Exploiting Seismic Waveforms

Exploiting Seismic Waveforms introduces a range of rece including the application of correlation techniques, heterogeneity and the extraction of structure and source ir inversion. It provides a full treatment of correlation meth signals and develops inverse methods for both sources and structure. Higher frequency components of seismograms are frequently neglected, or removed by filtering, but they contain information about seismic structure on scales that cannot be revealed by seismic tomography. Sufficient computational resources are now available for waveform inversion for 3-D structure to be a practical procedure and this book describes suitable algorithms and examples reflecting current best practice. Intended for students and researchers in seismology, this book provides a physical understanding of seismic waveforms and the way that different aspects of the seismic wavefield are revealed by the way that seismic data are handled.

Brian L.N. Kennett is Emeritus Professor of Seismology at the Australian National University. His research interests are directed towards understanding the structure of the Earth from seismological observations. He is the recipient of numerous awards and medals for his work and is a Fellow of the Australian Academy of Sciences and the Royal Society (London). He is the author of more than 320 research papers and nine other books.

Andreas Fichtner is Professor of Seismology and Wave Physics at ETH Zurich. His work is focused on the development of waveform tomography methods using high-performance computing, and their application in seismology and medical imaging. He is the recipient of several awards, including the 2011 Aki award of the American Geophysical Union, and is the editor or author of three other books.

Exploiting Seismic Waveforms

Correlations, Heterogeneity and Inversion

BRIAN L.N. KENNETT

Research School of Earth Sciences, The Australian National University

ANDREAS FICHTNER

Department of Earth Sciences, Swiss Federal Institute of Technology (ETH), Zurich

CAMBRIDGE
UNIVERSITY PRESS

CAMBRIDGE
UNIVERSITY PRESS

University Printing House, Cambridge CB2 8BS, United Kingdom

One Liberty Plaza, 20th Floor, New York, NY 10006, USA

477 Williamstown Road, Port Melbourne, VIC 3207, Australia

314–321, 3rd Floor, Plot 3, Splendor Forum, Jasola District Centre, New Delhi – 110025, India

79 Anson Road, #06–04/06, Singapore 079906

Cambridge University Press is part of the University of Cambridge.

It furthers the University's mission by disseminating knowledge in the pursuit of education, learning, and research at the highest international levels of excellence.

www.cambridge.org
Information on this title: www.cambridge.org/9781108830744
DOI: 10.1017/9781108903035

First published 2021

Printed in the United Kingdom by TJ Books Limited, Padstow Cornwall

A catalogue record for this publication is available from the British Library.

ISBN 978-1-108-83074-4 Hardback
ISBN 978-1-108-82878-9 Paperback

To Heather and Carolin

Contents

Preface

Over the last couple of decades, enhancements of computational power and denser seismic networks have made a considerable impact on seismological practice. As a result, it is now possible to examine and model complex structures at relatively high frequencies. Further, the development of correlation techniques has allowed the exploitation of ambient noise and reduced dependence on fortuitous placement of earthquakes. Waveform inversion techniques enable the exploitation of much more of the seismogram than hitherto, but the small-scale structures that determine the high-frequency character of seismograms remain beyond reach of direct investigation.

This book provides an account of the use of correlation concepts in a broad range of applications in seismology, and the use of higher-frequency waves to examine the finer-scale aspects of the heterogeneity of the Earth. One of the major objectives of seismology has always been to extract as much information as possible from seismograms about the seismic source and the structure of the Earth. The growth of computational power means that it is now possible to undertake direct calculations of the seismic wavefield for realistic three-dimensional models and to use these to invert for complex structure, so we provide a full discussion of the inversion of seismic waveforms.

In recent years the density of seismometers available in some parts of the world for earthquake studies has reached the point where signal enhancement using multi-sensor techniques can exploit experience gained in the exploration field. The work therefore endeavours to provide links between the applications of seismology to earthquakes, ambient noise, regional and global studies, and seismic exploration. With numerical simulation we can approach the complexity of unfiltered observed seismograms, and the best results come when the physical processes controlling the behaviour of the wavefield are well understood. Our aim in this book is therefore to provide a suitable background for the appreciation of recent developments in seismic wave analysis.

An introductory chapter discusses the background to the work and the way in which it builds on and integrates material from prior studies in seismology. This is followed by a summary of the structure of the book, indicating the nature of the topics to be covered and the way that they interact.

To keep the work in bounds, we have assumed a reasonable acquaintance with the principles of seismology, and provide a concise recapitulation of important results in Part I that are exploited in later parts.

The treatment in this book draws on the fundamentals developed in the two volumes of *The Seismic Wavefield* (Kennett, 2001; 2002), and thus does not attempt to provide a comprehensive treatment of basic topics. References to sections in these volumes are indicated using a section marker (e.g., § SWI:3.1.2). For equations the volume number is represented explicitly as in (SWII:17.2.5). Where reference is made to the book *Geophysical Continua* (Kennett & Bunge, 2008) the designator GC is employed (e.g., § GC:11.3.2). Occasional use is also made of *Seismic Wave Propagation in Stratified Media* (Kennett, 1983; 2009) designated SM.

Acknowledgments

We are grateful to many people who have been kind enough to provide material for figures including M. Afanasiev, N. Blom, C. Bösch, P. Boué, Y. Chen, A. El-Sharawy, M. de Kool, T.-K. Hong, E. Kissling, T. Meier, T.-S. Pham, K. Priestley, N. Rawlinson, A.M. Reading, J. Ritsema, K. Sager, E. Saygin, B. Tauzin, H. Tkalčić, and D. Thompson. Takashi Furumura kindly produced some new numerical simulations for the work, as well as contributing figures.

1

Introduction

The early advances in understanding the internal structure of the Earth and the nature of earthquakes were achieved with a very limited number of seismic stations. Yet, by 1914, not only the presence of a core was recognised by Oldham (1906) from his analysis of the travel times of seismic waves from 14 earthquakes across the globe, but also a good estimate of its radius had been made (Gutenberg, 1913).

Many advances came from painstaking analysis of large numbers of earthquake records at the same stations; such as the discovery of the presence of the inner core by Inge Lehmann from the records of the Copenhagen station (Lehmann, 1936) – fortunately the plentiful events on the Pacific Rim were at a suitable distance.

The basic theory of linear elastic waves was well established by the 1850's, but seismograms did not conform to the expectations for simple models such as a uniform half space with simple *P*, *S* and Rayleigh waves (Lamb, 1904). Love (1911) introduced the possibility of dispersion for observed seismograms, though the details remained sketchy because the necessary calculations for multilayered structures were difficult even with early digital computers (e.g., Ewing, Jardetsky & Press, 1957).

In this book we explore the aspects of seismic waves that have been made accessible by the dramatic increase in the availability and quality of seismic data across a broad range of frequencies and the computational power needed for modelling and analysis. Indeed we have reached a situation where numerical simulation is capable of generating seismograms that rival observations in their complexity. In consequence we still need suitable tools to disentangle the complex physical processes that control both observations and synthetics.

Our focus is on recent developments that reach beyond a radially stratified model of the Earth. We try to develop concepts that bridge scales and help develop understanding of the seismic wavefield in three-dimensions (3-D). We examine the application of correlations between seismic waveforms in a wide range of contexts, from the estimation of time and phase delays, through the exploitation of ambient noise and earthquake coda for structural studies, to receiver studies.

Improved sampling of the seismic wavefield has shed light on the smaller scales of heterogeneity. Rather than just simplifying seismograms by low-pass filtering, it is possible to utilise the higher-frequency components of the coda of seismic phase to gain information about structure.

Parametric stochastic models provide a convenient description of the smaller-scale features of heterogeneity via spatial correlation functions. We use such models to discuss the effects of heterogeneity on seismic waves and the way in which multiple scales of heterogeneity interact. The capacity to simulate realistic seismograms opens the opportunity for using waveform inversion methods to extract structural information directly from observed seismograms. This approach is facilitated by the use of adjoint techniques that reduce computational demands, and we show how waveform inversion can be used for both event records and correlation data.

1.1 Growth of Recording Networks

Since the recognition that waves from earthquakes could be detected thousands of kilometres away from the source, the development of earthquake seismology has had two major strands. The first strand comes from the deployment of seismographs in regions of earthquake activity with the aim of understanding the nature of local events and their implications for earthquake hazard. The second strand utilises seismic stations across the globe and exploits distant recordings to determine source characteristics of larger events and to understand the internal structure of the Earth. From the 1960s efforts to monitor nuclear testing led to coherent global networks and the development of seismic arrays.

The advent of digital recording encouraged the development of broadband sensors that could cover the full frequency range encompassed by seismic waves. Now it was no longer necessary to have two separate seismometer systems to focus on the signals above and below the dominant microseismic peak around 0.12 Hz. The versatility of the new instrumentation meant that it was possible to capture both local and global signals with a single sensor, and so markedly enlarge the scope of seismological studies. The result has been a fusion of the two strands of enquiry, with enhanced national networks also able to be used for global studies. The increased suite of permanent stations has been accompanied by extensive deployments of portable instrumentation directed towards structural studies.

The increase in the number of seismic stations in the world has been dramatic. For example for events with moment magnitude Mw 7 in the New Ireland region of Papua New Guinea, the number of stations reporting time readings to the International Seismological Centre increased from 324 in 1978, 639 in 1999, to 1255 in 2015. The volume of waveform data also continues to grow at a rapid rate: for Mw 7.7 events at a very similar location in this region, around 1200 stations were potentially available in 2000, compared with more than 3000 stations in 2016.

One of the side effects of the introduction of broadband sensors is that the spectrum of seismic noise is faithfully recorded, rather being suppressed by selective frequency response. Before long, methods for exploiting seismic noise were developed building on ideas from acoustics (e.g., Campillo & Paul, 2003). The stacking of correlation of seismic noise records between two different seismic stations over a substantial period of time was found to be equivalent to having virtual sources and receivers at each location. This advance opened up new ways of examining seismic structure, particularly with higher-frequency surface waves, since pairs of stations could be used without need for alignment to the great-circle path from a distant earthquake that was a major restriction on early structural studies (Knopoff, 1972). The exploitation of ambient noise correlation has become a major part of the toolkit of seismology with many different applications (see, e.g., Nakata, Gualtieri & Fichtner, 2019). Since the number of stations pairs in a network of N stations is $\frac{1}{2}N(N-1)$, very large numbers of virtual propagation paths can be explored via correlation methods.

One of the developments in the use of portable broadband instrumentation has been the creation of broad areal coverage through multi-year experiments with progressive movement of stations. An early example was the SKIPPY experiment in Australia that achieved reconnaissance coverage of the continent at 400 km station spacing over a period of 4 years (van der Hilst et al., 1994). With a limited number of stations, a sequence of 6-month deployments were made to progressively cover the continent from east to west, with deployments in the south in the austral summer to avoid the 'wet' in the tropics. The relatively short duration of each deployment was sufficient to collect numerous events in the earthquake belts around Australia for surface wave studies, but less effective for body wave work.

Subsequent more ambitious projects with much larger numbers of instruments have typically used more than 2 years recording at each site. The most comprehensive project was USArray in which around 400 stations in the Transportable Array were moved like a shutter blind across the continental United States over a period of 7 years from 2007–2014. At the end of their recording interval, stations were moved from the western limb of the array to the east to maintain continuous coverage. Following the successful deployment, the transportable stations were moved to the Alaskan region. Other major portable deployments have been the ongoing ChinArray in China, and the IberArray that spanned from Morocco to northern Spain in three deployments in the period 2009–2014. The AlpArray in Europe (2016-2019) has been a multi-national effort with 45 research institutions from 18 countries cooperating to link the permanent stations in the Alpine area with a broad-scale deployment of portable instruments, with the objective of studying the structure and evolution of the lithosphere in the entire Alpine region of Europe. The AlpArray experiment has included both land-based stations and sea-bottom seismometers (Figure 1.1).

Even without any major experiment, a combination of smaller experiments can

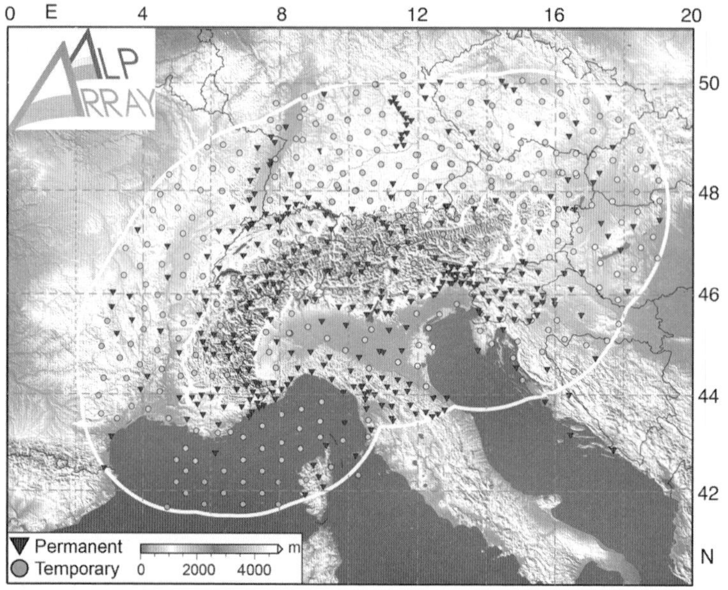

Figure 1.1 Configuration of the AlpArray experiment in Europe, covering the entire Alpine area with a combination of permanent and portable stations. [Courtesy of AlpArray Working Group.]

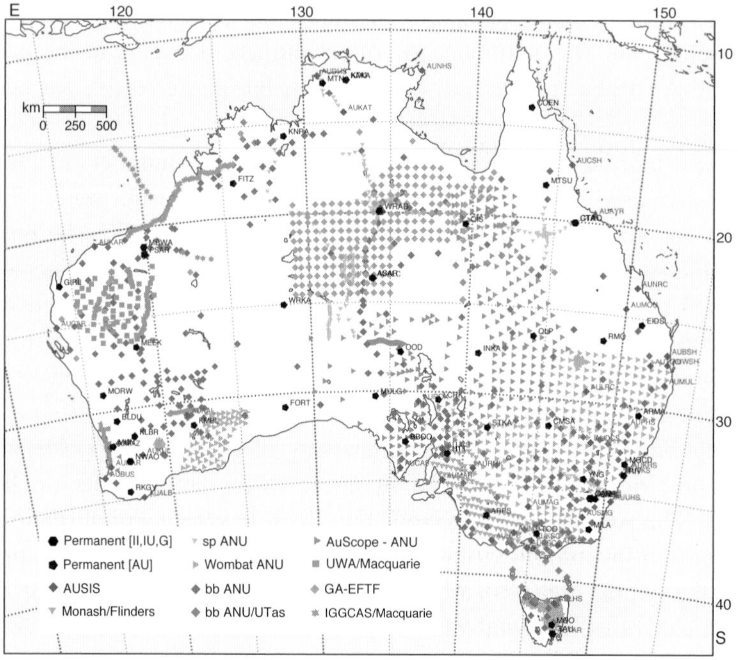

Figure 1.2 Station coverage in Australia from both permanent stations and portable deployments to the end of 2019. [Courtesy of AuScope.]

over time produce substantial coverage. In Australia, the early SKIPPY project was followed by a number of more localised broadband experiments, and a sequence of portable deployments in southeastern Australia, collectively known as the WOMBAT project. Additional recent deployments in northern and western Australia have produced reasonable coverage at the continental scale (Figure 1.2). Substantial areas of desert on the Australian continent remain undersampled, since access and logistics are difficult.

The progress from initially sparse sensors to dense networks has been even more marked in exploration seismology. The amount of energy reflected from depth is rather small, but can be enhanced if multiple coverage of the subsurface can be combined. Thus stacking procedures of increasing sophistication became an essential part of the reflection surveys. Where possible, multiple line profiles have been replaced by full surveys with an areal distribution of source and receivers and intensive processing of enormous data sets to extract three-dimensional structure. On land, the use of sensors linked by long cables is often now replaced by large numbers of autonomous sensors that allow more flexible configurations. Thousands of sensors are routinely deployed with the aim of securing sufficient sampling of the wavefield that the details of reflections are faithfully recorded

Both aspects of seismology are converging towards a goal of full rendering of the seismic wavefield over a broad range of frequencies, so that the maximum structural information can be extracted. We therefore provide links between the treatment of the reflection wavefield and techniques for the study of larger-scale structure.

1.2 Theoretical and Computational Developments

Theoretical developments in seismology have drawn heavily on the tools developed by mathematical physicists, exploiting separation of variables and the inter-relations between plane, cylindrical, and spherical waves. Lamb (1904) formulated the response of an infinite half-space to a surface source in terms of double integrals over wavenumber and frequency. The elegant work of Lapwood (1948) used steepest-descent and stationary-phase methods to examine the seismic phases arising from a buried source in a half-space. Pekeris (1948) also used stationary phase techniques and recognised the Airy phase in dispersive wave propagation associated with extrema in group slowness that tends to dominate in long-distance propagation. Cagniard (1939) developed a way of recasting the integral contributions for an individual arrival so that the time response was more readily obtained. This approach was simplified somewhat by de Hoop (1958), and the modified form has been extensively used in the generalised ray method (see, e.g., Aki & Richards, 2002).

In 1968, two articles appeared that showed the application of numerical seismograms to understanding active source experiments. Helmberger (1968) used a development of the wavefield in terms of generalised ray contributions

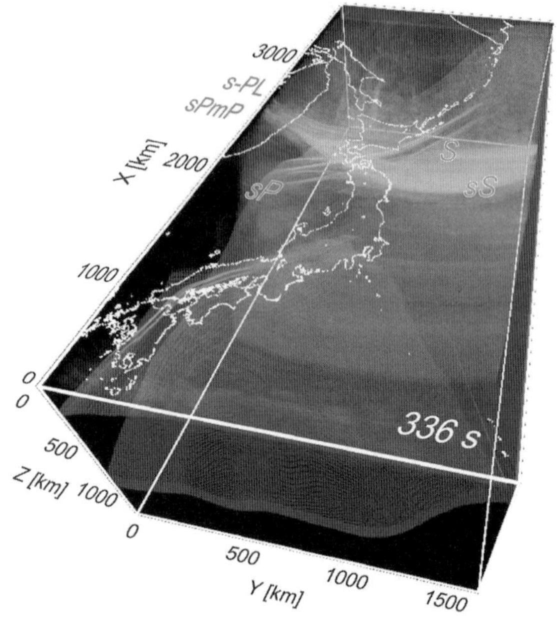

Figure 1.3 3-D modelling using the finite-difference technique for an event in the Sea of Okhotsk recorded across Japan. [Courtesy of T. Furumura.]

whose time signatures were summed to yield seismograms at various distances. Fuchs (1968) worked directly with the integral formulations for a multi-layered medium with numerical integration over wavenumber and a fast Fourier transform to recover the time response. Fuchs & Müller (1971) introduced a shallow zone with just transmission overlying layering for which the full reflection response was calculated. This *reflectivity* method was used to calculate a full record section for the interpretation of a refraction profile. Active source experiments provided early applications of synthetic seismograms because they had relatively dense sampling for the time. It was not long though before earthquake observations were being used and the structure employed extended through the upper mantle transition zone (Helmberger & Wiggins, 1971). Kennett (1975) pointed out the important role of attenuation on the character of the wavefield and the potential trade-offs between structure and zones of low Q.

Alongside the use of layered models, direct numerical solutions were developed for the underlying differential equations (e.g., Alterman & Loewenthal, 1972). Fully numerical approaches emerged at around the same time and by 1972 finite-difference techniques had been applied to simple two-dimensional (2-D) heterogeneous models (Boore, 1970), and a finite-element approach used for surface wave propagation across complex structures (Lysmer & Drake, 1972). In these years a major limitation for all the numerical procedures was the availability of computer memory, since even with the most powerful machines of the 1970s only relatively small grids could be used.

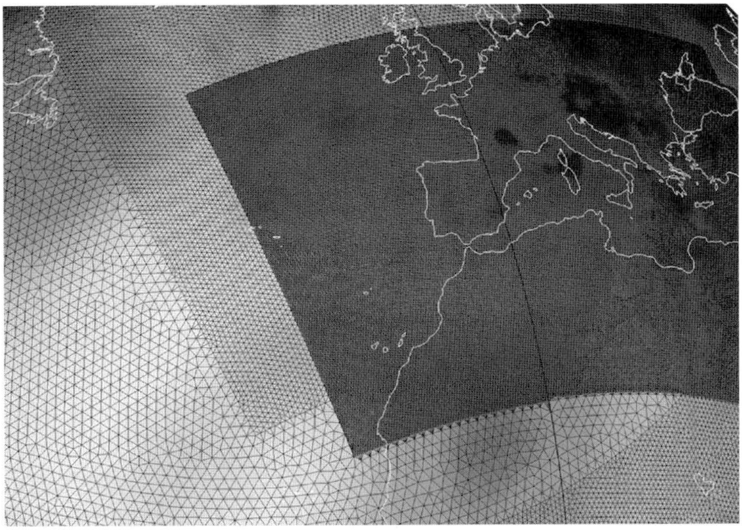

Figure 1.4 Variable-density grid employed with spectral element modelling.

For single processor systems, finite-difference methods tended to concentrate on higher-order difference schemes or pseudospectral methods (e.g., Furumura et al, 1998) since these minimised numerical dispersion for coarser grids. But, once parallel computation became fully established, it was recognised that simpler schemes, e.g., 4th order in space and 2nd order in time were more effective for inter-processor communications. The finite-difference approach for 3-D models exploits staggered grids where displacements and tractions are defined at different grid positions. These displaced grids minimise numerical dispersion effects, but at the cost of a complex rendering of interface conditions and the free surface. With thousands of compute nodes that have gigabytes of local memory, it has become possible to undertake finite-difference numerical simulations for relatively high frequencies across a large spatial domain (Figure 1.3). The computational effort for such complex calculations remains high, and care needs to be taken in constructing 3-D models. Two-dimensional simulations remain valuable for exploration of the effects of fine-scale heterogeneity.

The finite-element method exploits an integral equation formulation of elastic wave propagation and so automatically satisfies the free-surface boundary condition even in the presence of topography. The computational domain is divided into discrete elements so irregular geometries can be accommodated. With low-order polynomial representations in each element and explicit continuity conditions, the solution of the resulting system of linear equations requires considerable effort and numerical dispersion can be significant. As a result the direct finite-element approach has not been extensively used in seismic applications.

However, a variant of the finite-element technique has achieved widespread use. This spectral element method again uses non-overlapping elements that can be adapted to complex geometries. Within each element a high-order spectral representation is used. With a suitable choice of polynomial and quadrature procedure within each element, the effort of solving the requisite linear equations is much reduced. Since the free-surface condition is built in, the spectral element method is very effective for modelling surface wave propagation and similar phenomena (e.g., Komatitsch & Tromp, 2002a,b). Spectral element procedures can be adapted to a wide range of conditions, with substantial variations in the size of mesh as illustrated in Figure 1.4. The spectral representation carries the implicit assumption that the variation in wavespeed in each element is smooth. Thus rather small elements are needed for rapidly varying wavespeeds.

The advent of full 3-D computations opens the possibility of exploiting much more of the seismic wavefield than in the past. We can simulate entire seismic waveforms from events for moderate frequencies, and make use of such results in the inversion of observed seismograms. We still need good starting models, which can exploit the wide range of results obtained with travel-time tomography as well as earlier waveform inversions. The importance of scattered energy increases rapidly with frequency, and so the presence of unresolvable fine-scale structure is likely to impose limits on the applications of full waveform inversion.

1.3 Structure of the Book

In this book we provide a treatment of a broad range of topics that relate to the nature of seismic wave propagation in complex structures and the ways in which understanding of 3-D structure can be extracted from observations. We exploit recent developments, but draw on a range of techniques and modes of analysis. We endeavour to provide links between the applications of seismology to earthquakes, ambient noise, regional and global studies, and seismic exploration. We use both observations and numerical simulations to illustrate the topics and provide insight into the character of the seismic wavefield.

In Part I we develop the representations of the seismic wavefield needed for the later topics. In Chapter 2, we provide a discussion of the way that the higher-frequency body waves and surface wave components emerge from the normal mode spectra of the Earth to link global and local concepts. In Chapter 3, we consider integral representations of the wavefield and develop the concepts of reflection and transmission operators that allow insight gained in the discussion of stratified media to be transferred to a more general 3-D environment. This operator approach also provides a convenient link between larger-scale seismology and seismic reflection studies. Chapter 4 provides a discussion of the reflection field in an exploration context, which allows links to be drawn to current developments for exploiting teleseismic arrivals. We consider propagation issues using the operator development and show how this leads to understanding of migration procedures, with links to remapping of reflectivity as is currently being exploited in Marchenko techniques.

Part II is concerned with the exploitation of correlations in seismology in a wide range of circumstances. In Chapter 5 we introduce the concepts of waveform correlations, and the way in which they can be exploited to extract information from seismograms. We then consider the closely related topic of transfer functions between aspects of the wavefield, and this leads into a discussion of the ways by which seismograms can be compared – a topic of importance in the comparison of observations and simulations. We also consider the nature of receiver functions and the correlation of teleseismic signals at a receiver. Chapter 6 addresses the nature of the correlation wavefield and its relation to the group of techniques collectively known as *seismic interferometry*. We establish a direct representation of the cross-correlation of the seismic signals between two stations and show how, with a suitable distribution of sources, this correlation can provide a virtual source–receiver pair whose phase properties arise from differencing. We then discuss the concept of generalised interferometry with an arbitrary distribution of sources, and illustrate the way in which processing procedures can affect the nature of correlated signals. Having laid down the fundamentals, in Chapter 7 we examine the application of correlation procedures to the exploitation of the ambient noise field where the dominant component of the correlation field comes from surface waves, though body waves can be extracted in some circumstances.

In Chapter 8 we turn attention to applications of correlation techniques to the coda of seismic source signals where steeply travelling body waves are the main contributors to the correlation wavefield. The correlation wavefield emphasises seismic phases that are difficult to detect in direct excitation by a source and so can provide new information on internal structure, e.g., an improved estimate of the shear wavespeed in the inner core. In Chapter 9 we consider correlation methods applied to surface recordings with the objective of extracting information on subsurface reflectivity. We show how the auto-correlation of seismic signals can provide information on reflections without conversions, and can be exploited to provide indirect imaging of heterogeneous structure. Correlations between signals at different sensors can also be exploited in reflection work to provide virtual sources that provide new ways of imaging complex structure.

In Part III we examine the interaction of seismic waves with heterogeneity at all scales, with an emphasis on the influence of structure on multiple scales. We discuss ways in which numerical simulations and inversions can exploit data with differing station density to provide maximum resolution of structure. The finer scales of variation within the Earth lie beyond any capacity for direct imaging, but the scattered wavefield that they produce contains important information on structure. In Chapter 10 we contrast deterministic and stochastic representations of heterogeneity, and look at the way that ensemble results can be exploited for Earth structure that is time invariant. We also consider the way that effective media, with simpler structure, can be extracted from complex models by the process of wavespeed upscaling. In Chapter 11 we examine the ways in which the effects of heterogeneity can be handled. We first consider perturbations of the wavefield using Born series and show how such concepts can be combined with the use of reflection and transmission operators to provide a flexible treatment of structures with varying heterogeneity in different zones of the model. Although the various modes of surface waves propagate independently in simple structure, the presence of heterogeneity induces cross-coupling that modifies the wavefield. In Chapter 12 we examine the processes of scattering in the Earth, the various zones where it is important and the way that these influence observed seismograms. Guided waves in heterogeneous structures play an important role at high frequencies. We discuss examples from the propagation of deep earthquakes in subduction zone environments, and for the oceanic and continental lithosphere. Chapter 13 discusses the interaction of seismic waves with multiple scales of heterogeneity, particularly in the lithosphere.

Part IV introduces the concepts and developments needed for the inversion of seismic waveforms, with a discussion of algorithms and illustrations of practical inversions. Chapter 14 presents the basic inversion framework in a Bayesian context, leading into the formulation of the nonlinear inversion process in terms of optimisation of a measure of misfit. In Chapter 15 we provide a broad survey of methods for inversion relevant to waveforms, showing how inversions can be performed for models described by very large numbers of parameters. Chapter 16 first describes the way in which adjoint techniques allow the computation of derivatives in complex models and so enable practical non-linear inversion. This is followed by a discussion of sensitivity kernels associated with the variation of critical parameters for both structure and sources. We show how such kernels can provide insight into the nature of Earth structure and potential resolution. In Chapter 17 we describe the process of waveform inversion for earthquake data to extract 3-D structure, including computational aspects. We show how, as a model is improved, it becomes possible to incorporate further data and hence achieve model refinement. We examine the issues of practical resolution assessment, and validation of proposed models. In Chapter 18 we turn attention from event data to the exploitation of the correlation wavefield, demonstrating that it is possible to achieve joint inversion for noise sources and Earth structure and include a discussion of the relevant sensitivity kernels. The final chapter (Chapter 19) addresses a range of topics that hold considerable promise for future developments. We start by considering nested inversions that allow definition of heterogeneity across a wide range of length scales from local through regional to global. This is followed by discussion of adaptive numerical gridding, exploitation of data redundancy, the development of efficient random sampling methods for inversion, and the use of Hamiltonian Monte Carlo techniques for efficient searching of high-dimensional spaces.

Part I

Building the Seismic Wavefield

2

Stratified Media

In Part I we bring together a wide range of results on the nature of the seismic wavefield and its relation to the physical processes associated with wave propagation. These concepts will be exploited in later parts of the work. We concentrate on aspects of the seismic wavefield for frequencies above 10 mHz, so that it is natural to think in terms of body wave and surface wave components of the field.

This chapter presents a brief description of wave propagation in stratified media on both global and local scales. We draw on results from the first two volumes of *The Seismic Wavefield* (Kennett, 2001; 2002) and *Geophysical Continua* (Kennett & Bunge, 2008) to provide a concise treatment of the main results.

In our treatment of the correlation wavefield in Chapters 6 and 8 we exploit information for the full globe, and so need to establish results in a spherical geometry. We first establish the link between the general modal representation for the sphere and integral properties in frequency–slowness space that allow direct inclusion of the effects of attenuation. From there we examine the extraction of body wave and surface wave contributions, and their relation to the physical processes encountered in the passage of seismic waves between source and receiver.

2.1 From Normal Modes to Seismograms

The response of the whole Earth to excitation by a source can be represented in terms of a summation of contributions from normal modes. For long period deformation, seismic disturbances perturb the gravitational field so that seismic displacements are coupled to the gravitational potential. Such self-gravitational effects diminish rapidly with increasing frequency ω. Here we consider a non-rotating, spherical body with radial stratification (as in §GC:11.3). The additional complications imposed by Earth's rotation and 3-D structure are discussed in detail in Dahlen & Tromp (1998).

Although a sum of normal modes provides a complete description of global propagation phenomena, considerable effort is required to extract explicit expressions that relate to the body wave and surface components recognised in seismograms. We here concentrate on higher frequencies and bring together a range of relevant results that are distributed through other works (e.g., BenMenahem & Singh, 1981; Dahlen & Tromp, 1998). We show how the modal sum can be converted into an equivalent integral representation, from which both surface wave and body wave contributions can be extracted.

2.1.1 Normal Modes

The displacement $\mathbf{u}(\mathbf{x}, \omega)$ induced by a force system $\mathbf{f}(\mathbf{x})$ is determined by the equation of motion for frequency ω, (GC:11.3.24),

$$\mathfrak{H}(\mathbf{u}, \psi) + \rho_0 \omega^2 \mathbf{u} = -\mathbf{f}, \tag{2.1.1}$$

where ψ is the gravitational potential and ρ_0 the undisturbed density. The term $\mathfrak{H}(\mathbf{u}, \psi)$ includes both the gradient of the stress tensor $\boldsymbol{\tau}$ associated with the displacement $\mathbf{u}$ and gravitational coupling terms:

$$\mathfrak{H}_i(\mathbf{u}, \psi) = \partial_j \tau_{ij} + \rho_0[\partial_k u_k \partial_i \psi_0 - \partial_i \psi_1 - \partial_i(u_k \partial_k \psi_0)], \tag{2.1.2}$$

in terms of the unperturbed gravitational potential ψ_0 and the incremental effect from deformation ψ_1. We have used the compressed notation ∂_i for $\partial/\partial x_i$. We work in the frequency domain since we are then able to include the effects of seismic attenuation via the use of complex elastic moduli (§ SWI:8.3; § GC:11.1).

The displacement field must remain finite at the centre of the Earth, and have vanishing traction at the surface. At internal boundaries, displacement and traction are continuous at solid–solid interfaces, whilst fluid–solid boundaries have continuity of traction and displacement normal to the surface.

The normal modes of this system satisfy the equation of motion (2.1.1) and the full set of boundary conditions in the absence of a source, for a discrete set of eigenfrequencies $\{\omega_K\}$. The associated displacement eigenfunctions $\mathbf{u}_K^e(\mathbf{x}, \omega_K)$ are orthogonal and can be normalised so that

$$\int_{V_e} d^3\mathbf{x}\, \rho_0(r) \mathbf{u}_I^e(\mathbf{x}) \cdot [\mathbf{u}_J^e(\mathbf{x})]^* = \delta_{IJ}, \tag{2.1.3}$$

where the integral is taken over the whole volume of the Earth (V_e) and the superscript * indicates a complex conjugate.

2.1.2 Modal Sum

In the presence of a force system $\mathbf{f}(\mathbf{x})$ we can expand the displacement field as a sum of contributions from the suite of eigenfunctions

$$\mathbf{u}(\mathbf{x}, \omega) = \sum_{I=0}^{\infty} c_I\, \mathbf{u}_I^e(\mathbf{x}, \omega_K). \tag{2.1.4}$$

Exploiting the orthonormality of the eigenfunctions, we can express the displacement response induced by the force system $\mathbf{f}$ in the frequency domain as

$$\mathbf{u}(\mathbf{x}, \omega) = \sum_{K} \frac{\mathbf{u}_K^e}{\omega_K^2 - \omega^2} \int_{V_e} d^3\mathbf{x}\, \mathbf{f}(\mathbf{x}) \cdot [\mathbf{u}_K^e(\mathbf{x}, \omega_K]^*. \tag{2.1.5}$$

If we specialise to a point moment tensor $\mathbf{M}(\mathbf{x}_s)$ at location $\mathbf{x}_s$ (§ SWI:11.4) then

$$\int_{V_e} d^3\mathbf{x}\, \mathbf{f} \cdot [\mathbf{u}_K^e(\mathbf{x}, \omega_K)]^* = [\mathbf{e}_K^e(\mathbf{x}, \omega_K)]^* : \mathbf{M}(\mathbf{x}_s), \tag{2.1.6}$$

where $\mathbf{e}_K^e(\mathbf{x}, \omega_K)$ is the strain tensor associated with the Kth mode. The tensor contraction $[\mathbf{e}_K^e]^* : \mathbf{M}$ in component form is $[\mathbf{e}_K^e]_{pq}^* \mathbf{M}_{pq}$, with summation over repeated suffices.

Using the moment rate tensor $\mathfrak{M}(t) = \partial \mathbf{M}(t) \partial t$, which is only non-zero whilst earthquake displacement occurs, we can write the full time response as a convolution of modal contributions with the moment rate tensor

$$\mathbf{u}(\mathbf{x}, t) = \sum_{K} \int_{-\infty}^{\infty} dt' \frac{1}{\omega_K^2} \left\{ C_K(t - t') \mathbf{u}_K^e\, [\mathbf{e}_K^e]^* : \mathfrak{M}(\mathbf{x}_s, t) \right\}. \tag{2.1.7}$$

Here we have introduced the effect of mild anelasticity through the modal time term

$$C_K(t) = \left[1 - \cos(\omega_K t) e^{-i\omega_K t/2Q_K}\right] H(t), \tag{2.1.8}$$

where Q_K is the quality factor for the Kth mode.

For a spherical non-rotating Earth, the eigenfunctions in spherical polar coordinates (r, θ, ϕ) can be represented in terms of spherical harmonics (§ SWI:12.2). In terms of unit vectors $\hat{\mathbf{e}}_r$, $\hat{\mathbf{e}}_\theta$, $\hat{\mathbf{e}}_\phi$ associated with each of the spherical coordinates, the gradient on the surface of a spherical shell $r = const$ can be expressed as

$$\nabla_1 = \hat{\mathbf{e}}_\theta \partial_\theta + \hat{\mathbf{e}}_\phi \frac{1}{\sin\theta} \partial_\phi, \tag{2.1.9}$$

with the surface curl operator

$$\hat{\mathbf{e}}_r \wedge \nabla_1 = -\hat{\mathbf{e}}_\theta \frac{1}{\sin\theta} \partial_\phi + \hat{\mathbf{e}}_\phi \partial_\theta, \tag{2.1.10}$$

acting in a direction perpendicular to ∇_1.

The normalised surface harmonic

$$Y_l^m(\theta, \phi) = (-1)^m \left(\frac{2l+1}{4\pi} \right)^{\frac{1}{2}} \left[\frac{(l-m)!}{(l+m)!} \right]^{\frac{1}{2}} P_l^m(\cos\theta) e^{im\phi}, \tag{2.1.11}$$

where l, m are integers, $-l \leq m \leq l$, and $P_l^m(\cos\theta)$ is the associated Legendre function. We set

$$\mathcal{L} = [l(l+1)]^{\frac{1}{2}}, \tag{2.1.12}$$

and define a set of orthogonal vector spherical harmonics

$$\mathbf{P}_l^m = \hat{\mathbf{e}}_r Y_l^m(\theta, \phi),$$

$$\mathbf{B}_l^m = \frac{1}{Ł}\nabla_1 Y_l^m(\theta, \phi), \tag{2.1.13}$$

$$\mathbf{C}_l^m = \frac{1}{Ł}\hat{\mathbf{e}}_r \wedge \nabla_1 Y_l^m(\theta, \phi). \tag{2.1.14}$$

The displacement $\mathbf{u}$ and the radial component of traction $\boldsymbol{\tau}_r = (\tau_{rr}, \tau_{r\theta}, \tau_{r\phi})$ can be represented in terms of this set of vector spherical harmonics as:

$$\mathbf{u} = U(r)\mathbf{P}_l^m + V(r)\mathbf{B}_l^m + W(r)\mathbf{C}_l^m, \tag{2.1.15}$$

$$\omega^{-1}\boldsymbol{\tau}_r = P(r)\mathbf{P}_l^m + S(r)\mathbf{B}_l^m + T(r)\mathbf{C}_l^m, \tag{2.1.16}$$

with displacement terms (U, V, W) and traction terms (P, S, T) that depend on the orders l, m of the spherical harmonics. Since Y_0^0 is independent of θ and ϕ, $\mathbf{B}_0^0 = 0$, $\mathbf{C}_0^0 = 0$.

For a radially stratified Earth, the equation of motion and the gravitational coupling equations can be written in terms of a coupled set of first-order differential equations for the displacement and traction terms and the gravitational potential, involving only radial derivatives. For isotropic media or transversely isotropic media with a radial symmetry axis, these differential equations divide into two sets. The first pair of equations describes toroidal modes with *SH* wave behaviour, and just involves W, T. The second set of equations link U, V to P, S and the gravitational potential ψ. These equations describe spheroidal modes that include both *P* and *SV* wave propagation. The full suite of equations including gravitational terms are presented in § GC:11.3.2, and the high-frequency forms with self-gravitation neglected in § SWI:12.2. For a spherically symmetric Earth, the differential equations in the absence of sources are independent of the azimuthal order m, and so the normal modes are $(2l + 1)$ fold degenerate.

For each angular order l there is a suite of modes which for the toroidal modes are designated ${}_nT_l$ in order of increasing eigenfrequency $\omega_n, n = 0, 1, 2, 3, \ldots$. The index n is the radial order, which for these toroidal modes corresponds to the number of zero crossings in the displacement eigenfunction $W(r)$ as a function of radius. For spheroidal modes ${}_nS_l$, the index n does not correspond directly to the number of nodal surfaces in radius, because of the presence of additional modes corresponding to Stoneley waves at the core–mantle boundary and the inner core boundary.

2.1.3 Character of Modal Spectrum

The character of the patterns of normal modes of the Earth at higher frequencies are illustrated in Figure 2.1, where we show both toroidal (T) and spheroidal (S) modes for the Earth model PREM (Dziewoński & Anderson, 1981) in the frequency band from 30 mHz to 70 mHz. Rather than display the modes in terms of angular order l (cf. § SWII:24.4) we use spherical slowness $\wp$ that can be directly related to

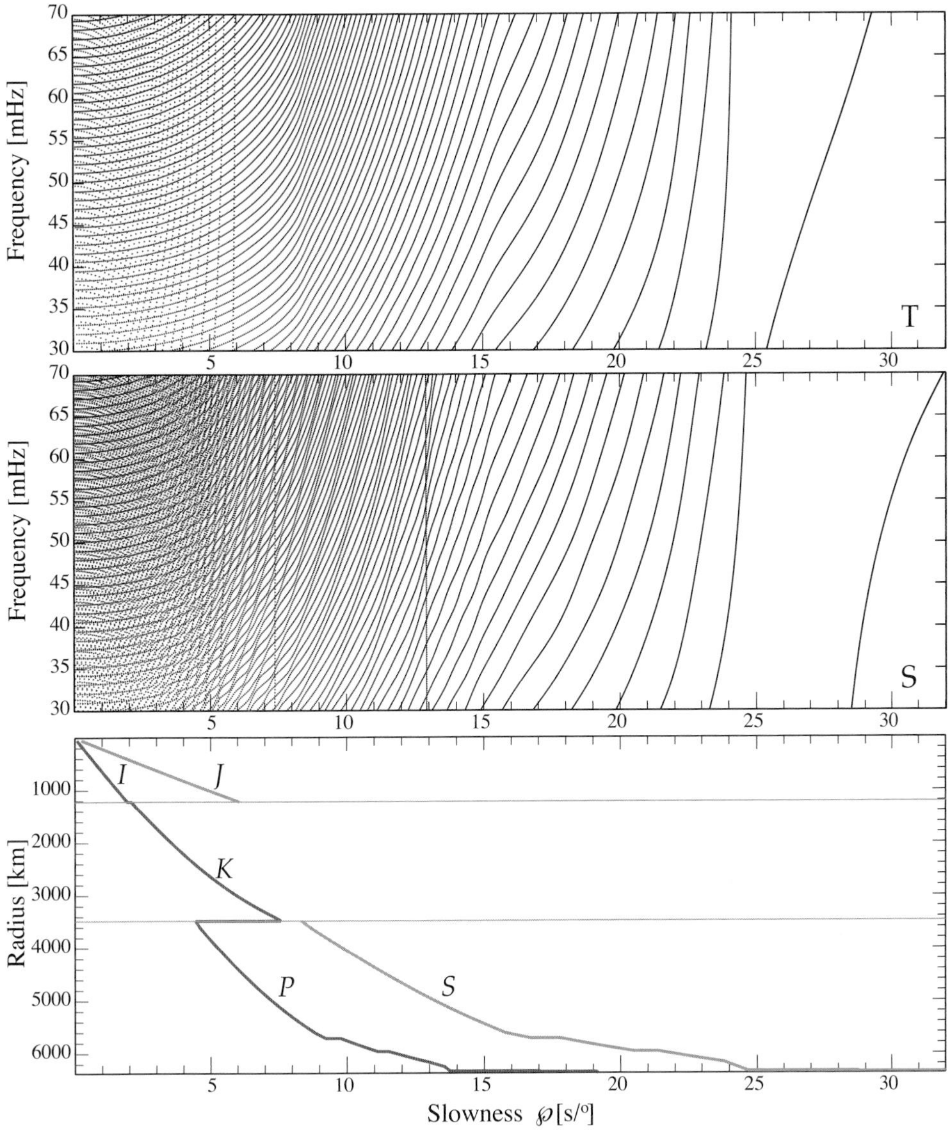

Figure 2.1 The normal modes for the radial model PREM (Dziewoński & Anderson, 1981) in the frequency band 30–70 mHz as a function of spherical slowness $\wp$. Top panel: toroidal modes (T). Centre panel: spheroidal modes (S). Bottom panel: the Earth model represented in terms of radius and spherical slowness.

the properties of the radial Earth model, as indicated in the bottom panel of Figure 2.1. For large angular order l, $\mathcal{L} = \omega\wp \sim l + \frac{1}{2}$, and thus we show $\wp = (l + \frac{1}{2})/\omega$ for each mode.

The toroidal modes in the upper panel of Figure 2.1 divide into two classes: modes trapped in the inner core (for $\wp < 6$ s/°), and modes trapped in the mantle and crust which span the full slowness range. *SH* waves cannot penetrate the fluid core. The regularly spaced modes at small slownesses represent multiple *ScS* waves reflected at the core–mantle boundary and the Earth's surface. For each radial order, the modal behaviour with frequency shows a slight inflection as the slowness passes 8.2 s/° and the *SH* waves are no longer reflected from the core–mantle boundary. For large slowness, the mode branches become well separated, and we will see later that these branches can be identified with surface waves.

The situation is more complex for the spheroidal modes displayed in the middle panel of Figure 2.1; now *P* and *SV* waves can interact in many different ways. We can still discern mode branches equivalent to *ScS* and inner core modes, but there are additional effects from *P* propagation in the fluid outer core as well as in the solid parts of the Earth, to produce *PKIKP* equivalent modes for small slowness. The presence of both *P* and *SV* waves give rise to complex propagation effects, so when an individual mode branch is tracked it partakes of different physical character in successive slowness ranges. This produces distinct changes in the slope of the mode branches with respect to frequency. Modulation of the spacing of the spheroidal mode branches between 5 s/° and 13 s/° reflect the presence of propagating *P* waves as well as *S*.

Additional disruptions to the spheroidal mode patterns are produced by the presence of modes corresponding to Stoneley waves on the core–mantle boundary and the inner core/outer core boundary that have almost constant slowness (7.3 s/° for the inner core boundary and 12.9 s/° for the core–mantle boundary. The Stoneley waves have amplitudes that decay exponentially away from the interfaces on both the solid and fluid side.

Although individual normal modes are plotted as discrete points in Figure 2.1, we see that much of the behaviour with frequency and slowness appears to be continuous. The individual mode branches correspond to different values of the radial order n, and for most branches the separation of modes in frequency is small so that they appear as distinct lines. Only the fundamental modes ($n = 0$) with the largest slowness are well separated from the rest, though the next few mode branches ($n = 1, 2, ..., 5$) have relatively broad spacing in slowness at fixed frequency. Once the radial order is above 10, the slowness spacing between mode branches is markedly reduced, and it is hard to think in terms of individual mode contributions. Indeed wavefield features such as long-period body waves arise from constructive interference between groups of modes.

2.1.4 From Sum to Integral

The dependence of the displacement $\mathbf{u}$ on azimuthal order m enters through the nature of seismic sources. For a point source the displacement field takes a simple form when the coordinate axis $\theta = 0$ is taken through the source location, so that θ can be interpreted as the epicentral distance Δ. The sum over azimuthal orders is then restricted to $m < 2$. For such a point source the spatial and temporal displacement response can be written as

$$\mathbf{u}(r,\Delta,\phi,t) = \frac{1}{2\pi}\int_{-\infty}^{\infty} d\omega\, e^{-i\omega t} \sum_{l=0}^{\infty} \tag{2.1.17}$$

$$\cdot \sum_{m=-2}^{2} [U(r,l,m,\omega)\mathbf{P}_l^m + V(r,l,m,\omega)\mathbf{B}_l^m + W(r,l,m,\omega)\mathbf{C}_l^m] .$$

The angular dependence of the vector spherical harmonics $\mathbf{P}_l^m, \mathbf{B}_l^m, \mathbf{C}_l^m$ enters through $P_l^m(\cos\Delta)e^{im\phi}$. The associated Legendre function $P_l^m(\cos\Delta)$ represents a stationary pattern on the sphere. In order to study specific seismic wave phenomena, it is convenient to make a decomposition into two oppositely directed travelling wave disturbances that interfere to produce the stationary $P_l^m(\cos\Delta)$ pattern (Nussenveig, 1965; Gilbert, 1976). Thus we set

$$P_l^m(\cos\Delta) = Q_{lm}^{(1)}(\Delta) + Q_{lm}^{(2)}(\Delta), \tag{2.1.18}$$

where asymptotically, for large l,

$$Q_{lm}^{(1,2)}(\Delta) \sim \frac{2(-l)^m}{(\pi l \sin\Delta)^{\frac{1}{2}}} \exp\left[\mp i\left((l+\tfrac{1}{2})\Delta + m\frac{\pi}{2} - \frac{\pi}{4}\right)\right]. \tag{2.1.19}$$

The convergence of the sum over angular order in (2.1.17) is slow for higher frequencies, with several thousand l-contributions needed to accurately represent body waves with frequencies above 0.1 Hz. As we have seen in Figure 2.1, the mode branches at higher frequencies appear to be continuous. This property can be exploited by converting the summation in (2.1.17) into an integral (§ SWI:15.4). Consider a contribution that is independent of azimuth (i.e., $m=0$). With the aid of the Watson transformation, a sum such as

$$\bar{f}(\omega) = \sum_{l=0}^{\infty} (l+\tfrac{1}{2})P_l(\cos\Delta)f_l, \tag{2.1.20}$$

can be converted into an equivalent form corresponding to a summation over multiple circuits of the globe

$$\bar{f}(\omega) = \sum_{q=\infty}^{\infty} (-1)^q \int_0^{\infty} dk\, kP_{k-\frac{1}{2}}(\cos\Delta)f(k)e^{2iqk\pi}, \tag{2.1.21}$$

where $f(k) = f_l$ when $k = l + \frac{1}{2}$. In (2.1.21) $f(k)$ has been extended as an even function of complex wavenumber k.

The standing wave representation (2.1.21) can be recast in terms of the travelling wave terms $Q^{(1,2)}$ introduced in (2.1.19). $Q^{(1)}(\Delta)$ is a wave that propagates in the direction of increasing Δ and $Q^{(2)}(\Delta)$ travels in the opposite sense. Away from the source ($\Delta = 0$) and its antipole ($\Delta = \pi$), the response spectrum can be written as

$$\bar{f}(\omega) = \sum_{q=1,3,5,\ldots}^{\infty} (-1)^{(q-1)/2} \int_0^{\infty} dk\, k Q^{(1)}_{k-\frac{1}{2}}(\Delta) f(k) e^{2i(q-1)k\pi}$$
$$+ \sum_{q=2,4,6,\ldots}^{\infty} (-1)^{q/2} \int_0^{\infty} dk\, k Q^{(1)}_{k-\frac{1}{2}}(\Delta) f(k) e^{2iqk\pi}. \qquad (2.1.22)$$

The terms $q = 1$ corresponds to direct propagation from the source along the minor arc and $q = 2$ to the major arc (in the opposite direction along the great circle). The remaining terms correspond to multi-orbit waves that have circled the Earth more than once before arriving at the receiver.

For a general point source when we consider just the first orbit, the representation for the vertical displacement (i.e., along the radius vector) takes the form

$$u_z(\Delta, \phi, \omega) = \sum_{m=-2}^{2} \int_0^{\infty} dk\, k P^m_{k-\frac{1}{2}}(\cos\Delta) f(k) e^{im\phi}. \qquad (2.1.23)$$

For high frequencies we can exploit the asymptotic representation of the Legendre function (e.g., Olver, 1974, §12.4; BenMenahem & Singh, 1981, Appendix H)

$$P^m_{k-\frac{1}{2}}(\cos\Delta) e^{im\phi} \sim \left[\frac{\Delta}{\sin\Delta}\right]^{\frac{1}{2}} J_m(kX) e^{im\phi}, \qquad (2.1.24)$$

in terms of the horizontal range $X = r_e\Delta$ at the Earth's surface $r = r_e$.

We thereby extract the approximate integral form

$$u_z(\Delta, \phi, \omega) \approx \left[\frac{\Delta}{\sin\Delta}\right]^{\frac{1}{2}} \sum_{m=-2}^{2} \int_0^{\infty} dk\, k J_m(kX) e^{im\phi} f(k), \qquad (2.1.25)$$

which recovers the expression for horizontally stratified media (§ SWI:15.2) in terms of Bessel functions. Now, however, the result is modulated by the scaling factor $[\Delta/\sin\Delta]^{1/2}$. This sphericity correction is close to unity for small epicentral distance Δ ($< 12°$).

On the Earth's surface the spherical slowness $\wp = pr_e$, where r_e is the radius of the Earth. The wavenumber $k = \omega p$ in terms of the horizontal slowness p. It is convenient to remap the variable of integration in expressions such as (2.1.25) to slowness p, since this provides a more direct relation to physical processes. For a general source, we can then write a representation of the surface displacement field in the frequency–slowness domain as (cf: § SWI:15.3)

$$\mathbf{u}_0(\Delta, \phi, \omega) = \left[\frac{\Delta}{\sin\Delta}\right]^{\frac{1}{2}} \omega^2 \int_0^{\infty} dp\, p \sum_m \mathbf{w}_0^T(p, m, \omega) \mathsf{T}_m(\omega p r_e \Delta), \qquad (2.1.26)$$

where T_m represents the tensor field of horizontal phase terms of order m, derived from the vector surface harmonics. The response from the source and propagation from source to receiver $\mathbf{w}_0(p, m, \omega) = [U(p, m, \omega), V(p, m, \omega), W(p, m, \omega)]$. The superscript ^{T} denotes the transpose.

When $\omega\wp\Delta = \omega p r_e \Delta \gg 1$ we can make a further asymptotic development in terms of travelling waves:

$$\mathbf{u}_0(\Delta, \phi, \omega) = \left[\frac{\Delta}{\sin\Delta}\right]^{\frac{1}{2}} \tfrac{1}{2}\omega|\omega| \int_0^\infty dp\, p \sum_m \mathbf{w}_0^T(p, m, \omega) \mathsf{T}_m^{(1)}(\omega p r_e \Delta), \quad (2.1.27)$$

where the contour of integration in the p plane goes just above the real axis at the origin to avoid a branch point. The asymptotic forms of the vector harmonic contributions $\mathsf{T}_m^{(1)}(\omega p r_e \Delta)$ have just an exponential dependence on range, and are directed along the orthogonal coordinate vectors $\hat{\mathbf{e}}_z = -\hat{\mathbf{e}}_r, \hat{\mathbf{e}}_\Delta, \hat{\mathbf{e}}_\phi$. Thus the tensor phase-field contribution is

$$\mathsf{T}_m^{(1)} \sim \begin{pmatrix} \hat{\mathbf{e}}_z \\ i\hat{\mathbf{e}}_r \\ -i\hat{\mathbf{e}}_\phi \end{pmatrix} \left(\frac{2}{\pi\omega p r_e \Delta}\right)^{\frac{1}{2}} \exp\left[i\omega p r_e \Delta - i(2m+1)\frac{\pi}{4} - i\sigma\frac{\pi}{2}\right], \quad (2.1.28)$$

with the same slowness and distance dependence for all three components of displacement. The asymptotic representation (2.1.28) is suitable for use away from the poles of the sphere ($\Delta = 0°$ and $180°$); the index σ is incremented by unity each time a pole is traversed.

2.2 Extraction of Surface Waves and Body Waves

The integration over slowness can be adapted to extract different aspects of the seismic wavefield. Consider the azimuthally symmetric part ($m = 0$) of the vertical displacement at the surface at epicentral distance Δ. We use a moment rate model for the source and separate its spectrum $S(\omega)$ from the response of the spherically stratified model $G(p, \omega)$. At frequency ω

$$u_{z0}(\Delta, \omega) = f(\Delta) S(\omega) \omega^{\frac{1}{2}} \int_0^{p_0} dp\, p^{\frac{1}{2}} G(p, \omega) e^{i\omega p r_e \Delta}. \quad (2.2.1)$$

The limit in the slowness integral p_0 is to be taken somewhat larger than p_{R0}, the slowness of the fundamental mode Rayleigh wave, i.e. $p_0 > 1.25\beta_0^{-1}$, where β_0 is the surface *S* wavespeed. For the sphere, the spreading factor for amplitude is given by $f(\Delta) = (1/\sin\Delta)^{1/2}$.

As we can see from Figure 2.1 (central panel), the fundamental Rayleigh mode branch in the spheroidal modes is well separated from the other branches for frequencies above 30 mHz. We can distort the slowness integral away from the real axis into the upper quadrant of the complex p plane so as to pick up the contribution from the fundamental mode Rayleigh wave (Figure 2.2). We break away from the real axis at p_c in the gap before the next mode branch is reached. For a slightly attenuative medium the pole corresponding to the fundamental mode lies just above the real p axis. There will be limited 'leaking-mode' additions from

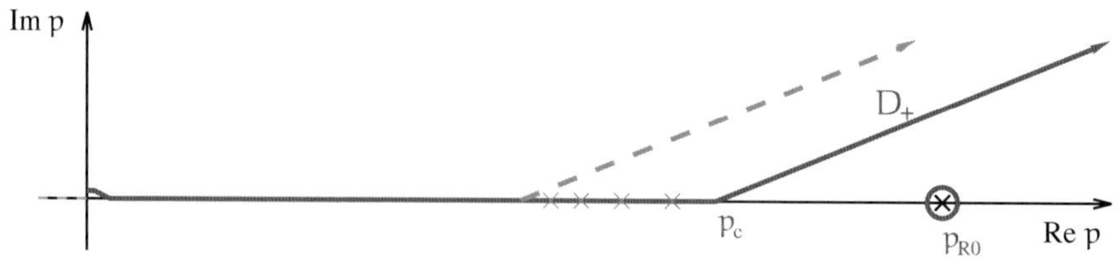

Figure 2.2 Deformation of the contour path in the complex slowness (p) domain, to pick up the fundamental mode Rayleigh pole, at fixed frequency ω. Moving the hinge point p_c to smaller values as indicated by the dashed lines, exposes higher mode poles and gives a stronger contribution from D_+.

the contour D_+, that will decay rapidly with increasing Δ. The body waves will be contained in the real-axis contribution out to p_c.

The slowness integral in (2.2.1) can be written as

$$u_{z0}(\Delta, \omega) = \omega^{\frac{1}{2}} S(\omega) f(\Delta) \left\{ \int_0^{p_c} dp\, p^{\frac{1}{2}} G(p, \omega) e^{i(\omega p r_e \Delta)} + H(p_{R0}(\omega), \omega) e^{i(\omega p_{R0}(\omega) r_e \Delta)} + \int_{D_+} dp\, p^{\frac{1}{2}} G(p, \omega) e^{i(\omega p r_e \Delta)} \right\}. \tag{2.2.2}$$

$H(p_{R0}(\omega), \omega)$ represents the residue contribution from the fundamental Rayleigh wave pole $p_{R0}(\omega)$, which for a spherically stratified model depends solely on the frequency ω and the Earth model. In this way we isolate the dominant surface wave contribution for the Rayleigh waves. The main body wave contributions come from the integral along the real slowness axis out to the hinge point p_c.

For the 30–70 mHz band a suitable choice for p_c is β_m^{-1}, i.e., the inverse of the uppermost mantle *S* wavespeed. Then crustal *S* contributions come from the fundamental mode Rayleigh wave, with some contributions from D_+. At higher frequencies, the surface waves penetrate less far into the Earth (§ SWI:16.3) and so, although there is a clear separation of the fundamental mode Rayleigh wave from the other modes, the hinge point p_c needs to be shifted to larger slowness values.

The original choice of hinge point p_c was chosen so that just the fundamental model pole was revealed. But, we can peel back the integration contour in the upper quadrant by progressively reducing p_c to reveal further discrete poles from the higher modes of the Rayleigh waves. In this way we can extract a sum of modal pole contributions, but now the contributions from the portion of the contour off the real axis D_+ become more significant especially at short epicentral distances. This process is indicated by the dashed line segment and pole markings in Figure 2.2.

A comparable development can be made for the Love wave contributions to the tangential component of motion from the toroidal mode spectrum (upper panel of Figure 2.1). The separation of the fundamental Love mode from the other mode branches is somewhat less, and so the 'leaking mode' contribution from D_+ can be more significant. For such *SH* wave propagation we can explicitly identify the sum of a suite of higher-mode contributions with multiple *SH* wave reflections from the surface – an example of *mode–ray duality* (see, e.g., § SWI:16.2; BenMenahem & Singh 1981, §8.4).

When we start from the normal mode field we include all aspects of the seismic wavefield at once. In many applications we wish to concentrate on particular aspects of the wavefield and so use some subset of the full response of the Earth. The extraction of modal contributions, as discussed above, enables the large amplitude surface wave portions of seismograms to be isolated. To look at specific body wave arrivals, we need to set up ways of exploiting the physical character of the wavefield, and use these in association with selective integration over slowness (cf. § SWI:15.3,16.1).

2.3 Description of Physical Processes

An effective way of relating the nature of the seismic wavefield to the processes that contribute to the passage of waves between source and receiver is in terms of the reflection and transmission of seismic waves. This approach is developed in detail in SWI:chapters 12–14, based on the exploitation of results in the slowness–frequency (p, ω) domain. Here we bring out some of the main results for the reflection and transmission properties, including the interaction between regions, and the inclusion of sources that will be of significance in later developments in this book. We show how systematic approximations to the nature of the seismic wavefield can be developed to emphasise specific features of the field.

2.3.1 *Reflection and Transmission Matrices*

In a uniform medium we are able to represent the displacement and traction fields in terms of a combination of upgoing and downgoing waves. Such wave elements will be reflected and transmitted at interfaces, and the process can be described by the introduction of reflection and transmission matrices **R**, **T** linking the different wave types. For isotropic (or radially anisotropic) media the 3×3 reflection and transmission matrices block partition into coupled *P–SV* terms and *SH* elements

$$\mathbf{R} = \begin{pmatrix} R^{PP} & R^{PS} & 0 \\ R^{PP} & R^{PS} & 0 \\ 0 & 0 & R^{HH} \end{pmatrix}, \qquad \mathbf{T} = \begin{pmatrix} T^{PP} & T^{PS} & 0 \\ T^{PP} & T^{PS} & 0 \\ 0 & 0 & T^{HH} \end{pmatrix} \tag{2.3.1}$$

We use the subscripts D, U to indicate the initial direction of the incident wave, so that $\mathbf{R}_D$ is produced by an incident downgoing wave. We also use the convention that, e.g., R^{PS} represents reflection from a *SV* wave into a *P* wave so that we can use the ordinary rules of matrix multiplication to generate component expressions.

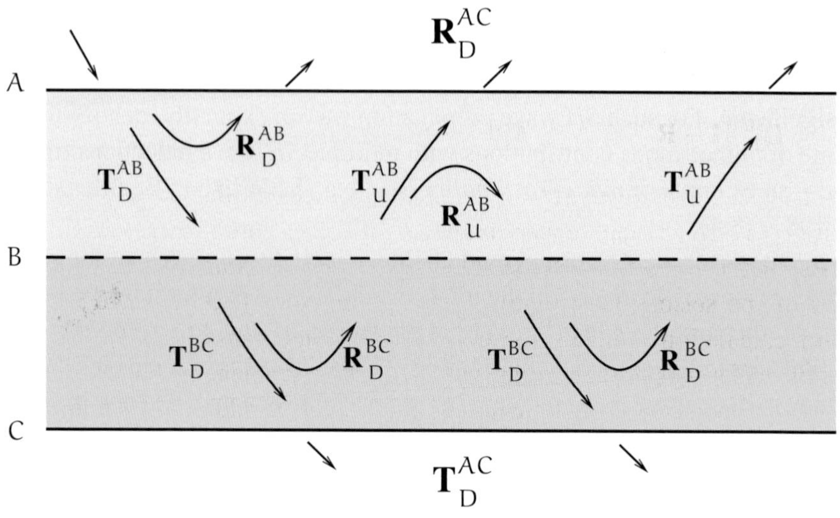

Figure 2.3 Building the reflection and transmission response for downward propagating waves impinging on a composite region AC in terms of the reflection and transmission matrices for upgoing and downward waves on the subdivisions AB and AC.

For a single interface in horizontally stratified media, the reflection and transmission coefficients depend only on slowness p, and this property applies asymptotically for high frequencies in the sphere.

The free surface plays a critical role in many wave phenomena. We introduce the free-surface reflection matrix $\mathbf{R}_F$, and the surface displacement transfer matrix $\mathbf{W}_F$ that converts the upgoing wavefield at the surface into displacement (§ SWI:13.1). There is coupling between the *P* and *SV* waves at the surface, but the factor W_F^{HH} for *SH* waves that are completely reflected at the surface is 2.

We can build the reflection and transmission response of a portion of the structure by bounding it by uniform media from which downgoing and upgoing waves can impinge on the structure. For a zone between radial surfaces A, B we introduce the set of matrices $\mathbf{R}_D^{AB}(p,\omega), \mathbf{T}_D^{AB}(p,\omega)$ for downward incident waves at A, and $\mathbf{R}_U^{AB}(p,\omega), \mathbf{T}_U^{AB}(p,\omega)$ for upward incident waves at B. Even for a uniform zone, frequency dependence is introduced through the phase delays in propagation.

We can then add in the effects of further structure in the region between B and C in terms of the reflection and transmission matrices for the region BC. The effect of a downward wave incident on the composite zone AC from A to C, can then be built by taking into consideration all the interactions with the upper and lower zones – as indicated schematically in Figure 2.3. The reflection and transmission matrices $\mathbf{R}_D^{AC}$, $\mathbf{T}_D^{AC}$ for the composite zone AC can be built from the reflection and transmission matrices for the zones AB and BC for both downward and upward incident waves. The addition rules take the form:

$$\mathbf{R}_{\mathrm{D}}^{\mathrm{AC}} = \mathbf{R}_{\mathrm{D}}^{\mathrm{AB}} + \mathbf{T}_{\mathrm{U}}^{\mathrm{AB}}\mathbf{R}_{\mathrm{D}}^{\mathrm{BC}}\left[\mathbf{I} - \mathbf{R}_{\mathrm{U}}^{\mathrm{AB}}\mathbf{R}_{\mathrm{D}}^{\mathrm{BC}}\right]^{-1}\mathbf{T}_{\mathrm{D}}^{\mathrm{AB}}, \tag{2.3.2}$$

$$\mathbf{T}_{\mathrm{D}}^{\mathrm{AC}} = \mathbf{T}_{\mathrm{U}}^{\mathrm{BC}}\left[\mathbf{I} - \mathbf{R}_{\mathrm{U}}^{\mathrm{AB}}\mathbf{R}_{\mathrm{D}}^{\mathrm{BC}}\right]^{-1}\mathbf{T}_{\mathrm{D}}^{\mathrm{AB}}. \tag{2.3.3}$$

where $\mathbf{I}$ is the identity matrix. The set of processes that involve successive reflections between the upper and lower zones are represented through $[\mathbf{I} - \mathbf{R}_{\mathrm{U}}^{\mathrm{AB}}\mathbf{R}_{\mathrm{D}}^{\mathrm{BC}}]^{-1}$. Such reverberation terms play an important role in many aspects of the seismic wavefield. The addition rules (2.3.2), (2.3.3) form the basis of calculation schemes for the reflection and transmission properties of stacks of uniform layers, or regions with velocity gradients (§ SWI:14.2) in the slowness–frequency (p, ω) domain. A similar set of addition relations for the reflection and transmission matrices $\mathbf{R}_{\mathrm{U}}^{\mathrm{AC}}$, $\mathbf{T}_{\mathrm{U}}^{\mathrm{AC}}$ can be developed for upgoing waves incident at C.

An important special case for incident upgoing waves is provided by a zone bounded above by the free surface. For the zone between B and the free surface we write $\mathbf{R}_{\mathrm{U}}^{\mathrm{fB}}(p, \omega)$ for reflection from beneath and $\mathbf{W}_{\mathrm{U}}^{\mathrm{fB}}(p, \omega)$ for the surface displacement transfer matrix from incident upgoing waves at B, which represents the combination of transmission to the free surface and amplification at the surface. When an additional zone is added below extending to the surface C, the matrices for the composite region are built in a similar way to (2.3.2), (2.3.3):

$$\mathbf{R}_{\mathrm{U}}^{\mathrm{fC}} = \mathbf{R}_{\mathrm{U}}^{\mathrm{BC}} + \mathbf{T}_{\mathrm{D}}^{\mathrm{BC}}\mathbf{R}_{\mathrm{U}}^{\mathrm{fB}}\left[\mathbf{I} - \mathbf{R}_{\mathrm{D}}^{\mathrm{BC}}\mathbf{R}_{\mathrm{U}}^{\mathrm{fB}}\right]^{-1}\mathbf{T}_{\mathrm{U}}^{\mathrm{BC}}, \tag{2.3.4}$$

$$\mathbf{W}_{\mathrm{U}}^{\mathrm{fC}} = \mathbf{W}_{\mathrm{U}}^{\mathrm{fB}}\left[\mathbf{I} - \mathbf{R}_{\mathrm{D}}^{\mathrm{BC}}\mathbf{R}_{\mathrm{U}}^{\mathrm{fB}}\right]^{-1}\mathbf{T}_{\mathrm{U}}^{\mathrm{BC}}. \tag{2.3.5}$$

In terms of propagating waves normalised to unit energy transport in the vertical direction, there are general symmetry properties for reflection and transmission (Kennett et al., 1978),

$$\mathbf{R}_{\mathrm{D}}^{\mathrm{AB}}(p, \omega) = [\mathbf{R}_{\mathrm{D}}^{\mathrm{AB}}(p, \omega)]^{\mathrm{T}}, \quad \mathbf{T}_{\mathrm{U}}^{\mathrm{AB}}(p, \omega) = [\mathbf{T}_{\mathrm{D}}^{\mathrm{AB}}(p, \omega)]^{\mathrm{T}}. \tag{2.3.6}$$

For a perfectly elastic medium, there are a number of other interrelations between the reflection and transmission matrices that involve complex conjugates (see. e.g., § SM:5.Appendix).

2.3.2 Inclusion of a Source

The concept of reflection and transmission processes can be linked to the presence of a source by decomposing the source radiation into an upgoing part U^{S} (heading towards the surface) and a downgoing part D^{S} that has to interact with structure before it can contribute to surface displacement. We can develop a representation for the medium response $\mathbf{w}_0(p, \omega)$ directly in terms of propagation elements above and below the source level (§ SWI:14.5.1). Each element of the propagation process can be described by the action of a reflection or transmission matrix in

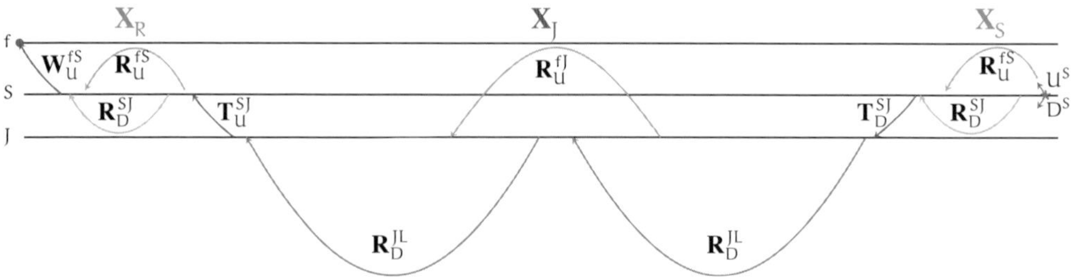

Figure 2.4 Schematic representation of the propagation elements that build the full seismic response for an Earth model for a source at level S, with a separation surface at level J beneath the source.

the slowness–frequency (p, ω) domain. For consistency with § SWI:14.5, we will designate the reflection matrix from below the level S as $\mathbf{R}_D^{SL}$. We can then express $\boldsymbol{w}_0(p, \omega)$ in the form

$$\boldsymbol{w}_0 = \mathbf{W}_U^{fS}\mathbf{U}^S + \mathbf{W}_U^{fS}[\mathbf{I} - \mathbf{R}_D^{SL}\mathbf{R}_U^{fS}]^{-1}\mathbf{R}_D^{SL}(\mathbf{D}^S + \mathbf{R}_U^{fS}\mathbf{U}^S). \tag{2.3.7}$$

The dependence of the surface response $\boldsymbol{w}_0(p, m, \omega)$ on angular order m enters solely through the source terms $\mathbf{D}^S$, $\mathbf{U}^S$. The reflection and transmission elements are independent of m.

The response due to excitation by a source can then be found from a slowness integral as in (2.1.27):

$$\mathbf{u}_0(\Delta, 0, \omega) = \left[\frac{\Delta}{\sin\Delta}\right]^{\frac{1}{2}} \tfrac{1}{2}\omega|\omega|M(\omega)\int dp\, p\sum_m \boldsymbol{w}_0^T(p, m, \omega)\mathsf{T}_m^{(1)}(\omega p r_e \Delta), \tag{2.3.8}$$

where the slowness range needs to be tailored to the portion of the wavefield of interest. Thus, surface waves will be found from the residue contributions at the poles of the reverberation operator $[\mathbf{I} - \mathbf{R}_D^{SL}\mathbf{R}_U^{fS}]^{-1}$, as discussed in Section 2.5. Body waves will be associated with smaller slownesses.

For many purposes it is desirable to allow for the influence of reduced seismic wavespeeds near the Earth's surface. We can achieve this by making an additional split in the stratification at a level J below the source (Figure 2.4), and thereby allow the propagation processes associated with shallow structure to be separated from those that involve interaction with the entire structure.

The radiation from the source initiates interaction with local structure, but this soon extends to the full model. Near the receiver, there is also the possibility of reverberative effects. The full wavefield for the situation with a separation level J can be written as:

$$\begin{aligned}\boldsymbol{w}_0 = {} & \mathbf{W}_U^{fS}\mathbf{U}^S + \mathbf{W}_U^{fS}[\mathbf{I} - \mathbf{R}_D^{SJ}\mathbf{R}_U^{fS}]^{-1}\mathbf{R}_D^{SJ}(\mathbf{D}^S + \mathbf{R}_U^{fS}\mathbf{U}^S) && (s)\\ & + \mathbf{W}_U^{fS}[\mathbf{I} - \mathbf{R}_D^{SJ}\mathbf{R}_U^{fS}]^{-1}\mathbf{T}_U^{SJ}[\mathbf{I} - \mathbf{R}_D^{JL}\mathbf{R}_U^{fJ}]^{-1}\mathbf{R}_D^{JL}\mathbf{T}_D^{SJ} && (2.3.9)\\ & \quad \times[\mathbf{I} - \mathbf{R}_U^{fS}\mathbf{R}_D^{SJ}]^{-1}(\mathbf{D}^S + \mathbf{R}_U^{fS}\mathbf{U}^S) && (d)\end{aligned}$$

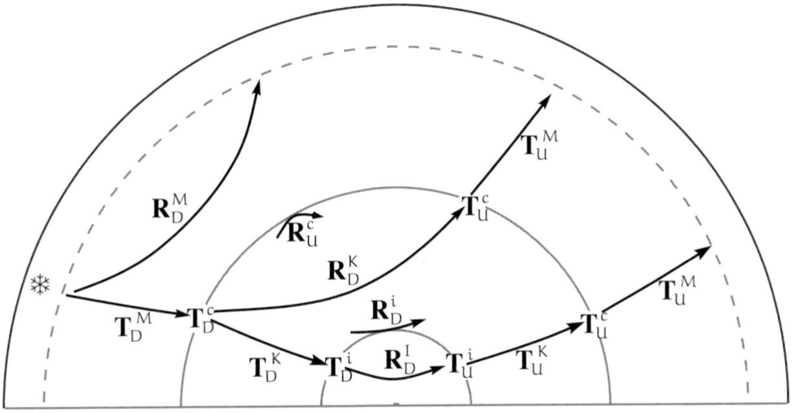

Figure 2.5 Representation of mantle and core propagation processes in terms of reflection and transmission matrices for the different regions and interfaces.

The first contribution (s) in (2.3.9) includes all propagation solely above the separation level J. $\mathbf{W}_U^{fS}\mathbf{U}^S$ is the direct wave from the source to the surface, and the second term in (s) includes the effects of downward radiation from the source $\mathbf{D}^S$ together with the surface 'ghost' $\mathbf{R}_U^{fS}\mathbf{U}^S$ that includes such surface reflected phases as *pP, sP, pS, sS*. The second contribution (d) describes all the processes that involve return of energy from beneath the separation level J.

We can recognise in the expression (2.3.9) the presence of reverberative effects through the whole structure through

$$\mathbf{X}_J = [\mathbf{I} - \mathbf{R}_D^{JL}\mathbf{R}_U^{fJ}]^{-1}. \tag{2.3.10}$$

The singularities of $\mathbf{X}_J(p, \omega)$ determine the position of the mode branches for the Earth model. We also have reverberations occurring near the source $\mathbf{X}_S$ and near the receiver $\mathbf{X}_R$.

$$\mathbf{X}_S = [\mathbf{I} - \mathbf{R}_U^{fS}\mathbf{R}_D^{SJ}]^{-1}, \qquad \mathbf{X}_R = [\mathbf{I} - \mathbf{R}_D^{SJ}\mathbf{R}_U^{fS}]^{-1}. \tag{2.3.11}$$

With these expressions we can write the shallow propagation terms from (2.3.9s) as

$$\boldsymbol{w}_{0s} = \mathbf{W}_U^{fS}\mathbf{U}^S + \mathbf{W}_U^{fS}\mathbf{X}_R\mathbf{R}_D^{SJ}(\mathbf{D}^S + \mathbf{R}_U^{fS}\mathbf{U}^S) \tag{2.3.12}$$

and the deep propagation contribution (2.3.9d) as

$$\boldsymbol{w}_{0d} = \mathbf{W}_U^{fS}\,\mathbf{X}_R\,\mathbf{T}_U^{SJ}\mathbf{X}_J\mathbf{R}_D^{JL}\mathbf{T}_D^{SJ}\,\mathbf{X}_S(\mathbf{D}^S + \mathbf{R}_U^{fS}\mathbf{U}^S). \tag{2.3.13}$$

As discussed in § SWII:26.1 the effect of all structure in the spherical Earth can be included by suitable representations of the energy return from depth $\mathbf{R}_D^{JL}$. The various elements associated with propagation through the major divisions of Earth structure with interactions with the major interfaces are indicated schematically in Figure 2.5. We use a designation linked to conventional seismological notation so that the superscripts c is linked to the core–mantle boundary, and i to the inner-core boundary. We use M for the mantle, K for the outer core and I for the inner core. For global studies it is convenient to split the response of the Earth at the core–mantle boundary so that

$$\mathbf{R}_{\mathrm{D}}^{\mathrm{JL}} = \mathbf{R}_{\mathrm{D}}^{\mathrm{M}} + \mathbf{T}_{\mathrm{U}}^{\mathrm{M}}\mathbf{T}_{\mathrm{U}}^{\mathrm{c}}[\mathbf{I} - \mathbf{R}_{\mathrm{D}}^{\mathrm{KI}}\mathbf{R}_{\mathrm{U}}^{\mathrm{c}}]^{-1}\mathbf{R}_{\mathrm{D}}^{\mathrm{KI}}\mathbf{T}_{\mathrm{D}}^{\mathrm{c}}\mathbf{T}_{\mathrm{D}}^{\mathrm{M}}, \tag{2.3.14}$$

where $\mathbf{R}_{\mathrm{D}}^{\mathrm{KI}}$ is the return from the entire structure beneath the core–mantle boundary, and $\mathbf{R}_{\mathrm{D}}^{\mathrm{M}}$ includes all mantle propagation and upperside core–mantle boundary reflections (*PcP*, *ScP*, *ScS*, *PcS*). The terms $\mathbf{T}_{\mathrm{D}}^{\mathrm{M}}$, $\mathbf{T}_{\mathrm{U}}^{\mathrm{M}}$ carry the waves through the mantle to and from the core–mantle boundary. At the core-mantle boundary the transmission matrices $\mathbf{T}_{\mathrm{D}}^{\mathrm{c}}$, $\mathbf{T}_{\mathrm{U}}^{\mathrm{c}}$ include the possibility of conversion between *P* in the core and *SV* in the mantle. Within the fluid outer core, we have only *P* wave propagation, and so the reflection matrices $\mathbf{R}_{\mathrm{D}}^{\mathrm{KI}}$ for the entire core and $\mathbf{R}_{\mathrm{U}}^{\mathrm{c}}$ for the underside reflection at the core-mantle boundary have a single *PP* entry. Conversion to *SV* waves is possible at the inner core boundary.

In the integration over slowness at fixed frequency, as in (2.1.27), the slowness components of $\boldsymbol{w}_0$ combine with the phase contribution from $\mathsf{T}^{(1)}$ to produce the displacement field at an epicentral distance Δ. Generally, the contribution (s) in (2.3.9) is most significant at shorter distances, and the return from depth (d) dominates at larger distances, notably because of the presence of surface waves from the poles of $\mathbf{X}_{\mathrm{J}}$.

2.3.3 Multiple Deep Reflections

We can recast the component (d) of (2.3.9), for wave phenomena that link below the separation level J, in terms of the reverberation terms introduced in (2.3.10), (2.3.11):

$$\boldsymbol{w}_{0\mathrm{d}} = \mathbf{W}_{\mathrm{U}}^{\mathrm{fS}}\,\mathbf{X}_{\mathrm{R}}\,\mathbf{T}_{\mathrm{U}}^{\mathrm{SJ}}\,\mathbf{X}_{\mathrm{J}}\mathbf{R}_{\mathrm{D}}^{\mathrm{JL}}\,\mathbf{T}_{\mathrm{D}}^{\mathrm{SJ}}\,\mathbf{X}_{\mathrm{S}}(\mathbf{D}^{\mathrm{S}} + \mathbf{R}_{\mathrm{U}}^{\mathrm{fS}}\mathbf{U}^{\mathrm{S}}). \tag{2.3.15}$$

Both the downgoing waves and their surface reflections interact with shallow structure at source and receiver ends and these effects modulate the deep reflections. The representation (2.3.15) will be found useful in our discussion of the reflection field in Chapter 4, and is helpful whenever attention is directed to multiple reflections. Attention can be drawn to just the first few reflections by expanding the reverberation term $\mathbf{X}_{\mathrm{J}}$ using the identity

$$(\mathbf{I} - \mathbf{A})^{-1} = \mathbf{I} + \mathbf{A}(\mathbf{I} - \mathbf{A})^{-1}. \tag{2.3.16}$$

Thus we can express the reflections from depth through

$$\mathbf{X}_{\mathrm{J}}\mathbf{R}_{\mathrm{D}}^{\mathrm{JL}} = \mathbf{R}_{\mathrm{D}}^{\mathrm{JL}} + \mathbf{R}_{\mathrm{D}}^{\mathrm{JL}}\mathbf{R}_{\mathrm{U}}^{\mathrm{fJ}}\mathbf{R}_{\mathrm{D}}^{\mathrm{JL}} + [\mathbf{I} - \mathbf{R}_{\mathrm{D}}^{\mathrm{JL}}\mathbf{R}_{\mathrm{U}}^{\mathrm{fJ}}]^{-1}\mathbf{R}_{\mathrm{D}}^{\mathrm{JL}}\mathbf{R}_{\mathrm{U}}^{\mathrm{fJ}}\mathbf{R}_{\mathrm{D}}^{\mathrm{JL}}, \tag{2.3.17}$$

which allows the identification of successive multiples.

If we introduce further substructure in the region below J as in the description of mantle and core processes in (2.3.14) then we can make use of a further identity

$$(\mathbf{I} - \mathbf{A} - \mathbf{B})^{-1} = (\mathbf{I} - \mathbf{A})^{-1} + (\mathbf{I} - \mathbf{A})^{-1}\mathbf{B}(\mathbf{I} - \mathbf{A} - \mathbf{B})^{-1}, \tag{2.3.18}$$

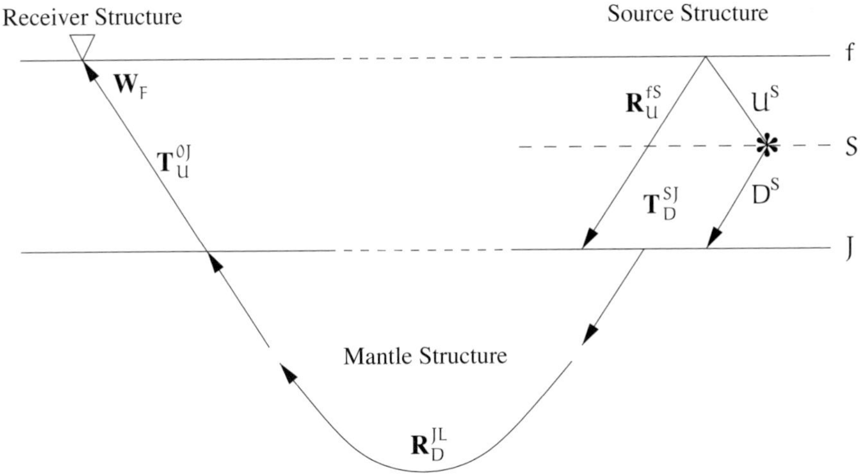

Figure 2.6 Schematic representation of teleseismic propagation processes with passage through the deep mantle, shallow reverberations associated with $\mathbf{X}_R$, $\mathbf{X}_S$ are not indicated.

and so successively produce an expansion that emphasises $(\mathbf{I} - \mathbf{A})^{-1}$, with a sequence of multiple powers of the combination $(\mathbf{I} - \mathbf{A})^{-1}\mathbf{B}(\mathbf{I} - \mathbf{A})^{-1}$ By appropriate choice of $\mathbf{A}$ and $\mathbf{B}$ we can concentrate attention on reverberations involving different parts of the structure.

For example, we can exploit the separation of mantle and core propagation in (2.3.14), by expressing $\mathbf{R}_D^{JL}$ in the form

$$\mathbf{R}_D^{JL} = \mathbf{R}_D^{M} + \mathbf{T}_U^{M}\mathbf{Y}^{C}\mathbf{T}_D^{M}, \quad \text{with} \quad \mathbf{Y}^{C} = \mathbf{T}_U^{c}[\mathbf{I} - \mathbf{R}_D^{KI}\mathbf{R}_U^{c}]^{-1}\mathbf{R}_D^{KI}\mathbf{T}_D^{c}, \tag{2.3.19}$$

so that mantle and core propagation processes are distinct. Then we may represent the reverberation operator $\mathbf{X}_J = [\mathbf{I} - \mathbf{R}_D^{JL}\mathbf{R}_U^{fJ}]^{-1}$ in a form where mantle reverberations are emphasised

$$\mathbf{X}_J = [\mathbf{I} - \mathbf{R}_D^{M}\mathbf{R}_U^{fJ}]^{-1} + [\mathbf{I} - \mathbf{R}_D^{M}\mathbf{R}_U^{fJ}]^{-1}\mathbf{T}_U^{M}\mathbf{Y}^{C}\mathbf{T}_D^{M}[\mathbf{I} - \mathbf{R}_D^{M}\mathbf{R}_U^{fJ}]^{-1} + \dots. \tag{2.3.20}$$

Each transmission element through the mantle is accompanied by the effects of reflection from the free surface back into the mantle.

2.4 Teleseismic Phases at a Receiver

For teleseismic arrivals, we can place attention on just those waves that have been returned once from depth so that from (2.3.15) we extract

$$w_0^{tel} = \mathbf{C}_R\mathbf{R}_D^{JL}\mathbf{C}_S = \mathbf{W}_U^{fS}\,\mathbf{X}_R\,\mathbf{T}_U^{SJ}\,\mathbf{R}_D^{JL}\,\mathbf{T}_D^{SJ}\,\mathbf{X}_S(\mathbf{D}^S + \mathbf{R}_U^{fS}\mathbf{U}^S). \tag{2.4.1}$$

The element $\mathbf{C}_S$ represents the source, and the multiples and conversions in the structure near the source, $\mathbf{R}_D^{JL}$ describes a single reflection below J, and the element $\mathbf{C}_R$ is the contribution from the structure in the neighbourhood of the receiver.

At teleseismic ranges between 3000 km and 9500 km from the source, the

PP, SS or HH elements of the mantle reflection matrix $\mathbf{R}_D^{JL}$ can normally be approximated in terms of a simple process of total reflection from the lower mantle so that, e.g.,

$$[\mathbf{R}_D^{JL}]_{PP} \sim \exp\{i\omega\tau_M(p) - i\tfrac{\pi}{2}\}, \tag{2.4.2}$$

where $\tau_M(p)$ is the phase delay for the mantle below the separation level z_J,

$$\tau_M(p) = 2\int_{z_J}^{Z_p} dz\, q_\alpha(z), \tag{2.4.3}$$

and the extra phase delay of $\exp(-i\pi/2)$ is associated with the turning point for a ray. Frequency-dependent compensation can also be included for the effects of attenuation (§ SWI:16.1.2). Depending on the distance range it may be necessary to also include core reflections and refractions that can interfere with the main *P* and *S* arrivals at larger teleseismic distances.

To recover the seismograms at a given range Δ we have to evaluate an integral of the form (2.1.27) over a restricted range of positive slowness

$$\mathbf{u}_0(\Delta, 0, \omega) = \left[\frac{\Delta}{\sin\Delta}\right]^{\frac{1}{2}} \tfrac{1}{2}\omega|\omega|M(\omega)\int_\Gamma dp\, p\, [\boldsymbol{w}_0^{tel}]^T\, T^{(1)}(\omega p r_e \Delta), \tag{2.4.4}$$

for a suitable interval Γ. In (2.4.4) we have extracted a common source spectrum $M(\omega)$. For simplicity the summation over angular order m associated with source radiation is suppressed. For many purposes it is adequate to assume that the arrival at a teleseismic receiver can be treated as a plane wave with the geometric slowness for the wave type under consideration, e.g., p_{gP} for a *P* wave. This is equivalent to making a saddle-point approximation for the slowness integral in (2.4.4). A more accurate treatment is to undertake a limited slowness integral around p_g with a weighting function to avoid numerical arrivals from the truncation of the slowness integral (§ SWI:16.1.2), as in the work of Marson-Pidgeon & Kennett (2000).

With the simpler, single-slowness representation, the character of a teleseismic arrival at a station can be expressed in terms of incident plane waves with a single slowness component p_g. We separate the contributions that are local to the receiver from those imposed by the source and the propagation path to the local structure. Thus we represent the surface displacement as

$$\boldsymbol{w}_0^{tel}(p_g, \omega) = S_I(p_g, \omega)\mathbf{Z}(p_g, \omega)\mathbf{C}_R(p_g, \omega)\Phi(p_g, \omega), \tag{2.4.5}$$

where $\Phi(p_g, \omega)$ represents the phase propagation effects from the source, $\mathbf{Z}(p_g, \omega)$ includes amplitude contributions prior to arrival at the zone below the station and $S_I(p_g, \omega)$ the combined effects of excitation and the instrumental response. $\mathbf{C}_R(p_g, \omega)$ describes the effect of structure local to the station.

We take a transmission zone from a level $z = z_J$ below the base of the crust to the surface at $z = 0$, as sketched in Figure 2.7. The local structural term $\mathbf{C}_R(p, \omega)$

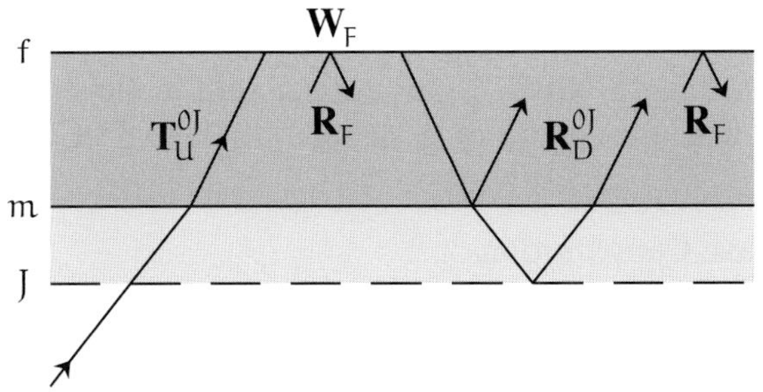

Figure 2.7 Transmission of plane waves through the zone from level J to the surface. The lower boundary is taken below the base of the crust at m. The action of the transmission and reflection matrices is represented schematically.

can then be written as

$$\begin{aligned} \mathbf{C}_R(p_g,\omega) &= \mathbf{W}_F(p_g)\mathbf{T}_U^{FJ}(p_g,\omega) \\ &= \mathbf{W}_F(p_g)\left[\mathbf{I}-\mathbf{R}_D^{0J}(p_g,\omega)\mathbf{R}_F(p_g)\right]^{-1}\mathbf{T}_U^{0J}(p_g,\omega), \end{aligned} \tag{2.4.6}$$

in terms of the transmission matrix through the zone 0J, $\mathbf{T}_U^{0J}$; the reflection matrix at the free surface, $\mathbf{R}_F$; reflection back from 0J, $\mathbf{R}_D^{0J}$; and the amplification of ground motion at the free surface, represented by $\mathbf{W}_F$. We also introduce the modified reflection matrix

$$\mathbf{R}_D^{FJ}(p,\omega) = \left[\mathbf{I}-\mathbf{R}_D^{0J}(p,\omega)\mathbf{R}_F(p)\right]^{-1}\mathbf{R}_D^{0J}(p,\omega), \tag{2.4.7}$$

which includes the effects from reflection at the free surface, and reverberations above the level J.

The symmetry property of the reflection matrix (2.3.6) corresponds to symmetry in conversion between *P* and *SV* waves. For a perfectly elastic medium we have a further property for propagating waves,

$$[\mathbf{R}_D^{0J}(p)]^{T*}\mathbf{R}_D^{0J}(p) + [\mathbf{T}_D^{0J}(p)]^{T*}\mathbf{T}_D^{0J}(p) = \mathbf{I}, \tag{2.4.8}$$

where the star denotes a complex conjugate. The modified reflection and transmission matrices also satisfy the relation

$$\begin{aligned} \mathbf{I} + \mathbf{R}_F(p)[\mathbf{R}_D^{FJ}(p,\omega)]^* + [\mathbf{R}_D^{FJ}(p,\omega)]^T[\mathbf{R}_F(p)]^T \\ = [\mathbf{T}_U^{FJ}(p,\omega)]^*[\mathbf{T}_U^{FJ}(p,\omega)]^T, \end{aligned} \tag{2.4.9}$$

where transmission symmetry has been employed to represent the right-hand side in terms of upward transmission. Frasier (1970) was the first to obtain a relation equivalent to (2.4.9) for the *P–SV* system, and a convenient derivation is provided by Ursin (1983). As we shall see in Section 5.2.1 the relation (2.4.9) forms the basis for extracting reflection information from the auto-correlation of transmitted waves from distant events recorded at the surface.

A more detailed treatment of the propagation effects near a receiver is provided in § SWII:28.1 in a discussion of the nature of receiver functions. The representation used in § SWII:28.1 introduces a split in the stratification at a level K, above J, to isolate reverberations in shallow structure near the receiver.

2.5 Surface-Wave Contributions

For a stratified medium, the representation of the seismic response to a source as a function of angular frequency (ω) and slowness (p) has singularities in the form of poles in the complex ω–p domain corresponding to those combinations of frequency and slowness for which it is possible to satisfy simultaneously the conditions on the wavefield for vanishing traction at the surface, and decay at great depth.

In these circumstances the reverberation operator for the stratified structure is singular and the pole locations are to found from the multiple roots of

$$\det[\mathbf{I} - \mathbf{R}_D^{JL}(p,\omega)\mathbf{R}_U^{fJ}(p,\omega)] = 0, \tag{2.5.1}$$

in terms of the reflection response above a level J including the free surface $\mathbf{R}_U^{fJ}(p,\omega)$ and that from below this level $\mathbf{R}_D^{JL}(p,\omega)$. Note that since the level J is arbitrary a wide range of different forms of equation have the same pole locations in slowness–frequency (p,ω) space; the residue contributions from this common set of poles constitute the surface-wave field.

The modal contribution to the surface displacement for Rayleigh waves at range $X = r_e\Delta$ and azimuth ϕ from the source can be represented as the sum of residue contributions at the poles of the dispersion relation (2.5.1) linked to the phase contribution from the vector surface harmonics:

$$\mathbf{u}_S(r,\phi,\omega) = \tfrac{1}{2}i\omega^2 \sum_m e^{im\phi}\left[\sum_{j=0}^{N(\omega)} p_j \operatorname*{Res}_{\omega,p=p_j}[\mathbf{w}_0^T(p,m,\omega)]\mathbf{T}_m^{(1)}(\omega p_j X)\right]. \tag{2.5.2}$$

Here $\mathbf{w}_0(p,m,\omega)$ represents the spectral response for the coupled *P–SV* wave system in the ω–p domain and $N(\omega)$ is the number of Rayleigh modes extracted for frequency ω. The angular variation in radiation from the source is represented through the summation over the angular order m of the vector surface harmonics contained in the tensor field $\mathbf{T}_m^{(1)}$. The azimuthal variation in the spectral representation of the response comes from the expansion of the source contribution, which can be conveniently represented in terms of discontinuities in displacement (S_W) and traction (S_T) across the source level $z = z_s$, (§ SWI:14.4):

$$\mathbf{S}(p,m,\omega,z_s) = [\mathbf{S}_W^m(z_s), \mathbf{S}_T^m(z_s)]^T. \tag{2.5.3}$$

At a pole corresponding to a surface-wave mode, a single eigen-displacement field $\mathbf{u}^e$ satisfies both the surface and radiation conditions. The residue at the pole

$H(p, \omega)$ can be extracted in terms of the source jumps and this eigen-displacement field

$$\operatorname*{Res}_{\omega, p=p_j} [\mathbf{w}_0^T(p, m, \omega)] = \frac{g_j}{2\omega I_j}[\mathbf{t}_j^{eT}(z_s)\mathbf{S}_W^m - \mathbf{u}_j^{eT}(z_s)\mathbf{S}_T^m]\mathbf{u}_j^e(0), \tag{2.5.4}$$

where g_j is the group slowness for the particular mode, and I_j is related to the kinetic energy content in the eigenfunctions

$$I_j = \int_0^\infty dz\, \rho(z)\, \mathbf{u}_j^{eT}(z)\mathbf{u}_j^e(z). \tag{2.5.5}$$

The residue contribution has a convenient separation into a part which is linked to the source $[\mathbf{t}_j^{eT}(z_s)\mathbf{S}_W^m - \mathbf{u}_j^{eT}(z_s)\mathbf{S}_T^m]$, and a displacement contribution at the receiver $\mathbf{u}_j^e(0)$.

With this expression for the residue contribution from each mode, the surface displacement can be written as

$$\mathbf{u}_S(r, \phi, \omega) = i\omega \sum_m e^{im\phi} \left[\sum_{j=0}^{N(\omega)} \frac{p_j g_j}{4I_j} \mathcal{S}_R^m(p_j)\mathcal{R}(p_j)\mathbf{T}_m^{(1)}(\omega p_j X) \right]. \tag{2.5.6}$$

The source contribution ($\mathcal{S}_R$) and receiver contribution ($\mathcal{R}$) to the displacement can be expressed in ways which have more direct physical content. The source dependence can be written in terms of the radiation components

$$\mathcal{S}_R^m(p_j) = [\mathbf{t}_j^{eT}(z_s)\mathbf{S}_W^m - \mathbf{u}_j^{eT}(z_s)\mathbf{S}_T^m] = -i[\mathbf{U}^m(z_s) + \mathbf{R}_D^{SL}\mathbf{D}^m(z_s)], \tag{2.5.7}$$

where $\mathbf{U}^m(z_s)$ is the upgoing radiation from the source in angular order m and $\mathbf{D}^m(z_s)$ is the corresponding downgoing radiation. We note that since (2.5.7) involves the processes of reflection from beneath the source level through $\mathbf{R}_D^{SL}$, the contribution involving the source is not localised at the source itself, but will involve the full propagation path. The receiver term includes the processes of transmission from the source level to the surface

$$\mathcal{R}(p_j) = \mathbf{u}_j^e(0) = \mathbf{W}_F[\mathbf{I} - \mathbf{R}_D^{0S}\mathbf{R}_F]^{-1}\mathbf{T}_U^{0S}, \tag{2.5.8}$$

where $\mathbf{W}_F$ allows for the amplification of displacement due to interaction with the free surface and the instrument response. $\mathbf{R}_F$ is the free-surface reflection matrix. The reflection and transmission terms $\mathbf{R}_D^{0S}$, $\mathbf{T}_U^{0S}$ involve the region between the source and the surface. For long propagation paths or higher frequencies, we can use asymptotic approximations for the vector surface harmonics (2.1.28). The representation of multi-modal Rayleigh waves then takes the form

$$\mathbf{u}_S(X, \phi, t) = \begin{pmatrix} i\mathbf{e}_z \\ -\mathbf{e}_r \end{pmatrix} \sum_m e^{im\phi} \int_{-\infty}^{\infty} d\omega\, e^{-i\omega t}$$
$$\sum_{j=0}^{N(\omega)} \sqrt{\frac{2\omega p_j}{\pi X}} \frac{g_j}{4I_{Rj}} \mathcal{S}_R^m(p_j)\, \mathcal{R}_R(p_j)\, e^{[i\omega p_j X - i(2m+1)\pi/4]}. \tag{2.5.9}$$

This asymptotic representation separates the phase contribution from the passage to the range X from the decay in amplitude associated with the spreading out the waves across the surface. We should remember that there is an implied dependence on frequency ω for the phase and group slowness p_j, g_j, and all the terms such as $\mathcal{S}$, $\mathcal{R}$ which depend on the properties of individual mode branches. Although we have identified source and receiver terms above, we have already seen that these have *non-local* properties.

The equivalent representation for Love waves is

$$\mathbf{u}_H(X, \phi, t) = \mathbf{e}_\phi \sum_m e^{im\phi} \int_{-\infty}^{\infty} d\omega\, e^{-i\omega t} \sum_{l=0}^{M(\omega)} \sqrt{\frac{2\omega p_l}{\pi X}\frac{g_l}{4I_{Ll}}}\, \mathcal{S}_L^m(p_l)\, \mathcal{R}_L(p_l)\, e^{[i\omega p_l X - i(2m+1)\pi/4]}. \quad (2.5.10)$$

In both (2.5.9) and (2.5.10) the major phase contribution comes from the horizontal propagation term, but there will also be some influence from the source and instrumental response.

The fundamental mode contributions separate from the higher modes by their larger slowness, and often the first and second higher modes are also distinct for larger propagation distances. The remainder of the higher modes are equivalent to multiple *S* arrivals (*SS*, *SSS*, ...), with some coupling to *P* for Rayleigh waves. This mode–ray correspondence becomes apparent if we make an expansion of the reverberation term $[\mathbf{I} - \mathbf{R}_D^{JL}(p,\omega)\mathbf{R}_U^{fJ}(p,\omega)]^{-1}$ that controls dispersion using the relation (2.3.16):

$$[\mathbf{I} - \mathbf{R}_D^{JL}\mathbf{R}_U^{fJ}]^{-1} = \mathbf{I} + \mathbf{R}_D^{JL}\mathbf{R}_U^{fJ} + \mathbf{R}_D^{JL}\mathbf{R}_U^{fJ}\mathbf{R}_D^{JL}\mathbf{R}_U^{fJ}[\mathbf{I} - \mathbf{R}_D^{JL}\mathbf{R}_U^{fJ}]^{-1}. \quad (2.5.11)$$

With each successive term in the expansion an extra contribution $\mathbf{R}_D^{JL}\mathbf{R}_U^{fJ}$ is introduced, which corresponds to an additional reflection from below the level J and from the zone above J including the free surface. The remainder term always contains the reverberations that yield the modal contributions. For *SH* waves, the *S* multiples appear directly. For *SV* waves a further decomposition of the reverberation term by wave types is required, and there is potential for modulation by shallow *P* wave propagation.

3

Laterally Varying Media

The use of stratified models provides much insight into the nature of the seismic wavefield. However, heterogeneity is pervasive on all scales and produces complexity in the wavefield and the character of seismograms. Direct numerical methods such as finite-difference and finite-element techniques provide estimates of the field, but do not identify physical processes. In this chapter we consider integral (weak) forms of the elastodynamic equations that allow us to extract representations for the displacement field in terms of surface integrals and volume contribution from real, or effective, sources. These expressions will see utility in our later discussion of correlation fields in Part II.

With the aid of the integral techniques, we develop a system of reflection and transmission operators for portions of the medium. Each propagation process can be associated with a specific operator or built up from the cumulative action of other operators. The algebra of these operators parallels that of the reflection and transmission matrices in stratified media, and so we can build a wide range of results for complex media.

This operator approach gives a full description of the propagation processes in complex media and so enables us to clearly identify and assess approximations for both body wave and surface wave propagation. For stratified media, the propagation of individual horizontal slowness components is independent, but in the presence of heterogeneity coupling is introduced.

3.1 Convolutions and Correlations

Although most of our mathematical development is made in the frequency domain, when we wish to exploit actual seismograms we need to convert to the time domain. In this context we will encounter two important styles of relations between different time series, correlations, and convolutions. Convolutional models are common for successive physical processes and play an important role in the representation of seismograms. Correlation properties have become of major importance in recent years as a means of exploiting noise and other less regarded features of the seismic wavefield.

The cross-correlation $C_{u,v}(t)$ of two time series $u(t)$, $v(t)$ of length T provides a measure of their similarity

$$C_{u,v}(t) = \int_{-T}^{T} d\tau\, u(\tau)v(\tau + t). \tag{3.1.1}$$

If the two signals are scaled but shifted in time, the correlation will peak at the time shift.

A closely related quantity, the convolution of the two time series $D_{u,v}(t)$ differs in the sign of t:

$$D_{u,v}(t) = u \star v(t) = \int_{-T}^{T} d\tau\, u(\tau)v(\tau - t). \tag{3.1.2}$$

If we take the Fourier transform with respect to time to convert into the frequency (ω) domain

$$\bar{C}_{u,v}(\omega) = \bar{u}(\omega)\bar{v}(\omega)^*, \tag{3.1.3}$$

where as in (2.1.3) the superscript * denotes the complex conjugate. The transform of the convolution D is just the product of the Fourier transforms of u, v:

$$\bar{D}_{u,v}(\omega) = \bar{u}(\omega)\bar{v}(\omega). \tag{3.1.4}$$

A further useful quantity is the auto-correlation of a time series, i.e., the correlation of a time series with itself,

$$C_{u,u}(t) = \int_{-T}^{T} d\tau\, u(\tau)u(\tau + t). \tag{3.1.5}$$

$C_{u,u}$ always has a peak at zero time lag, indicating the power in the signal, but features can emerge at later times where, e.g., there are repetitive processes with time delay. The Fourier transform of $C_{u,u}(t)$ is just the power spectrum $|\bar{u}(\omega)|^2$.

Both convolutions and correlations play an important role in integral representations for the seismic wavefield. Correlation results are linked to conservation of energy and so apply to perfectly elastic media. In contrast, the convolutional forms are of general application and apply in the presence of attenuation.

3.2 Integral Representations

We develop integral forms for the seismic wavefield that allow us to extract representation theorems for displacement, and propagation invariants.

Consider two elastic displacement fields $\mathbf{u}^{(1)}$, $\mathbf{u}^{(2)}$ with associated stress-tensor fields $\boldsymbol{\tau}^{(1)}$, $\boldsymbol{\tau}^{(2)}$. In Cartesian components, the equations of motion in the frequency domain are

$$\partial_j \tau_{ij}^{(1)} - \rho\omega^2 u_i^{(1)} = -f_i^{(1)}, \qquad \partial_j \tau_{ij}^{(2)} - \rho\omega^2 u_i^{(2)} = -f_i^{(2)}, \tag{3.2.1}$$

where we have written $\partial_j = \partial/\partial x_j$. The stress tensor $\tau_{ij} = c_{ijkl}\partial_k u_l$, and the elastic modulus tensor has the symmetries

$$c_{ijkl} = c_{jikl} = c_{ijlk} = c_{klij}. \tag{3.2.2}$$

We form the scalar product of each equation of motion with the alternate displacement field, and then subtract to eliminate the explicit frequency dependence. Then, integrating over a volume V in which the displacements and stresses are continuous,

$$\int_V d^3\mathbf{x}\left(u_i^{(1)}\partial_j\tau_{ij}^{(2)} - u_i^{(2)}\partial_j\tau_{ij}^{(1)}\right) = -\int_V d^3\mathbf{x}\left(u_i^{(1)}f_i^{(2)} - u_i^{(2)}f_i^{(1)}\right). \tag{3.2.3}$$

Now, using the divergence theorem, we find that over the bounding surface ∂V of V

$$\int_{\partial V} d^2\mathbf{x}\, n_j\left(u_i^{(1)}\tau_{ij}^{(2)} - u_i^{(2)}\tau_{ij}^{(1)}\right) = -\int_V d^3\mathbf{x}\left(u_i^{(1)}f_i^{(2)} - u_i^{(2)}f_i^{(1)}\right), \tag{3.2.4}$$

where $\mathbf{n}$ is the normal vector to ∂V. The other terms cancel because of the properties of the elastic moduli (3.2.2).

In terms of the traction vector $\mathbf{t}$ on the surface ∂V, we can rewrite (3.2.4) in the more compact form

$$\int_{\partial V} d^2\mathbf{x}\left(u_i^{(1)}t_i^{(2)} - u_i^{(2)}t_j^{(1)}\right) = -\int_V d^3\mathbf{x}\left(u_i^{(1)}f_i^{(2)} - u_i^{(2)}f_i^{(1)}\right), \tag{3.2.5}$$

where $t_i = n_j\tau_{ij}$. Although we have derived (3.2.5) under the assumption of continuous stress and displacement fields, we may extend the result to include regions containing material discontinuities. With conditions of welded contact, both displacement and traction will be continuous across such a discontinuity. The additional integrals introduced on the two sides of the discontinuity in the application of the divergence theorem will then cancel.

When the medium is perfectly elastic, $c_{ijkl} = c^*_{ijkl}$, and the complex conjugate $\mathbf{u}^{(2)*}$ satisfies the same equation of motion as $\mathbf{u}^{(2)}$. As a result we have a further relation,

$$\int_{\partial V} d^2\mathbf{x}\left(u_i^{(1)}t_i^{(2)*} - u_i^{(2)*}t_j^{(1)}\right) = -\int_V d^3\mathbf{x}\left(u_i^{(1)}f_i^{(2)*} - u_i^{(2)*}f_i^{(1)}\right), \tag{3.2.6}$$

in the absence of attenuation.

3.2.1 Representation Theorems

We introduce the Green's tensor $\mathbf{G}(\mathbf{x}, \boldsymbol{\xi}, \omega)$. The jp component $G_{jp}(\mathbf{x}, \boldsymbol{\xi}, \omega)$ of the tensor represents the displacement in the jth direction at $\mathbf{x}$ due to a unit force in the pth direction at $\boldsymbol{\xi}$. The particular form of the Green's tensor will depend on the boundary conditions that it satisfies. We also introduce the stress tensor $H_{ijp} = c_{ijkl}\partial_k G_{lp}$ that corresponds to the force system $\delta_{ip}\delta(\mathbf{x} - \boldsymbol{\xi})$. We can then use the relation (3.2.5) for the displacement field $\mathbf{u}(\mathbf{x}, \omega)$ with $\mathbf{u}^{(2)}$ selected as the Green's tensor

$$\begin{aligned}\Theta(\mathbf{x})u_p(\mathbf{x}, \omega) = &\int_V d^3\boldsymbol{\xi}\, G_{jp}(\mathbf{x}, \boldsymbol{\xi}, \omega) f_j(\boldsymbol{\xi}, \omega) \\ &+ \int_{\partial V} d^2\boldsymbol{\xi}\, [G_{qp}(\mathbf{x}, \boldsymbol{\xi}, \omega) t_q(\boldsymbol{\xi}, \omega) - u_q(\boldsymbol{\xi}, \omega) H_{qp}(\mathbf{x}, \boldsymbol{\xi}, \omega)],\end{aligned} \tag{3.2.7}$$

where

$$\Theta(\mathbf{x}) = 1, \mathbf{x} \in V, \quad \Theta(\mathbf{x}) = 0, \mathbf{x} \notin V, \tag{3.2.8}$$

and H_{qp} is the traction on the surface ∂V associated with the Green's tensor. For homogeneous boundary conditions on ∂V (e.g., vanishing traction at the free surface), the surface integral vanishes and the Green's tensor has the symmetry

$$G_{jp}(\mathbf{x}, \boldsymbol{\xi}, \omega) = G_{pj}(\boldsymbol{\xi}, \mathbf{x}, \omega). \tag{3.2.9}$$

As a result the representation (3.2.8) can be recast with a receiver at $\mathbf{x}$ and a source at $\boldsymbol{\xi}$,

$$\begin{aligned}\Theta(\mathbf{x})u_k(\mathbf{x}, \omega) = &\int_V d^3\boldsymbol{\xi}\, G_{kq}(\mathbf{x}, \boldsymbol{\xi}, \omega) f_q(\boldsymbol{\xi}, \omega) \\ &+ \int_{\partial V} d^2\boldsymbol{\xi}\, [G_{kq}(\mathbf{x}, \boldsymbol{\xi}, \omega) t_q(\boldsymbol{\xi}, \omega) - u_q(\boldsymbol{\xi}, \omega) H_{kq}(\mathbf{x}, \boldsymbol{\xi}, \omega)].\end{aligned} \tag{3.2.10}$$

We note that we have a set of products of Fourier transforms, and hence (3.2.10) is a convolutional form in the time domain.

A comparable development can be made from the relation (3.2.6), for a perfectly elastic medium, again taking $\mathbf{u}^{(2)}$ as the Green's tensor $\mathbf{G}(\mathbf{x}, \boldsymbol{\xi}, \omega)$. This produces a second representation

$$\begin{aligned}\Theta(\mathbf{x})u_k(\mathbf{x}, \omega) = &\int_V d^3\boldsymbol{\xi}\, G^*_{kq}(\mathbf{x}, \boldsymbol{\xi}, \omega) f_q(\boldsymbol{\xi}, \omega) \\ &+ \int_{\partial V} d^2\boldsymbol{\xi}\, \left[G^*_{kq}(\mathbf{x}, \boldsymbol{\xi}, \omega) t_q(\boldsymbol{\xi}, \omega) - u_q(\boldsymbol{\xi}, \omega) H^*_{kq}(\mathbf{x}, \boldsymbol{\xi}, \omega)\right].\end{aligned} \tag{3.2.11}$$

This form is only valid for perfectly elastic media, and involves products such as $G^*(\omega)u(\omega)$, and hence correlations in the time domain.

When we apply (3.2.11) to the Green's function itself we extract the relation

$$\begin{aligned}G_{ij}(\mathbf{x}_1, \mathbf{x}_2, \omega) - G^*_{ij}(\mathbf{x}_1, \mathbf{x}_2, \omega) = \\ \int_{\partial V} d^2\boldsymbol{\xi}\, \left[G^*_{iq}(\mathbf{x}_2, \boldsymbol{\xi}, \omega) H_{qj}(\mathbf{x}_1, \boldsymbol{\xi}, \omega) - G_{iq}(\mathbf{x}_1, \boldsymbol{\xi}, \omega) H^*_{qj}(\mathbf{x}_2, \boldsymbol{\xi}, \omega)\right].\end{aligned} \tag{3.2.12}$$

This result relates the Green's tensor and its time reversal to correlations of displacement and traction fields on the surface ∂V.

3.2.2 Effect of Unmodelled Structure

As discussed in § SWI:11.1, 11.3, a source is effectively introduced into the equations of motion wherever our model for the stress tensor departs from the actual physical stress. We can account for such effects by introducing an equivalent force distribution $\mathbf{e}(\mathbf{x}, \omega)$ to represent the mismatch, and so introduce an extra contribution in the representation theorems (3.2.10) and (3.2.11).

For example, consider the situation where we use the perfectly elastic result (3.2.11) for an attenuative medium, where the complex elastic modulus tensor is written as

$$c_{ijkl}(\mathbf{x}, \omega) = c'_{ijkl}(\mathbf{x}, \omega) + ic''_{ijkl}(\mathbf{x}, \omega), \tag{3.2.13}$$

with the attenuative component in c''_{ijkl}. In this case the additional contribution to (3.2.11) takes the form

$$u''_p(\mathbf{x}, \omega) = i \int_V d^3\boldsymbol{\xi}\, c''(\boldsymbol{\xi}, \omega)\partial_j u_i(\boldsymbol{\xi}, \omega)\partial_l G^*_{kp}(\mathbf{x}, \boldsymbol{\xi}, \omega), \tag{3.2.14}$$

where $\mathbf{G}^*$ is the Green's tensor for just the real part of the elastic moduli. This distributed effective source thus depends on the gradients of both the displacement field itself and the complex conjugate of the Green's tensor. Now, rather than a direct representation for the displacement, we have an integral equation for $\mathbf{u}$.

The representation results (3.2.10) and (3.2.11) require the Green's tensor for the zone being studied in full complexity, including all heterogeneity effects. What is the consequence of using a Green's tensor for a simpler medium, e.g., one that excludes all small-scale heterogeneity? We write the equation of motion in the form

$$\partial_i(c^o_{ijkl}\partial_k u_l) + \rho^o\omega^2 u_j = -f_j + \mathcal{L}'\{u_i\}, \tag{3.2.15}$$

where c^o_{ijkl}, ρ^o are taken for a simplified structure. $\mathcal{L}'\{u_i\}$ represents the effects of deviations from this structure and so appears as an equivalent source. If we take $\mathbf{G}^o$ as the Green's tensor for the simplified structure, satisfying homogeneous boundary conditions on ∂V, then the displacement field takes the form

$$\begin{aligned} u_k(\mathbf{x}, \omega) = &\int_V d^3\boldsymbol{\xi}\, G_{kq}(\mathbf{x}, \boldsymbol{\xi}, \omega) f_q(\boldsymbol{\xi}, \omega) \\ &- \int_V d^3\boldsymbol{\xi}\, G_{kq}(\mathbf{x}, \boldsymbol{\xi}, \omega)\mathcal{L}'\{u_q\}(\boldsymbol{\xi}, \omega). \end{aligned} \tag{3.2.16}$$

This is again an integral equation for the displacement field $\mathbf{u}$. As discussed in § SWII:30.2, when the heterogeneous contribution is small the most common procedure is to make a perturbation expansion. The displacement field $\mathbf{u}^o$ for the smooth medium is used as a first estimate in the second integral on the right-hand side of (3.2.16), and then further refinement can be achieved by employing updated estimates for $\mathbf{u}$ in an iterative manner, often known as a Born series.

Often the sequence is terminated after the first stage, and this 'single scattering'

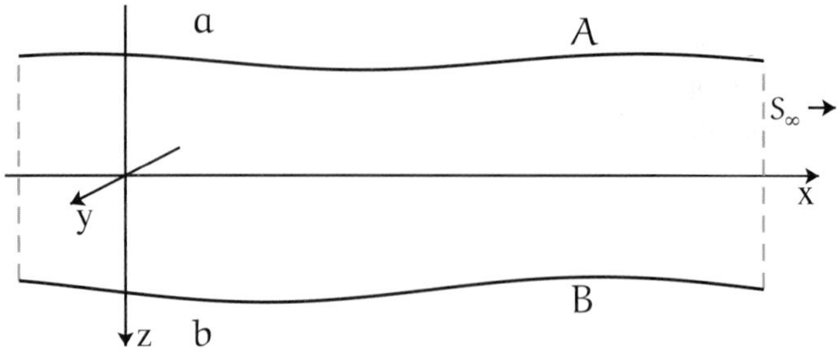

Figure 3.1 Quasi-stratified medium with surfaces A and B chosen to conform to the nature of the model. The zone above A is designated a, and b denotes the region below B.

result forms the basis of many studies. A comprehensive treatment of such scattering issues in simple media is given by Sato, Fehler & Maeda (2012).

3.2.3 Propagation Invariants

It is useful to be able to isolate portions of the elastodynamic field with specific character; this can be achieved by exploiting propagation invariants. Consider now a quasi-stratified medium in which the predominant variation in elastic parameters is in the z direction (Figure 3.1), with surfaces A and B spanning the whole x–y plane. Depending on the nature of the model, a convenient choice for such surfaces would be to follow material discontinuities or contours in elastic wavespeeds.

We consider a situation in the absence of sources. Choosing the surface ∂V in (3.2.5) to include both A and B with completion surfaces S_∞ at infinity, we have

$$\int_A d^2\mathbf{x}\left(u_i^{(1)}t_i^{(2)} - u_i^{(2)}t_j^{(1)}\right) = \int_B d^2\mathbf{x}\left(u_i^{(1)}t_i^{(2)} - u_i^{(2)}t_j^{(1)}\right), \tag{3.2.17}$$

provided that $\mathbf{u}^{(1)}$, $\mathbf{u}^{(2)}$ decay sufficiently rapidly as S_∞ is moved to infinity. Such would be the case if they were derived from sources lying outside the region ∂V. The surface integrals are thus an invariant of the field.

Since the integrals in (3.2.17) will vanish identically when $\mathbf{u}^{(1)}$ and $\mathbf{u}^{(2)}$ are the same field, we see that any part of $\mathbf{u}^{(2)}$ which is a multiple of $\mathbf{u}^{(1)}$ makes no contribution to the invariant. With suitable choice of $\mathbf{u}^{(1)}$ we can therefore isolate desired characteristics in the field $\mathbf{u}^{(2)}$.

For a perfectly elastic medium, we can make a similar treatment based on (3.2.6) to establish the invariant result

$$\int_A d^2\mathbf{x}\left(u_i^{(1)}t_i^{(2)*} - u_i^{(2)*}t_j^{(1)}\right) = \int_B d^2\mathbf{x}\left(u_i^{(1)}t_i^{(2)*} - u_i^{(2)*}t_j^{(1)}\right), \tag{3.2.18}$$

which is equivalent to conservation of wavefield energy.

3.3 Reflection and Transmission Operators

We now show how we can build a description of physical processes using reflection and transmission operators for portions of the medium. The reflection and transmission operator development may be used for both elastic and acoustic waves, and for solid models can provide a systematic means of following interconversions between *P* and *S* waves.

3.3.1 A Single Interface

As a preliminary to investigating the seismic response of an entire model we introduce the concept of reflection and transmission operators for a single interface and then consider more complex configurations. In order to introduce the concept of a reflection operator we start with the familiar case of a flat interface, and then look at the modifications imposed when the reflection surface has topography.

A flat interface

We consider a plane horizontal interface at the level z_I, lying between two uniform solid media. An incident downward travelling wave $D^a(\mathbf{x}, t)$ which impinges on the interface will generate an upward reflected wave $U^a(\mathbf{x}, t)$ in the same medium. We can make a plane-wave decomposition of each of these two wavefields, and the upgoing wave components will be related to the downgoing components by the action of the reflection coefficients at the interface z_I (§ SWI:13.2). In terms of frequency ω and horizontal slowness $\mathbf{p} \equiv (p_x, p_y)$, the downward travelling field at the interface z_I may be split into plane waves as

$$D(\mathbf{x}_h, z_I, \omega) = \frac{1}{2\pi} \int\!\!\int dp_x\, dp_y\, \omega^2 \bar{D}(\mathbf{p}, z_I, \omega) \exp[i\omega \mathbf{p}.\mathbf{x}_h], \tag{3.3.1}$$

where $\mathbf{x}_h \equiv (x, y)$ is the position vector in the horizontal plane. The reflected wave field at z_I may then be built up from the slowness components $\bar{D}(\mathbf{p}, z_I, \omega)$ by the action of the interfacial reflection coefficients $\mathbf{R}_D^I$ for the downward travelling incident wave. Thus,

$$U(\mathbf{x}_h, z_I, \omega) = \frac{1}{2\pi} \int\!\!\int dp_x\, dp_y\, \omega^2 \mathbf{R}_D^I(\mathbf{p}) \bar{D}(\mathbf{p}, z_I, \omega) \exp[i\omega \mathbf{p}.\mathbf{x}_h]. \tag{3.3.2}$$

$\mathbf{R}_D^I(\mathbf{p})$ includes the reflection coefficients for slowness $\mathbf{p}$ and is independent of frequency. For acoustic waves $\mathbf{R}_D^I$ is just a scalar, but for elastic problems $\mathbf{R}_D^I$ is a 3×3 matrix of coefficients. As noted in Chapter 2, in the case of isotropic and transversely isotropic media $\mathbf{R}_D^I$ factors into a 2×2 matrix coupling *P* and vertically polarised *S* waves (*SV*) and a scalar coefficient for horizontally polarised *S* waves (*SH*). For each slowness the reflected field is obtained by multiplying the component of the incident field by the appropriate reflection coefficient.

When the spatial response (3.3.2) is reconstructed via a Fourier integral over horizontal wavenumber, we recognise that there is an alternative representation of

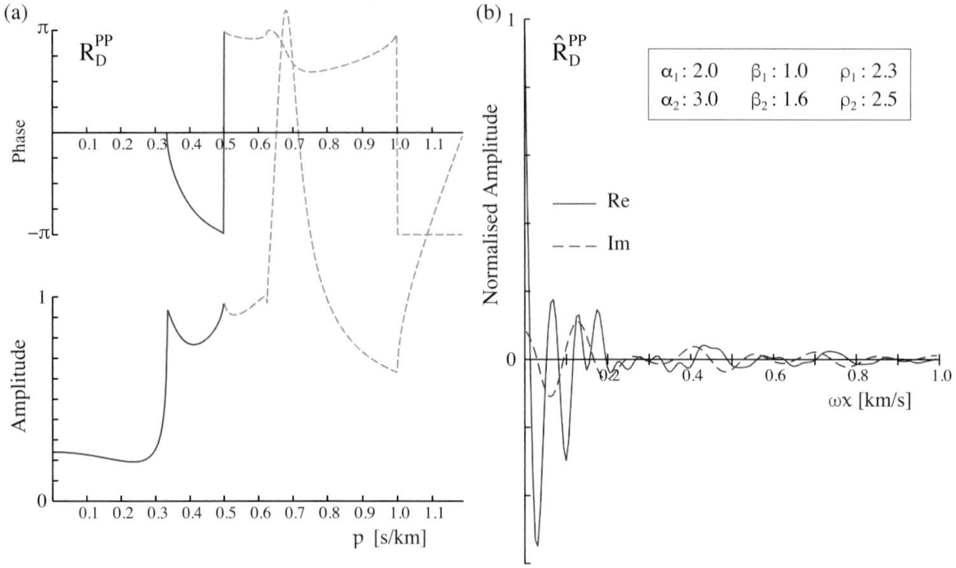

Figure 3.2 (a) The phase and amplitude behaviour of the PP reflection coefficient between two elastic media as a function of horizontal slowness p; evanescent *P* waves are indicated by dashed lines. (b) Spatial behaviour of the real (Re) and imaginary (Im) parts of the corresponding complex reflection operator.

the wave field as a spatial convolution between the incident downward wave field and a reflection term which describes the spatial action of the reflection process:

$$\mathsf{U}(\mathbf{x}_h, z_I, \omega) = \omega^2 \hat{\mathbf{R}}_D^I(\omega \mathbf{x}_h) \star \mathbf{D}(\mathbf{x}_h, z_I, \omega), \tag{3.3.3}$$

where the spatial representation of the reflection $\hat{\mathbf{R}}_D^I$ is obtained by the inverse Fourier transform of the reflection coefficients $\mathbf{R}_D^I(\mathbf{p})$ with respect to horizontal slowness,

$$\hat{\mathbf{R}}_D^I(\mathbf{x}_h) = \frac{1}{2\pi} \int\!\!\int d^2\mathbf{p}\, \mathbf{R}_D^I(\mathbf{p}) \exp[i\mathbf{p}.\mathbf{x}_h]. \tag{3.3.4}$$

The explicit frequency dependence in $\hat{\mathbf{R}}_D^I(\omega \mathbf{x}_h)$ in (3.3.3) arises from the fact that the reflection coefficients at the interface, $\mathbf{R}_D^I(\mathbf{p})$, are independent of frequency.

The representation (3.3.3) will include all upward travelling waves generated at the interface. $\hat{\mathbf{R}}_D^I(\mathbf{x}_h)$ will therefore be built up from the action of both pre-critical and post-critical reflections and head waves. Although the main concentration of the spatial reflection term $\hat{\mathbf{R}}_D^I(\mathbf{x}_h)$ is near the origin, there will be a long tail to include the decay of the head waves with distance.

We illustrate the behaviour of the spatial reflection operator $\hat{\mathbf{R}}_D^I$ in Figure 3.2 for the case of the *P*-wave reflection coefficient (PP) from an interface between two elastic media. In the Figure 3.2(a) we display the slowness dependence of the amplitude and phase of the reflection coefficient. Since the inversion integral is carried out over all slownesses we have to extend the definition of the reflection coefficient into the zone where *P* waves are evanescent in the upper medium (i.e., $p > 1/\alpha_1$). The onset of evanescence is marked by a change to dashed lines in Figure 3.2(a), for larger slownesses there is no direct physical significance to be attached to the reflection coefficients because the evanescent waves do not transport energy in the *z*-direction. The result of the inverse transformation over slowness to construct $\hat{R}_{PP}(\omega x)$ is displayed in Figure 3.2(b). We have plotted both the real and imaginary parts of this complex function. The real part describes the reflection in phase with the incident wave, which is of greatest interest in reflection work. The imaginary part of the reflection introduces a 90° phase shift, and is associated with the wide-angle reflections and head waves. As expected, $\hat{R}_{PP}(\omega x)$ has its greatest values near the origin and after an initial rapid drop there is an abrupt change in the gradient of the envelope around $\omega x = 0.15$ km/s; for larger arguments the amplitude decays fairly slowly. The structure in this tail reflects the details of the slowness behaviour illustrated in Figure 3.2(a). The oscillatory behaviour of these spatial reflection terms, as we move away from the geometrical reflection point, is in accord with a Fresnel zone picture of the pattern of reflection.

In a ray treatment of the reflection process we would assume a point-wise operator acting just at the reflection point, and so the spatial reflection operator would be concentrated at the origin. However, from Figure 3.2 we see that for any finite frequency a larger region is involved in reflection. In our illustrative example the principal contribution comes from a zone about $2000/\omega$ m wide for angular frequency ω, i.e., about a sixth of a wavelength across. A merit of the representation (3.3.3) for the reflected wavefield is that it shows how the effective size of the zone which contributes to the reflection scales inversely with frequency.

Let us now introduce the reflection operator R_D^I for this interface, defined such that the upgoing field U^a is to be found by the action of R_D^I on the incident field:

$$\mathsf{U}(\mathbf{x}_h, z_I, \omega) = \mathsf{R}_D^I[\mathbf{D}(\mathbf{x}_h, z_I, \omega)]. \qquad (3.3.5)$$

For this planar interface comparison of (3.3.3) and (3.3.5) shows that we must identify R_D^I with the spatial convolution operation $\omega^2\hat{\mathbf{R}}_D^I(\omega\mathbf{x}_h)\star$. In a similar way, we can introduce the transmission operator T_D^I for downward incident waves as a spatial convolution with the inverse Fourier transform of the plane-wave transmission coefficients as a function of horizontal position.

Once the interface deviates from a plane we lose the spatial convolution properties, although for a slowly undulating surface with small slopes (3.3.3) will often provide an effective representation of the reflected field.

Thomson (2012) has investigated reflection from a planar interface in the 2-D space–time domain rather than space–frequency as discussed above. The temporal operator $R(x,t)$ is a generalized function with a complicated form at the spatial origin, consistent with the complexity seen in Figure 3.2. The $R(x,t)$ response is V-shaped in the (x,t) plane, with linear arms given by the wave cones for the two media involved. These arms carry $H(t)t^{-1/2}$ signals, where $H(t)$ is the Heaviside step function.

A non-planar interface

For reflection from a non-planar interface $h_I(\mathbf{x}_h)$ we can extend the idea of a reflection operator $\mathbf{R}_D^I$ connecting the ingoing and outgoing fields at the interface. We consider waves emerging from a point source $\mathbf{x}_s$ and received at a point $\mathbf{x}_r$ in the uniform region a above the interface. In terms of a generic point $\boldsymbol{\xi}$ on the interface, the ingoing field will be a linear combination of the Green's tensor $\mathbf{G}^a(\boldsymbol{\xi}, \mathbf{x}_s)$ and its spatial derivatives. The outgoing (reflected) field will depend on $\mathbf{G}^a(\mathbf{x}_r, \boldsymbol{\xi})$. Following the treatment of Kennett (1984a) we can construct the wavefield at the receiver point as a combination of the waves propagating directly from the source and waves reflected from the surface. The effect of the irregular surface is included by introducing a set of apparent sources $\mathbf{R}_D$ along the interface, which will depend on the incident wave:

$$\mathbf{V}(\mathbf{x}_r) = \mathbf{D}^a(\mathbf{x}_r, \mathbf{x}_s) + \int_I d^2\boldsymbol{\xi}\, \mathbf{G}^a(\mathbf{x}_r, \boldsymbol{\xi}) \cdot \mathbf{R}_D(\boldsymbol{\xi}; \mathbf{D}^a), \tag{3.3.6}$$

where the integral is to be taken over the interface $I : h_I(\mathbf{x}_h)$, and $\mathbf{D}^a$ is the downgoing field which would be present in the absence of the interface. With the aid of the representation theorem (3.2.10), the source terms $\mathbf{R}_D(\boldsymbol{\xi}; \mathbf{D}^a)$ can be found as the solution of an integral equation which depends on the incident field $\mathbf{D}^a$ and the Green's tensors for the two media abutting the interface:

$$\mathbf{L}(\mathbf{x}_r; \mathbf{D}^a) = \int_I d^2\boldsymbol{\xi}\, \mathbf{K}(\mathbf{x}_r, \boldsymbol{\xi}) \cdot \mathbf{R}_D(\boldsymbol{\xi}; \mathbf{D}^a) \tag{3.3.7}$$

The integral kernel $\mathbf{K}$ depends on the difference in the expressions for the Green's tensors for propagation in the regions a and b on the two sides of the irregular interface I:

$$K_{kp} = \int_I d^2\boldsymbol{\eta}\, [G^b_{kq}(\mathbf{x}_r, \boldsymbol{\eta}) H^a_{qp}(\boldsymbol{\eta}, \boldsymbol{\xi}) - H^b_{kq}(\mathbf{x}_r, \boldsymbol{\eta}) G^a_{qp}(\boldsymbol{\eta}, \boldsymbol{\xi})], \tag{3.3.8}$$

where the integration is taken over the whole surface, and $\mathbf{H}$ is the traction on the surface I associated with the Green's tensor $\mathbf{G}$. The dependence on the incident field appears in $\mathbf{L}$:

$$L_k(\mathbf{x}_r; \mathbf{D}^a) = \int_I d^2\boldsymbol{\eta}\, [G^b_{kq}(\mathbf{x}_r, \boldsymbol{\eta}) E^a_q(\boldsymbol{\eta}, \mathbf{x}_s) - H^b_{kq}(\mathbf{x}_r, \boldsymbol{\eta}) D^a_q(\boldsymbol{\eta}, \mathbf{x}_s)], \tag{3.3.9}$$

where $\mathbf{E}^a$ is the traction produced by the incident wave. The combination of terms in $\mathbf{L}$ is such that any part of the incident field which resembles the transmitted field in medium b is removed and so the reflected contribution is left.

In general, the solution of (3.3.9) has to be sought numerically via a discretisation of the interface and the reduction of the integral equation (3.3.9) to a set of linear equations. The kernel $\mathbf{K}$ depends only on the properties of the media adjoining the interface, and so many different incident fields can be considered without the need for recomputing the inverse of $\mathbf{K}$. In the special case when the interface is flat, the solution can be found by transform methods and we recover the representation (3.3.9).

Once the integral equation (3.3.7) has been solved to find the force terms R_D, equation (3.3.6) enables us to determine the upward field at the receiver in terms of the downward field $[\mathsf{D}^a]$ at the interface. This enables us to identify the action of the reflection operator $\mathsf{R}_\mathrm{D}^\mathrm{I}$ for the reflection process from the interface as equivalent to the integral term in (3.3.6). Thus, we can write

$$\mathsf{U}^a(\mathbf{x}_r) = \mathsf{R}_\mathrm{D}^\mathrm{I}[\mathsf{D}^a] = \int_\mathrm{I} \mathrm{d}^2\xi\, \mathsf{G}^a(\mathbf{x}_r, \xi)\mathsf{R}_\mathrm{D}(\xi; \mathsf{D}^a). \tag{3.3.10}$$

As for the flat case, the full reflective effect is distributed over the interface. In a ray treatment we would use a point operator at the geometrical reflection point. For all finite frequencies, the full representation (3.3.10) gives a concentration of the reflection process near this point, but allows for the possibility of diffraction effects. As in the flat case, there will be a decaying oscillatory pattern in the force distribution as ξ moves away from the reflection point.

Effective approximate results for the reflection from an irregular surface can be obtained by making simple approximations for the form of the incident wave and the Green's functions for the media on the two sides of the interface, but retaining the full surface integral. This will give an approximate integral equation for the secondary forces R_D, but the reflected field will be well represented.

For a truncated reflector, the use of approximate reflection operators in the form of an integral over the reflecting surface gives a much better treatment for the diffracted energy than the quick approximation of scaling the results for a rigid screen (see, e.g., Trorey, 1970). The use of the reflection operator allows for the angle dependence of the reflection from an interface between two dissimilar media, and for elastic models both *P* and *S* wave diffraction can be modelled at the same time.

In transmission through an interface we take a similar representation of the reflected field to (3.3.6), by working with a secondary force system T_D along the interface. The transmitted field depends on the Green's function for the lower medium G^b, and the secondary sources are again to be determined by an integral equation (Kennett, 1984a). The action of the transmission operator for the interface can once more be identified with a surface integral which represents the propagation away from the secondary sources; so that the transmitted field

$$\mathsf{D}^b(\mathbf{x}_r) = \mathsf{T}_\mathrm{D}^\mathrm{I}(\mathsf{D}^a) = \int_\mathrm{I} \mathrm{d}^2\xi\, \mathsf{G}^b(\mathbf{x}_r, \xi)\mathsf{T}_\mathrm{D}(\xi; \mathsf{D}^a). \tag{3.3.11}$$

When an interface is smooth and has small slopes, it is possible to use simple approximations for the transmitted field. The full transmission operator development is however required when the radius of curvature of the surface is small, so that strong focussing and defocussing of the transmitted waves is introduced.

3.3.2 The Effect of a Heterogeneous Region

For a single interface we have established a means of generating the reflected and transmitted fields for a given incident wave. The method we have used – with a representation by secondary force systems – can be also extended to regions of a model, provided that its internal propagation characteristics are well known (Kennett, 1984a).

Operators for a heterogeneous region

Consider a heterogeneous region bounded above by the surface $A : z = f_A(\mathbf{x}_h)$ and below by the surface $B : z = f_B(\mathbf{x}_h)$, as in Figure 3.1. We will take the source at $\mathbf{x}_S$ to lie in the region a above the surface A and, as for a single interface, try to represent the reflected and transmitted field via surface integrals of secondary force terms. We will also denote the zone below the surface B as b.

For reflection with an observation point $\mathbf{x}_R$ lying above A we will look for the total displacement field in the form

$$u_k(\mathbf{x}_R) = D^a_k(\mathbf{x}_R, \mathbf{x}_S) + \int_A d^2\boldsymbol{\xi}\, G^a_{kp}(\mathbf{x}_R, \boldsymbol{\xi}) R_p(\boldsymbol{\xi}, \mathbf{x}_S). \tag{3.3.12}$$

Here D^a is the displacement radiated directly from the source, and as in our discussion of a single interface $\mathbf{G}^a$ is the Green's tensor for an unbounded medium with the properties of the region a. In transmission, in region b, we represent the displacement field as

$$u_k(\mathbf{x}_R) = \int_B d^2\boldsymbol{\xi}\, G^b_{kp}(\mathbf{x}_R, \boldsymbol{\xi}) T_p(\boldsymbol{\xi}, \mathbf{x}_S) \tag{3.3.13}$$

and again we take $\mathbf{G}^b$ to be the Green's tensor for an unbounded medium with the properties of the region b.

In order to determine the force distribution $\mathbf{R}$ we apply the representation theorem to the region b, which for reflection excludes both source and receiver points. Thus we have

$$0 = \int_B d^2\boldsymbol{\eta}\, [G^b_{kq}(\mathbf{x}_R, \boldsymbol{\eta}) t_q(\boldsymbol{\eta}) - H^b_{kq}(\mathbf{x}_R, \boldsymbol{\eta}) u_q(\boldsymbol{\eta})], \tag{3.3.14}$$

and we may now make use of the propagation invariant (3.2.17) to transfer this relation to the surface A. Now we have

$$0 = \int_A d^2\boldsymbol{\eta}\, [\bar{G}^b_{kq}(\mathbf{x}_R, \boldsymbol{\eta}) t_q(\boldsymbol{\eta}) - \bar{H}^b_{kq}(\mathbf{x}_R, \boldsymbol{\eta}) u_q(\boldsymbol{\eta})], \tag{3.3.15}$$

where $\bar{\mathbf{G}}^b$ represents the displacement field obtained by extrapolating the displacement and tractions corresponding to the outgoing Green's tensor for the medium b from the surface B into the heterogeneous region.

We now use the form (3.3.12) for the displacement fields on the surface A in the integral (3.3.15), and are able to derive an integral equation for the force distribution **R** in terms of invariant integrals of the Green's tensors and the source radiation. The components of **R** are to be found from the equation

$$L_k(\mathbf{x}_R, \mathbf{x}_S) = \int_A d^2\boldsymbol{\eta}\, K_{kp}(\mathbf{x}_R, \boldsymbol{\xi}))R_p(\boldsymbol{\xi}, \mathbf{x}_S), \tag{3.3.16}$$

with

$$L_k(\mathbf{x}_R, \mathbf{x}_S) = \int_A d^2\boldsymbol{\eta}\, [\bar{G}^b_{kq}(\mathbf{x}_R, \boldsymbol{\eta})h_q(\boldsymbol{\eta}) - \bar{H}^b_{kq}(\mathbf{x}_R, \boldsymbol{\eta})g_q(\boldsymbol{\eta})], \tag{3.3.17}$$

and kernel

$$K_{kp}(\mathbf{x}_R, \boldsymbol{\xi}) = \int_A d^2\boldsymbol{\eta}\, [\bar{G}^b_{kq}(\mathbf{x}_R, \boldsymbol{\eta})H^a_{qp}(\boldsymbol{\eta}, \mathbf{x}_S) - \bar{H}^b_{kq}(\mathbf{x}_R, \boldsymbol{\eta})G^a_{qp}(\boldsymbol{\eta}, \mathbf{x}_S)]. \tag{3.3.18}$$

Provided that we exclude zones containing the source and receiver points, the two integrals (3.3.17) and (3.3.18) are invariant, and so may be evaluated over any convenient surface, with suitable extrapolation of the fields occurring in the integrand.

For the transmission problem we also make use of the representation theorem. Now we consider the region a containing the source, but excluding the observation point $\mathbf{x}_R$. Then

$$0 = D^a_k(\mathbf{x}_R, \mathbf{x}_S) + \int_A d^2\boldsymbol{\eta}\, [G^a_{kq}(\mathbf{x}_R, \boldsymbol{\eta})t_q(\boldsymbol{\eta}) - H^a_{kq}(\mathbf{x}_R, \boldsymbol{\eta})u_q(\boldsymbol{\eta})], \tag{3.3.19}$$

where $D^a_k(\mathbf{x}_R, \mathbf{x}_S)$ is the field which would be observed if both $\mathbf{x}_R$ and $\mathbf{x}_S$ lay in material with the properties of the region a. As in the reflection case we may use the invariance properties of the integral to transfer it to the surface B by extrapolating the Green's tensor values $\mathbf{G}^a$. Thus

$$0 = D^a_k(\mathbf{x}_R, \mathbf{x}_S) + \int_B d^2\boldsymbol{\eta}\, [\bar{G}^a_{kq}(\mathbf{x}_R, \boldsymbol{\eta})t_q(\boldsymbol{\eta}) - \bar{H}^a_{kq}(\mathbf{x}_R, \boldsymbol{\eta})u_q(\boldsymbol{\eta})], \tag{3.3.20}$$

in terms of the extrapolated field $\bar{\mathbf{G}}^a$. With this representation at the surface B we can derive an integral equation for the surface forces **T** in the form

$$-D^a_k(\mathbf{x}_R, \mathbf{x}_S) = \int_B d^2\boldsymbol{\xi}\, \tilde{K}_{kp}(\mathbf{x}_R, \boldsymbol{\xi})T_p(\boldsymbol{\xi}, \mathbf{x}_S), \tag{3.3.21}$$

as in the case of a single interface. The kernel $\tilde{\mathbf{K}}$ takes the form of an integral over the surface B,

$$\tilde{K}_{kp}(\mathbf{x}_R, \boldsymbol{\xi}) = \int_B d^2\boldsymbol{\eta}\, [\bar{G}^a_{kq}(\mathbf{x}_R, \boldsymbol{\eta})H^b_{qp}(\boldsymbol{\eta}, \mathbf{x}_S) - \bar{H}^a_{kq}(\mathbf{x}_R, \boldsymbol{\eta})G^b_{qp}(\boldsymbol{\eta}, \mathbf{x}_S)], \tag{3.3.22}$$

but since it is an invariant can be transferred to other surfaces between A and B.

We see therefore that with suitable knowledge of the wave propagation in the region AB, we are able to construct the secondary force distributions **R** and **T** from equations (3.3.16) and (3.3.21). This enables us to construct the reflection ($\mathbf{R}^{AB}$) and transmission ($\mathbf{T}^{AB}$) operators for the region AB in the form

$$\mathsf{R}_k^{AB}[\mathsf{D}^a] = \int_A d^2\xi\, G_{kp}^a(\mathbf{x}_R, \xi) R_p(\xi, \mathbf{x}_S), \tag{3.3.23}$$

$$\mathsf{T}_k^{AB}[\mathsf{D}^a] = \int_B d^2\xi\, G_{kp}^b(\mathbf{x}_R, \xi) T_p(\xi, \mathbf{x}_S). \tag{3.3.24}$$

With uniform media on the two sides of the region AB we can introduce representations of Green's tensors $\mathbf{G}^a$, $\mathbf{G}^b$ in terms of the different wave types. The reflection and transmission between different wave types can then be represented by matrix force systems.

For horizontal stratification the results we have just obtained reduce, under Fourier transformation over the horizontal coordinates $\mathbf{x}_h$, to the forms presented in § SM:5.2.2, which also employ propagation invariants. Although we originally introduced the expressions (3.3.12) and 3.3.13) to describe the reflected and transmitted fields outside the region AB, once we have found the force systems R and T we can extend the range of these representations into the region AB. This requires that we use the extrapolation of the Green's tensors $\mathbf{G}^a$, and $\mathbf{G}^b$, and the source radiation $\mathbf{D}^a$ into the heterogeneous region. The action of the operators generates fields that depend on both the source and receiver positions $\mathbf{x}_S$, $\mathbf{x}_R$. But, once the source is fixed, the $\mathbf{x}_R$ dependence dominates.

One way of constructing the reflection and transmission response from a heterogeneous region is to use a plane-wave decomposition of the wavefield and allow for transfer of energy between different wavenumbers induced by departures from stratification. Kennett et al. (1990) have established the general properties of such solutions by using certain properties of the wavefield which remain invariant with depth even in the presence of heterogeneity. Calculation schemes using coupled wavenumbers work well for models which consist of irregular surfaces separated by homogeneous regions (see, e.g., Geli et al., 1988; Koketsu et al., 1991).

For weak perturbations in material properties about a horizontally stratified medium, an approximate procedure may be used. With a Born series development for the scattered field from the inhomogeneities as in Section 3.2.2 (see, e.g., Hudson, 1968; Clayton & Stolt, 1981) we get a representation in terms of a volume distribution of source terms. To first order, these terms depend on the wavefield in the background medium and are linear in the size of the inhomogeneity. The reflection and transmission operators can then be built up by combining the weak scattering from the heterogeneity with the major effects of the horizontal stratification.

Kennett (1986) has shown how the coupled wavenumber and first-order Born approximation can be combined to give a recursive development which enables the reflection and transmission properties to be evaluated in a way that allows for

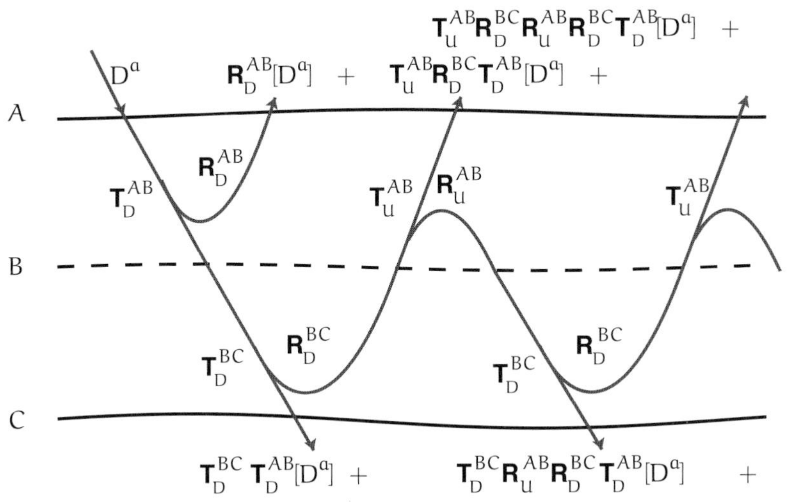

Figure 3.3 Configuration of a region AC divided into two parts AB and BC with a schematic representation of the construction of the reflection and transmission operators for the region AC from the operators for AB and BC.

multiple interactions between different portions of the model which internally have only weak heterogeneity effects (see Section 11.1.4).

Combining reflection and transmission operators

How then do we build up the response of a composite region, if we know the reflection and transmission operators for its constituent parts? The answer lies in following the cumulative propagation processes, and may be illustrated by reference to Figure 3.3.

Consider a region bounded above by an interface A, below by an interface C and divided into two parts by an interface B. The zones above A, and below C are assumed to be uniform. We can build up the action of the overall reflection operator on an incident downward field by looking at the classes of propagation processes in turn, combining the relevant reflection and transmission operators.

The reflected wavefield returned from AB alone is given by

$$\mathbf{R}_D^{AB}[\mathrm{D}^a], \tag{3.3.25}$$

but this is only part of the total reflected field returned from AC, since we have the possibility of transmission through AB and reflection from BC to give a further contribution to the wavefield of

$$\mathbf{T}_U^{AB}\mathbf{R}_D^{BC}\mathbf{T}_D^{AB}[\mathrm{D}^a]. \tag{3.3.26}$$

In transmission, the first set of interactions between the regions AB and BC produce

$$\mathsf{T}_D^{BC}\mathsf{T}_D^{AB}[\mathsf{D}^a] + \mathsf{T}_D^{BC}\mathsf{R}_U^{AB}\mathsf{R}_D^{BC}\mathsf{T}_D^{AB}[\mathsf{D}^a]. \tag{3.3.27}$$

The process of reflection and transmission from the regions AB and BC will continue alternately to generate the remainder of the reflected and transmitted field for the region AC.

We can now bring together all the contributions to the reflected field. We started with the reflection from AB (3.3.25) and then added the reflection from BC with two-way transmission through AB (3.3.26). Subsequent terms involve a double reflection: from above by AB and from below from BC. The total effect is

$$\mathsf{R}_D^{AC}[\mathsf{D}^a] = \mathsf{R}_D^{AB}[\mathsf{D}^a]+\mathsf{T}_U^{AB}\mathsf{R}_D^{BC}\mathsf{T}_D^{AB}[\mathsf{D}^a]+\mathsf{T}_U^{AB}\mathsf{R}_D^{BC}\mathsf{R}_U^{AB}\mathsf{R}_D^{BC}\mathsf{T}_D^{AB}[\mathsf{D}^a]+..., \tag{3.3.28}$$

where we have represented the cumulative action of operators by a composite operator. We can therefore identify the reflection operator R_D^{AC} with a sequence of reflection terms

$$\mathsf{R}_D^{AC} = \mathsf{R}_D^{AB} + \mathsf{T}_U^{AB}(\mathsf{R}_D^{BC} + \mathsf{R}_D^{BC}\mathsf{R}_U^{AB}\mathsf{R}_D^{BC} + \mathsf{R}_D^{BC}\mathsf{R}_U^{AB}\mathsf{R}_D^{BC}\mathsf{R}_U^{AB}\mathsf{R}_D^{BC} + ...)\mathsf{T}_D^{AB}. \tag{3.3.29}$$

The entire sequence of possible propagation effects can be represented via a single formal expression (Kennett, 1984b):

$$\mathsf{R}_D^{AC} = \mathsf{R}_D^{AB} + \mathsf{T}_U^{AB}\mathsf{R}_D^{BC}[\mathsf{I} - \mathsf{R}_U^{AB}\mathsf{R}_D^{BC}]^{-1}\mathsf{T}_D^{AB}. \tag{3.3.30}$$

where I is the identity operator. The inverse operator $[\mathsf{I} - \mathsf{R}_U^{AB}\mathsf{R}_D^{BC}]^{-1}$ represents the cumulative effect of the entire sequence of internal reverberations in AC, since we have the identity

$$[\mathsf{I} - \mathsf{R}_U^{AB}\mathsf{R}_D^{BC}]^{-1} = \mathsf{I} + \mathsf{R}_U^{AB}\mathsf{R}_D^{BC} + \mathsf{R}_U^{AB}\mathsf{R}_D^{BC}\mathsf{R}_U^{AB}\mathsf{R}_D^{BC}[\mathsf{I} - \mathsf{R}_U^{AB}\mathsf{R}_D^{BC}]^{-1} \tag{3.3.31}$$

in which we can recognise those parts of the field we have already encountered.

In transmission we obtain a similar result,

$$\mathsf{T}_D^{AC} = \mathsf{T}_D^{BC}(\mathsf{I} + \mathsf{R}_U^{AB}\mathsf{R}_D^{BC} + \mathsf{R}_U^{AB}\mathsf{R}_D^{BC}\mathsf{R}_U^{AB}\mathsf{R}_D^{BC} + ...)\mathsf{T}_D^{AB}. \tag{3.3.32}$$

The overall transmission operator T_D^{AC} can also be expressed in terms of the same reverberation operator

$$\mathsf{T}_D^{AC} = \mathsf{T}_D^{BC}[\mathsf{I} - \mathsf{R}_U^{AB}\mathsf{R}_D^{BC}]^{-1}\mathsf{T}_D^{AB}. \tag{3.3.33}$$

The reflection and transmission operators act to their right on earlier wavefield constructs. The operators cannot be commuted. Thus the physical content is obtained by reading from right to left as may be seen by comparison with Figure 3.3. Kennett (1984b) gives an alternative derivation of the composite operator results for both reflection and transmission in which the reverberation operator (3.3.31) is introduced directly.

If a further zone is added to extend to D, we can use the same approach to build the operators for AD. The composition relations for the reflection and transmission operators have the same form as (3.3.30) and (3.3.33), but now in terms of $\mathbf{R}_D^{AC}$, $\mathbf{R}_D^{CD}$ etc.

The addition rules for reflection and transmission operators (3.3.30, 3.3.33) are identical to those previously established for reflection matrices in the frequency/slowness domain for the case of horizontally stratified media (Section 2.3.1), as we would expect from our discussion of the flat interface problem. We thus can establish a *correspondence principle* for the reflection and transmission operators, so that we can translate any formulation for reflection and transmission matrices in a stratified medium directly into an operator equivalent.

Wavetype representations

Up to this point we have considered the reflection and transmission operators as entities acting on the entire incident field on a region. But, within the uniform half-spaces we have taken to border the heterogeneous zone (a, b), we can decompose the wavefield into three independent wavetypes. In isotropic media we would recognise these as *P* waves, *SV* waves, and *SH* waves, but in anisotropic media the eigenvector decomposition may not have such a readily identifiable physical character. For definiteness we will label the three wave types by P, S and H. In passage through a heterogeneous zone, we can get inter-conversions between the wavetypes so that *P-SV* and *SH* become coupled.

We partition the reflection operators by wavetype so that,

$$\mathbf{R}_D^{BC} = \begin{pmatrix} [\mathbf{R}_D^{BC}]^{PP} & [\mathbf{R}_D^{BC}]^{PS} & [\mathbf{R}_D^{BC}]^{PH} \\ [\mathbf{R}_D^{BC}]^{SP} & [\mathbf{R}_D^{BC}]^{SS} & [\mathbf{R}_D^{BC}]^{SH} \\ [\mathbf{R}_D^{BC}]^{HP} & [\mathbf{R}_D^{BC}]^{HS} & [\mathbf{R}_D^{BC}]^{HH} \end{pmatrix}, \tag{3.3.34}$$

where $\mathbf{R}^{PS}$ is the operator which generates an upgoing *P* wave from a downgoing *S* wave. The full suite of *P* and *S* wave interactions can then be followed by compounding the partitioned forms for the operators. These operators combine as if matrices were being multiplied, so that

$$[\mathbf{R}_D^{BC}\mathbf{T}_D^{AB}]^{PP} = [\mathbf{R}_D^{BC}]^{PP}[\mathbf{T}_D^{AB}]^{PP} + [\mathbf{R}_D^{BC}]^{PS}[\mathbf{T}_D^{AB}]^{SP} + [\mathbf{R}_D^{BC}]^{PH}[\mathbf{T}_D^{AB}]^{HP}. \tag{3.3.35}$$

The interaction of multiple reflections and conversions can be determined by making an expansion of the reverberation operator (3.3.31). Kennett (1986) has shown that when coupling between *P* and *S* is restricted to a single conversion it is possible to obtain a very useful approximation for the reverberation operator (3.3.31) in which the *P* and *S* wave reverberation sequences can be identified separately,

$$[\mathbf{I} - \mathbf{R}_U^{AB}\mathbf{R}_D^{BC}]^{-1} = \tag{3.3.36}$$
$$\begin{bmatrix} [\mathsf{X}^{PP}]^{-1} & [\mathsf{X}^{PP}]^{-1}\mathsf{Y}^{PS}[\mathsf{X}^{SS}]^{-1} & [\mathsf{X}^{PP}]^{-1}\mathsf{Y}^{PH}[\mathsf{X}^{HH}]^{-1} \\ [\mathsf{X}^{SS}]^{-1}\mathsf{Y}^{SP}[\mathsf{X}^{PP}]^{-1} & [\mathsf{X}^{SS}]^{-1} & [\mathsf{X}^{SS}]^{-1}\mathsf{Y}^{SH}[\mathsf{X}^{HH}]^{-1} \\ [\mathsf{X}^{HH}]^{-1}\mathsf{Y}^{HP}[\mathsf{X}^{PP}]^{-1} & [\mathsf{X}^{HH}]^{-1}\mathsf{Y}^{HS}[\mathsf{X}^{SS}]^{-1} & [\mathsf{X}^{HH}]^{-1} \end{bmatrix}.$$

Here $[\mathsf{X}^{PP}]^{-1}$ is the reverberation operator for *P* waves alone:

$$[\mathsf{X}^{PP}]^{-1} = [\mathsf{I} - [\mathbf{R}_U^{AB}]^{PP}[\mathbf{R}_D^{BC}]^{PP}]^{-1}, \tag{3.3.37}$$

with a similar interpretation for $[\mathsf{X}^{SS}]^{-1}$, $[\mathsf{X}^{HH}]^{-1}$.
The cross terms such as Y^{SP} allow for one conversion between wavetypes:

$$\mathsf{Y}^{SP} = [\mathbf{R}_U^{AB}]^{SP}[\mathbf{R}_D^{BC}]^{PP} + [\mathbf{R}_U^{AB}]^{SS}[\mathbf{R}_D^{BC}]^{SP} + [\mathbf{R}_U^{AB}]^{SH}[\mathbf{R}_D^{BC}]^{HP}. \tag{3.3.38}$$

With such expansions of the full reverberation operator it is relatively easy to follow the main conversion processes, recognising that the interactions between *P* and *SV* waves are likely to be stronger than those with *SH* waves.

3.3.3 Free-Surface Reflections

The idea of a reflection operator may also be used for waves incident from below on a region bounded by a free surface. To distinguish such operators, we will use the notation $\mathbf{R}_U^{fS}$ for reflection from the region between the interface S and the free surface f. The downgoing waves D^S returned from the region bounded below by S are then related to the upgoing waves U^S at S by

$$\mathsf{D}^S = \mathbf{R}_U^{fS}[\mathsf{D}^S], \tag{3.3.39}$$

The closest analogue to transmission in this case is the relation of the surface disturbance to the incident field at S. We can describe this by the action of a transfer operator W_U^{fS} (Kennett 1984a) so that the surface displacement can be found from

$$\boldsymbol{w}(\mathbf{x}_f) = \mathsf{W}_U^{fS}[\mathsf{U}^S]. \tag{3.3.40}$$

The two free-surface operators $\mathbf{R}_U^{fS}$ and W_U^{fS} play an important role in the description of the seismic wavefield generated by a source.

3.3.4 Operator Representations for the Full Wavefield

In order to be able to make use of the reflection and transmission operator development of the seismic wave field, we have to introduce the effects of a source. Our description must take care of the attributes of the source, and the way in which it interacts with the medium in its immediate vicinity.

The source at level S can be regarded as placed in a uniform medium, which allows a clear separation of the radiation into upgoing $[\mathsf{U}^S]$ and downgoing $[\mathsf{D}^S]$ parts.

The direct radiation to the surface from upgoing waves is

$$\mathsf{W}_U^{fS}\mathsf{U}^S. \tag{3.3.41}$$

Whereas the net downgoing radiation, including the underside reflection at the free surface, is

$$\mathsf{D}^S + \mathbf{R}_U^{fS}\mathsf{U}^S, \tag{3.3.42}$$

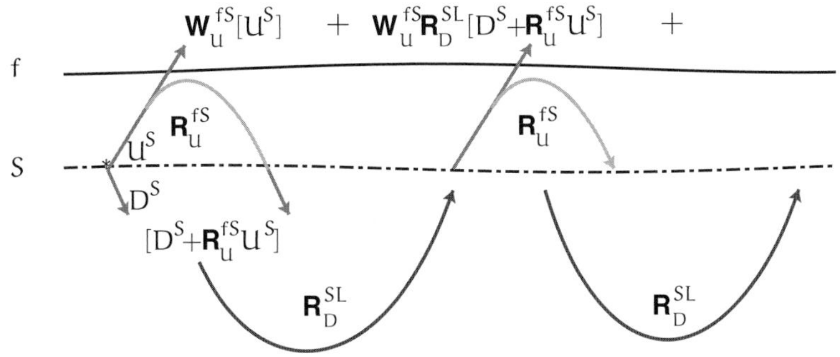

Figure 3.4 Schematic representation of the construction of the wavefield generated by a source at level S in terms of the reflection and transmission operators for the zone above the source to the surface fS and the region below S.

which is reflected back from below S and produces a surface contribution

$$\mathsf{W}_U^{fS}\mathsf{R}_D^{SL}[\mathsf{D}^S + \mathsf{R}_U^{fS}\mathsf{U}^S]. \tag{3.3.43}$$

When we allow for multiples generated at the free surface, each successive multiple contributes $\mathsf{R}_U^{fS}\mathsf{R}_D^{SL}$ from reflection above and below the source level S. Such multiples are indicated schematically in Figure 3.4.

For the seismic disturbance $\boldsymbol{w}_o$ at the surface the result is a propagation sequence, including free-surface multiples, of the form

$$\boldsymbol{w}_o = \mathsf{W}_U^{fS}\mathsf{U}^S + \mathsf{W}_U^{fS}(\mathsf{I} + \mathsf{R}_D^{SL}\mathsf{R}_U^{fS} + \mathsf{R}_D^{SL}\mathsf{R}_U^{fS}\mathsf{R}_D^{SL}\mathsf{R}_U^{fS} + \ldots)\mathsf{R}_D^{SL}[\mathsf{D}^S + \mathsf{R}_U^{fS}\mathsf{U}^S], \tag{3.3.44}$$

which by analogy with (3.3.29, 3.3.30) we can express formally as

$$\boldsymbol{w}_o = \mathsf{W}_U^{fS}\mathsf{U}^S + \mathsf{W}_U^{fS}[\mathsf{I} - \mathsf{R}_D^{SL}\mathsf{R}_U^{fS}]^{-1}\mathsf{R}_D^{SL}[\mathsf{D}^S + \mathsf{R}_U^{fS}\mathsf{U}^S]. \tag{3.3.45}$$

The inverse operator $[\mathsf{I} - \mathsf{R}_D^{SL}\mathsf{R}_U^{fS}]^{-1}$ includes all multiple interactions between the regions above and below the source. In the representation (3.3.45) for the wavefield we combine the reverberation operator for the entire structure with reflection from below the source level. Since

$$[\mathsf{I} - \mathsf{R}_D^{SL}\mathsf{R}_U^{fS}]^{-1}\mathsf{R}_D^{SL} = \mathsf{R}_D^{SL}[\mathsf{I} - \mathsf{R}_U^{fS}\mathsf{R}_D^{SL}]^{-1}, \tag{3.3.46}$$

we can view the reverberation sequence as occurring on either the source or receiver sides of the main reflection R_D^{SL}.

We have thus been able to use operator concepts directly to produce a representation of the seismic wavefield split at the source level. As expected this is consistent with the comparable result for a stratified medium (§ SWI:14.5, SM:7.2). We can then explore the range of different forms for the seismic response, e.g. with a split at a surface J, as in (2.3.9), to extract the most suitable form for the situation at hand. Approximations to the field can be made by truncating the expansion of the reverberation operators, and this allows us to examine the influences of multiple reflections, especially from the free surface.

3.3.5 *Effective Sources*

We can exploit the correspondence between the results for stratified media and the reflection and transmission operators to extract results for the effective source if the reference level is changed. We build on the development in § SM:9.2 to produce expressions for transferring the effect of a source to both a deeper and a shallower level.

When the new level J lies below the true source we exploit the reflection and transmission properties of the region SJ to express the upward and downward radiation components at J in terms of the original radiation in the form

$$\begin{aligned} \mathsf{U}^{J} &= [\mathbf{T}_{U}^{SJ}]^{-1}(\mathsf{U}^{S} + \mathbf{R}_{D}^{SJ}\mathsf{D}^{S}), \\ \mathsf{D}^{J} &= \mathbf{T}_{D}^{SJ}\mathsf{D}^{S} + \mathbf{R}_{U}^{SJ}[\mathbf{T}_{U}^{SJ}]^{-1}(\mathsf{U}^{S} + \mathbf{R}_{D}^{SJ}\mathsf{D}^{S}). \end{aligned} \tag{3.3.47}$$

The contribution $[\mathbf{T}_{U}^{SJ}]^{-1}$ arises because we are attempting to move upgoing waves back along their propagation path. However, the combination $\mathbf{C}_{D}^{J} = \{\mathsf{D}^{J} + \mathbf{R}_{U}^{fJ}\mathsf{U}^{J}\}$ corresponding to the net downward radiation at the level J is free of such terms,

$$\mathbf{C}_{D}^{J} = \mathsf{D}^{J} + \mathbf{R}_{U}^{fJ}\mathsf{U}^{J} = \mathbf{T}_{D}^{SJ}[\mathsf{I} - \mathbf{R}_{U}^{fS}\mathbf{R}_{D}^{SJ}]^{-1}\{\mathsf{D}^{S} + \mathbf{R}_{U}^{fS}\mathsf{U}^{S}\}. \tag{3.3.48}$$

The right-hand side of (3.3.48) can be recognised as the total downward radiation at the level J produced by a source at the level S; this expression allows for reverberation between the surface and J, in the neighbourhood of the source.

For a source which lies *below* the desired level for the equivalent source K, the upward and downward radiation components at K are

$$\begin{aligned} \mathsf{D}^{K} &= (\mathbf{T}_{D}^{KS})^{-1}(\mathsf{D}^{S} + \mathbf{R}_{U}^{KS}\mathsf{U}^{S}), \\ \mathsf{U}^{K} &= \mathbf{T}_{U}^{KS}\mathsf{U}^{S} + \mathbf{R}_{U}^{KS}\mathsf{D}^{K}. \end{aligned} \tag{3.3.49}$$

The net upward radiation $\mathbf{C}_{U}^{K}$ at K, allowing for reflection from below K down to a base level L, now has the equivalent representations

$$\begin{aligned} \mathbf{C}_{U}^{K} &= \mathsf{U}^{K} + \mathbf{R}_{D}^{KL}\mathsf{D}^{J}, \\ &= \mathbf{T}_{U}^{KS}[\mathsf{I} - \mathbf{R}_{D}^{SL}\mathbf{R}_{U}^{KS}]^{-1}(\mathsf{U}^{S} + \mathbf{R}_{D}^{SL}\mathsf{D}^{S}). \end{aligned} \tag{3.3.50}$$

The second expression includes the full interactions of the waves from the source with the region below K. We may emphasise reflections at this level by writing

$$\mathbf{C}_{U}^{K} = \mathbf{T}_{U}^{KS}\{[\mathsf{I} - \mathbf{R}_{D}^{SL}\mathbf{R}_{U}^{KS}]^{-1}\mathbf{R}_{D}^{SL}(\mathsf{D}^{S} + \mathbf{R}_{U}^{KS}\mathsf{U}^{S}) + \mathsf{U}^{S}\}. \tag{3.3.51}$$

The surface displacement field in this case can be found from

$$w_0 = \mathbf{W}_{U}^{fK}[\mathsf{I} - \mathbf{R}_{D}^{KL}\mathbf{R}_{U}^{fK}]^{-1}\mathbf{C}_{U}^{K}, \tag{3.3.52}$$

and there is no simple separation of a shallow propagation term.

3.4 Body-Wave Approximations

The reflection and transmission operators for specific zones of a model can, in principle, be constructed from the solutions of integral equations as in Section 3.3.3. However, the composition of multiple operators rapidly leads to very complex results. Thus the operators are best regarded as a conceptual framework for understanding the structure of the seismic wavefield derived from direct numerical methods. Operator representations also provide a means of assessing the implications of approximations made in simplifying models, and the impact of procedures used in seismogram analysis. Approximate methods attempt to isolate near independent portions of the wavefield, and to handle additional complexity in structure via coupling. Such methods work best when there are a limited number of major zones in a model. When there are many entities, keeping track of all multiples and conversions requires careful tabulation, and the computational effort can outweigh the gains in the simplicity of the main results.

3.4.1 Coupled Wavenumbers

The relative simplicity of handling wave propagation in stratified media, is associated with the fact that different horizontal slowness components propagate independently. The presence of lateral variations induces coupling between slowness components. Consider a medium with lateral variation superimposed on vertical variation, e.g.,

$$\alpha(x,z) = \alpha_0(z) + \alpha_1(x,z). \tag{3.4.1}$$

The propagation equations can be written in terms of the displacement–traction vector **b** as

$$\frac{\partial}{\partial z}\mathbf{b}(x,z,\omega) = \mathbf{A}_0(z,\omega)\mathbf{b}(x,z,\omega) + \mathbf{A}_1(x,z,\omega)\mathbf{b}(x,z,\omega), \tag{3.4.2}$$

where $\mathbf{A}_0(z,\omega)$ is the operator matrix for a laterally homogeneous medium (§ SWI:12.1) and $\mathbf{A}_1(x,z,\omega)$ includes all the extra terms associated with horizontal variation in material properties. After a Fourier transform with respect to x to convert to the wavenumber domain,

$$\frac{\partial}{\partial z}\bar{\mathbf{b}}(k,z,\omega) = \bar{\mathbf{A}}_0(k,z,\omega)\bar{\mathbf{b}}(k,z,\omega) + \int_{-\infty}^{\infty} d\xi\, \bar{\mathbf{A}}_1(k-\xi,z,\omega)\bar{\mathbf{b}}(\xi,z,\omega), \tag{3.4.3}$$

where the effect of heterogeneity appears through a convolution in wavenumber (Kennett, 1972) modulated by the wavenumber transform of the horizontal variations in $\bar{\mathbf{A}}_1(k,z,\omega)$. The effects of such wavenumber coupling can be handled with a discretisation of the wavenumber domain (e.g., Koketsu et al., 1991). The coupled wavenumber procedure has the merit that the expressions we have obtained for different portions of the wavefield can be used directly by substituting the appropriate reflection and transmission supermatrices, in the discrete wavenumber domain, in place of the operators.

Such a wavenumber coupling system can accommodate heterogeneity on a wide variety of scales by simply changing the span of the major lobe in the wavenumber scattering. Multiple scattering can be simulated by the cascading of first-order approximations for different regions, and such results can give valuable clues as to the likely effects of different classes of heterogeneity (Kennett, 1986). In Section 11.1.4 we discuss the way in which such representations can be combined with operator techniques to provide insight into the action of heterogeneity on the wavefield.

3.4.2 Modified Ray Theory

The simplest representations of reflection and transmission processes are provided by working directly with ray theory (SWI:Chapter 9), applying reflection and transmission coefficients when a ray impinges on an interface. A local plane wave is taken as incident on the tangent plane at the intersection point. Geometrical spreading is included between interfaces.

More sophisticated methods for handling wave propagation in laterally heterogeneous media such as the Maslov method (Chapman & Drummond, 1982) and Gaussian beams (Červeny et al., 1982) aim to find localised solutions around ray paths that asymptotically, at least, can be treated as independent. The modification of the slowness of a beam with position avoids the introduction of coupling effects, but these modified ray methods can only easily cope with relatively smooth parameter distributions. An extensive discussion of modified ray methods is given in Chapter 10 of Chapman (2004).

3.4.3 Interfaces

For interfaces between uniform media improved representations of the reflection and transmission processes for complex geometries can be made using the displacement representation results introduced in Section 3.2.1. The contribution from reflection at an interface K can be expressed in terms of the displacement at the surface, and the elastic moduli $\mathbf{c}^-$ on the upper side of the surface as

$$u_p^r(\mathbf{x},t) = \int_K dS(\xi) n_j(\xi) c^-_{ijkl}(\xi)[G_{pi}(\mathbf{x},t;\xi,0) * \partial_l u_k^r(\xi,t;\mathbf{x}_s) - u_i^r(\xi,t;\mathbf{x}_s) * \partial_l G_{pk}(\mathbf{x},t;\xi,0)]. \quad (3.4.4)$$

The convolutional representation (3.4.4) for the reflected part of the field is only useful if we can obtain satisfactory approximations for the displacement field on K. When the curvature of the surface K does not vary rapidly, we can adopt a local tangent plane approximation for the generation of the reflected field. In other words, we locally treat the incident wavefront as plane and impinging on a flat boundary (the tangent plane) in order to estimate the reflected field contribution

$u^r(\xi, t; \mathbf{x}_s)$ to appear in the surface integral in (3.4.4). With this approximation we can rewrite (3.4.4) as

$$u_p^r(\mathbf{x}, t) = \int_K dS(\xi)\, n_j(\xi) c_{ijkl}^-(\xi) \partial_l G_{pk}(\mathbf{x}, t; \xi, 0) * R_{iq}(\xi) u_q^{in}(\xi, t; \mathbf{x}_s), \qquad (3.4.5)$$

where $\mathbf{u}^{in}$ is the displacement that would be present in the absence of the interface. The matrix $R(\xi)$ will be built up from the reflection coefficients for plane elastic waves on the tangent plane and inclination factors for the different wave types. Although we have used a local approximation for $\mathbf{u}^r$ on the boundary in (3.4.5), the representation of the total reflected field involves contribution from the whole surface and will be more accurate than the internal approximation. A similar form can be used in transmission, with a transmission operator $T(\xi)$; we then need to make use of the moduli $\mathbf{c}^+$ on the further side of the interface and a Green's function suitable for the second medium.

The representation of the reflected wavefield in (3.4.5) is equivalent to the Kirchhoff techniques used in acoustics, and can be regarded as an approximate form of Huyghen's principle. Any suitable form for the Green's tensor can be employed in (3.4.5) so that, for example, the effects of strong internal multiples in the region above K could be included by using a slowness–frequency integral to set up the Green's tensor for a layered model. The Green's tensor also appears implicitly in the expression for the incident field on the boundary $\mathbf{u}^{in}$ since this will be generated by the action of the force system at $\mathbf{x}_s$ on the medium above K.

The representation (3.4.5) thus takes the form of a Green's tensor description of the propagation from the source to the boundary through $\mathbf{u}^{in}$, a reflection operation depending on the local characteristics of the interface $\mathbf{R}$ and a Green's tensor description of the propagation from the boundary to the receiver through $G(\mathbf{x}, t; \xi, 0)$. The results can be cast in a form in which the contribution to the seismogram at each time can be recognised (Cao & Kennett, 1989). For reflections, such an approach has several merits. Firstly, it is possible to isolate the reflected contributions for a given time interval. Secondly, the time surfaces for a given reflector, which are an intermediate product of the approach, have a direct correspondence to horizontal slice techniques used in some forms of three-dimensional reflection processing.

3.5 Surface Waves in Varying Media

For a stratified medium, with a one-dimensional velocity model depending on depth alone, we have no horizontal variation in seismic properties and the propagation path will automatically lie along the great circle between source and receiver. For a realistic medium we have to account for the presence of three-dimensional variations in seismic properties and therefore need to know to what extent we can rely on the results of calculations for stratified media for interpretation. We first consider the influence of smooth horizontal variations in structure, and then turn our attention to the effect of a sharp transition zone between distinct structures.

3.5.1 Smoothly Varying Structure

For a smoothly varying medium where the seismic parameters depend weakly on the horizontal coordinates, Woodhouse (1974) has demonstrated that the propagation characteristics are governed by the vertical structure beneath each point on the propagation path. The path itself is governed by the variations in phase slowness, and the incremental phase $\phi(\omega)$ along the path is simply the integral of the phase slowness

$$\phi(\omega) = \omega \int_{path} \mathbf{p}_j.\mathbf{dr}, \tag{3.5.1}$$

where $\mathbf{p}_j$ is the horizontal slowness vector for the mode and $\mathbf{dr}$ is an element along the path. This style of phase representation is subject to the same limitations as the asymptotic approximation for the vector harmonics in the stratified case. We therefore require that for a propagation range X, the combination $\omega p X \gg 1$, for the slowness p of all modes considered.

For structural models with slow variations in seismic parameters, we can adapt the formulation for a stratified medium by envisaging the modal field to be launched in a uniform region with the source properties. Then the field of surface waves propagates retaining the identity of the individual modes, and finally generates the observable displacements using the receiver structure.

The propagation component can be further simplified if the propagation path is close to a great circle, by using the averaged structure which produces the same incremental phase. Thus, for range X, we require

$$\phi(\omega) = \omega \int_{path} \mathbf{p}_j.\mathbf{dr} = \omega p_j^{av} X, \tag{3.5.2}$$

where p_j^{av} is the slowness for the averaged structure.

Using the averaged structure to represent the main propagation characteristics, we can represent an approximation to the surface wave part of the wavefield via a close analogue to the forms for a stratified medium, (2.5.9)–(2.5.10). For example, for Love waves we have

$$\mathbf{u}_H(X, \phi, t) = \mathbf{e}_\phi \sum_m e^{im\phi} \int_{-\infty}^{\infty} d\omega\, e^{-i\omega t}$$
$$\sum_{l=0}^{M(\omega)} \sqrt{\frac{2\omega p_l^{av}}{\pi X}\frac{g_l^{av}}{4 I_l}}\, \mathcal{S}_L^m(p_l^s)\, \mathcal{R}_L(p_l^r)\, e^{[i\omega p_l^{av} X - i(2m+1)\pi/4]}, \tag{3.5.3}$$

where the source, receiver, and path contributions are evaluated in different structures, as indicated by the superscripts for the relevant slownesses. The requirement that has to be imposed on these separate structures is that there is a one-to-one correspondence between the mode shapes as a function of depth. The relationship between different modal sets can be quantified by examining the expansion of the members of one set of modes, e.g. for the source structure, in terms of the orthogonal set of eigenfunctions for the other models. For the simple approximation (3.5.3) to the wavefield to be valid, we require that the expansion of mode j (for the source structure) in terms of the other mode sets (for path and receiver) is dominated by the same mode j, with no significant admixture of the neighbouring modes j $\pm$ 1. At the stage where the influence of the neighbouring modes on the expansion becomes noticeable, the shape of the modal eigenfunction no longer corresponds to its equivalent in the other modal sets.

We have noted above (Section 2.5) that the notional source contribution is not local in its dependence on structure. However, if the modal eigenfunctions of the propagation segment and the source structure are well matched, then (3.5.3) should be a good approximation for the surface wave portion of the field. The level of eigenfunction match will depend on the nature of the differences in structure and the modal order. For each situation there will be a frequency band within which this simplified propagation approximation can be used. The lower frequency limit will be dictated by the requirement that the product of frequency and path length (ωX) be large enough to allow the use of the phase representation for the dominant propagation term. The upper frequency limit will depend on the character of the structural variations and mode number. The effects of changing velocity structure are less severe than moving a major boundary. This means that the relative depths of the Moho between structures becomes important: too large a shift will change the eigenfunction shapes and mean that coupling between modes has to be introduced to compensate.

For frequencies less than 30 mHz, different crustal structures can be used at the source, near the receiver, and along the propagation path, provided that the change in crustal thickness is not mode than 10 km between contiguous structures. This restrictions ensures that there will be a good match in the behaviour of the eigenfunctions with depth. In waveform inversions for structure in surface wave tomography, it should then be possible to employ a single set of modes across a region and account for perturbations smaller than 5% for lithospheric S wavespeed along individual propagation paths.

For continental scale paths (1000–4000 km), the path-averaged approximation should be suitable for waveform fitting in the range 10–30 mHz. The lower limit comes from the use of the asymptotic approximations in (3.5.3), and the upper one from the influence of heterogeneity on the modal content of the seismograms.

For structures with stronger structural variation we can build the response in terms of a modal sum, working either with a fixed set of modes from a reference structure or *local modes* associated with the vertical structure beneath a point. In each case the influence of propagation through varying structure is introduced by coupling between modes, so that there is amplitude transfer between modal contributions described by a suite of evolution equations for the modal coefficients, as discussed in Section 11.2.

For fully three-dimensional structures, a mode coupling treatment has to take into account conversion between Love and Rayleigh modes of propagation (Kennett, 1998). The extent of such coupling depends strongly on the horizontal gradients in wavespeed (Section 11.2.2). For complex situations, it is generally most efficient to make a full 3-D simulation of the wavefield using finite-element or finite-difference techniques, though even modal concepts provide useful aids to the interpretation of the results.

3.5.2 Propagation Across Structural Boundaries

Often surface waves have crossed major structural boundaries, such as the transition from an oceanic to a continental regime. Such waves carry with them information about the structures through which they have passed. We can understand such situations in terms of a redistribution of energy between modes, working with the structures on the two sides of the transition.

If we consider a source in a nearly stratified region (e.g., an oceanic region), we can produce a good representation of the surface wave field in that region by using approximations such as (3.5.3) to include some allowance for near-source or near-receiver heterogeneity. However, once the receiver moves outside this zone into a different structure, we have to include an allowance for the transition between the source and receiver structures. When we consider long-period waves (< 0.06 Hz), the shear wavelengths will generally be much longer than the width of the transition, so that a simplified model of an abrupt transition can be used to include the major effects of the structural change.

Within the oceanic zone surrounding the source, the wavefield should not be strongly affected by the presence of the structural transition. Thus we can project the wavefield up to the boundary using the seismic properties of this zone. The superposition of different modal contributions can be used to build up the profile of seismic displacements and tractions at the boundary, as a function of depth.

If the source and receiver lie well away from the boundary, we can use the ray formulation of Levshin (1985) to consider propagation to the structural boundary, transmission with refraction due to the contrast in structures, and then propagation in the receiver structure to the recording location. In this high-frequency ray approximation the continuity conditions for displacement and traction have only to be applied at the point where the ray crosses the boundary. Gregersen & Alsop (1974) have introduced a convenient approximation for long-period propagation in which any diffraction processes or other local effects at the boundary are ignored. The displacement and traction field calculated for the source structure is expanded in terms of the modes of the receiver structure and these contributions are then propagated to the receiver.

We consider a source at a distance X^S from the boundary, and then construct the vertical and radial components of the displacement profile as a function of depth in the frequency domain, as a sum of Rayleigh modes,

$$\mathbf{u}_S^s(X^s, \phi, z, \omega) = \begin{pmatrix} i\mathbf{e}_z \\ -\mathbf{e}_r \end{pmatrix} \sum_m e^{im\phi} \sum_{j=0}^{N^s(\omega)} \sqrt{\frac{2\omega p_j^s}{\pi X^s}\frac{g_j^s}{4I_{Rj}^s}}\, \mathcal{S}_R^m(p_j^s) w_j^e(p_j^s, z) e^{[i\omega p_j^s X^s - i(2m+1)\pi/4]}; \quad (3.5.4)$$

with a comparable expression for the tangential component of displacement in terms of Love modes. Similar representations can be written for the traction components in terms of the requisite eigenfunctions. These displacement and traction profiles are then to be expanded into the modes of the receiver structure by using the orthogonality properties of the modal set (including both Love and Rayleigh polarisations)

$$i\int_0^\infty dz\, \left[\mathbf{u}_j^{eT}(-p_j, z, \omega)\hat{\mathbf{t}}_k^e(p_k, z, \omega) - \hat{\mathbf{t}}_j^{eT}(-p_j, z, \omega)\mathbf{u}_k^e(p_k, z, \omega)\right] = \delta_{jk}, \quad (3.5.5)$$

where $\hat{\mathbf{t}}$ is the traction acting on the vertical boundary. Gregersen (1978) has shown the need for allowing both Love and Rayleigh waves for oblique incidence on a ocean–continent transition. Vaccari et al. (1989) have shown that for models with uniform layers an analytic integration can improve the efficiency of the computation of modal coefficients, via integrals of the type

$$C_j = i\int_0^\infty dz\, \left[\mathbf{u}_j^{eT}(-p_j, z, \omega)\hat{\mathbf{t}}^s(z, \omega) - \hat{\mathbf{t}}_j^{eT}(-p_j, z, \omega)\mathbf{u}^s(z, \omega)\right]. \quad (3.5.6)$$

However, for more general structural models, numerical integration cannot be avoided. Once the modal coefficients (3.5.6) have been evaluated, we can construct the displacement at the receiver at a distance X^r from the boundary by

$$\mathbf{u}_S(X^r, 0, \omega) = \begin{pmatrix} i\mathbf{e}_z \\ -\mathbf{e}_r \end{pmatrix} \frac{1}{J} \sum_{j=0}^{N^r(\omega)} \sqrt{\frac{2\omega p_j^r}{\pi X^r}\frac{g_j^r}{4I_j^r}}\, w_j^e(p_j^s, z) C_j(\omega)\, e^{[i\omega p_j^r X^r - i\pi/4]}, \quad (3.5.7)$$

where J represents the geometrical spreading along the refracted ray path (Levshin, 1985).

The two-stage computation procedure involves propagation in both source and receiver structures, with transfer coefficients evaluated via the continuity of displacement and traction at the point of refraction. The resemblance of the modes on the two sides of the boundary will be strongest for modes at lower frequencies, so transmission can then be quite strong. As the frequency increases, the discrepancies between the modes become progressively more pronounced, and the transmission efficiency drops. In Figure 3.5 we show the mode shapes for the

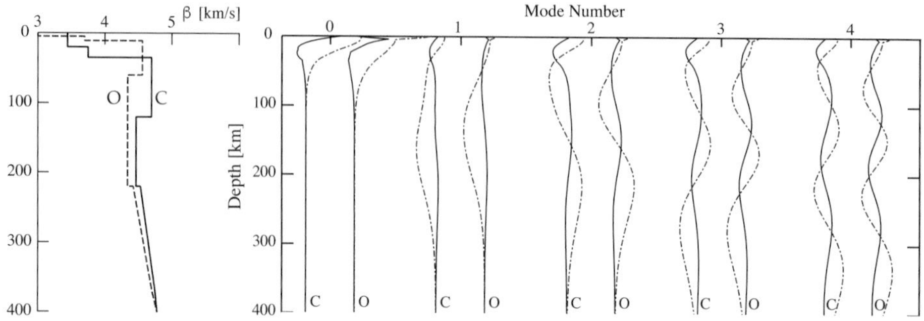

Figure 3.5 Comparison of mode shapes for Rayleigh waves as a function of depth for continental (C) and oceanic (O) models, at a frequency of 60 mHz. The shear wavespeed profiles with depth are shown at the left. The vertical component of the modal eigenfunctions is indicated with a solid line and the radial component with a dashed line.

continental and oceanic models introduced in § SWI:16.3. These velocity models only differ in the top 400 km. The differences between the shapes of the modal eigenfunctions in the two structures at 60 mHz (period 16.67 s) tend to extend over a larger range of depth as the mode number increases. Even for the fundamental mode at this intermediate frequency, the mismatch in mode shapes extends well below the crust. The fundamental mode in the oceanic structure penetrates deeper because the lithospheric lid is thinner.

The differences between modal shapes for the Rayleigh waves diminish with decreasing frequency, and so at 25 mHz (period 40 s) there are modest differences in the crust but the mantle behaviour is very similar (Maupin, 2007). This means that the fundamental mode Rayleigh wave would most likely cross the continental margin with little coupling to other modes. However, the situation is quite different for Love waves, the fundamental mode shapes at 25 mHz differ strongly. For the continent, energy is concentrated in the upper 50 km, but for oceanic structure much of the energy is carried in the low-velocity zone below 50 km depth. The only way that a match could be achieved would be with coupling to the higher modes.

4

The Reflection Field

Although early studies using man-made seismic sources exploited refracted waves, the use of reflected waves was soon found to be more effective for delineating structure. With the aim of better resolution of the complex structures that host petroleum resources, there has been a steady trend to larger numbers of high-fidelity recorders for each source. This means that most aspects of the seismic wavefield near the source can be captured in detail.

Early sparse recordings along individual lines were replaced by dense recordings across long spreads, and then surface grids to allow direct access to three-dimensional structure. The nature of the seismic wavefield in reflection work is strongly influenced by the character of the seismic source used to generate the seismic waves and the receivers used to record the field. In order to maximise the sensitivity to energy travelling near the vertical, arrays of sources and receivers are commonly used. Such arrays impose directivity related to their geometrical configuration on the basic characteristics of the wavefield. The directivity varies with frequency and horizontal slowness and any array design needs to balance conflicting demands. For example, the suppression of horizontally travelling waves, such as ground roll, can be achieved by summing the outputs from a long sub-array of receivers, at the expense of horizontal resolution in the recorded traces. Modern deployments with very large numbers of autonomous sensors mean that processing works directly on the seismic wavefield, so that its characteristics need to be well understood.

In this brief survey, we look at the nature of the reflected wavefield and the ways in which it can be modified by acquisition and processing in the pursuit of understanding of structure at depth. *The Leading Edge*, published by the Society of Exploration Geophysicists, provides topical special issues relating to many aspects of exploration seismology that reach to the current state of the art.

4.1 General Representations

In (2.3.9) we have established a representation of the full seismic wavefield for a stratified medium with a source at level S and a split in the structure at a level J. This leads to an equivalent operator development that allows a treatment of a more complex medium. We have a separation of the field into two parts. Firstly, shallow propagation

$$\boldsymbol{w}_{o}^{s} = \mathsf{W}_{U}^{fS}\mathsf{u}^{S} + \mathsf{W}_{U}^{fS}(\mathsf{I} - \mathsf{R}_{D}^{SJ}\mathsf{R}_{U}^{fS})^{-1}\mathsf{R}_{D}^{SJ}\,[\mathsf{D}^{S} + \mathsf{R}_{U}^{fS}\mathsf{u}^{S}], \tag{4.1.1}$$

confined to the region above the level J. Secondly, the deeper reflections of principal interest

$$\boldsymbol{w}_{o}^{d} = \mathsf{W}_{U}^{fS}\mathsf{X}_{R}\mathsf{T}_{U}^{SJ}\mathsf{X}_{J}\mathsf{R}_{D}^{JL}\mathsf{T}_{D}^{SJ}\mathsf{X}_{S}\,[\mathsf{D}^{S} + \mathsf{R}_{U}^{fS}\mathsf{u}^{S}], \tag{4.1.2}$$

where X_S, X_R represent shallow reverberation operators, and X_J the effect of multiple reflections through the entire structure. The case of a surface source is obtained by moving S up to the free surface. Although the expressions (4.1.1)–(4.1.2) are valid for any separation surface J lying beneath the source, it is advantageous to choose this surface to achieve the maximum separation between the different classes of propagation. For marine models, it is therefore desirable to take J a little below the sea bed itself so that sea bed interactions are included in the shallow terms, even though this means that R_{U}^{fJ} itself will include contributions at the sea bed as well as from the sea surface.

The objective of many aspects of reflection seismology is to extract a simple reflection response of the type

$$\boldsymbol{w}_{o}^{ref} = \mathsf{W}_{U}^{fS}\mathsf{T}_{U}^{SJ}\mathsf{R}_{D}^{JL}\mathsf{T}_{D}^{SJ}[\mathsf{D}^{S}], \tag{4.1.3}$$

with the split level J close to the surface. Even R_{D}^{JL} will contain internal multiples that have the potential to obscure the primary reflections that delineate structure. This is why the ultimate goal remains direct extraction of structure from the reflection signal rather than imaging.

Many developments in reflection seismology can be readily expressed through adaptations of the operator formulation. For example, the matrix operators invoked by Berkhout (1983, 1997) can be regarded as special cases of the reflection and transmission operators developed here.

The nature of the reflection wavefield is strongly conditioned by the nature of the available sources and sensors, so we examine such effects for both land and marine recording in the next two sections.

4.2 Land Seismic Profiling

Although explosive sources are still sometimes used in reflection seismic work on land, most reflection surveys now use surface seismic sources. For small-scale surveys weight-drop techniques can be used to generate seismic energy, but the dominant type of source is some form of vibrator.

In the vibroseis technique the vibrating plate is anchored to the ground by the weight of the vehicle and a vibration signal applied which sweeps across a range of frequencies. The resulting signals are cross-correlated with the signal applied to the plate to produce a recorded trace equivalent to a source with a limited

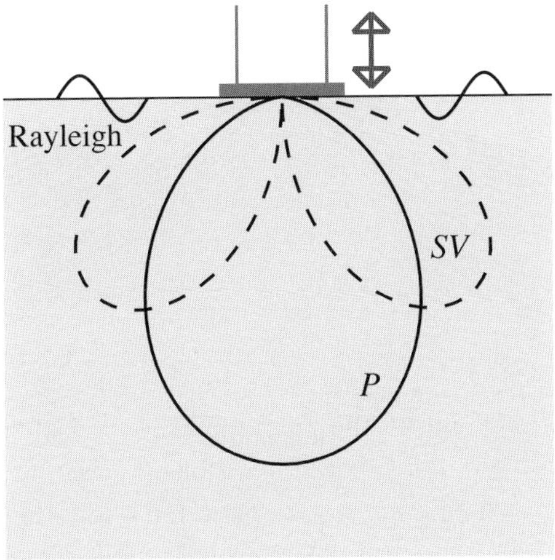

Figure 4.1 Schematic representation of the action of a vibrator source, with generation of both *SV* waves and Rayleigh waves in addition to *P* waves.

duration pulse. The interaction of the vibrator with the surface is quite complex, and nonlinear phenomena may well occur (Baeten & Ziolowski 1990).

A simple theoretical model that enables us to see the basic radiation characteristics of the vibrator source is a force system imposed at the free surface of a half-space (Figure 4.1). The reaction of the ground can be reasonably well approximated by a force dipole (vertical for standard vibrators, horizontal for *SH* wave sources).

A vertical vibrator generates substantial *P*-wave radiation with the largest radiation oriented vertically, but also produces *SV* radiation with a propagation angle around 45 degrees to the vertical. In addition, a large fraction of the radiated energy is carried in the form of fundamental mode Rayleigh waves to create ground roll. Because the source lies at the surface in the zone of greatest variation in elastic properties, there can be significant local variations in the quality of seismic coupling and the resulting radiation patterns. The strong ground roll component from a vibrator source normally requires the use of geophone arrays to enhance the *P*-wave signal. Horizontal vibrators can produce significant *SH* waves, but also generate *P* and *SV* waves (Pugin & Yilmaz, 2019).

For land seismic profiles, the array of sources moves steadily through a pre-deployed set of sensors, so that the available coverage of the wavefield is normally two-sided. An example of the recording from a single shot point using an array of three vibrators is presented in Figure 4.2. Only part of the recording of the 10 km spread is shown; recording extended to 20 s two-way time for this full crustal survey in northern Australia. In the record section in Figure 4.2 there are clear reflections from depth in the 5 km span closest to the sources. The early parts of the records display a complex pattern of refractions and shallow wide-angle

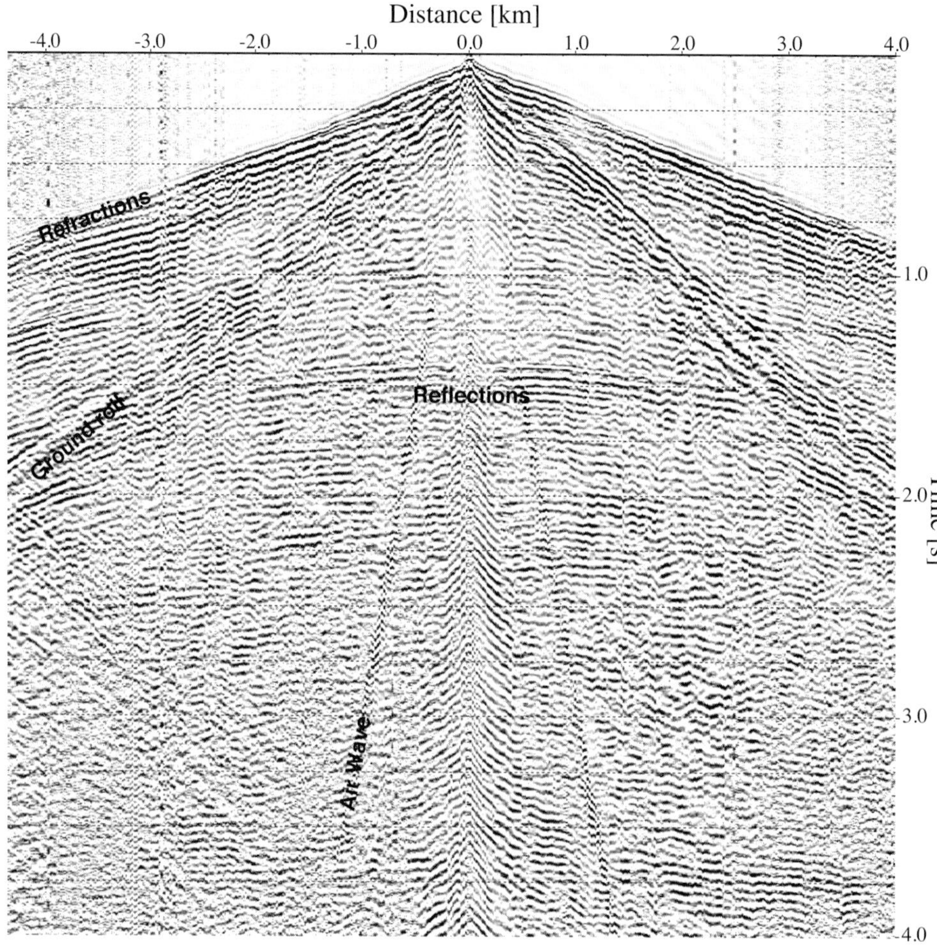

Figure 4.2 Illustration of part of a spread for a single shot point using an array of vibrator sources, from full crustal reflection profiling in northern Australia with recording to 20 s two-way time. [Courtesy of Geoscience Australia.]

reflections. Prominent ground roll cuts across the record. There is also a distinct air wave that has travelled from the vibrators to the sensor above the ground with a wavespeed around 330 m/s. In such profiling, there is a high fold of ground cover. The stacking of many sources enhances the weaker reflections from the base of the crust at about 14 s two-way time in this location.

The reception and recording of the seismic wavefield imposes much of the character of reflection seismic records. Whatever the actual transducer, the seismic signals undergo amplification and filtering before being recorded and this process can be used to reduce the significance of unwanted arrivals such as ground roll. However, the reduction of bandwidth imposed by filtering also reduces the potential resolution available.

The geophones used in land work are frequently sensitive to only a single component of ground velocity. A full vectorial representation of the surface

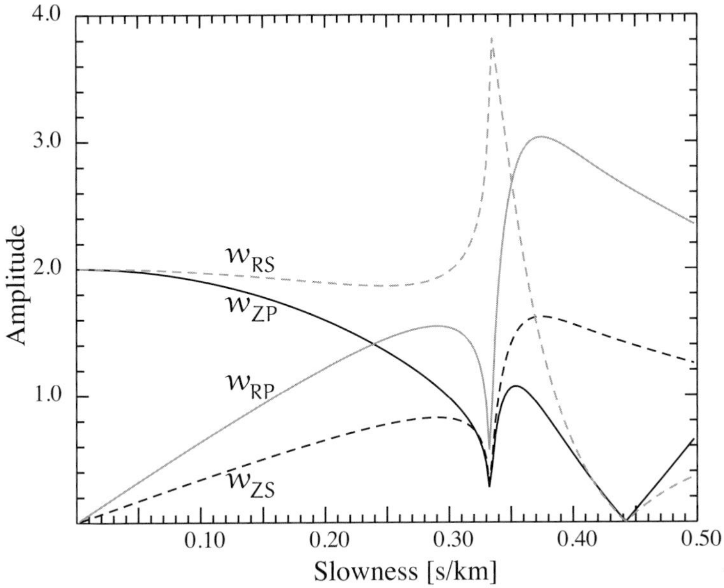

Figure 4.3 Surface amplification factors for incident upgoing waves for a medium with surface properties $\alpha_o = 3.0$ km/s; $\beta_o = 1.6$ km/s. w_{ZP}, w_{RP} are the amplification factors for vertical and radial components for incident *P* waves, and w_{ZS}, w_{RS} for incident *SV* waves.

wavefield then requires separate geophones for three orthogonal components of motion, which are recorded separately and combined as necessary at the processing stage. Some three-component geophone systems include local processing to suppress, e.g., ground roll, but have the disadvantage that the original records cannot be recovered.

Since the geophones are normally planted just at the surface they record the combination of the incident and reflected wavefields at the surface with amplification effects that differ for different wavetypes. For *SH* waves the surface amplitude is just double the incident wave amplitude and there is no dependence on slowness. However, for *P* and *SV* waves the surface amplitudes depend on the surface wavespeeds α_o, β_o and the slowness p (§ SWI:13.1). The slowness dependence of the amplification factors for the vertical and radial ground velocities for incident *P* and *SV* waves is illustrated in Figure 4.3. The amplification of *P* waves on the vertical component reduces steadily as the angle of incidence increases up to the onset of evanescence for *P* at a slowness of 0.333 s/km. *SV* waves have less effect, but become more significant for larger slownesses.

Groups of geophones are commonly deployed as arrays to discriminate against ground roll, and such arrays can also help to enhance coherent signal from depth. For high-resolution work single geophones are often used and the coherent noise removed by careful frequency–wavenumber filtering. In areal surveys designed to extract 3-D structure, particularly with individual node recording, it can be difficult to discriminate against ground roll arriving from different directions, and additional processing-stage removal is often necessary.

An alternative style of sampling of the seismic wavefield is provided by *distributed acoustic sensing* using fibre optic cables (e.g., Parker et al., 2014). In this system, a stream of coherent laser pulses are sent along a fibre, and are scattered back by irregularities within the fibre. The position is determined by the time of transit of the reflected pulse. Changes in the reflected amplitude indicate modifications of the radial strain along the fibre. The spatial resolution is determined by the length of the laser pulse, about 10 m resolution can be achieved with a 100 ns pulse. The temporal resolution depends on the length of the fibre optic cable, since no new pulse can be transmitted until energy reflected at the end of the cable gets back to the transmitter. This presents no problems for the seismic energy range, with sampling available up to 1000 Hz. The practical limitation on the length of cables, without some form of optoelectronic booster system, is of the order of tens of kilometres. With the relatively fine spacing of the effective sensors there is the possibility of recording much of the seismic wavefield without aliasing any of the arrivals. Distributed acoustic sensing systems have begun to play a major role in vertical seismic profiling, since the fibre systems can be readily deployed in drill holes without the complications of anchoring needed for geophone strings. Further, the fibre systems are able to work effectively in the harsh high-temperature environment of deep holes. To optimise the strain sensitivity, the fibre optic cable can be wound in a helix around a mandrel or the hole casing (see, e.g., Wuestefeld & Wilks, 2019).

4.3 Marine Seismic Profiling

4.3.1 Marine Sources

The dominant type of seismic source used in current marine operations is the airgun which discharges air under very high pressure into the water. The gas bubble produced by the discharge oscillates and migrates towards the surface, giving rise to a more complex wavetrain than a simple pulse. Further, the sea surface acts as a strong reflector so that the effective source for radiation is a dipole consisting of the original airgun source and its negative mirror image, an equal height above the surface. Constructive interference between the source and its 'ghost' occurs when the source depth is an integral number of half wavelengths and this leads to prominent peaks in the spectral response.

Although it is possible to make very large airguns, the replenishment time for a large air chamber is long and so the firing rate is slow. An equivalent energy release can be achieved by an array of smaller guns and many different array designs have been implemented. If the individual guns are well separated, there is no interaction between different sources and the total radiation can be determined by summing the effects produced by each of the sources. Commonly, guns are fired in clusters and then the full nonlinear interaction between neighbouring sources

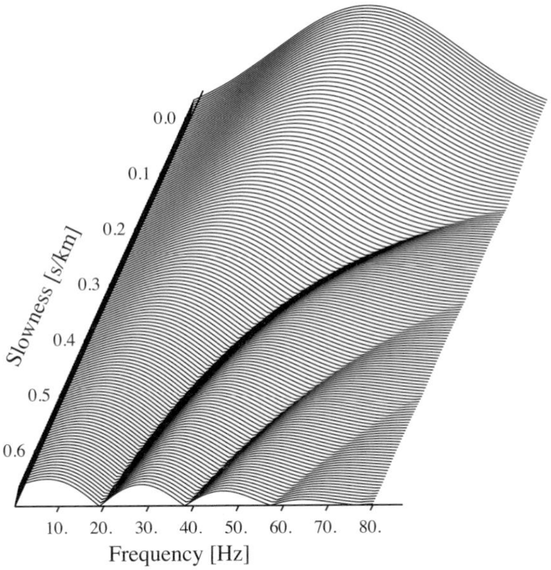

Figure 4.4 The array response for a group of 15 equally weighted hydrophones, 5 m apart as a function of frequency and slowness for a Gaussian source spectrum (centred at 40 Hz with a width of 30 Hz).

has to be taken into account (Parkes & Hatton, 1986). An airgun array can have substantial areal dimensions, e.g. 15× 25 m, but can usually be represented to adequate fidelity by a point source with a directivity with strong frequency and slowness dependence. The spacing and timing of the guns are arranged to enhance near-vertical radiation, but the in-line horizontal dimension is insufficient to provide suppression of near-horizontally travelling waves with frequencies less than around 40 Hz. Such energy remains a substantial contributor to the recorded signals.

4.3.2 Marine Recording

The transducer used in marine recording streamers is a piezoelectric hydrophone which is sensitive to pressure changes in the water. The transducer does not have any directional sensitivity and all the dependence on the direction of propagation is imposed by combining the signals from many transducers. The hydrophones are evenly spaced along the streamer and the response of any sub-array can be determined by linear superposition (see Figure 4.4). In order to reduce the effects of guided wave energy, especially in shallow water, the output of many transducers can be combined. However with such long sub-arrays there will be strong correlation between successive recorded traces and horizontal resolution will be reduced by smearing. The streamers are constructed to be neutrally buoyant at a suitable depth, and so receive both upgoing energy from the reflection process and its 'ghost' reflected in the sea surface.

For an array of N receivers at a spacing Δx, the response to a plane wave with frequency ω and slowness p can be calculated summing the effects of the phase delays across the hydrophone group

$$A_N(\omega, p) = \sum_{j=0}^{N-1} w_j \exp[i\omega p\, j\Delta x], \tag{4.3.1}$$

where the w_j are the weights applied to the individual hydrophone inputs. For equal weighting the amplitude of the output is

$$|A_N(\omega, p)| = \sin(\tfrac{1}{2}N\omega p\Delta x)/\sin(\tfrac{1}{2}\omega p\Delta x). \tag{4.3.2}$$

This response is displayed in Figure 4.4 for a 15-element group, with modulation also by a Gaussian source spectrum. Note the presence of zeroes in the receiver array response, which will have the effect of suppressing certain combinations of slowness and frequency. Near-vertically travelling waves are little affected, but as the slowness increases the array response can significantly modify the apparent behaviour of the wavefield.

On land, direct 3-D sampling can be achieved with an areal array of sensors recording each source. The equivalent in marine work is a network of bottom nodes, commonly with both hydrophones and geophones, with a moving source towed by a ship on a track designed for maximal coverage. Nevertheless, most marine profiling is accomplished using a single vessel and multiple simultaneous streamers are used to cover a broad swath of the sea floor. The particular configuration depends on the ship, and careful track is made of the streamer positions to facilitate processing.

4.3.3 Representation of Marine Records

The general representations introduced in Section 4.1 allow for the possibility of a source at depth, but have the receiver placed at the surface. For marine work, the receiver will always lie at some depth. For a streamer, the depth will be comparable to that of the source, but could lie slightly above or below the source level. With a bottom node, the receiver will be well below the source level but, depending on the mode of deployment, may lie on the sea bed or embedded into the sea floor.

We take the split level J to lie just below the sea bed, and consider an upgoing wave $[\mathsf{U}^J]$ impinging on this level. There will be transmission through the sea floor to the receiver level described by $\mathbf{T}_U^{RJ}$, accompanied by reflection from the sea surface $\mathbf{R}_U^{fR}$. Energy can then reverberate around the receiver level R, with reflection at both the sea bed and surface. Both upgoing and downgoing waves will arrive at the receiver level and we represent the conversion to, e.g., pressure with operators $\mathbf{V}_U^R$ and $\mathbf{V}_D^R$, with the subscript indicating the sense of propagation.

The composite effect of these receiver side processes then takes the form

$$\boldsymbol{w}_R = (\mathbf{V}_U^R + \mathbf{V}_D^R\mathbf{R}_U^{fR})[\mathbf{I} - \mathbf{R}_D^{RJ}\mathbf{R}_U^{fR}]^{-1}\mathbf{T}_U^{RJ}[\mathsf{U}^J], \tag{4.3.3}$$

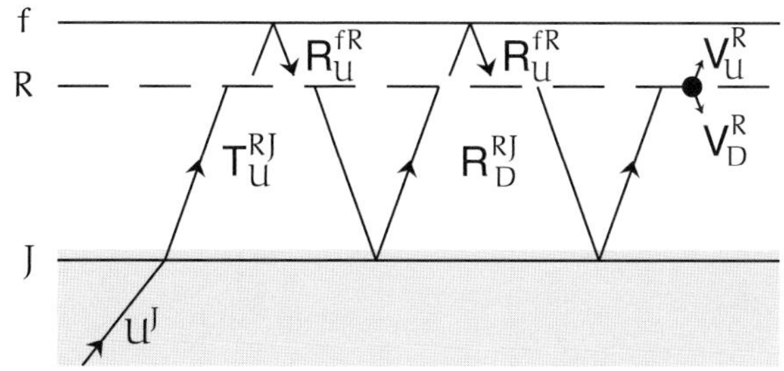

Figure 4.5 Schematic representation of receiver side propagation processes in passage from the split level J through the sea bed to the receiver.

in which we can recognise $(\mathbf{V}_U^R + \mathbf{V}_D^R \mathbf{R}_U^{fR})$ as including the receiver 'ghost' from reflection at the sea surface. These elements are represented schematically in Figure 4.5.

The full response at the receiver including the reflections from depth can then be written as

$$\boldsymbol{w}_R = (\mathbf{V}_U^R + \mathbf{V}_D^R \mathbf{R}_U^{fR})[\mathbf{I} - \mathbf{R}_D^{RJ} \mathbf{R}_U^{fR}]^{-1} \mathbf{T}_U^{RJ} \mathbf{X}_J \mathbf{R}_D^{JL} \mathbf{T}_D^{SJ} [\mathbf{I} - \mathbf{R}_U^{fS} \mathbf{R}_D^{SJ}]^{-1} [\mathbf{D}^S + \mathbf{R}_U^{fS} \mathbf{U}^S], \quad (4.3.4)$$

where we have shown both the source and receiver side shallow reverberations explicitly. The full set of reflections from depth are described by

$$\mathbf{X}_J \mathbf{R}_D^{JL} = [\mathbf{I} - \mathbf{R}_D^{JL} \mathbf{R}_U^{fJ}]^{-1} \mathbf{R}_D^{JL}. \quad (4.3.5)$$

As noted above the reverberation term $\mathbf{X}_J$ will include interaction with the sea bed as well as the water surface, and so the nature of multiples is strongly sensitive to the nature of the sea bed.

The representation (4.3.4) can be used for both pressure and vertical velocity for a geophone sitting on the sea floor, with $\mathbf{R}_U^{fR}$ just involving *P*-wave propagation in the water. The difference between the pressure and velocity forms comes in the nature of the receiver terms $\mathbf{V}_U^R$, $\mathbf{V}_D^R$. Pressure is omni-directional and so the two terms have the same sign. But, for vertical velocity the up- and down-going terms are of opposite sign. This difference can be exploited to help reduce the influence of shallow receiver side multiples, by combining pressure and velocity records.

For a bottom node embedded in the sea bed, the effective position of the velocity sensor will be below the sea floor. The form (4.3.4) can still be used but $\mathbf{R}_U^{fR}$ will include shear elements from passage through the sea floor. The representation for a pressure sensor in the water on the upper side of the sea floor will then not be directly comparable to that for the embedded geophone.

4.3.4 Influence of Sea Bed Structure

In an acoustic approximation the influence of shear waves on the reflection seismic wavefield is neglected. Although such calculations can predict the correct travel times for *P*-wave propagation, they can be very misleading with respect to the

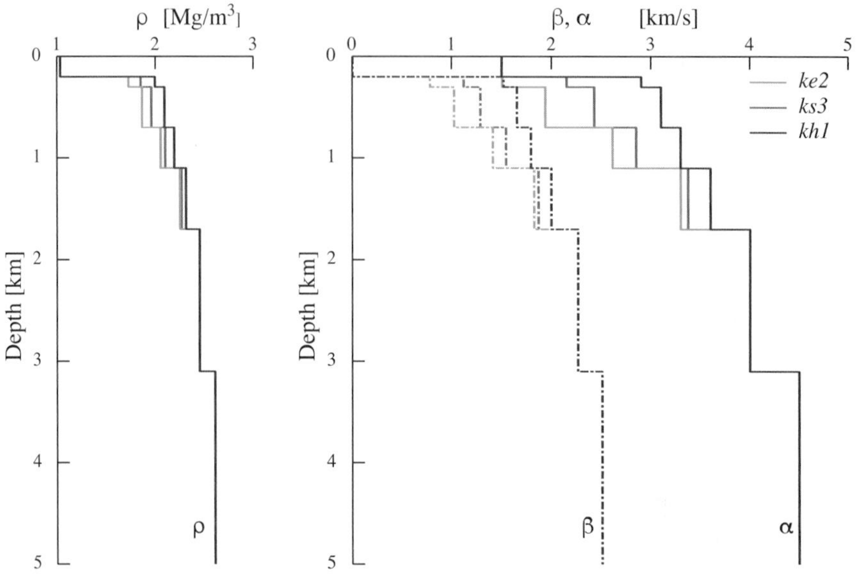

Figure 4.6 *P* wavespeed (α), *S* wavespeed (β) and density (ρ) profiles with depth for the three models *ke2*, *ks3*, and *kh1* used to illustrate the effect of changes in structure near the sea bed.

distribution of seismic amplitudes, especially at larger offsets. At the sea bed, the acoustic approximation predicts total reflection for *P* waves incident beyond the critical angle. However, once the presence of shear waves in the solid material below the sea floor is taken into account, *P* waves in the water incident on the sea bed beyond the critical angle can give rise to transmitted *SV* waves, with a consequent major change in the propagation pattern. Such effects are very important for areas with high velocities at the sea floor, as commonly occurs in tropical waters, such as the north-west shelf of Australia. The character of the water-borne noise in these conditions depends on whether the shear wavespeed at the sea bed lies above or below the *P* wavespeed in the sea water above. For high shear velocities, two distinct sets of critically reflected multiples can be produced to give a very energetic noise train trapped in the water column that contains much high-frequency energy. Conversion of *P* to *S* in transmission at the sea floor can often be important and give rise to significant arrivals on the outer traces from long marine cables. Further, the conversion of energy to *S* waves reduces the energy available for *P*-wave multiples and dramatically reduces the influence of water-bottom multiples compared with a purely acoustic situation (see, e.g., Kennett, 1996).

When the sea bed is soft with gradational properties, the efficiency of conversion is very low (see, e.g., Kim and Seriff, 1992). However, once the material at the sea bed is more competent, transmission through the sea floor with conversion between *P* waves in the water and *S* beneath can be quite efficient, especially for

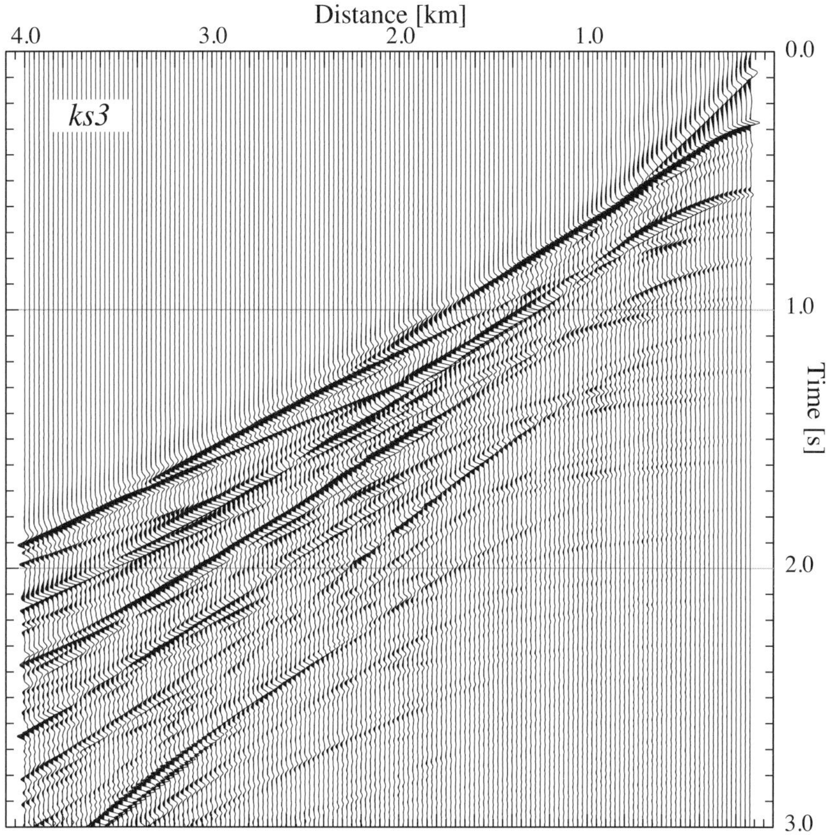

Figure 4.7 Pressure record section for model *ks3* for a delta function source and a frequency band 5–70 Hz. The source is placed at 10 m depth, and the receiver strings are set at 15 m depth. Up to nine surface multiples and all conversions between *P* and *S* waves are included.

wide-angle propagation. The *S* waves can then be reflected back by the structure beneath the sea floor to give rise to *PSSP* arrivals on the outer traces of marine shot gathers. The moveout of such converted reflections differs from the *P*-wave arrivals but it can be quite difficult to separate *PSSP* arrivals from the unconverted waves.

We illustrate the variations in reflection character using a sequence of models that have a fixed structure in the deeper part of the model, with a variable wavespeed gradient for *P* and *S* in the upper part of the model, as illustrated in Figure 4.6.

In Figure 4.7 we show synthetic shot gathers for the model *ks3*, which is representative of medium-hard sediments with a good contrast at the sea bed. The calculated record sections use an array of five sources at 10 m depth with 10 m spacing, and a sequence of receiver arrays at 15 m depth each consisting of 12 elements at 2 m spacing. Near-source and near-receiver ghosts were included in the theoretical seismogram calculation, and up to nine multiples were allowed in the water layer. The time signature of the source is a band-passed delta function filtered with a sine ramp from 5 Hz to 7 Hz, a flat spectrum from 7 Hz to 60 Hz, and a sine ramp down to zero at 70 Hz. The models include an allowance for attenuation, with $Q_p = 500$ and $Q_s = 250$ in each of the solid layers. Full conversion between *P* and *S* is allowed at each interface and the pressure records include all the conversions and multiples within the solid layers. The *P* reflections are clear on the near offset traces, but *PSSP* reflections are prominent at larger offsets, e.g., beyond 2 km with a two-way time greater than 2 s. Each of the primary reflections (both *P* and *PSSP*) is accompanied by a set of water-layer multiples, which help to build up the complexity of the records on the outer traces. The primary *PSSP* reflections are substantially larger than the corresponding *P* reflections for offsets beyond 1000 m, and their presence will help to degrade the effectiveness of processing procedures such as stacking.

The differences in the character of the conversion process at the sea bed can be most readily understood by looking at the patterns of reflection and transmission coefficients. From Snell's law the *P* and *S* waves will travel at different angles of propagation to the vertical, but will share a common horizontal slowness. For the marine seismic situation the range of slownesses is from 0 s/km (vertical incidence) to 0.1667 s/km (horizontal *P*-wave propagation). The reflection pattern at the sea bed is strongly influenced by the value of the *S* wavespeed. If the *S* wavespeed is less than 1.5 km/s (the *P* wavespeed in the water) there is a single critical slowness equal to the reciprocal of the *P* wavespeed beneath the sea floor (Figure 4.8a); beyond critical, most of the energy is converted to *S* in transmission rather than being reflected at the water bottom (Tatham & Stoffa, 1976). Once the *S* wavespeed beneath the sea bed exceeds 1.5 km/s, there are two critical points associated with the refraction of both *P* and *S* waves beneath the sea floor (Figure 4.8b). Sharp increases in the reflection coefficient for *P* waves occur at the critical slownesses for both *P* and *S*. For larger slownesses than the critical point for *S* waves, reflection from the sea bed is total. As a result, all the energy from the source for slownesses between 0.571 s/km and 0.667 s/km is trapped in the water column.

In Figure 4.9 we show synthetic shot gathers for the two models whose sea bed conditions are explored in Figure 4.8. The model *ke2* has a *P* wavespeed at the sea bed only just above the wavespeed in the water. In consequence the *P*-wave critical point is displaced to large offsets, and reflected waves from the second interface become quite important. The dominant *P*-wave multiples in Figure 4.9(a) come from this second interface. However, there is a large contrast for shear wavespeed at the sea bed, even though the *S* wavespeed beneath the sea floor is around 0.9 km/s. As a consequence, conversion from *P* waves in the water to *S* waves beneath the sea bed makes a major contribution to the wavefield, with noticeable *PSSP* reflections on the far offsets. The prominent train of energy travelling close to

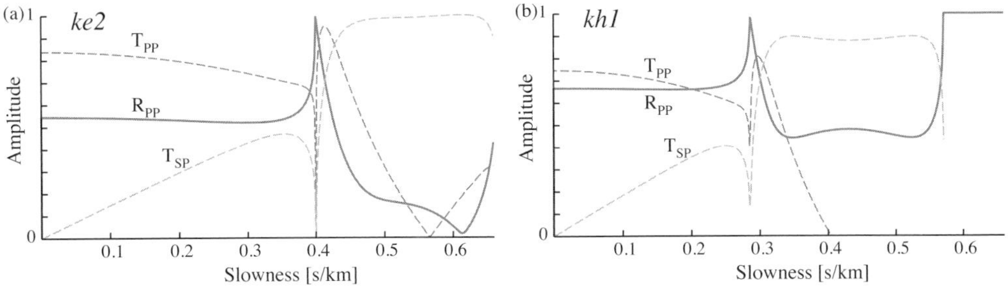

Figure 4.8 The pattern of reflection and transmission coefficients for incident *P* waves on the sea floor for (a) the model *ke2* with a low shear wavespeed beneath the sea bed, and (b) the model *kh1* with shear wavespeed above the *P* wavespeed in the water.

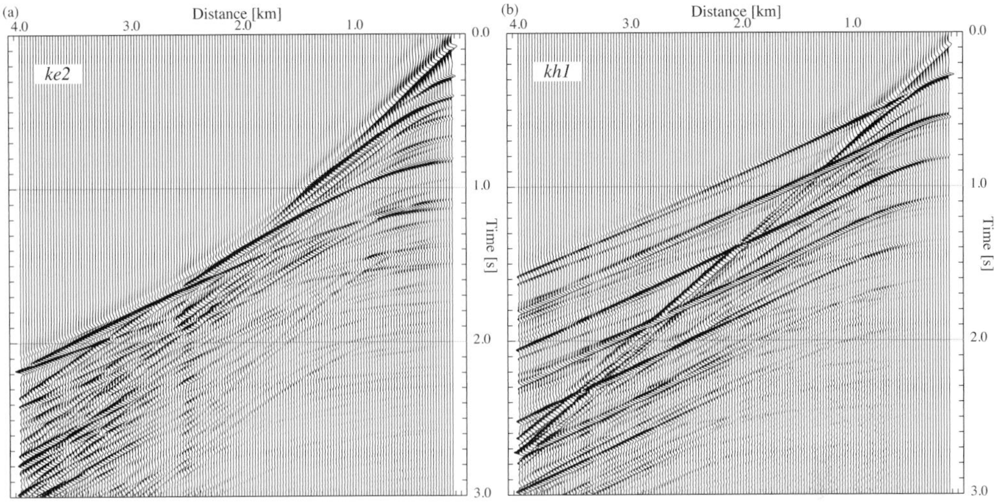

Figure 4.9 Pressure record sections using the same source–receiver configuration as in Figure 4.7 for (a) the model *ke2* with moderately stiff sediments, and (b) the model *kh1* with a hard water bottom.

1.5 km/s, comprises *P*-wave multiples reinforced by shallow converted *S*-wave reflections.

As the *P* wavespeed at the sea bed increases, the separation of the critical reflection points diminishes, modifying the pattern of the wavefield. When the shear wavespeed at the sea bed lies below the *P* wavespeed in the water, there is no band of slownesses for which total reflection occurs. As a result, water-borne energy with velocities close to 1.5 km/s is broken up into a sequence of separate bursts, with amplitude maxima just beyond the critical points for the successive water-layer multiples, as in Figure 4.7 for the model *ks3*. With a long airgun waveform these bursts of energy tend to coalesce to give a nearly continuous train.

A larger P-wavespeed contrast at the sea floor leads to a more significant role for P-wave refractions and their water-layer multiples. A sharp change in the appearance of the guided energy in the water column occurs once the S wavespeed rises above the P wavespeed in the water (1.5 km/s), as in the model *kh1*, since then total reflection occurs at the sea bed for a band of slownesses. Such a situation can readily arise with a hard carbonate layer at the sea bed, as is often found in tropical waters such as the Northwest Shelf of Australia. On the synthetic shot gather (Figure 4.9b) for the model *kh1*, the total reflection shows up as a relatively strong arrival, travelling apparently at the P wavespeed in the water with a rich high-frequency content. This phase cuts across the very prominent water-layer multiples of the shallow P-wave reflections, which tend to obscure primary reflections from depth. *PSSP* reflections generated by conversion at the sea floor are not large but have a distinctive moveout, and can be discerned where there is constructive interference with the P-wave multiples.

Arrivals with apparent velocities close to the P wavespeed in water are largely built up from the interference of multiple critical reflections, and thus are strongly influenced by the horizontal positions of the critical points, which vary with the depth of water and the wavespeeds beneath the sea bed. The pattern of multiple critical points compresses towards shorter distances with both increasing wavespeed at the sea bed and decreasing water depth.

4.3.5 Stacking below the Sea Bed

The standard procedures for the stacking of seismic waves and the estimation of seismic velocities are based on the normal moveout (nmo) relations for a model of a single uniform layer overlying the reflector of interest. For a multi-layered model, the interval velocity for the individual layers can be extracted sequentially from the stacking velocities for the various layers. For large offsets, the time–offset trajectories start to deviate from a hyperbola, and such issues are particularly severe for the converted *PSSP* waves whose largest amplitudes occur at large offsets. The situation can be rectified by introducing a two-layer model in which the water layer is represented explicitly. Not only does this give a better rendering of the moveout behaviour, it also means that apparent velocities below the sea bed can be associated directly with their mode of propagation.

For a situation with a single layer with wavespeed v_1 and thickness h_1, the reflection trajectory in time–offset (t–x) space is the simple hyperbola

$$\mathrm{v}_1^2 \mathrm{t}^2 - \mathrm{x}^2 = 4\mathrm{h}_1^2. \tag{4.3.6}$$

This hyperbolic trajectory is the basis of the estimation of seismic velocities from the moveout of reflections via stacking. For a two-layer scenario, there is no analytic result. The parametric equations for reflections from the bottom boundary,

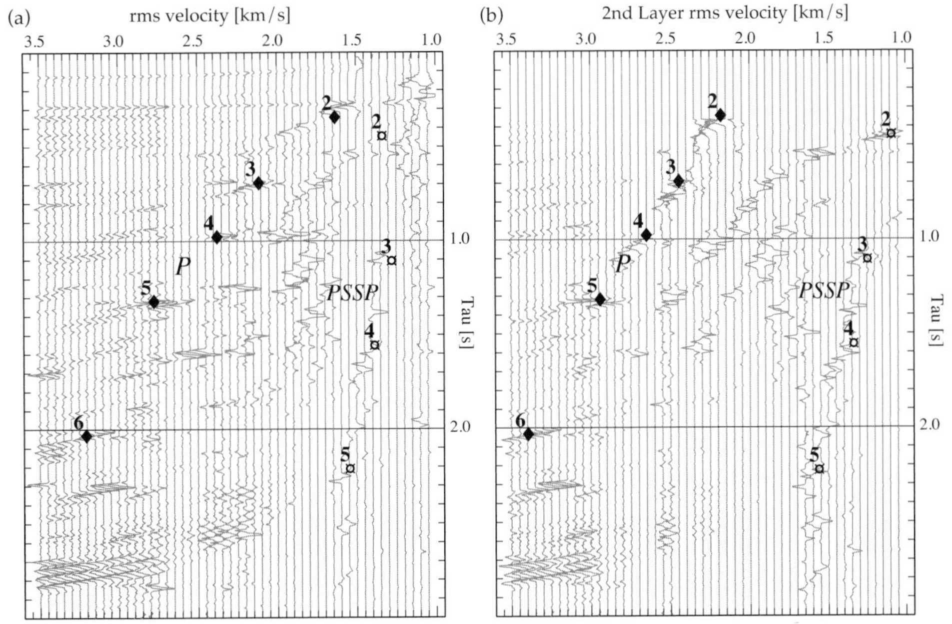

Figure 4.10 Velocity estimation for model *ks3* using semblance-enhanced stacks: (a) conventional rms estimates, (b) two-layer estimates. The values for *P* reflections are shown with solid symbols, and those for *PSSP* reflections with open symbols.

below the sea bed, in terms of horizontal slowness p are

$$x = 2h_1 p v_1/[1 - v_1^2 p^2]^{1/2} + 2h_2 p v_2/[1 - v_2^2 p^2]^{1/2},$$
$$t = 2h_1/v_1[1 - v_1^2 p^2]^{1/2} + 2h_2/v_2[1 - v_2^2 p^2]^{1/2}. \quad (4.3.7)$$

To describe the stacking trajectory starting at the vertical two-way time

$$t = 2h_1/v_1 + 2h_2/v_2, \quad (4.3.8)$$

we have to be able to find the slowness p corresponding to a wave that arrives at offset x. As described by Kennett (1993), this can readily be achieved with a simple shooting scheme to bracket the requisite slowness p_x. Now the apparent velocity in the second layer is used for scanning over stacking trajectories.

We illustrate the application of this approach to results from the model *ks3* (Figures 4.6, 4.7) in Figure 4.10. Stacking has been applied applied with no spatial gain and a linear gain in time. Muting excludes refracted arrivals travelling faster than 1.5 km/s. In panel (a) we display the results for an root-mean-square (rms) velocity scan for the single-layer case, and in panel (b) we show the equivalent results with scanning over the second-layer velocity v_2 with fixed parameters for the water layer. In each panel we mark the locations of the expected rms velocities. Those for *P*-wave reflections are shown with numbered solid symbols, and the equivalent locations for the *PSSP* reflections (with conversion at the sea floor) are marked by open symbols.

Most of the *S* wavespeeds below the sea floor in *ks3* are less than the *P* wavespeed in the water (1.5 km/s). As a result the effective velocities for *PSSP* reflections in the single layer case initially drop below 1.5 km/s before rising again for the deepest reflectors. In the two-layer case we are only concerned with the velocities below the sea bed for both *P* and *S* waves. As a result the effective velocities for the second layer increase monotonically for each wave type.

For the single-layer case (Figure 4.10a) the simple stacking procedure provides a good rendering of the behaviour of the *P* reflections. The presence of the *PSSP* reflections can be distinguished, although amplitude variations as a function of offset, including changes of sign, tend to reduce the stacked amplitude. Because all the *PSSP* rms velocities lie close to 1.5 km/s, it is rather difficult to get constraints on *S*-wave interval velocities.

The corresponding two-layer stacking analysis is displayed in Figure 4.10(b); this takes about four times as much computing as the conventional stack in Figure Figure 4.10(a). The stack displays start 0.1 s after the two-way time through the water (i.e., 0.367 s). The *P* reflectors can be clearly followed as a clear set of amplitude maxima that are better defined than before, because the reflection trajectories are more accurately modelled. Because the two-layer stack has a more direct physical correspondence to the marine case than conventional stacking, it is relatively easy to predict the effect of multiples. Water-layer multiples occur 0.267 s after the equivalent primaries. For *P* reflections the multiples have reduced stacking velocities because of the extra time spent in the water layer. The *PSSP* reflections have stacked well, with a distinct set of stacking maxima, even though amplitude variation with offset reduces the effectiveness of the stack compared with the *P* reflections. Multiples for *PSSP* are displaced to slightly higher effective velocity because the *S* wavespeed is less than the *P* wavespeed in the water down to the third reflector.

Similar analysis can be applied to any other situation where there is a natural physical boundary that separates the propagation zone into two parts, e.g., situations with deep weathering or permafrost in land surveys.

4.4 Influences on Amplitudes

In order to understand the action of the processing applied to the seismic reflection field it is useful to review the range of factors which control the amplitude of a reflector seen on a stacked seismic section. These include the generation of elastic waves by the source, their interaction with the subsurface followed by reception at the surface, and subsequent processing. Each of these stages imposes its characteristic influence on the appearance of a reflection event.

(a) Source directivity and array effects

We have considered the nature of seismic sources in Sections4.2 and 4.3. We have seen that the character of the seismic radiation is controlled by the nature of the source and its configuration relative to the surface. The effective radiation pattern changes with source depth, and is also modified when multiple sources are deployed in an array. Where the individual sources do not interact, the resulting radiation can be estimated by linear superposition, but for marine arrays a more complex nonlinear representation is required.

(b) Receiver arrays

The most faithful reproduction of the seismic wavefield is achieved by taking the seismic records from a single receiver, but this has the disadvantage that large amplitude energy travelling near horizontally is retained on the seismic traces. Such energy is normally suppressed by groups of receivers whose outputs are combined into a single seismic trace, but as noted in Section 4.2 this imposes a selective frequency and slowness filtering on the data, which influences components well away from those for which the suppression is designed. Appropriate wavenumber filtering will have have less effect on the amplitudes of near vertically travelling waves than those at shallower angles.

(c) Near-surface velocity variations

One of the most complex zones in the Earth is the immediate vicinity of the surface (or just below the sea bed in marine work). Small-scale variations in seismic wavespeeds are common and can have a strong focussing or defocussing influence on the radiated seismic wavefield. It is very difficult to estimate the effects of this near-surface variability, which will be most marked for high frequencies for which scattering processes will be important.

With explosive sources on land it is often possible to get below the worst of the near-surface heterogeneity at the source, but a vibrator source lies on top of this zone. In each case the receiver sits just in the heterogeneity zone. The principal limitation on high-resolution work arises from this near-surface zone.

(d) The reflection path

In order for the seismic energy to reach a given reflector it must penetrate the material above this reflector. There will be transmission losses as the waves pass through the bedding both on the way to and on the way back from the reflector. A strong reflection zone will act as a barrier to transmission and leave a region of muted reflections beneath. Such features are often observed in the presence of localised regions of strong cyclic bedding.

In addition, the wavefronts spread in passage through the medium; this pattern of 'geometrical spreading' will depend on the way the velocity distribution varies with depth and will be affected by any velocity anisotropy. There will be two stages of such spreading: the first from the source to the reflector, and the second from the reflector (which can be viewed as a virtual source) to the receiver. If the waves pass through a region of strong attenuation there can also be significant phase distortion of the incident waveform.

(e) The nature of the reflector

The principal control on the amplitude of a reflected wave arises from the nature of the reflector which turns around the incident energy. The material contrasts at individual interfaces are rarely large and coherent reflections often arise from the combined effects of many thin beds. Such reflected energy is built up from a complex interference pattern of primary reflections and internal multiples in the fine bedding, which depend heavily on the thicknesses and arrangements of the beds. Horizontal changes in the character of the local bedding can have significant effects on the waveform interference pattern and so change the apparent reflector strength.

Even where we have a single interface acting as a reflector the character of the observed reflections will depend strongly on the shape of the reflector. When the curvature of the reflector is greater than that of the incident wavefront, we will have the possibility of the formation of subsurface focussing with possible phase distortion of the observed reflections.

Conversions to S waves become more significant at larger offsets, where wide-angle reflections from shallow structure can be important. As the angle of incidence on a reflector increases, the influence of shear wavespeed becomes more significant even where direct conversion is small. The hidden effect of shear properties is to give an apparent attenuation of wider-angle reflections relative to a purely acoustic situation.

(f) Seismic processing

Many of the commonly applied processes in generating stacked seismic records have a nonlinear component whose influence on the amplitudes of reflection is not easy to predict without detailed analysis.

In the standard common mid-point stacking procedure a time-varying moveout stretch is applied to the traces. This stretch has a particularly dramatic effect on shallow reflectors. The post-stack result will therefore include transfer of energy between higher-frequency and lower-frequency components.

Another operation whose influence is pervasive is the ‘muting’ of shot gathers: by removing the residual horizontal travelling energy with a pair of numerical scissors, much of the shallow reflected energy is also lost, with the result that the relative amplitude of shallow and deep reflectors will be altered.

It should also be remembered that most migration procedures are not amplitude preserving. Such techniques move energy to the appropriate spatial location but the weighting procedures are often based on simple approximations for acoustic wave propagation which achieve amplitude fidelity over a limited range of propagation angles. The comparison of observed and calculated amplitudes is therefore best conducted on time sections before migration.

Many modelling techniques employed to simulate seismic reflections are based on the use of acoustic wave theory, to minimise computational cost. Such methods cannot allow for the possibility of conversion of energy to shear waves. For near-normal incidence such conversion tends to be small but cannot be entirely neglected, especially for reflectors with complex shapes.

4.5 Structural Challenges

Current applications of reflection seismic methods are focussed on the delineation of structures in complex environments. In petroleum seismology, the elucidation of structural traps such as those linked to salt features penetrating sedimentary sequences necessitate 3-D data acquisition and analysis. Reflectors may arise from distinct contrasts in wavespeed and density, but also may indicate a change in the style of sedimentation without significant physical contrasts.

Near-surface heterogeneity can produce considerable complications for reflection surveys for sedimentary structures, since they prevent significant energy penetrating to depth. Hard carbonates at the sea floor in tropical waters pose problems from high-amplitude multiple trains in the sea water that obscure reflected energy (Section 4.3.4). Carbonates can also pose a major problem for land surveys in karstic terrains, where near-surface conditions vary rapidly and localised cavities in the subsurface scatter and impede the propagation of seismic waves to and from the targets. Similar issues arise with shallow volcanic layers overlying sedimentary structure because the strong contrast between the basalts and sediments reduces energy propagation to deeper horizons. Further, internal scattering in the basalts and rough boundaries compound the transmission issues. Lower frequencies are less affected by small-scale structure, but reduce potential resolution. Multiple basalt layers are particularly problematic, because of strong multiples in the basalt/ash sequence.

Even when energy has penetrated to depth, the elucidation of complex structures with steep dips can be problematic, since reflected energy may be deflected outside the available receiver array. Imaging in and around salt bodies can be particularly challenging (Jones & Davison, 2014) because of complex geometries and variable physical properties, including anisotropy associated with salt flow. With strongly inclined interfaces conversions between P and S waves are likely, and can be misinterpreted.

Equally challenging circumstances arise in the application of relection techniques to hard-rock situations, where physical contrasts are reduced and many structures are far from horizontal – indeed steep dips are to be expected. Here again the origin of reflectors is hard to unravel, and may represent textural features imposing local anisotropic variations rather than simple wavespeed contrasts. Nevertheless, reflection results have frequently been of considerable value in delineating local contrasts at the mining camp scale, and in providing indications of controlling structures for mineralisation at a larger scale.

Full-crustal reflection profiling reveals the presence of many different styles of structure extending through the crust, though the steepest may well be missed because reflected energy from them does not arrive within the span of the recording

(a)

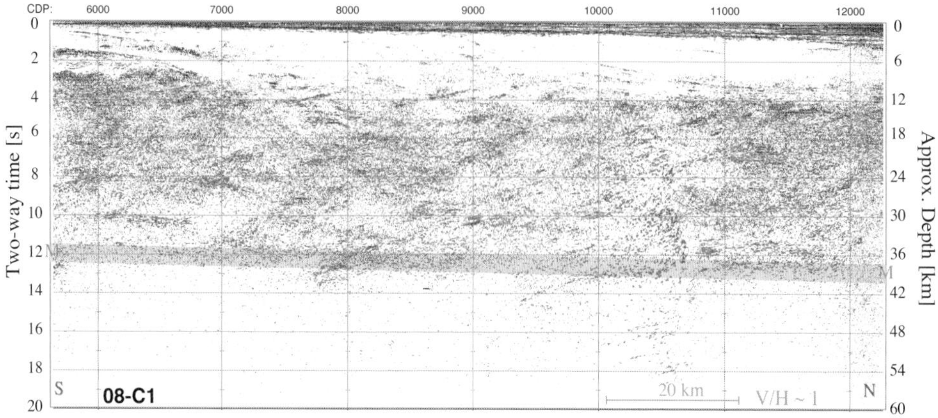

(b)

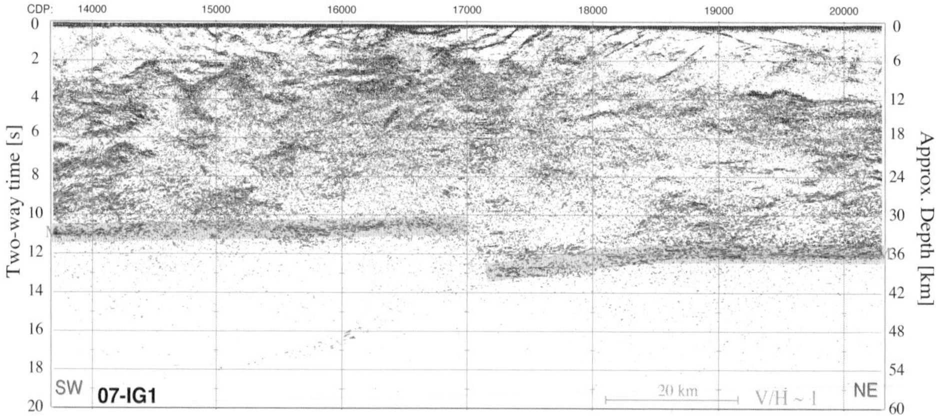

Figure 4.11 Reflection results from full-crustal surveys using an array of three vibrator sources presented at close to true scale: (a) from the Curnamona Province of the South Australian Craton showing complex structure close to the crust–mantle boundary (shaded zone M); (b) from northern Queensland, beneath sedimentary cover, with indications of a jump in the reflection Moho possibly linked to ancient subduction.

array. Examples of segments of reflection profiles in Australia are shown in Figure 4.11, from a cratonic environment in Western Australia and likely Precambrian crust hidden beneath a later sedimentary basin in northern Queensland. These reflection results have been acquired along long nearly linear traverses, and processed with 2-D migration, since there is little control out of line, even though structure is undoubtedly 3-D. For such profiles, careful analysis of the effects of the highly heterogeneous near-surface is needed to achieve good imaging of the structure at depth. Wavespeed variations are small, and even with long spreads of more than 4 km on each side of the source array it is hard to exploit moveout for velocity analysis for features below 8 km depth. The record sections shown in Figure 4.11 include corrections for dip at depth based on dip–moveout analysis.

The segments of reflection profiles are displayed as migrated time sections to 20 s two-way time; an approximate depth conversion can be made by using a mean crustal stacking velocity of 6 km/s and this depth scale is indicated on the figures. In a number of parts of the continent it has been possible to calibrate the depth conversion using results from receiver functions exploiting the reflections and conversions following the onset of the *P* waves on the records from distant earthquakes. The typical difference between the receiver function estimates of the depth to the Moho and those derived from the reflection analysis is around 1 km.

4.6 Processing and Migration

The application of reflection seismic methods to increasingly complex structures poses challenges for concepts originally developed for simpler configurations. With the aid of the operator representation of the medium response developed above, we are able to examine the action of commonly used seismic processing operations as they interact with the seismic wavefield. Each of the major processing steps can itself be represented in operator form. A particularly important class of such operators arises in migration. Here we are able to specify an ideal pre-stack migration scheme and also to investigate the extent to which feasible schemes can give an adequate representation of the ideal.

Because we are dealing with the cumulative action of physical propagation processes, the order in which the reflection and transmission operators act is very important. Even for purely acoustic energy, the operators are non-commutative and their order cannot be varied without deleterious effects.

When we attempt to unravel the seismic wavefield in order to determine a structural model, we must also be careful about the order in which the processing operations are applied. The use of the operator formalism provides a clear specification of the optimum order in a processing sequence.

4.6.1 Reflection Processing

With the formal representation (2.3.9) for the seismic response of a model to excitation by a source, we can investigate the way in which major data processing procedures interact with the seismic wavefield. The effect of the use of receiver groups in the field will be to modify the operators $\mathsf{W}_{\mathsf{U}}^{\mathsf{fS}}$ for actual recordings with smearing over the horizontal span of the group.

In order to reduce the contaminating influence of the shallow terms (4.1.1) the standard approach is to mute out the beginning of the records in each shot spread. The time trajectory of the mute is usually chosen to eliminate the largest early energy, at the expense of information on the shallow velocity structure from wide-angle reflections (e.g., Schulz, 1982; Harding, 1985). Where guided wave energy is important, as in shallow marine work or in some land operations with pronounced ground roll, it may be necessary to resort to frequency–wavenumber filtering (e.g., see Christie et al., 1983), or more delicate filtering in intercept time–slowness space (Kennett & Harding, 1984). We can represent the action of muting and filtering by an operator $\mathcal{M}$. In an ideal situation we would have just removed the portion (4.1.1) of the response, but the deeper reflections will not emerge unscathed. We will have

$$\mathcal{M}\mathbf{w}_{o}^{ref} = \mathcal{M}\mathsf{W}_{U}^{fS}\mathbf{X}_{R}\mathbf{T}_{U}^{SJ}\mathbf{X}_{J}\mathsf{R}_{D}^{JL}\mathsf{T}_{D}^{SJ}\mathbf{X}_{S}[\mathsf{D}^{S} + \mathsf{R}_{U}^{fS}\mathsf{U}^{S}], \tag{4.6.1}$$

where, e.g., the outer parts of reflection hyperbolae may be lost. The action of $\mathcal{M}$ cannot be represented as just a modification of the reflection R_{D}^{JL}, since it does not commute with the other operators in (4.6.1). The reverberation sequences $\mathbf{X}_{R}$, $\mathbf{X}_{S}$ will be modified since the beginning of the process will be removed. The long-lag multiple sequence $\mathbf{X}_{J}$ corresponding to surface multiples from deeper interfaces will be less affected by the action of the mute. The expressions (4.1.1) and (4.1.2) represent the seismic wavefield at surface receivers for a single shot, but can be regarded as one of an ensemble of realizations for different shots. For a *single shot gather*, data processing operators will therefore act from the left on $\mathbf{w}_{o}^{ref}$, as for muting in (4.6.1). Whereas operations on *single receiver gathers* involve a suite of shots and must act on the source terms. We can represent the action of such processes by an operator acting from the right as $\mathbf{w}_{o}^{ref}\mathcal{Z}$.

There is some reciprocity in the situation with source or receiver gathers, but unfortunately there is no easy way to represent the conventional common mid-point gather without using a higher-dimensional operator configuration (cf. Berkhout, 1997). The conventional common mid-point trace gather mixes combinations of many shot–receiver pairs, and has no direct relation to the propagation pattern, except for horizontally stratified media where there is no distinction between the gathers. Our interest is of course in situations where there is horizontal change to provide the conditions necessary for resource accumulations.

We will concentrate on the use of source gathers in a pre-stack mode, since this will provide the most convenient description of the action of processing and migration operations. However, it will be possible to recast most steps in terms of a common receiver gather.

Although some processing steps can be applied at any stage, in many cases the optimum effect requires that a specific sequence be followed. For example, let us consider the action of wavelet extraction and stacking. With a single source gather we can attempt to extract a source wavelet by deconvolution because, if the directivity of the source array is not too strong, we will have a similar source time function on each trace. This deconvolution operator $\mathcal{V}$ will commute with the muting process and the reflection operators (if it is one-sided), so that it has only to act on the effective source radiation $\mathsf{D}^{S} + \mathsf{R}_{U}^{fS}\mathsf{U}^{S}$:

$$\mathcal{V}\mathcal{M}\mathbf{w}_{o}^{ref} = \mathcal{M}\mathsf{W}_{U}^{fS}\mathbf{X}_{R}\mathsf{T}_{U}^{SJ}\mathbf{X}_{J}\mathsf{R}_{D}^{JL}\mathsf{T}_{D}^{SJ}\mathbf{X}_{S}\mathcal{V}[\mathsf{D}^{S} + \mathsf{R}_{U}^{fS}\mathsf{U}^{S}]. \tag{4.6.2}$$

If we now perform a stack to correct for nmo across this gather, the effect will be to enhance the primary reflections from below J and to reduce surface multiples. We can expand the reverberation sequence to emphasise the primary return by using the identity

$$\mathbf{X}_J \mathbf{R}_D^{JL} = \mathbf{R}_D^{JL} + \mathbf{X}_J \mathbf{R}_D^{JL} \mathbf{R}_U^{fJ} \mathbf{R}_D^{JL}. \tag{4.6.3}$$

With a stacking operator $\mathcal{S}$ we can now write the result of stacking after a wavelet extraction in a form which attempts to separate the long-lag surface multiples:

$$\begin{aligned} \mathcal{S}\mathcal{V}\mathcal{M}\boldsymbol{w}_o^{ref} = \mathcal{M}\mathbf{W}_U^{fS}\mathbf{X}_R\mathbf{T}_U^{SJ}\mathcal{S}\mathbf{R}_D^{JL}\mathbf{T}_D^{SJ}\mathbf{X}_S\mathcal{V}[\mathbf{D}^S + \mathbf{R}_U^{fS}\mathbf{U}^S] \\ +\mathcal{M}\mathbf{W}_U^{fS}\mathbf{X}_R\mathbf{T}_U^{SJ}\mathcal{S}\,\mathbf{X}_J\mathbf{R}_D^{JL}\mathbf{R}_U^{fJ}\mathbf{R}_D^{JL}\mathbf{T}_D^{SJ}\mathbf{X}_S\,\mathcal{V}[\mathbf{D}^S + \mathbf{R}_U^{fS}\mathbf{U}^S]. \end{aligned} \tag{4.6.4}$$

and in optimum circumstances $\mathcal{S}$ will largely remove the reverberation operator $\mathbf{X}_J$ from the second term, and $\mathcal{S}\mathbf{R}_D^{JL}$ will be close to $\mathbf{R}_D^{JL}$ itself. We have assumed above that $\mathcal{S}$ would commute with the short-lag multiple operators $\mathbf{X}_R, \mathbf{X}_S$ because there would not be significant time separation between such multiples and the primaries. This is a reasonable approximation for the situation where shallow multiples decay rapidly in amplitude with time, but would not be appropriate in the case of a hard water bottom when shallow reverberation will both be of large amplitude and slow decay in time. This set of processes is summarised in Figure 4.12(a).

When wavelet extraction is applied after stacking so that we work with $\mathcal{V}\mathcal{S}\mathcal{M}\boldsymbol{w}_o$ rather than $\mathcal{S}\mathcal{V}\mathcal{M}\boldsymbol{w}_o$ as in (4.6.4), the multiple suppression is not affected. But now, the stretching of the outer traces in the gather under nmo correction during stacking distorts the spectrum and after summation there will be no consistent wavelet for the whole trace. For common mid-point gathers the situation is compounded by including different shots and receiver response in the stacking process.

In practice, of course, it may be necessary to resort to post-stack wavelet extraction because the statistical improvement for noisy traces outweighs the distortions produced in the result. Similarly, pre-stack equalisation procedures, e.g., for land sources, can repair many of the problems with different shots and receivers.

If we attempt to suppress short-lag multiples $\mathbf{X}_S$, $\mathbf{X}_R$ using some form of predictive deconvolution, we can no longer assume that the operator is independent of location since the true periodicity in the multiples occurs only at fixed slowness. The requisite operators will not commute with $\mathbf{X}_S$ and $\mathbf{X}_R$, and dereverberation operators should strictly be applied to both source and receiver gathers to remove all shallow multiple effects (see below).

4.6.2 Migration and Inversion

The aim of a migration procedure is to attempt to reverse the pattern of the propagation effects which generated the recorded seismograms, with the object of building an image of the subsurface structure. The accuracy of reconstruction depends on the approximations made in particular algorithms and the extent to which the velocity distribution in the model is known.

How then does migration fit into the general operator scheme? The key lies

$$(a)\ \boldsymbol{w}_{o}^{ref} \xrightarrow{\text{mute}} \mathcal{M}\boldsymbol{w}_{o}^{ref} \xrightarrow{\text{decon}} \mathcal{V}\mathcal{M}\boldsymbol{w}_{o}^{ref} \underset{\text{stack}}{\xrightarrow{\text{nmo}}} \mathcal{S}\mathcal{V}\mathcal{M}\boldsymbol{w}_{o}^{ref} \equiv \bar{\boldsymbol{w}}_{o}^{ref}$$

$$(b)\ \bar{\boldsymbol{w}}_{o}^{ref} \underset{\text{shallow effects}}{\xrightarrow{\text{remove}}} \mathsf{X}_{R}^{-1}(\mathsf{W}_{U}^{fS})^{-1}\boldsymbol{w}_{o}^{ref} \xrightarrow{\text{deghost}} \mathsf{X}_{R}^{-1}[(\mathsf{W}_{U}^{fS})^{-1}\boldsymbol{w}_{o}^{ref}\mathsf{E}^{-1}]\mathsf{X}_{S}^{-1} \equiv \mathsf{V}$$

$$(c)\ \mathsf{V} \xrightarrow{\text{redatum}} (\mathbf{T}_{U}^{SJ})^{-1}\mathsf{V} \equiv \mathsf{V}^{J} \underset{\text{suppression}}{\xrightarrow{\text{multiple}}} \mathfrak{P}(\mathsf{V}^{J}) \xrightarrow{\text{migration}} \mathfrak{M}[\mathfrak{P}(\mathsf{V}^{J})]$$

Figure 4.12 Summary of reflection processing operators: (a) early stage to stacking; (b) near-source/near-receiver effects; (c) redatuming followed by multiple suppression and migration with nonlinear operations.

in recognising the assumption which is built into most migration schemes that previous processing has left the ideal case of a reflection operator $\hat{\mathbf{R}}_{D}^{JL}$ with only primary reflections included, acting on a broad band source signal with no ghosts. Ambient noise and residual multiples will contaminate the actual solution, but let us see what is needed in principle.

We discuss an ideal pre-stack migration procedure, which shows the way that possible processing steps can be related to the physical character of the wave propagation. We start from the expression (4.6.3) for the deep reflections and assume that those waves whose propagation is confined to the shallowest regions have been removed. Reinstating the full form for the long-lag multiple operator X_{J}, we can write (4.1.2) as

$$\boldsymbol{w}_{o}^{ref} = \mathsf{W}_{U}^{fS}\mathsf{X}_{R}\mathsf{T}_{U}^{SJ}(\mathsf{I}-\mathsf{R}_{D}^{JL}\mathsf{R}_{U}^{fJ})^{-1}\mathsf{R}_{D}^{JL}\mathsf{T}_{D}^{SJ}\mathsf{X}_{S}[\mathsf{D}^{S}+\mathsf{R}_{U}^{fS}\mathsf{U}^{S}]. \tag{4.6.5}$$

We recall that X_{R} and X_{S} are operators representing the effect of reverberations in the shallow structure: X_{R} near the receiver and X_{S} near the source. With an ensemble of sources and receivers we wish to extract the reflections from depth and so we need to disentangle the near-receiver and near-source effects from the available records.

By working with a set of records from a common source we can try to remove receiver coupling effects and the shallow reverberations: formally, this means we need to construct

$$\mathsf{X}_{R}^{-1}(\mathsf{W}_{U}^{fS})^{-1}\boldsymbol{w}_{o}^{ref}. \tag{4.6.6}$$

Because we are working with a common source gather, the process of removing reverberations requires the action of an operator from the left. For land surveys, Kennett (1991) has shown that the inverse of W_{U}^{fS} can be constructed when three-component recordings are available; under the assumption that the incoming wavefield at each instant can be approximated by a set of different wavetypes with a common horizontal slowness. The result of applying $(\mathsf{W}_{U}^{fS})^{-1}$ would be a set of record sections for each wavetype *P*, *SV*, and *SH*. This would have the advantage that most of the residual ground roll energy would appear on the two shear-wave components. If transverse energy can be ignored, it would be possible to obtain estimates of the *P* and *SV* wavefields from vertical and radial component recordings. When only single component recordings are available an approximate correction can be made for the effect of the free surface on *P* waves, but it is not possible to separate any *S*-wave contributions (and so there would not be any significant ground roll suppression). For marine recordings, $(\mathsf{W}_{\mathrm{U}}^{\mathrm{fS}})^{-1}$ can also be constructed under a similar assumption that only a single horizontal slowness component contributes to the record at any time. This assumption will rarely be exactly satisfied in the offset–time domain, but both the receiver coupling operator $(\mathsf{W}_{\mathrm{U}}^{\mathrm{fS}})^{-1}$ and dereverberation operator $\mathbf{X}_{\mathrm{R}}^{-1}$ can be applied directly in the slowness–time domain. The data then have to be re-transformed to offset–time to allow further processing.

In order to remove effects associated with the source we have to use common receiver gathers. If the source wavelet is consistent across many sources we can aim to improve time resolution by deconvolution with an operator $\mathcal{V}$ such that the original downgoing wave from the source D^{S} is converted to a tighter pulse form d^{S}. When the near-surface is close to horizontal stratification the surface reflection can be removed in the slowness–time domain to yield an new effective source term e^{S}. The combination of deconvolution and surface 'ghost' removal can be summarised by the action of an operator E^{-1} so that

$$[\mathsf{D}^{\mathrm{S}} + \mathbf{R}_{\mathrm{U}}^{\mathrm{fS}}\mathsf{U}^{\mathrm{S}}]\mathsf{E}^{-1} = \mathsf{e}^{\mathrm{S}}, \tag{4.6.7}$$

and because we are dealing with a common receiver gather the operator acts from the right. If we wish to try to remove the effect of source directivity (beyond the removal of the 'ghost') multiple source experiments are required for land recordings in order to provide sufficient information to construct the inverse matrix operators that are needed to allow simultaneous treatment of the *P* and *S* wavefields. The inverse of the reverberation operator $\mathbf{X}_{\mathrm{S}}^{-1}$ can be applied in the slowness–time domain after the deconvolution process on the assumption that it does not affect the new effective source e^{S}. The migration operators will act in the offset–time domain and so we need to re-transform once again, and at this stage we will have constructed

$$\mathsf{V} = \mathbf{X}_{\mathrm{R}}^{-1}[(\mathsf{W}_{\mathrm{U}}^{\mathrm{fS}})^{-1}\mathbf{w}_{0}^{\mathrm{ref}}\mathsf{E}^{-1}]\mathbf{X}_{\mathrm{S}}^{-1}, \tag{4.6.8}$$

where we have to recognise that the actual processing operations will only represent approximations to the true operators. This set of operations to remove near-source and near-receiver effects is summarised in Figure 4.12(b).

If we were able to invert the operators exactly, the wavefield V would represent

$$\mathsf{V} = \mathbf{T}_{\mathrm{U}}^{\mathrm{SJ}}(\mathsf{I} - \mathbf{R}_{\mathrm{D}}^{\mathrm{JL}}\mathbf{R}_{\mathrm{U}}^{\mathrm{fJ}})^{-1}\mathbf{R}_{\mathrm{D}}^{\mathrm{JL}}\mathbf{T}_{\mathrm{D}}^{\mathrm{SJ}}\mathsf{e}^{\mathrm{S}}, \tag{4.6.9}$$

where the wavefield V is separated into the *P* and *S* waves arriving at the receiver. With the two stages of processing for shallow effects we will still retain the long-lag multiples in $(\mathsf{I}-\mathsf{R}_{\mathsf{D}}^{\mathsf{JL}}\mathsf{R}_{\mathsf{U}}^{\mathsf{fS}})^{-1}$ and transmission from the source level down to the separation level J. The ideal effective source e^{S} would correspond to isotropic radiation for a particular wavetype with a delta-function time dependence.

In any practical situation, we would have an imperfect knowledge of V since the inverse reverberation operations $\mathsf{X}_{\mathsf{S}}^{-1}$, $\mathsf{X}_{\mathsf{R}}^{-1}$, at least, will be incomplete with finite amounts of data. However, once we have found V we can follow the prescription suggested by Kennett (1979) and elaborated by Berkhout (1983), and aim to remove the long-lag surface multiples in a single operation. If we have an adequate knowledge of the model to construct an accurate estimate of $\mathbf{R}_{\mathsf{U}}^{\mathsf{fJ}}$ and the transmission operators $\mathbf{T}_{\mathsf{U}}^{\mathsf{SJ}}$, $\mathbf{T}_{\mathsf{D}}^{\mathsf{SJ}}$ we can shift the effective source depth to J:

$$\mathsf{V}^{\mathsf{J}} = (\mathsf{T}_{\mathsf{U}}^{\mathsf{SJ}})^{-1}\mathsf{V} = (\mathsf{I}-\mathsf{R}_{\mathsf{D}}^{\mathsf{JL}}\mathsf{R}_{\mathsf{U}}^{\mathsf{fJ}})^{-1}\mathsf{R}_{\mathsf{D}}^{\mathsf{JL}}\mathsf{e}^{\mathsf{J}}, \tag{4.6.10}$$

where we have absorbed the downward transmission term into e^{J}. We can now view (4.6.10) as an equation for the operator $\mathsf{R}_{\mathsf{D}}^{\mathsf{JL}}$ and after rearrangement we have

$$(\mathsf{I}-\mathsf{R}_{\mathsf{D}}^{\mathsf{JL}}\mathsf{R}_{\mathsf{U}}^{\mathsf{fJ}})\mathsf{V}^{\mathsf{J}} = \mathsf{R}_{\mathsf{D}}^{\mathsf{JL}}\mathsf{e}^{\mathsf{J}}, \tag{4.6.11}$$

and thus

$$\mathsf{R}_{\mathsf{D}}^{\mathsf{JL}}[\mathsf{R}_{\mathsf{U}}^{\mathsf{fJ}}\mathsf{V}^{\mathsf{J}} + \mathsf{e}^{\mathsf{J}}] = \mathsf{V}^{\mathsf{J}}. \tag{4.6.12}$$

We can extract $\mathsf{R}_{\mathsf{D}}^{\mathsf{JL}}$ as

$$\mathsf{R}_{\mathsf{D}}^{\mathsf{JL}} = \mathsf{V}^{\mathsf{J}}[\mathsf{R}_{\mathsf{U}}^{\mathsf{fJ}}\mathsf{V}^{\mathsf{J}} + \mathsf{e}^{\mathsf{J}}]^{-1} \tag{4.6.13}$$

and we can partially stabilise the estimate by considering the action of $\mathsf{R}_{\mathsf{D}}^{\mathsf{JL}}$ on the effective source e^{J}:

$$\mathsf{R}_{\mathsf{D}}^{\mathsf{JL}}\mathsf{e}^{\mathsf{J}} = \mathsf{V}^{\mathsf{J}}[\mathsf{R}_{\mathsf{U}}^{\mathsf{fJ}}\mathsf{V}^{\mathsf{J}} + \mathsf{e}^{\mathsf{J}}]^{-1}\mathsf{e}^{\mathsf{J}}. \tag{4.6.14}$$

We would normally need to expand the operator inverse in (4.6.14) and this will lead to a sequence of terms with alternating sign. With imperfect knowledge of V^{J} it can be difficult to control the numerical stability of this free-surface multiple removal process.

Nevertheless, the estimate of the action of $\mathsf{R}_{\mathsf{D}}^{\mathsf{JL}}\mathsf{e}^{\mathsf{J}}$ obtained from (4.6.14) is the starting point for the development of a migration procedure intended to improve the positioning of reflectors by exploiting knowledge of the velocity structure and the character of the reflected wavefield. Let us start at the top of the model and assume that we have sufficient information about the large-scale structure of the model to enable us to find any requisite reflection and transmission operators. Consider a small slice of the reflection response by splitting at a surface K corresponding to a small time step down the seismograms. We can split up the operator $\mathsf{R}_{\mathsf{D}}^{\mathsf{JL}}$ to expose the properties of the slice JK by using (3.3.30):

$$\mathsf{R}_{\mathsf{D}}^{\mathsf{JL}} - \mathsf{R}_{\mathsf{D}}^{\mathsf{JK}} = \mathsf{T}_{\mathsf{U}}^{\mathsf{JK}}(\mathsf{I}-\mathsf{R}_{\mathsf{D}}^{\mathsf{KL}}\mathsf{R}_{\mathsf{U}}^{\mathsf{JK}})^{-1}\mathsf{R}_{\mathsf{D}}^{\mathsf{KL}}\mathsf{T}_{\mathsf{D}}^{\mathsf{JK}}. \tag{4.6.15}$$

For a small enough slice $\mathsf{R}_{\mathsf{D}}^{\mathsf{JK}}\mathsf{e}^{\mathsf{J}}$ will be dominated by primaries and can be directly identified. The reflection sequence for the zone below K, i.e. $\mathbf{R}_D^{KL}e^J$, can be found by downward continuation and multiple removal. We construct

$$Q = (\mathbf{T}_U^{JK})^{-1}(\mathbf{R}_D^{JL}e^J - \mathbf{R}_D^{JK}e^J)(\mathbf{T}_D^{JK})^{-1} \tag{4.6.16}$$

by downward continuation on both source and receiver gathers – the 'double square root' procedure applies for a uniform medium (see, e.g., Berkhout, 1983). We then have to extract the action of $\mathbf{R}_D^{KL}$ from Q as

$$\mathbf{R}_D^{KL}e^J = Q[\mathbf{R}_U^{JK}Q + e^J]^{-1}e^J, \tag{4.6.17}$$

where we have to assume we have sufficient knowledge about upward propagation in the region JK. When internal multiples in the slice are not important we can use Q as a direct estimator of $\mathbf{R}_D^{KL}e^J$. Once again errors will cascade to produce instability in the estimate of $\mathbf{R}_D^{KL}e^J$, so that the base wave velocity distribution needs to be known well.

Once one slice has been extracted, the whole process can be repeated to give a reflection image with correctly positioned reflectors. This ideal pre-stack migration procedure depends on the accuracy of the downward continuation operators. Errors tend to build up as one attempts to move deeper in the section. In principle, if the density is known and the Born approximation is adequate the amplitudes in $\mathbf{R}_D^{JK}$ could be interpreted in terms of velocity variations, so that an inversion is achieved at the same time.

In practice, of course, we have limited information and thus cannot realise the ideal scheme we have just outlined (Figure 4.12), especially with regard to full inversion. However, the operator approach has the merit of laying bare the bones of the technique and revealing the most likely sources of error.

Practical migration schemes currently operate on stacked seismic data so that it is more difficult to establish a direct relation to the physical propagation processes. However, once again we can examine the way in which processing and propagation are related by working with the operator approach. Although we have based our discussion on surface seismic methods, similar ideas can be applied to vertical seismic profiling, and operator methods provide a convenient link between the two techniques.

Beyond migration

Rather than image the structure at depth by processing the recorded wavefield the ultimate goal is to extract structural information directly. This requires the capacity to model the elastic wavefield directly for complex models, and the need to make allowance for all the pre-processing designed to enhance the reflection signal. Current *full waveform inversion* schemes are only effective if the underlying velocity model is well constrained from prior processing, but inversion can then add critical detail. Successful schemes for reflection work have been applied both in the frequency domain, working from lower to higher frequencies (Pratt, 1999), and with time-domain implementations (e.g., Virieux & Operto,

2009). A general treatment of waveform inversion is provided in Part IV of this work, particularly Chapters 15, 16, and 17.

A variety of current approaches to full waveform inversion for reflection data are compared and discussed by Brittan & Jones (2019). Often, purely acoustic equations are used for forward modelling rather than a fully elastic model, since the computational costs are substantially reduced. Such an approach may be adequate where only small offsets are considered, but the examples in Section 4.3.4 indicate that systematic error can readily accrue from the neglect of the influence of shear-wave structure on the reflection field, since the amplitudes of P reflections are modified. A persistent problem in applications of waveform inversion is the problem of cycle-skipping, where the alignment of the observations and the calculated seismograms are offset and inversion gets locked into a local minimum in misfit rather than the true global minimum. Such problems can be reduced by starting with the lowest available frequencies so that a good initial alignment can be achieved. Advances in source and receiver technology have improved the recording of lower frequencies in recent years, but the majority of data has limited low-frequency content.

A further issue highlighted by Brittan & Jones (2019) is the estimation of an appropriate source wavelet for use in modelling. The extraction of a suitable wavelet is particularly difficult for shallow-water situations and for land surveys. Signal-to-noise problems can be then be significant for the lowest frequencies because of interference with surface waves, despite the efforts to suppress them through processing.

4.6.3 Remapping Reflectivity

Alongside efforts to extract structural information by waveform inversion, there has been increasing interest in using the recorded wavefield to produce *virtual sources* that provide different illumination of the subsurface. Most effort has gone into methods based on seismic interferometry, whereby cross-correlation of specific portions of the wavefield turns receivers into virtual sources. This approach is discussed in Section 9.4 in the general context of correlation methods for receiver studies.

Here we show what is needed to transfer the surface recorded wavefield to that corresponding to a source at depth, and how this can be accomplished with the aid of the focussing functions introduced by Wapenaar et al. (2014) in the context of developing Marchenko equations.

We consider an elastic medium with surface recording of just vertical ground motion, and approximate the subsurface response by working with just the PP component of the propagator operators. The response to excitation to a surface source can then be written in the approximate form, based on (2.3.7),

$$w_{zo} = \mathsf{W}^{f0}[\mathsf{I} - \mathsf{R}_{\mathsf{D}}^{0L}\mathsf{R}^{f0}]^{-1}(\mathsf{U}_{\mathsf{P}}^{0} + \mathsf{R}_{\mathsf{D}}^{0L}\mathsf{D}_{\mathsf{P}}^{0}). \tag{4.6.18}$$

We allow for the effect of near-surface structure with the operators W^{f0}, R^{f0}. The

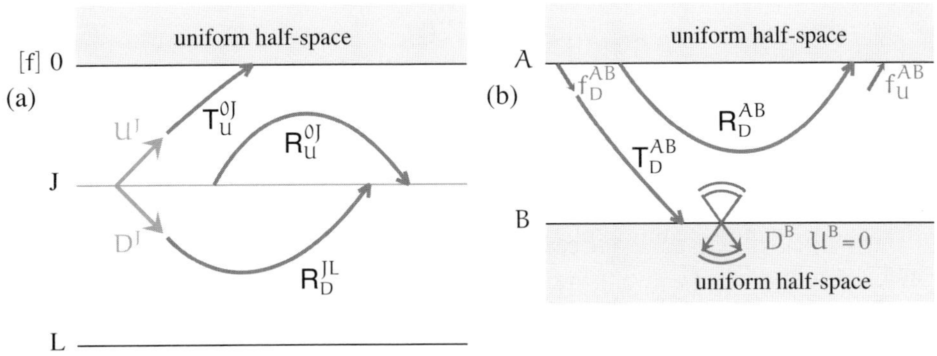

Figure 4.13 (a) Configuration of medium after removal of free surface effects, with virtual source at level J, The main propagation elements are illustrated. (b) Downward focussing function for the region AB with convergence to just downward wave propagation in the uniform half-space below B, with the significant propagation elements indicated.

effective *P*-wave content at the surface is given by $\mathsf{P}_0 = [\mathsf{W}^{f0}]^{-1} w_{zo}$. Our interest is in the reflection from the subsurface R_D^{0L}, and we can solve (4.6.18) for this quantity in the form

$$\mathsf{R}_D^{0L} = [\mathsf{P}_0 - \mathsf{U}_P^0][\mathsf{D}_P^0 + \mathsf{R}^{f0}\mathsf{P}_0]^{-1}. \tag{4.6.19}$$

This expression for the approximate suppression of surface multiples is equivalent to that presented by Kennett (1979). R_D^{0L} corresponds to the situation with the free surface removed and the zone above replaced by a uniform half-space, with continuity of properties across $z = 0$ (Figure 4.13a).

Now we have extracted the reflectivity of the medium as seen from the surface R_D^{0L}, we can look at how this relates to a subsurface level J, with the aid of the composition rule (3.3.30):

$$\mathsf{R}_D^{0L} = \mathsf{R}_D^{0J} + \mathsf{T}_U^{0J}\mathsf{R}_D^{JL}[\mathsf{I} - \mathsf{R}_U^{0J}\mathsf{R}_D^{JL}]^{-1}\mathsf{T}_D^{0J}. \tag{4.6.20}$$

We can project the reflection response to the level J by rearranging (4.6.3) to the form

$$\mathsf{R}_D^{JL}[\mathsf{I} - \mathsf{R}_U^{0J}\mathsf{R}_D^{JL}]^{-1} = [\mathsf{T}_U^{0J}]^{-1}\{\mathsf{R}_D^{0L} - \mathsf{R}_D^{0J}\}[\mathsf{T}_D^{0J}]^{-1} = \mathsf{Y}, \tag{4.6.21}$$

which includes the reverberation between the upper and lower parts of the structure. Since surface multiples have been suppressed these will be purely internal multiples. The reflectivity from below J is R_D^{JL}, which can be extracted from (4.6.21) as

$$\mathsf{R}_D^{JL} = \mathsf{Y}[\mathsf{I} + \mathsf{R}_D^{0J}\mathsf{Y}]^{-1} = [\mathsf{I} + \mathsf{Y}\mathsf{R}_D^{0J}]^{-1}\mathsf{Y}. \tag{4.6.22}$$

If we consider a source at level J in the inhomogeneous half-space (Figure

4.13a), the *P*-wave contribution at level 0 takes the form:

$$\begin{aligned} P^{0J}(0) &= G_U^{0J}(0)U_P^J + G_D^{0J}(0)D_P^J, \\ &= T_U^{0J}\{I + [I - R_U^{0J}R_D^{JL}]^{-1}R_U^{0J}\}U_P^J + T_U^{0J}R_D^{JL}[I - R_U^{0J}R_D^{JL}]^{-1}D_P^J, \end{aligned} \tag{4.6.23}$$

where U_P^J, D_P^J are the upward and downward radiation from the source. We see that the expressions for $G_U^{0J}(0)$, $G_D^{0J}(0)$ are closely linked to that for the reflectivity R_D^{0L}, and we can make this explicit with the introduction of focussing functions f_U^{0J} and f_D^{0J}.

Consider a region AB with a scenario as in Figure 4.13(b) where there are only downgoing waves in the uniform half-space below B. We can use the results for effective sources from Section 3.3.5 to establish suitable source strengths at A to produce no upgoing radiation at B. From (3.3.47) with $U^B = 0$, we require

$$0 = [T_U^{AB}]^{-1}(U^A + R_D^{AB}D^A), \qquad D^B = T_D^{AB}. \tag{4.6.24}$$

These effective sources create a wavefield with the appropriate character.

Applying this approach to the region 0J, we can establish a downward focussing field f^{0J} producing only downgoing waves at J:

$$f_U^{0J}(0) = R_D^{0J}[T_D^{0J}]^{-1}, \qquad f_D^{0J}(0) = [T_D^{0J}]^{-1}. \tag{4.6.25}$$

In the case of a 1-D stratified medium these relations reduce to those presented by Slob et al. (2014). Consider now the action of the focussing field f^{0J} on the reflectivity R_D^{0L},

$$\begin{aligned} R_D^{0L}f_D^{0J}(0) &= R_D^{0J}[T_D^{0J}]^{-1} + T_U^{0J}R_D^{JL}[I - R_U^{0J}R_D^{JL}]^{-1}, \\ &= f_U^{0J}(0) + G_U^{0J}(0). \end{aligned} \tag{4.6.26}$$

Thus the downward radiation reaching the level 0 from a source at level J can be represented as

$$G_D^{0J}(0) = R_D^{0L}f_D^{0J}(0) - f_U^{0J}(0). \tag{4.6.27}$$

This is a general relation that holds for dissipative media, and has a direct analogue for the fully elastic case. Equation (4.6.27) is one of the standard relations used in the development of Marchenko equations for reflection work (e.g., Wapenaar et al., 2014). In the time domain, the product in the frequency domain in (4.6.27) turns into a temporal convolution. With an explicit representation of the reflection operator, we obtain

$$G_D^{0J}(\mathbf{x}_0, \mathbf{x}_J) = \int_0 d\xi_0\, R_D^{0L}(\xi_0, \mathbf{x}_0, t) \star f_D^{0J}(\xi_0, \mathbf{x}_J, t) - f_U^{0J}(0)(\xi_0, \mathbf{x}_J, t), \tag{4.6.28}$$

for a source at $\mathbf{x}_J$, and recording at $\mathbf{x}_0$ with integration along the full surface 0.

An equivalent representation of the *P* waves radiated upward from J can be made for perfect elasticity. In this case we work with the complex conjugates of f_U^{0J} and

f_D^{0J}, which corresponds to using time reversal in the time domain. Then we can obtain the relation

$$G_U^{0J}(0) = [f_D^{0J}(0)]^* - R_D^{0L}[f_U^{0J}(0)]^*, \tag{4.6.29}$$

which is only valid for propagating waves; all evanescent waves must be excluded. There is again a direct elastic analogue of (4.6.29). It is only in the situation without any dissipation that time reversal reproduces amplitude and phase behaviour, so that backward downward propagation can be exactly equivalent to upward travel. For propagating waves in perfectly elastic medium, unitary relations link the reflection and transmission operators (cf. § SM:5.Appendix). The time-domain equivalent of (4.6.29) is

$$G_U^{0J}(\mathbf{x}_0, \mathbf{x}_J) = f_D^{0J}(0)(\xi_0, \mathbf{x}_J, -t) - \int_0 d\xi_0 \, R_D^{0L}(\xi_0, \mathbf{x}_0, t) \star f_U^{0J}(\xi_0, \mathbf{x}_J, -t). \tag{4.6.30}$$

The complex conjugates of the downward focussing field can be linked to a scenario of upward focussing, where incident fields on J from a non-reflecting half-space converge to a focus at 0 (e.g, Wapenaar et al., 2014) since

$$[f_U^{0J}(0)]^{*T} = [T_D^{0J*}]^{-1} R_D^{0J*} = -R_U^{0J}[T_D^{0J}]^{-1}. \tag{4.6.31}$$

An alternative derivation of elastic results with focussing functions has been made by da Costa Filho et al. (2014), using the propagation invariants introduced in Section 3.2.

The coupled equations (4.6.27), (4.6.29), for the downward- and upward-components of the Green's function, form the starting point for the application of Marchenko equations to extract estimates of the focussing function $\mathbf{f}^{0J}$, and hence to construct the response for a virtual *P* source at J. The concepts have their origin in quantum scattering inversion. The applications in reflection seismology are under active development because they allow mapping to a source at depth without detailed knowledge of the velocity structure in the subsurface (Lomas & Curtis, 2019).

The assumption used in the estimation of the focussing function elements is that the inverse transmission function f_D^{0j} can be separated into a direct arrival $\mathcal{T}_D^{inv}$ and a coda M^{0J} of reflection events produced by interaction with the structure above J:

$$f_D^{0J}(0)(\mathbf{x}_0, \mathbf{x}_J, t) = \mathcal{T}_D^{inv}(\mathbf{x}_0, \mathbf{x}_J, t) + M^{0J}(\mathbf{x}_0, \mathbf{x}_J, t). \tag{4.6.32}$$

An iterative development is then made which exploits the causality properties of the propagation elements G_U^{0J}, G_D^{0J} using a careful windowing to construct coupled equations for f_U^{0J}, f_D^{0J} (e.g., van der Neut et al., 2015). The first approximation to f_D^{0J} is taken as just the direct arrival, and the iterations progressively add detail to the coda function M^{0J}. The dominant internal multiples in the wavefield emerge in the first few iterations with reasonable amplitudes and can be refined by further iteration. The windowing conditions work well in media with modestly curved interfaces, but can be upset by isolated heterogeneity. The accuracy of the iterative scheme depends on the aperture of the recording array and will diminish at the edges.

At each iteration, the current focussing functions can be used to estimate G_U^{0J}, G_D^{0J} from (4.6.27), (4.6.29). van der Neut et al. (2015) point out that for G_D^{0J} the amplitudes of true reflections are updated at each iteration, whereas for G_U^{0J} such physical events emerge immediately but further iterations are needed to remove artefacts. As noted above, the expression for G_D^{0J} emerges directly, but more manipulation is needed to produce the representation of G_U^{0J}; these differences clearly influence the reconstruction process.

Once the response of a virtual source at level J has been established, a variety of approaches can be used to employ the estimates in imaging (e.g., Wapenaar et al., 2014; Lomas & Curtis, 2019). Marchenko imaging is built on the concept of obtaining the reflection response from the combination of the upgoing and downgoing wavefields at an arbitrary depth level. For example, a cross-correlation can be made between the downgoing field G_D^{0J} and the estimate of the direct arrival $\mathcal{T}_D^{\text{inv}}$; the zero-lag component emphasises those locations where the fields are most similar, which is expected to be in the neighbourhood of reflectors. The advantage of such an approach over conventional methods using back-projection of the surface wavefield is that internal multiples are included in the Marchenko results, so that many artefacts are suppressed.

Part II

Correlation Wavefields

5

Correlations and Transfer Functions

Correlation procedures play an important role in the determination of time and phase delays in the analysis of seismic signals. The cross-correlation function has the same phase as the transfer function between two signals, and so there is a close link to deconvolution. Receiver function methods exploit the relationships between different components of seismograms to exploit conversions between wavetypes, and, although commonly implemented by deconvolution, can also exploit correlation. Supplementary information is provided by auto-correlations that represent the local reflection response for a single wavetype.

Transfer functions between seismograms yield a measure of their similarity, and can be exploited to track the local evolution of the wavefield across a group of stations. The phase of the transfer function matches that of cross-correlation.

An important role for the comparison of seismograms comes in the matching of observed waveforms with synthetics. The best results are obtained when both amplitude and phase characteristics are captured. Even when there is only a modest similarity between the traces under comparison, the transfer function can provide a useful measure of fit.

5.1 Correlations for Time and Phase Delay

Cross-correlation methods have been extensively used for the analysis of longer-period records where seismic phases tend to be emergent, and there is no sharp arrival to be picked. Correlations of observations with suitable constructed synthetics can provide direct estimates of time shifts from a specified Earth model. Multi-channel cross-correlations also provide an effective means of estimating relative time delays between stations for arrivals from a single source as, e.g., for teleseismic arrivals across an areal array. An alternative procedure, when waveforms are less consistent, is provided by adaptive stacking.

In the two-station method for the analysis of surface-wave dispersion, the fundamental mode portion of the seismograms at two stations lying close to a common great-circle path are correlated to estimate the phase difference in multiple frequency bands for the traverse between the two stations.

5.1.1 Cross-Correlation for Time Shifts

Consider the cross-correlation of two segments of time series $u(t)$, $v(t)$. If $v(t)$ is derived from $u(t)$ by scaling and a time-shift, $v(t) = cu(t - t_o)$, and the cross-correlation $C_{u,v}(t)$ (3.1.1) will itself be a scaled and time-shifted version of the auto-correlation of $u(t)$ with a peak at time t_o. When the resemblance in the time series is strong, but not a direct reproduction, there will still be a maximum in the cross-correlation at the time shift with the closest correspondence between the time segments.

The *cross-correlation time shift* $\mathcal{T}$ is defined as the time when the cross-correlation between $u(t)$ and $v(t)$,

$$C_{u,v}(\tau) = \int_t dt\, u(t)v(t+\tau), \tag{5.1.1}$$

attains its global maximum. Thus $\mathcal{T} > 0$ when $v(t)$ is delayed compared to $u(t)$, and $\mathcal{T} < 0$ when $v(t)$ is advanced. This approach allows time shifts to be directly linked to a reference trace, by suitable choice of $v(t)$.

The reference trace can be constructed by, e.g., modal summation for a specified Earth model with the extraction of a suitable seismogram segment, or by a simpler process concentrating on an individual phase. Thus, Bolton & Masters (2001) have constructed reference pulse forms for teleseismic arrivals by convolving a delta function at the appropriate time with the impulse response of the instruments and an attenuation operator. Bolton & Masters (2001) rely on a visual correlation of the synthetic pulse to the first swing of the observed waveform, using an interactive graphical process that takes a few seconds per measurement. The use of just the first swing in the visual correlation process minimises effects from depth phase interference that become more obvious later in the waveform. More commonly, such a correlation is carried out in some class of automated process. For shallower events the combination of the main *P* and *S* phases and their associated depth phases form a complex interference pattern that changes with the frequency band employed. This led Sigloch & Nolet (2006) to combine the process of inversion for the source and correlation time estimation in an iterative scheme to measure apparent times in multiple frequency bands. An alternative is to undertake a source inversion as in § SWI:6.4, and then use these parameters to construct synthetic reference traces.

For higher-frequency arrivals, picking on individual traces using, e.g., the auto-regressive approach described in § SWII:29.2.1, can be quite effective for limited amounts of data. Such procedures do not take into account the properties of nearby stations and so the full information in the wavefield is not exploited. When many stations record similar arrivals, an effective scheme is to look at the cross-correlation of all pairs of traces, and then extract a self-consistent set of time delays between the stations (Vandecar & Crosson, 1990). In this multi-channel cross-correlation procedure, the time delay $\mathcal{T}_{ij}$ between stations i and j is measured from the maximum of the normalised cross-correlation $C_{ij}(t)$ between the traces

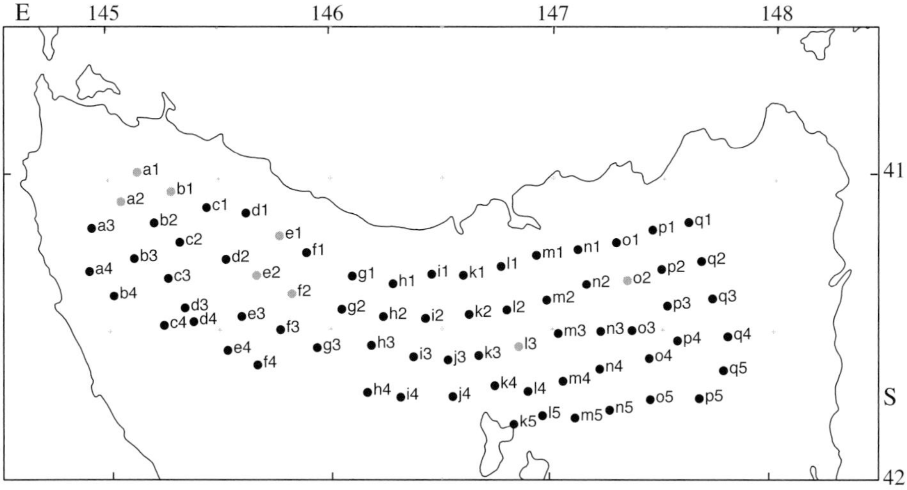

Figure 5.1 Configuration of the seismic array in northern Tasmania, used in the examples of time-delay estimation illustrated in Figures 5.2– 5.4. Stations shown in grey had broadband instrumentation, the remainder short-period seismometers.

$u_i(t)$ and $u_j(t)$ over a specified segment of time (t_b, t_e) around the arrival of the phase of interest:

$$C_{ij}(t) = \int_{t_b}^{t_e} ds\, u_i(s) u_j(s+t). \tag{5.1.2}$$

The width of the maximum peak provides a measure of the confidence interval in the measurement. The best results are achieved when all stations have comparable instrumentation, and a strong similarity in waveform over a reasonable band of frequencies. The position of the maximum in the correlation provides a measure of the time shift $\mathcal{T}_{ij}$ between the ith and jth traces for the arrival of the specified phase. By working with the full set of pairs of stations, a set of estimates of the time differences between the stations is established:

$$\mathcal{T}_{ij} = T_i - T_j, \qquad i, j = 1, \dots, N. \tag{5.1.3}$$

Since $\mathcal{T}_{ij} = -\mathcal{T}_{ji}$, it is only necessary to carry out the cross-correlations for $j > i$ so there are $N(N-1)/2$ equations. Because of the influence of noise and local conditions the measured time differences are not necessarily additive. With the addition of an additional constraint, e.g., the requirement that the sum of the relative arrival times vanishes,

$$\sum_{i=1}^{N} T_i = 0, \tag{5.1.4}$$

the set of over-determined equations is no longer singular and can be solved by

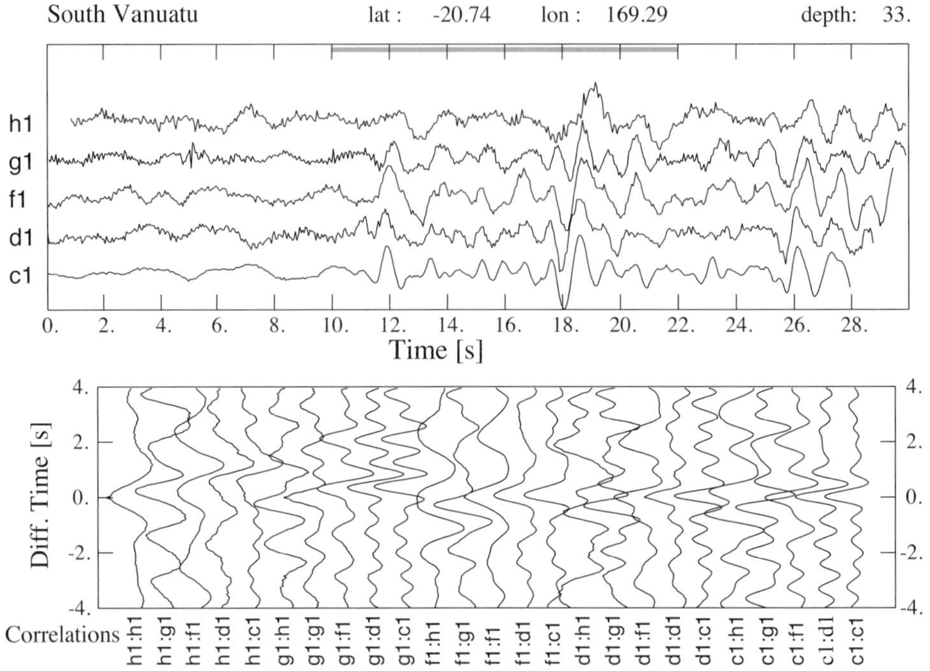

Figure 5.2 A group of seismograms from the west-central part of the seismic array shown in Figure 5.1 for an event in southern Vanuatu (20.74°S, 169.29°E, depth 33 km), together with the cross-correlation of every pair of records for the segment indicated in grey in the upper panel.

least squares. The simplicity of the equations, including (5.1.4), allows an explicit solution for the relative times

$$T_i = \frac{1}{N} \sum_{i \neq j}^{N} \mathcal{T}_{ij} \tag{5.1.5}$$

with a formal estimate for the variance of T_i:

$$\sigma_i^2 = \frac{1}{N-2} \sum_{i \neq j}^{N} [\mathcal{T}_{ij} - (T_i - T_j)]^2. \tag{5.1.6}$$

Here $\mathcal{T}_{ij}$ is the measured quantity and T_i, T_j are the derived values from (5.1.5).

As an example of this approach we consider a set of seismograms extracted from an extensive deployment of portable seismic recorders in northern Tasmania (Figure 5.1). Noise levels and local conditions were quite variable across this dominantly agricultural area, and so although there are many similarities in the seismograms the wavelet for a particular phase can show noticeable variation. In the upper panel of Figure 5.2 we show five traces from the west-central part of the deployment for a shallow event in southern Vanuatu. Time corrections for the

AK135 model (Kennett et al., 1995) have been applied to achieve approximate alignment of the *P* wave. These records show more variation than would be considered desirable for cross-correlation measurements, but nevertheless effective results can be obtained.

In the lower panel of Figure 5.2 we show the full suite of correlations between all the traces, for the time interval spanning the arrival of *P* and the subsequent depth phases. We include the auto-correlation traces, e.g., c1:c1 for which the peak is necessarily at zero time, but the width of this peak widens for poorer quality traces such as h1:h1. In the patterns of cross-correlation we can readily pick up the relative time shifts between the stations. Correlations between the noisier traces can show significant secondary maxima, but the relevant peak is normally readily identifiable.

5.1.2 Adaptive Stacking

In circumstances where the variability between stations is larger than is desirable for correlation measurements, an alternative procedure based on adaptive stacking can provide high-quality estimates of relative arrival times for a seismic phase across a network (Rawlinson & Kennett, 2004). The stack of the full suite of aligned records is compared with each trace, and the time corrections are iteratively updated to improve the relative time estimates, as in Figure 5.3.

For an extended network, the first step is to compensate for the time dependence on epicentral distance. For this we use the AK135 model for the desired phase, as in the multi-channel cross-correlation example above. After the initial alignment using this set of time shifts, linear and quadratic stacks of the records are calculated for the time segment around the selected phase. For N stations with records $\{u_i(t)\}$, the linear stack $V_l(t)$ and the quadratic stack $V_q(t)$ are constructed as

$$V_l(t) = \frac{1}{N}\sum_{i=1}^{N} u_i(t - t_i^c), \qquad V_q(t) = \frac{1}{N}\sum_{i=1}^{N} u_i(t - t_i^c)^2. \tag{5.1.7}$$

The linear stack $V_l(t)$ represents an estimate of the typical waveform across the array and the quadratic stack $V_q(t)$ is an indicator of the spread in alignment between stations. With these initial corrections, the optimum match of the ith trace with the stacked trace $V_l(t)$ is obtained by using a direct search over time shift τ to minimise an L_p measure of misfit

$$\mathcal{P}_i^p(\tau) = \sum_{j=1}^{M} |V_l(t_j) - u_i(t_j - t_i^c - \tau)|^p, \tag{5.1.8}$$

where M is the number of samples in the trace segment. The full search means that the minimum of the $\mathcal{P}_i^p$ misfit measure, can be definitely located with an associated time shift τ_i. The allowed range for $\{\tau_i\}$ can be restricted to a specified interval (e.g., $-1 < \tau_i < 1$ s) to reduce the computational effort. It should be noted that

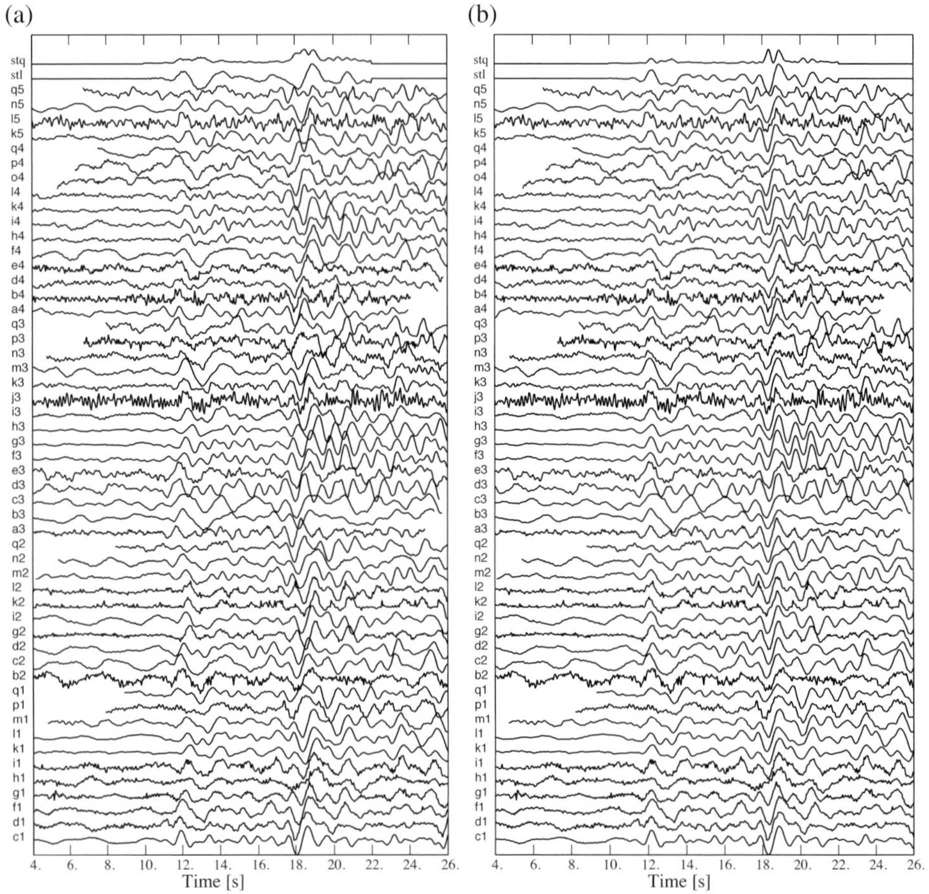

Figure 5.3 Adaptive stacking example using an event from southern Vanuatu (located at 169.29°E, 20.74°S and 33 km depth): (a) initial trace alignment with shifts from just the AK135 reference model; (b) final trace alignment achieved after five iterations of adaptive stacking. The stack traces are stl - linear stack, stq - quadratic energy stack.

the emphasis is on finding the time shift τ_i with minimum misfit of the trace to the stack – the quality of fit is not taken into consideration.

Once the time shifts $\{\tau_i\}$ have been estimated for all the traces, composite time corrections $\{t_i^c + \tau_i\}$ are applied to each trace to improve alignment. The linear and quadratic stacks are then recalculated with the revised time adjustments. The new stacked trace $V_l'(t)$ represents an improvement on the initial stack. The alignment procedure is then repeated for each station trace, using the new stack, to produce an improved estimate of the residuals $\{\tau_i\}$ from values predicted by the propagation model $\{t_i^c\}$. The process is then iterated until accurate and stable trace alignment is achieved.

Rawlinson & Kennett (2004) have found that using an L_3 measure in (5.1.8) is particularly effective in achieving trace alignment for teleseismic data. Experiments with different choices for p indicate that the benefits of larger p come

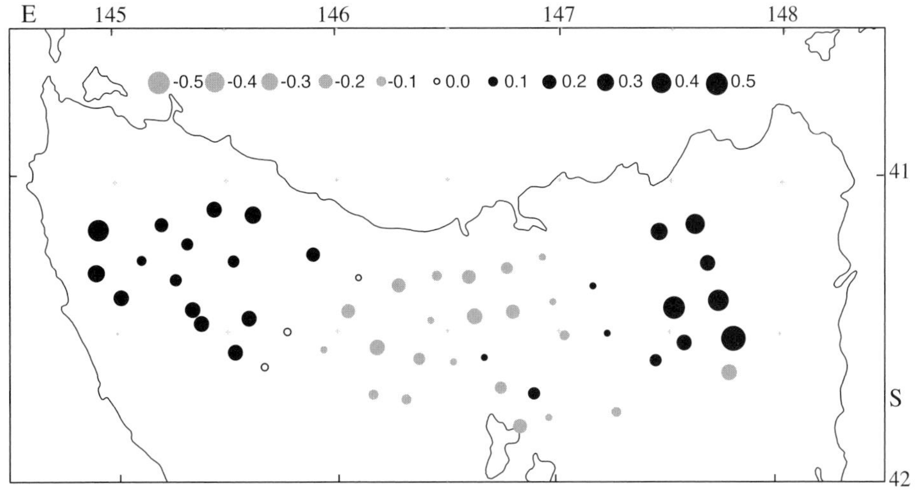

Figure 5.4 Patterns of time shifts (in seconds) between the initial and final stages of adaptive stacking for the southern Vanuatu event illustrated in Figure 5.3.

from the strong penalties imposed on even slight time offsets from the minimum of $\mathcal{P}_i^p$. This leads to more rapid convergence of the iterative process with very stable results achieved in two to three iterations, even when poor traces are included. A comparable approach can be used with trace correlation (e.g., Bungum & Husebye, 1971), but a correlation misfit measure has a broader maximum so convergence is slower and can be hijacked by noisy traces.

The width of the minimum in the misfit measure can be used as an estimator for the accuracy of the time shifts, since the width decreases as the similarity between the stack trace and the individual record improves. For the ith trace, with minimum misfit $\mathcal{P}_I^3(\tau_i)$, the uncertainty in the estimated residual τ_i can be estimated as the smallest difference $|T_n - \tau_n|$ that results in $\mathcal{P}_3(T_n) \geq \epsilon\mathcal{P}_3(\tau_n)$. ϵ is a scaling factor chosen *a priori*. For example, $\epsilon = 1.15$ will result in an error estimate that equals the time shift required to increase $\mathcal{P}_3$ by 15% from its minimum value.

The adaptive stacking procedure provides a set of residuals $\{\tau_i\}$ at each station, relative to the predictions of the propagation model used for initial alignment. Thus with the use of the AK135 model, the pattern of recovered residuals across the array of stations is relative to this model

The high signal to noise for the stacked trace $V_l(t)$ aids the picking of the onset of motion corresponding to the reference point, and hence the absolute time for the particular phase at that point. Absolute times for the full set of traces can then be recovered by applying the composite time corrections $\{t_i^c + \tau_i\}$ for each of the stations.

This adaptive stacking procedure is illustrated with teleseismic data from the full suite of stations in the northern Tasmanian network shown in Figure 5.1. This network included a number of different types of seismometers and recording systems. The response function for the various seismometers were unified as far as possible in pre-processing, but some relative distortions of the waveform remain.

We again show the southern Vanuatu event but now use all seismograms (Figure 5.3). The *P* complex is quite long for this relatively shallow event because of interference with the depth phases (*pP*, *sP*). The signal to noise is moderate across the network, but there are a few poor traces. The initial alignment of the traces using the AK135 travel times from the event location is displayed in Figure 5.3(a). The pattern of white space reflects the geometry of the array. The imperfect alignment indicates the presence of lateral variation in structure beneath the network.

After five iterations of the adaptive stacking procedure, the traces are now very well aligned (Figure 5.3b), and the time shifts that needed to be applied at the individual stations form a coherent pattern of anomalies across the array (Figure 5.4). Although the adaptive scheme only requires three iterations to converge, the computation effort is small and a few extra iterations can used to ensure the stability of the solution. After the iteration even the noisiest traces (e.g., stations j3 and l5) are well aligned. A comparison of the quadratic stack at the initial stage (Figure 5.3a) and final stage (Figure 5.3b) clearly indicates the improved alignment achieved across the network.

5.1.3 Surface-Wave Dispersion

Consider a reference seismogram $\mathbf{u}_r(t)$ and a comparator seismogram $\mathbf{u}_c(t)$. We can define a transfer operator $\mathfrak{T}_{cr}$ that maps one seismogram into the other,

$$\mathbf{u}_c(t) = \mathfrak{T}_{cr}\mathbf{u}_r(t). \tag{5.1.9}$$

For full three-component seismograms the operation $\mathfrak{T}_{cr}$ is a tensor.

When we work with comparisons between single components of the seismograms, we can employ a scalar transfer operator

$$u_c(t) = \mathfrak{T}_{cr}u_r(t). \tag{5.1.10}$$

This operator $\mathfrak{T}_{cr}$ can be implemented via a convolutional integral so that

$$u_c(t) = \int d\tau\, T_{cr}(t-\tau)u_r(\tau) = T_{cr}(t) \star u_r(t), \tag{5.1.11}$$

in terms of a time-domain transfer function $T_{cr}(t)$. Thus in principle we can extract the transfer function by deconvolution of the comparator trace by the reference trace. In the frequency domain this deconvolution can be represented as a spectral division

$$\bar{T}_{cr}(\omega) = \frac{\bar{u}_c(\omega)}{\bar{u}_r(\omega)} = \frac{\bar{u}_c(\omega)\bar{u}_r^*(\omega)}{\bar{u}_r(\omega)\bar{u}_r^*(\omega)} = \frac{\bar{u}_c(\omega)\bar{u}_r^*(\omega)}{|\bar{u}_r(\omega)|^2} = \frac{\bar{C}_{c,r}(\omega)}{\bar{C}_{r,r}(\omega)}. \tag{5.1.12}$$

The phase of $\bar{u}_c(\omega)/\bar{u}_r(\omega)$ is the same as for the cross-correlation $C_{c,r}$, but the amplitude is modulated by the inverse of the power spectrum of $\bar{u}_r(\omega)$, i.e., the transform of the auto-correlation of $u_r(t)$. Such spectral division usually requires numerical tempering to avoid the influence of very small values in the denominator for certain frequencies.

We now apply these concepts to the estimation of surface-wave dispersion, making use of the representation for the propagation of surface-wave trains through slowly varying structure as in Section 3.5.1. If we take two stations along a common great circle through the source, the azimuth at the source is the same and so the excitation terms are in common. However, the modal contributions will have evolved in different ways in the passage to the separate locations. In the frequency domain, using (3.5.3), we can express the multi-mode surface-wave contribution for either Love or Rayleigh waves in the form

$$\bar{u}(X,\phi,\omega) = \sum_{j=0}^{M} \sqrt{\frac{2\omega p_j^{av}}{\pi X}\frac{g_j^{av}}{4I_j}}\, \hat{\mathcal{S}}(p_j^s)\,\mathcal{R}(p_j^r)\, e^{i\omega p_j^{av} X}, \tag{5.1.13}$$

where we have absorbed the azimuthal effects in the composite source term $\hat{\mathcal{S}}$. All of the slownesses (p^s, p^r, p_{av}) are functions of frequency, and it is the dominant path dependent component we seek to extract. Working with a common component of displacement, the multi-mode transfer function between stations at epicentral distances X_1, X_2 takes the form

$$\frac{\bar{u}(X_2,\phi,\omega)}{\bar{u}(X_1,\phi,\omega)} = \frac{\sum_{j=0}^{M} \sqrt{\frac{2\omega p_j^{av}}{\pi X_2}\frac{g_j^{av}}{4I_j}}\, \hat{\mathcal{S}}(p_j^s)\,\mathcal{R}(p_j^r)\, e^{i\omega p_j^{av} X_2}}{\sum_{j=0}^{M} \sqrt{\frac{2\omega p_j^{av}}{\pi X_1}\frac{g_j^{av}}{4I_j}}\, \hat{\mathcal{S}}(p_j^s)\,\mathcal{R}(p_j^r)\, e^{i\omega p_j^{av} X_1}}. \tag{5.1.14}$$

Only if the fundamental mode ($j = 0$) can be isolated does this relationship simplify to provide a direct relation to propagation characteristics:

$$\frac{\bar{u}^0(X_2,\phi,\omega)}{\bar{u}^0(X_1,\phi,\omega)} = \sqrt{\frac{X_1}{X_2}\frac{\mathcal{R}_2(p_0^r)}{\mathcal{R}_1(p_0^r)}}\, e^{i\omega p_0^{av}(X_2-X_1)}, \tag{5.1.15}$$

where we include the effects of geometrical spreading, and allow for modest differences in the receiver behaviour at the two sites. The phase of the transfer function (5.1.15) reflects the propagation of the fundamental mode between the two sites, and is just what we would get by the simpler process of cross-correlation.

A merit of the cross-correlation approach is that the contribution of the large-amplitude fundamental mode to the cross-correlation is enhanced by product of the amplitude spectra $\bar{u}^0(X_1,\omega)\bar{u}^0(X_2,\omega)$, at the expense of noise components.

Thus if we are able to measure the differential phase $\Delta\varphi_{12}^0(\omega)$ between the fundamental modes at the two stations as a function of frequency ω, we can estimate the phase velocity $c(\omega) = 1/p_{av}(\omega)$ from

$$c(\omega) = \frac{\omega(X_2 - X_1)}{\Delta\varphi_{12}^0(\omega) + 2n\pi}, \tag{5.1.16}$$

where n has to be chosen to unwrap the ambiguity in phase. Slight deviations from a great-circle do not introduce bias in the result, but noise in the seismograms will contaminate the dispersion estimates. Corrections may need to be made for any differences in instrumental response between the stations that will introduce a slight phase shift.

It is generally easier to isolate the fundamental mode Rayleigh waves than for Love. This is because the fundamental Rayleigh mode has a greater separation in dispersion characteristics from the higher modes (see, e.g., § SWII:16.3.3). Indeed, the fundamental Love mode has a close affinity to the higher modes, and interference becomes likely at frequencies below 15 mHz.

Two-station phase velocity measurements were of considerable importance in establishing differences between upper mantle structures in different regions (see, e.g., Knopoff, 1972). With a limited distribution of seismic stations across the globe, finding a source and a pair of stations that were sufficiently close to a great circle to allow effective measurements was not easy, and restricted path coverage. As a result, single-station approaches were introduced that require better knowledge of the source, to allow the estimation of initial phase, but which are not as geographically restricted.

As much denser networks of seismic stations have been established in many regions, the possibilities of finding suitable great-circle combinations has improved. Soomro et al. (2016) describe an automated approach to the estimation of fundamental mode Rayleigh and Love wave dispersion across Europe using both single-station and two-station methods. They have used a combination of frequency-domain and time-domain filtering with cross-correlation of the fundamental mode contributions to the seismograms. In frequency, they have employed a suite of Gaussian filters around target frequencies $\{\omega_n\}$

$$F(\omega, \omega_n) = \exp(-\alpha_f[1 - (\omega/\omega_n)]^2); \tag{5.1.17}$$

the width of the Gaussian filter depends linearly on frequency to optimise the time–frequency resolution: $\alpha_f = \gamma_f^2 \omega_n \Delta t$, where Δt is the sampling interval in the time domain. The parameter γ_f is typically taken to be around 16. Following cross-correlation of the waveforms at the two stations, windowing in the time domain becomes easier because dispersion is reduced from that in the original seismogram. This means that windows are easier to position and can be narrower:

$$w(t) = \exp[-\omega_n^2 (t - t_{max})^2 / 4\alpha_w], \tag{5.1.18}$$

where t_{max} is the time of the maximum amplitude of the cross-correlation after filtering. The width of the Gaussian window is again chosen as a linear function of frequency: $\alpha_w = \gamma_w^2 \omega_n \Delta t$. Because the effects of dispersion are more pronounced for longer paths, γ_w needs to increase with inter-station distance from 20 for 400 km separation to 50 for 3000 km separation.

The process of a cross-correlation estimate of phase velocity dispersion between two stations is illustrated in Figure 5.5. The chosen segment of the phase velocity curve shown in a heavy solid line is reasonably close to a background model,

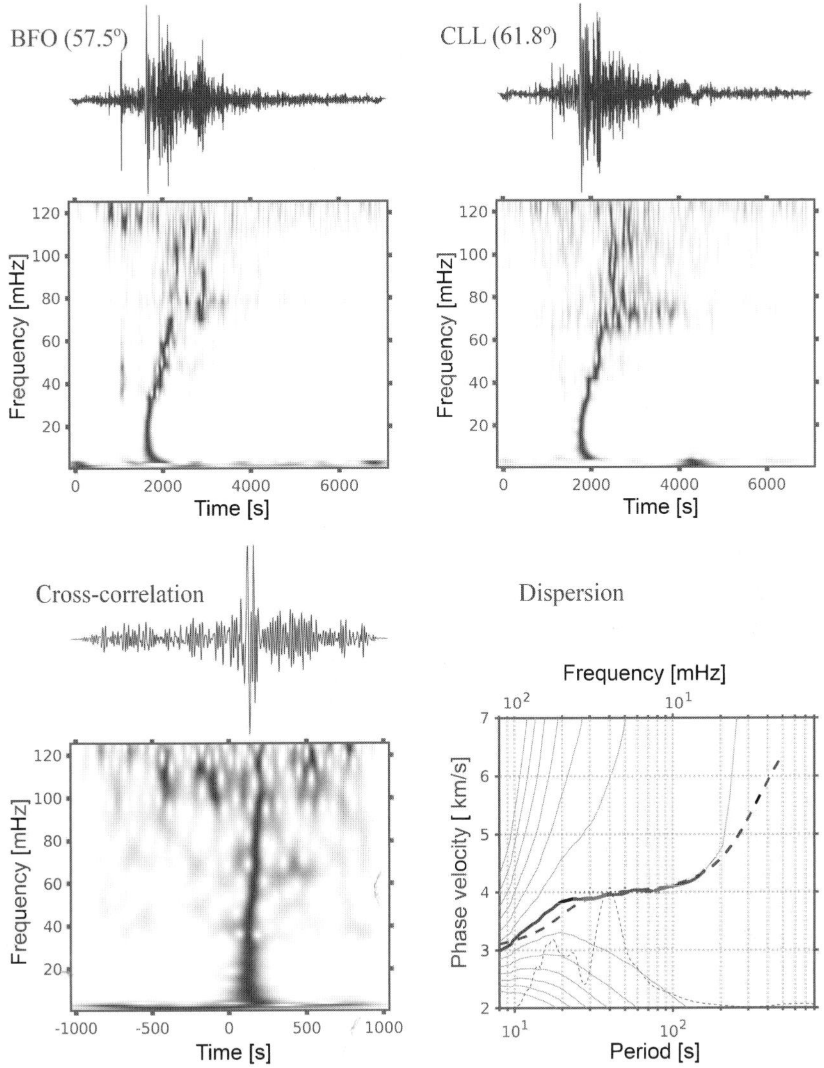

Figure 5.5 Inter-station estimation of phase velocity for Rayleigh waves between the station pair BFO (Black Forest Observatory, Germany) and CLL (Collmberg, Germany) for an event on 2000 October 27. The 2-hour long seismograms and the time–frequency representations of their waveforms are shown in the top panels. Lower left: the cross-correlation function and its time–frequency representation. Lower right: phase-velocity curves with an ambiguity of 2π (light lines), plotted together with the reference model (black dashed); heavy solid line: the manually selected portion of the dispersion curve that is accepted as the measurement. [Courtesy of A. El-Sharkawy.]

whereas other values for the phase-wrap compensation are clearly discordant. Despite complexities in the seismograms at higher frequencies (> 60 mHz) due to scattering, consistent dispersion can still be extracted.

With large data sets recorded by permanent and temporary stations, such as the more than 1000 stations in Europe, automated procedures represent the only practicable route. But, in order to make such a cross-correlation process work effectively, a number of hurdles have to be overcome (Soomro et al., 2016): (1) resolving the 2π ambiguity in phase; (2) selection of a suitable frequency range to provide reliable measurements; (3) identification of smooth segments of the dispersion curve and rejection of rough behaviour; (4) detection of quality problems caused by, e.g., erroneous station response information or timing problems; and (5) rejection of outliers. Each of these elements needs careful consideration and are described in detail by Soomro et al. (2016).

Modest deviations from the great circle (preferably less than 5°) can be tolerated, and this helps with event selection. With many years of available data, multiple measurements become accessible for many pairs of stations and so the consistency of dispersion estimates can provide good control on the accuracy of the final results.

The approximation of surface-wave propagation in a weakly varying medium that we have used in establishing the representation (5.1.13) is equivalent to assuming a plane-wave front progressing along the great circle. As the frequency increases, the wavelength diminishes, and so surface waves become more sensitive to the structures through which they progress, and the plane-wave approximation ceases to be adequate. This is the reason for the difficulty of measuring high-frequency dispersion except between rather close stations.

Similar issues arise when an attempt is made to exploit the dispersion across a number of stations in a network. The incoming wave front from a teleseismic event may well be quite complex. Forsyth & Li (2005) make the approximation that the incoming wavefield is composed of two interfering plane waves and use the properties of the seismograms to estimate their parameters. More complex schemes attempt to estimate the shape of the wavefront as it progresses across the array, as well as making local dispersion estimates (e.g., Friederich et al., 1994; Pedersen et al., 2006). Pedersen et al. (2015) present measurements of the deviation of fundamental mode Rayleigh waves from the great circle across the LAPNET array in northern Finland, and compare with estimates from synthetic seismograms for a fully three-dimensional Earth. Deviations are reasonably well simulated for lower frequencies (around 10 mHz) where deviations are typically around 3°. The agreement diminishes at higher frequency as the influence of the crust becomes more important and the synthetics underestimate the deviations of about 9° at 50 mHz.

5.2 Receiver Functions and Transfer Functions

The concept of a transfer function introduced in the context of the estimation of surface-wave dispersion in Section 5.1.3 has many other applications. We here consider two further ways in which transfer functions can be employed.

Conventional receiver function studies exploit the transfer functions between different components of a seismogram recorded at the same receiver. Additional information is available from auto-correlation of traces.

For distributed stations with a common source, the transfer function between a reference station and the others can provide insight into the nature of the wavefield via the moveout of peaks in the transfer function.

5.2.1 Receiver Functions

With receiver functions we endeavour to concentrate attention on the structural effects just beneath a recording station, and this can be achieved by extracting the transfer function between different components of a seismogram for the same time interval. In practice this means deconvolving one component by another in a way designed to enhance specific aspects of the wavefield.

In Section 2.4 we introduced a representation for the response at a teleseismic receiver (2.4.5) in terms of the ray geometric slowness $\mathrm{p_g}$:

$$\boldsymbol{w}_0^{\mathrm{tel}}(\mathrm{p_g},\omega) = \mathrm{S_I}(\mathrm{p_g},\omega)\mathbf{Z}(\mathrm{p_g},\omega)\mathbf{C}_{\mathrm{R}}(\mathrm{p_g},\omega)\Phi(\mathrm{p_g},\omega), \tag{5.2.1}$$

where $\mathrm{S_I}(\mathrm{p_g},\omega)$ represents the combined effects of excitation and the instrumental response, $\Phi(\mathrm{p_g},\omega)$ the phase acquired on propagation to the base of the structure beneath the station, and $\mathbf{Z}(\mathrm{p_g},\omega)$ the corresponding amplitude effects. $\mathbf{C}_{\mathrm{R}}(\mathrm{p_g},\omega)$ describes the effect of structure local to the station.

All components of the teleseismic arrival will share the same source and propagation effects. Thus extracting the transfer function between different components recorded at the same receiver should isolate the influence of structure beneath the receiver through the different elements of $\mathbf{C}_{\mathrm{R}}(\mathrm{p_g},\omega)$.

Deconvolution of receiver elements

A wide variety of different approaches have been taken to implement the deconvolution to construct such transfer functions in both the time and frequency domain. The character of the resulting *receiver function* depends on the components employed. A detailed examination of the character of receiver functions is given in § SWII:28.2, in terms of the reverberation processes in the crust and uppermost mantle.

A common choice for incident *P* waves is to deconvolve the radial component of motion by the vertical component to enhance *P* to *S* conversions in transmission to the receiver. This process produces a strong initial peak representing the direct *P* arrival and its interaction with very shallow structure. There is therefore some advantage in working with combinations of the components to achieve greater separation of the incident wavetype and its conversions. This can be achieved by a rotation in the vertical plane (Vinnik, 1977; Kind et al. 1995) or a transformation of the components to remove the effect of the free surface and so separate *P*, *SV*, and *SH* components (Kennett, 1991; Reading et al., 2003).

These approaches exploit the common slowness p for the arrivals in segment of seismograms that are being compared. For incident waves with slowness p

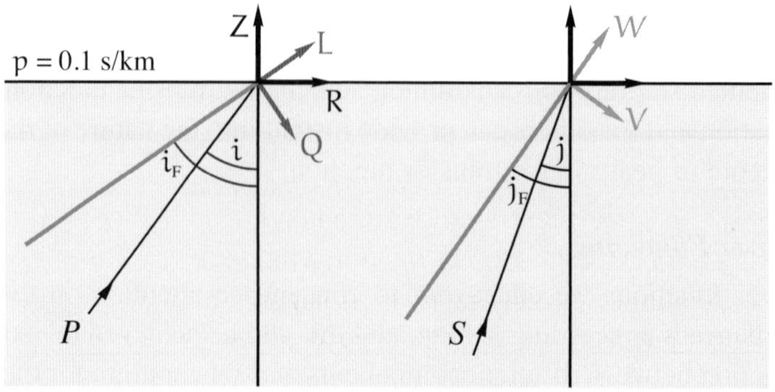

Figure 5.6 Configuration of the LQT and VWT coordinate systems at the free surface of a medium with α_0 = 5.8 km/s, β_0 = 3.4 km/s, for incident slowness p = 0.1 s/km. The T axis points out of the page.

impinging on the free-surface, the angles of incidence at the free surface for *P* and *S* waves with slowness p are related by Snell's law,

$$p = \frac{\sin i}{\alpha_0} = \frac{\sin j}{\beta_0}, \tag{5.2.2}$$

where α_0, β_0 are the *P* and *S* wavespeeds at the surface. The vertical slownesses

$$q_{\alpha 0} = \frac{\cos i}{\alpha_0} = (\alpha_0^{-2} - p^2)^{1/2}, \quad q_{\beta 0} = \frac{\cos j}{\beta_0} = (\beta_0^{-2} - p^2)^{1/2}. \tag{5.2.3}$$

In an unbounded medium, the components of motion for a *P* wave of amplitude P would be

$$Z_u = P\cos i, \quad R_u = P\sin i, \quad \text{so that} \quad \tan i = R_u/Z_u. \tag{5.2.4}$$

At the free surface we need to include the amplification factors for the displacement component (§ SWI:13.1),

$$Z = P\cos i \cdot C_1, \qquad R = P\sin i \cdot C_2. \tag{5.2.5}$$

The apparent angle of incidence i_F as seen from the relation of the components at the free surface is

$$\tan i_F = \frac{R}{Z} = \frac{R_u \cdot C_2}{Z_u \cdot C_1} = F\tan i = \tan 2j, \tag{5.2.6}$$

where the surface amplification factor

$$F = \frac{C_2}{C_1} = \left(\frac{\beta_0}{\alpha_0}\right) \frac{2\cos i \cos j}{1 - 2\beta_0^2 \sin^2 i/\alpha_0^2} = \left(\frac{\beta_0}{\alpha_0}\right) \frac{2\cos i \cos j}{1 - 2\sin^2 j}. \tag{5.2.7}$$

The LQT transformation (Vinnik, 1977) is based on a rotation in the vertical plane to align the L component along the apparent direction of propagation of *P* with the Q component perpendicular (Figure 5.6). The T component rests purely in the horizontal plane. Thus we can construct L and Q as

$$\begin{aligned} L &= -Z\cos i_F + R\sin i_F, \\ Q &= \;\; Z\sin i_F + R\cos i_F. \end{aligned} \tag{5.2.8}$$

The *P* motion should therefore be concentrated on L and the Q component should contain converted *SV* energy. This is useful for receiver function studies because the direct *P* pulse seen on the R component is suppressed on the Q component.

A similar development may be made for incident *S* waves with amplitude S. The apparent angle of incidence for the *S* wave is thus given by

$$\tan j_F = F\tan j. \tag{5.2.9}$$

The transformation to the VWT set (Yuan et al., 2006) again involves a rotation in the vertical plane. The V coordinate lies perpendicular to the apparent direction of propagation of *S*, i.e., in the expected direction for *SV* waves, the *W* component is at right angles. The transformation to V and *W* takes a similar form to the *P* wave case:

$$\begin{aligned} W &= -Z\cos j_F + R\sin j_F, \\ V &= \;\; Z\sin j_F + R\cos j_F. \end{aligned} \tag{5.2.10}$$

In this case the conversions from *S* to *P* are to be sought on the *W* component, and again the direct *S* pulse will be suppressed.

An alternative to these component rotations is make a transformation from the displacements into the amplitudes of *P* and *S* waves. The relation between the motion components at the surface (Z, R, T) to the incident *P* and *S* waves at slowness p (P, S, H) is the inverse of the surface amplification equations and has a surprisingly simple form (Kennett, 1991; § SWI:13.1):

$$\begin{pmatrix} P \\ S \\ H \end{pmatrix} = \begin{pmatrix} i(2\beta_0^2p^2-1)/q_{\alpha0}\epsilon_{\alpha0} & 2\beta_0^2p/\epsilon_{\alpha0} & 0 \\ 2\beta_0^2p/\epsilon_{\beta0} & i(2\beta_0^2p^2-1)/q_{\beta0}\epsilon_{\beta0} & 0 \\ 0 & 0 & 1/2 \end{pmatrix} \begin{pmatrix} Z \\ R \\ T \end{pmatrix}, \tag{5.2.11}$$

where

$$\epsilon_{\alpha0} = 1/(2\rho q_{\alpha0})^{1/2}, \quad \epsilon_{\beta0} = 1/(2\rho q_{\beta0})^{1/2}, \tag{5.2.12}$$

using fully normalised wavevectors (§ SWI:12.1).

Reading et al. (2003) have employed this transformation to extract the S and P wavevectors, and then deconvolve S with P to extract a receiver function in which the incident *P* wave peak is suppressed and *S* conversions are clarified. In Figure 5.7 we illustrate the difference between the conventional radial receiver function in which R is deconvolved by Z, and the result using the wavevectors. The wavevector result is simpler and gives improved results in inversion for one-dimensional structure because the inversion scheme does not have to expend

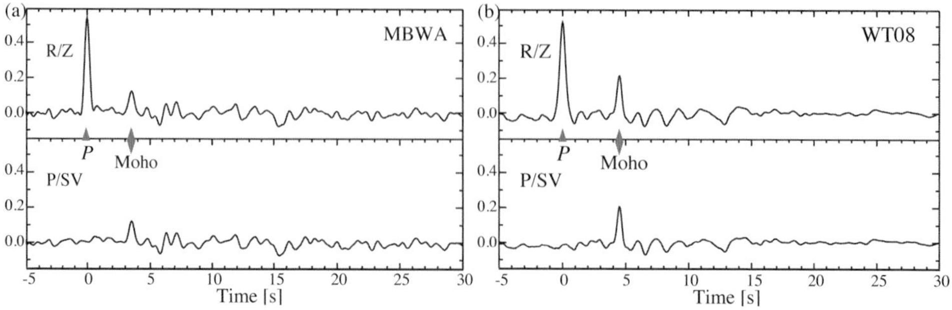

Figure 5.7 Receiver functions constructed from teleseismic observations at stations in Western Australia using the conventional approach and the wavevector transformation. The upper trace for each station is the radial RF (R/Z) and the lower trace is that using the wavevectors (P/SV). The results for MBWA come from a six-event stack, and WT08 from a nine-event stack. [Courtesy of A.M. Reading.]

effort to match the details of the *P* pulse. The peak corresponding to conversion at the Moho stands out very clearly in the wavevector result.

Using the wavevector transformation (5.2.11) the full effect of the free surface is removed, and so receiver functions for *SV* waves using S and those for *SH* waves using H are directly comparable in amplitude. This means that the size of the *SH* contribution can be used as a measure of the departure from locally horizontal stratification due, e.g., to dipping interfaces or the general influence of 3-D structure. The transformation requires a reasonable estimate for the surface wavespeeds α_0, β_0 but the elements are only a slowly varying function of the physical properties. Indeed because we are dealing with incident waves with wavelengths of several kilometres even at 1 Hz, the values to be used for α_0, β_0 should be a little higher than those immediately at the surface. The LQT rotation has similar properties to the transformation for suppressing the initial *P* pulse, but does not achieve the same balance of *SV* and *SH* contributions. Kind et al. (1995) advocate the use of an empirical estimate of the rotation angle based on minimising the Q component at the onset time of *P*.

When looking at deeper structure using *P*-wave (*Ps*) receiver functions, converted phases from depth can be masked by crustal multiples that arrive in the same time interval. As a result many workers have made use of *S*-wave (*Sp*) receiver functions to look at structures in the lithospheric mantle and below (e.g., Yuan et al., 2006). One of the merits of using incident *S* is the separation of converted *P* from crustal multiples, since these *P* conversions arrive as precursors to the *S* phase and the multiples follow *S*. However, the attenuation of *S* in propagation to teleseismic distances is much greater than for *P* ($t_P^* \sim 1, t_S^* \sim 4$), so that structural resolution is markedly reduced because of the relatively low-frequency band that can be used with teleseismic *S*. In consequence, interpretation of *Sp* receiver functions has tended to concentrate on rather simple structures with a few discrete discontinuities.

Correlation of receiver elements

An alternative approach to the use of transfer functions between components as in receiver functions is to exploit the correlation properties between the seismogram components. For correlations between dissimilar components, the phase of the cross-correlation is the same as for a transfer function. However, rather than cancelling the effect of the source and the path, their influence is squared by correlation. This means that somewhat careful filtering is needed to get good results. Yet, the correlation approach provides additional information from the auto-correlation of the segments of the seismogram around the *P* and *S* arrivals that allow direct examination of the *P* and *S* reflectivity. This information is complementary to the cross-component results that emphasise conversions.

We can demonstrate the relation of the correlation properties to reflectivity by exploiting the representation of the teleseismic response (5.2.1) for a single slowness component. We construct the tensor auto-correlation of the surface motion in the frequency domain as

$$[\mathbf{w}_0^{\text{tel}}(\mathrm{p},\omega)]^*[\mathbf{w}_0^{\text{tel}}(\mathrm{p},\omega)]^{\mathrm{T}} = \\ \mathrm{S}_{\mathrm{I}}^*(\mathrm{p},\omega)\mathbf{Z}^*(\mathrm{p},\omega)\mathbf{C}_{\mathrm{R}}^*(\mathrm{p},\omega)\Phi^*(\mathrm{p},\omega)\Phi^{\mathrm{T}}(\mathrm{p},\omega)\mathbf{C}_{\mathrm{R}}^{\mathrm{T}}(\mathrm{p},\omega)\mathbf{Z}^{\mathrm{T}}(\mathrm{p},\omega)\mathrm{S}_{\mathrm{I}}(\mathrm{p},\omega). \tag{5.2.13}$$

In the tensor auto-correlation (5.2.13) we can recognise the spectrum of the combined excitation and instrument response $\mathrm{S}_{\mathrm{I}}^*(\omega)\mathrm{S}_{\mathrm{I}}(\omega)$. The propagation phase term $\Phi^*(\mathrm{p},\omega)\Phi^{\mathrm{T}}(\mathrm{p},\omega)$ will reduce to the identity in the far-field from any source, since the components are orthogonal and the phase cancels out between the two complex conjugate terms. We are then left with the far-field contribution,

$$[\mathbf{w}_0^{\text{tel}}(\mathrm{p},\omega)]^{\mathrm{T}}[\mathbf{w}_0^{\text{tel}}(\mathrm{p},\omega)]^* = \\ \mathbf{Z}^*(\mathrm{p},\omega)\mathbf{C}_{\mathrm{R}}^*(\mathrm{p},\omega)\mathbf{C}_{\mathrm{R}}^{\mathrm{T}}(\mathrm{p},\omega)\mathbf{Z}^{\mathrm{T}}(\mathrm{p},\omega)|\mathrm{S}_{\mathrm{I}}(\mathrm{p},\omega)|^2. \tag{5.2.14}$$

We now can recognise the spectrum of the transmission term $\mathbf{C}_{\mathrm{R}}^*(\mathrm{p},\omega)\mathbf{C}_{\mathrm{R}}^{\mathrm{T}}(\mathrm{p},\omega)$ modulated by external amplitude effects. From the relation (2.4.9) for a perfectly elastic medium we can express this transmission spectrum as

$$\mathbf{C}_{\mathrm{R}}^*(\mathrm{p},\omega)\mathbf{C}_{\mathrm{R}}^{\mathrm{T}}(\mathrm{p},\omega) = \\ \mathbf{W}_{\mathrm{F}}^*(\mathrm{p})\mathbf{W}_{\mathrm{F}}^{\mathrm{T}}(\mathrm{p}) \\ + \mathbf{W}_{\mathrm{F}}^*(\mathrm{p})\left\{\mathbf{R}_{\mathrm{F}}^*(\mathrm{p})[\mathbf{R}_{\mathrm{D}}^{\mathrm{FJ}}(\mathrm{p},\omega)]^* + [\mathbf{R}_{\mathrm{D}}^{\mathrm{FJ}}(\mathrm{p},\omega)]^{\mathrm{T}}\mathbf{R}_{\mathrm{F}}^{\mathrm{T}}(\mathrm{p})\right\}\mathbf{W}_{\mathrm{F}}^{\mathrm{T}}(\mathrm{p}), \tag{5.2.15}$$

in terms of the reflection response $\mathbf{R}_{\mathrm{D}}^{\mathrm{FJ}}(\mathrm{p},\omega)$ including the effect of the free surface. For a weakly attenuative medium we can envisage that (5.2.15) will remain a reasonable approximation, but cannot be exact.

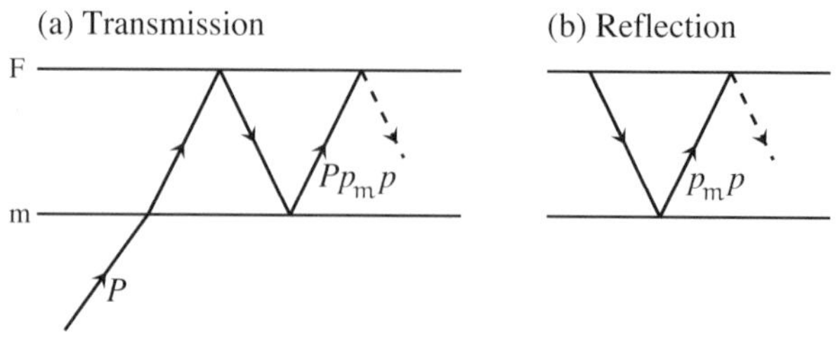

Figure 5.8 Equivalence of transmission and reflection results in the presence of a free surface (F) and reflector (m), the dashed path indicates further surface reflections.

Recognising that the Fourier transform of a product produces a convolution in the time domain, we obtain

$$\mathcal{A}[\mathbf{w}_0(p,t)] = \mathcal{A}[S_I(p,t)] * \mathcal{A}[\mathbf{C}_R(p,t)] * \hat{\mathbf{Z}}(p,t), \tag{5.2.16}$$

where we have written $\mathcal{A}[\ ]$ for the auto-correlogram. The last term $\hat{\mathbf{Z}}(p,t)$ in (5.2.16) summarises the amplitude effects on the wave field before it arrives at the base of the local structure. Whatever the nature of the propagation before energy arrives at the level z_J on its way to the surface, the first two terms on the right-hand side of (5.2.16) will be present, whereas $\hat{\mathbf{Z}}(p,t)$ will be variable. We see that the auto-correlation of the surface displacement at a particular station contains scaled information on the reflection response of the structure beneath the station including free-surface reverberations, but this is convolved with the combined effects of excitation by distant sources and the instrumental response.

When we concentrate attention on just the vertical component of ground motion, the dominant terms in $\mathcal{A}[\mathbf{C}_R(p,t)]$ will arise from *P* waves with some conversions for arrivals further away from the vertical. But, the full tensor auto-correlation includes both cross-terms between components with conversion between *P* and *S* and auto-correlations for the vertical, radial, and transverse components that concentrate on a single wavetype. In the spectral domain,

$$[\mathbf{w}_0(p,\omega)]^*[\mathbf{w}_0(p,\omega)]^T = \begin{pmatrix} Z^*Z(p,\omega) & Z^*R(p,\omega) & Z^*T(p,\omega) \\ R^*Z(p,\omega) & R^*R(p,\omega) & R^*T(p,\omega) \\ T^*Z(p,\omega) & T^*R(p,\omega) & T^*T(p,\omega) \end{pmatrix}. \tag{5.2.17}$$

When we construct the response at an individual station, we combine the estimates of $\mathcal{A}[\mathbf{w}_0]$ from many teleseismic events, with compensation for differential moveout between the different slowness arrivals. The stacking will emphasise the coherent contribution from $\mathcal{A}[S_I] * \mathcal{A}[\mathbf{C}_R]$.

Thus when we examine the auto-correlation of a single component we get a modulated version of the reflection response beneath the station, including the effects of the free surface for the appropriate wavetype. The cross-component terms represent a scaled version of the corresponding receiver function.

The process of auto-correlation emphasises the arrivals that have a systematic

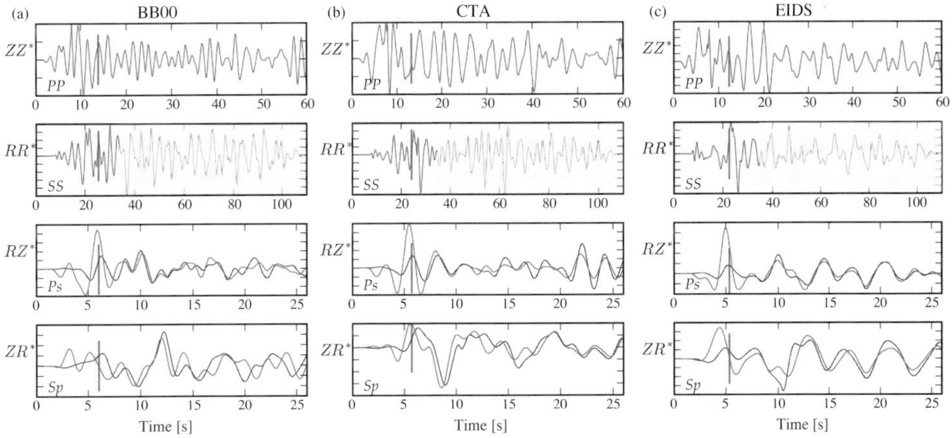

Figure 5.9 Tensor correlation results for a set of stations in northern Australia: (a) BBOO, (b) CTA, (c) EIDS. For each station four sub-panels display *P* and *SV* auto-correlograms as well as the *Ps* and *Sp* cross-correlograms. The vertical lines denote the time of the Moho reflection or conversion, converted from depth using an approximate *P* wavespeed of 6.0 km/s and *S* wavespeed of 3.3 km/s. The heavier lines in the lower two panels for each station show the *Ps* and *Sp* receiver functions for comparison with correlation results.

pattern of delays. In the case of upward transmission through the crust, waves reflected back from the free surface such as $Pp_{\mathrm{m}}p$ will have the equivalent delay pattern to a simple reflection from the crust–mantle interface $p_{\mathrm{m}}p$ (Figure 5.8). We can then think of the stacking process as enhancing such internal crustal reflections, which are common to a wide class of incident waves. The equivalence of the auto-correlation of transmission with reflection, was recognised for purely vertically travelling waves by Kunetz & d'Erceville (1962) and Claerbout (1968). The results were extended to obliquely travelling *P–SV* waves by Frasier (1970).

Figure 5.8 shows schematically how reflection results emerge from the auto-correlation of the surface response to transmitted waves. Each wave impinging from below on the free surface will produce a train of reflections, with the same pattern of time behaviour. The auto-correlation process reinforces this set of common reflection elements.

Sun & Kennett (2016) have made explicit use of the portions of the seismograms normally used for receiver function analysis to construct tensor correlation time series for both incident *P* and *S* waves. The surface-generated multiples (as in Figure 5.8a) are generally hidden in the coda of the *P* or *S* arrivals and are difficult to identify directly, but can be extracted by their consistent auto-correlation properties.

An example of this approach is illustrated in Figure 5.9 from teleseisms recorded at stations in northern Australia. The horizontal components of the seismogram have been reoriented, to work directly with the *SV* and *SH* arrivals on the radial and transverse components to the great-circle path.

The various events have differing slowness and so moveout corrections are applied to bring all the different styles of response to the equivalent of vertical incidence. The standard approach for converted waves is used for the *Ps* and *Sp* contribution. The trajectories of reflected arrivals in the time–slowness domain are approximately elliptical and when we stack the contributions from events at different epicentral distances we need to correct to a common time base, typically that for vertical incidence (zero slowness). Thus for the auto-correlations the moveout correction with slowness is

$$\tau_0 = \tau(p)/(1 - v_0^2 p^2)^{1/2} \approx \tau(p)/(1 - \tfrac{1}{2} v_0^2 p^2), \qquad (5.2.18)$$

where τ_0 is the vertical reflection time, $\tau(p)$ is the reflection time for slowness p, and v_0 is the average velocity above an interface. In this application the AK135 model is used for the moveout correction that requires both a time shift and a slight waveform stretch.

Prior to auto- and cross-correlation, all waveforms on both the vertical (Z) and radial (R) components are bandpass filtered with corners at 0.1–5 Hz. After auto-correlation and cross-correlation, an extra bandpass filter of 0.5–4 Hz for *P* and 0.5–3 Hz for *S* is employed. To suppress the large-amplitude peak at zero time in the auto-correlations, a one-sided Hanning window function is used over the first 5 s of the traces.

The upper two panels for each station in Figure 5.9 show the *PP* and *SS* reflectivity estimated from the stacked auto-correlograms for the vertical and radial components. Only the shallow part of the *SS* reflectivity can be expected to stack effectively and the *SS* response is masked after 35 s two-way time.

In the lower two panels of Figure 5.9, we compare the converted responses obtained by cross-correlation with the *Ps* and *Sp* receiver functions for the same time windows constructed by deconvolution in the frequency domain. For all stations, the *Ps* results agree well with the receiver function for incident *P* waves. However, there are differences in the details of the amplitude behaviour, which are likely to be associated with the differing treatment of source and station terms. We note that the amplitudes of the receiver function around 5 s are much weaker than the corresponding correlation responses. Similar variations are found between the *Sp* correlation response and the receiver function constructed from incident *S* waves.

5.2.2 Seismogram Transfer Functions

A further application of transfer functions is to the evolution of the seismic wavefield from a common source. We consider the records from a number of stations, using one station as a reference. The resulting suite of transfer functions can aid in the identification of different seismic phases.

In the case of two identical seismograms, the time-domain transfer function $T_{cr}(t)$ introduced in (5.1.11) takes the form of a delta function modulated by the available bandwidth $\delta_B(t)$. As differences between the comparison seismogram

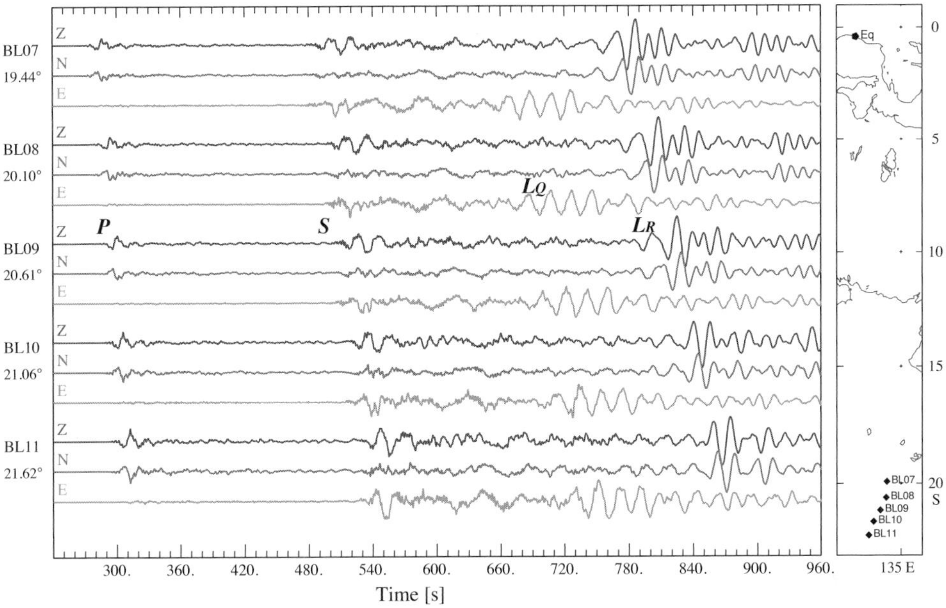

Figure 5.10 Three-component seismograms for an event on 2009 January 3 (Mw 7.7, 17 km deep) at the Birdshead of New Guinea almost to the north of the line of stations in the BILBY experiment in central Australia. The components are not rotated but are nearly naturally polarised. The configuration of the source and the portable broadband stations are shown in the right-hand panel.

$u_c(t)$ and the reference seismogram $u_r(t)$ increase, the transfer function $T_{cr}(t)$ acquires a distinctive character depending on the nature of the discrepancies between the traces. Where, for instance, there is a notable difference for a narrow frequency band, the transfer function will show a strong oscillation at those frequencies. In a similar way, if there is significant misfit in the long-period components, the transfer function will be elongated in time with long-period oscillations.

We illustrate the nature of the transfer function between seismograms by using part of a nearly linear array of portable broadband instruments in central Australia that covers a modest distance range and a narrow range of azimuths from a source to the north in the Birdshead of New Guinea. The influence of the source will be very similar at all stations, and so the differences in the seismograms come from path and near-receiver effects.

The source lies nearly due north of the stations and so the records are nearly naturally polarised, with the EW component approximately transverse to the propagation path. The general characteristic of the seismograms at this group of five stations are similar (Figure 5.10). However, because the stations span the epicentral distance range around 20° there are noticeable changes in the nature of the *P* and *S* phases associated with the influence of triplications in the arrivals

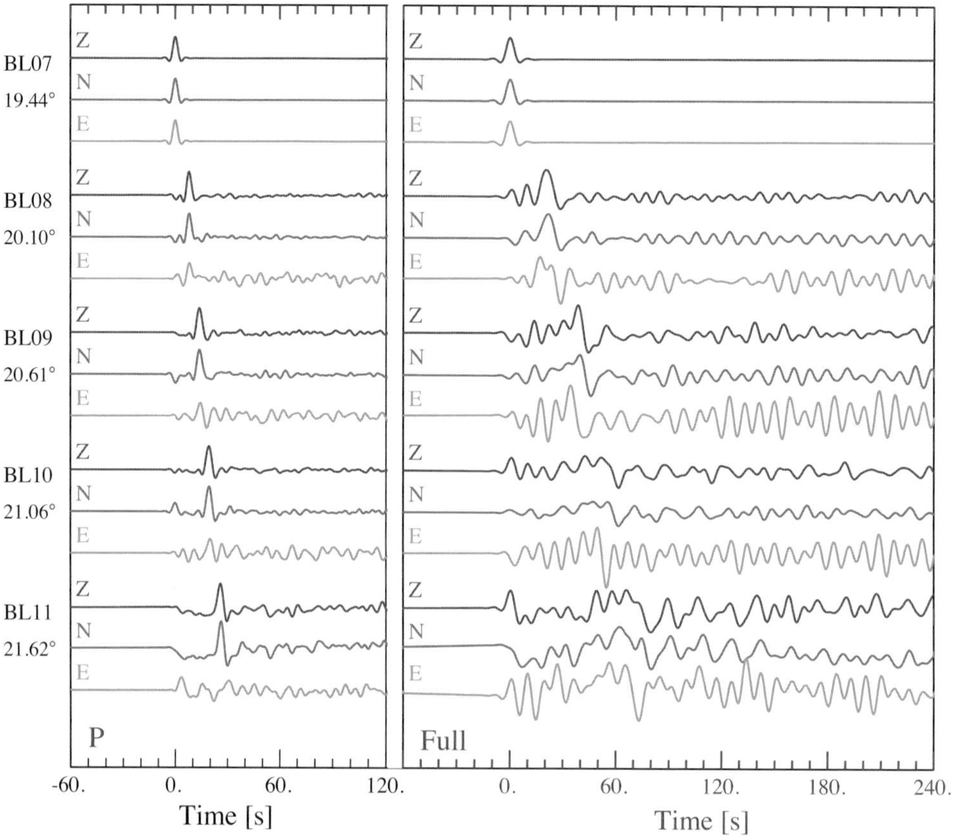

Figure 5.11 Three-component transfer functions for the Mw 7.7 event in New Guinea on 2009 January 3 for a group of stations in the BILBY deployment. The northern station BL07 is used as the reference trace. The left-hand panel shows the transfer function calculated from 240 s of the *P* wavetrain with a low-pass filter below 0.2 Hz, whereas the right-hand panel shows the transfer function calculated for the full seismogram (720 s) with a low-pass filter below 0.125 Hz. In each case the filtering was carried out with a four-pole Butterworth filter.

returned from the upper-mantle transition zone. Distinct moveout is readily apparent in the surface waves for both the Love wave (*LQ*) on the EW components and the Rayleigh waves (*LR*) on the vertical (Z) and NS components. Between *S* and the fundamental mode surface waves there is a complex signal that can be interpreted either in terms of interference of multiple *S* arrivals or higher mode surface waves.

The northern station BL07 is taken as the reference. Transfer functions are then constructed to the other stations. The deconvolution is performed with spectral division in the frequency domain using a small bias in the denominator, followed by back transformation to time. In Figure 5.11, we show the transfer functions for each of the three components at each station as a function of time. The reference trace at BL07 is included, so we can see the nature of the recovered band-passed delta-function response. Two separate constructions of transfer functions are displayed for different parts of the wavefield. The left-hand panel of Figure 5.11 concentrates on the *P*-wave arrivals; a window length of 240 s is used to construct the transfer function followed by low-pass filtering below 0.2 Hz. The right-hand window employs the full 720 s of record illustrated in Figure 5.10 to construct the transfer function, and so includes *P*, *S*, and the surface waves with a low-pass filter below 0.125 Hz.

As the distance from the reference station increases, the difference between the observations and the reference seismograms from BL07 becomes larger, and so the transfer operators are more complex. In the left-hand panel of Figure 5.10 progressive time offsets develop in the transfer function for the *P* waves with a shift in the timing of the main pulse on the vertical and NS components. The more distant stations also have a change in the shape in the *P* pulse associated with varying interactions with the mantle triplications. This change is reflected in the development of a negative lobe following the main peak. The orientation of the EW component, almost perpendicular to the propagation path, is unfavourable for recording *P* waves. There is not a well-defined peak in the EW transfer function, but some slight effect at the expected time offset.

In the right-hand panel of Figure 5.11 we display the transfer functions for the full seismic traces shown in Figure 5.10. Now, except for the closest station BL08, there is not a single well-defined peak in the transfer functions. This arises because of the competing influences of *P* waves, *S* waves, and the surface wave field. Each of these wave types have a different move-out with station separation, so there is not a single shifted broadened delta function but a complex of contributions, with a broad range of frequency components. In particular we see longer periods become stronger in the full transfer functions for the stations to the south, reflecting the more elongated surface-wave signals at larger distances from the source.

With suitable windowing of the seismic trace we can thus use transfer functions to track specific classes of arrival. Interference from waves with a different character will degrade the peak in the transfer function, or if they are strong enough produce a new peak with a different moveout.

5.3 Comparison of Seismograms

It is frequently necessary to assess how closely two seismograms resemble each other. The most familiar example would be the comparison of observed and simulated seismograms in structural or source studies. Other applications include time-lapse monitoring to detect changes in physical properties along a predefined propagation path, and the assessment of subtle changes of structure via the spatial variation of seismograms.

In each case, it is desirable to quantify the level of fit between the different seismograms, and then use this information to monitor change or develop improved models for structure and source character.

5.3.1 General Considerations

A commonly used measure of the misfit between a reference seismogram $\mathbf{u}_r(t)$ and a comparator seismogram $\mathbf{u}_c(t)$ is the L_2 norm

$$\mathcal{P}_{\mathbf{u}} = \int_T dt\, w(t)[\mathbf{u}_r(t) - \mathbf{u}_c(t)]^2, \tag{5.3.1}$$

where $w(t)$ is a weighting function. Such measures have been used by, e.g., Tarantola (1984) and Igel et al. (1996) in the context of full waveform inversion. Although it is simple, the measure $\mathcal{P}_{\mathbf{u}}$ has a number of problematic properties. Firstly, the numerical value of $\mathcal{P}_{\mathbf{u}}$ is dominated by the largest amplitude portions of the seismograms and mismatch in the lower amplitude portions of the traces can be masked. This problem can be countered to some extent by suitable choice of the weighting function $w(t)$. Secondly, even a slight time shift between otherwise identical seismograms leads to a significant misfit estimate.

To provide a measure of the coincidence in time, we can exploit the cross-correlation time shift introduced in (5.1.1). We construct $\mathcal{T}_{cr}$ as the time when the cross-correlation between $u_c(t)$ and $u_r(t)$,

$$C_{c,r}(\tau) = \int_t dt\, u_c(t) u_r(t+\tau), \tag{5.3.2}$$

is maximised. We then use

$$\mathcal{P}_{\mathcal{T}} = \tfrac{1}{2}\mathcal{T}_{cr}^2, \tag{5.3.3}$$

as a measure of the similarity in time of arrival of the main energy between the segments of the seismogram under comparison. $\mathcal{P}_{\mathcal{T}}$ is the misfit measure normally used in travel-time tomography exploiting finite-frequency observations.

In a similar way we can compare amplitudes. For a segment of a seismogram, we introduce the amplitude measure

$$\mathcal{A} = \left(\int_T dt\, [u(t)]^2 \right)^{\frac{1}{2}}. \tag{5.3.4}$$

We can then construct a robust amplitude misfit measure between the reference and comparator traces as

$$\mathcal{P}_{\mathcal{A}} = \tfrac{1}{2}\frac{(\mathcal{A}_c - \mathcal{A}_r)^2}{\mathcal{A}_r^2}. \tag{5.3.5}$$

The misfit measures for travel time and amplitude introduced in (5.3.3) and (5.3.5) can only be applied where single seismic phases can be isolated and the comparator and reference traces have already strong similarity. In the common situation where several phases interfere, e.g., the effect of depth phases for shallow sources, neither $\mathcal{P}_{\mathcal{T}}$ or $\mathcal{P}_{\mathcal{A}}$ give physically meaningful results. Similar issues arise when the comparator trace represents a somewhat distorted version of the reference trace, as can occur in the comparison of observations and synthetic seismograms from dispersion of waveforms due to the presence of 3-D heterogeneity.

5.3.2 Time and Frequency Response

In order to overcome the limitations of the simple time and amplitude misfit measures it is necessary to take into account the full behaviour of the seismic waveforms over the frequency bands of interest. This requires simultaneous consideration of phase and amplitude information. It is advantageous to have separate measures of fit for phase and amplitude because they carry different information about Earth structure. The variations in seismic wavespeeds provide the main controls on phase, whereas amplitudes are sensitive to wavespeed gradients and attenuation. The process of comparison also needs to be applicable to any part of the seismogram rather than dependent on any specific representation. For longer-period waves at smaller epicentral distances seismic phases tend to overlap and have no distinct identity. The arrivals between *S* and the fundamental mode surface waves are commonly complex, and can only be described by interference of multiple *S* waves or higher modes of the surface waves.

Fichtner et al. (2008) introduced misfit measures in time–frequency space so that the evolution of the phase and amplitude behaviour can be effectively tracked. The nature of the frequency content as a function of time can be extracted by sliding a window function $h(t)$ along the trace and examining the resulting Fourier transform. From $u_r(t)$ we construct

$$\tilde{u}_r(t,\omega) = \frac{1}{\sqrt{2\pi}} \int d\tau\, h(t-\tau) u_r(t) e^{-i\omega t}. \tag{5.3.6}$$

With this representation we can extract an envelope and phase representation for both the comparator and reference traces as

$$\tilde{u}_c(t,\omega) = |\tilde{u}_c(t,\omega)| e^{i\varphi_c(t,\omega)}, \qquad \tilde{u}_r(t,\omega) = |\tilde{u}_r(t,\omega)| e^{i\varphi_r(t,\omega)}. \tag{5.3.7}$$

A convenient form for the sliding window $h(t)$ is a Gaussian

$$h(t) = (\pi\sigma)^{-1/4} e^{-t^2/\sigma^2}, \tag{5.3.8}$$

for which the time–frequency resolution is maximised (Fichtner, 2011), and $\tilde{u}_r(t,\omega)$ is a *Gabor transform*. Many other alternatives can be used for $h(t)$. A wavelet transform is also an alternative to the windowed Fourier transform.

From the time–frequency response (5.3.7) we can construct an envelope misfit measure

$$\mathcal{P}_e = \int dt \int d\omega\, W_e(t,\omega) [|\tilde{u}_c(t,\omega)| - |\tilde{u}_r(t,\omega)|]^2, \tag{5.3.9}$$

and a corresponding phase misfit measure

$$\mathcal{P}_p = \int dt \int d\omega\, W_p(t,\omega) [\varphi_c(t,\omega) - \varphi_r(t,\omega)]^2, \tag{5.3.10}$$

where W_e and W_p are positive weighting functions. The difference in envelopes $|u_c| - |u_r|$ measures the amplitude discrepancies as a function of time and

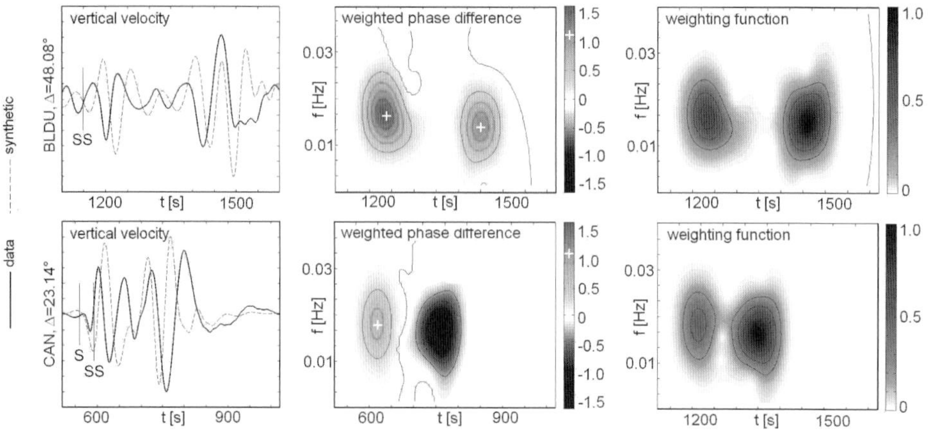

Figure 5.12 *Left*: Vertical-component velocity seismograms from an event in the Loyalty Islands (2007 March 25; 20.60°S, 169.12°E, depth 41 km) region recorded at the stations BLDU (30.61°S, 116.71°E, upper row) and CAN (-35.32°S, 148.99°E, lower row). The data are plotted as solid lines and the synthetics as dotted lines. Both data and synthetics are low-pass filtered with a cutoff frequency $f_c = 0.02$ Hz. *Centre*: Weighted phase differences $W_p(t,\omega)\Delta\varphi(t,\omega)$ corresponding to the seismograms on the left. Contour lines are plotted at multiples of 20% of the maximum value. *Right*: Weighting functions $W_p(t,\omega)$ used to compute the weighted phase differences $\Delta\varphi(t,\omega)$.

frequency. The phase shift $\Delta\varphi = \varphi_c - \varphi_r$ can be interpreted as a time shift $\Delta t = \Delta\varphi/\omega$ at frequency ω.

A suitable choice for the parameter σ that controls the character of the time–frequency transform is the dominant period of the data, after appropriate filtering (Fichtner et al., 2008). This choice means that there is a close correspondence between $\Delta\varphi$ and the concept of a phase difference as a time shift between comparable oscillations in the comparator and reference traces.

The weighting function for the envelope W_e is not critical, but can be used to emphasise certain aspects of seismograms, such as the higher modes, to enhance depth resolution in tomographic studies. However, W_p needs to be chosen to suppress phase difference contributions for those portions of time–frequency space where a physically meaningful measurement cannot be made. Such circumstances include segments where the signal is at or below the noise level $n(t,\omega)$. If a threshold can be found for the time–frequency variation of the noise, a suitable choice of W_p for semi-automatic application is (Fichtner, 2011)

$$\tilde{W}_p(t,\omega) = 1 - e^{-|u_r(t,\omega)|^2/n(t,\omega)^2}. \tag{5.3.11}$$

The time-–frequency domain misfits $\mathcal{P}_e$ and $\mathcal{P}_e$ are only useful when there is some degree of similarity between the seismograms being compared. It may be necessary to undertake preliminary low-pass filtering to improve such similarity.

Figure 5.12 illustrates measurements of the phase misfit for vertical-component waveforms from an event in the Loyalty Islands region (2007 March 25; latitude

20.60°S, longitude 169.12°E, depth 41 km) recorded at the stations CAN in southeastern Australia and BLDU in Western Australia. The phase misfit is a multi-frequency measurement for the frequency band 5–20 mHz. The advance of the observed waveforms with respect to the synthetic waveforms at station BLDU maps to a phase difference that is positive throughout the time–frequency range of interest (Figure 5.12, upper panels). Particular regions in the time–frequency space are emphasised or suppressed through the weighting function plotted in the upper right panel. At station CAN (Figure 5.12, lower panels), both a phase advance (positive phase difference) and a phase delay (negative phase difference) can be observed.

5.3.3 Use of Transfer Functions

The transfer function between seismograms introduced in Section 5.1.3 provides a means of comparison of seismograms that does not depend on close similarity. The transfer function T_{cr} connects the reference trace to the comparator trace via a convolutional equation

$$u_c(t) = \int d\tau\, T_{cr}(t-\tau) u_r(\tau) = T_{cr}(t) \star u_r(t). \tag{5.3.12}$$

The simplest representation of the transfer function comes in the frequency domain where the transform $\bar{T}_{cr}(\omega)$ can be expressed as a spectral division (5.1.12). This representation also indicates that the transfer function T_{cr} can be viewed as the deconvolution of the cross-correlation between the comparator and the reference traces by the auto-correlation of the reference trace.

In the context of waveform inversion we would employ the observed data u_{obs} as the reference trace ($u_r = u_{obs}$) and the corresponding synthetic seismogram u_{synth} constructed for a specific Earth model and source representation as the comparator, $u_c = u_{synth}$ (Kennett & Fichtner, 2012). This sets up a fixed reference trace for a set of iterations, and so it is possible to track the evolution of the transfer operator over the course of an inversion. When the data and the simulation match, the transfer operator $\mathfrak{T}_{cr}$ reduces to the identity operator $\mathfrak{I}$. The goal of waveform inversion can therefore be expressed as minimising the difference between the transfer operator $\mathfrak{T}_{cr}$ and the identity operator $\mathfrak{I}$.

Consider now an envelope and phase representation in the frequency domain for the comparator and reference seismograms ,

$$\bar{u}_c(\omega) = |\bar{u}_c(\omega)| e^{i\varphi_c(\omega)}, \quad \bar{u}_r(\omega) = |\bar{u}_r(\omega)| e^{i\varphi_r(\omega)}; \tag{5.3.13}$$

as well as the transfer function $\bar{T}_{cr}$ between them,

$$\bar{T}_{cr}(\omega) = |\bar{T}_{cr}(\omega)| e^{i\delta\varphi(\omega)}. \tag{5.3.14}$$

We can then express the relationships between the seismograms as

$$|\bar{u}_c(\omega)| e^{i\varphi_c(\omega)} = \bar{T}_{cr}(\omega) |\bar{u}_r(\omega)| e^{i\varphi_r(\omega)}. \tag{5.3.15}$$

Now take the natural logarithm of (5.3.15),

$$\ln|\bar{u}_c(\omega)| + i\varphi_c(\omega) = \ln\bar{T}_{cr}(\omega) + \ln|\bar{u}_r(\omega)| + i\varphi_r(\omega). \tag{5.3.16}$$

On separating the real and imaginary parts of (5.3.16) we obtain

$$\ln|\bar{u}_c(\omega)| - \ln|\bar{u}_r(\omega)| = \ln|\bar{T}_{cr}(\omega)|, \quad \varphi_c(\omega) - \varphi_r(\omega) = \delta\varphi(\omega). \tag{5.3.17}$$

If there is a full correspondence between the comparator and reference traces then both the logarithmic envelope $\ln|\bar{T}_{cr}(\omega)|$ and the transfer-function phase $\delta\varphi(\omega)$ will be zero for all frequencies.

When working in the time domain, we can make a comparable separation of envelope and phase characteristics by working with the analytic signal

$$\hat{T}_{cr}(t) = T_{cr}(t) + i\mathcal{H}T_{cr}(t), \tag{5.3.18}$$

where $\mathcal{H}T_{cr}(t)$ is the Hilbert transform of the transfer function. We can then construct an instantaneous envelope $(T_{cr}(t)^2 + \mathcal{H}T_{cr}(t)^2)^{1/2}$ and instantaneous phase $\tan^{-1}\{\mathcal{H}T_{cr}(t)/T_{cr}(t)\}$, or stabilised quantities derived by averaging over a number of time points.

One criterion for a fit between observed and simulated seismograms in the frequency domain can be the simultaneous minimisation of the absolute values of $\ln|\bar{T}_{cr}(\omega)|$ and $\delta\varphi(\omega)$, since both will be zero if a perfect fit is achieved. This can be viewed as a Pareto set problem, or alternatively one can employ a weighted combination of the two terms. Because of the different influence of noise, we can expect the amplitude and phase terms to behave in different ways.

In general the frequency-dependent phase of the frequency-domain transfer function $\bar{T}_{cr}(\omega)$ can be determined with much higher fidelity than the amplitude. When comparing observations at different stations, issues arise with local site effects, instrumental calibration, and orientation. In inversion there are additional theoretical complications for the calculation of synthetic seismogram amplitudes, such as the influence of focussing and defocussing due to complex structure, attenuation, and anisotropy.

We note that the use of the transfer function naturally employs the logarithm of amplitude as a direct counterpart to phase. The logarithm helps to mute the extraneous influences, but still logarithmic amplitude will be less well controlled than phase.

The adaptive waveform inversion approach of Warner & Guasch (2015), applied to seismic reflection data, represents an application of the transfer-function concept. They develop a set of convolutional filters designed to transform the predicted data into the observed data. Their formulation of the inversion problem requires iterative updates of their Earth model to force these Wiener filters toward zero-lag delta functions. Thus the Weiner filters represent an implementation of the transfer function T_{cr}.

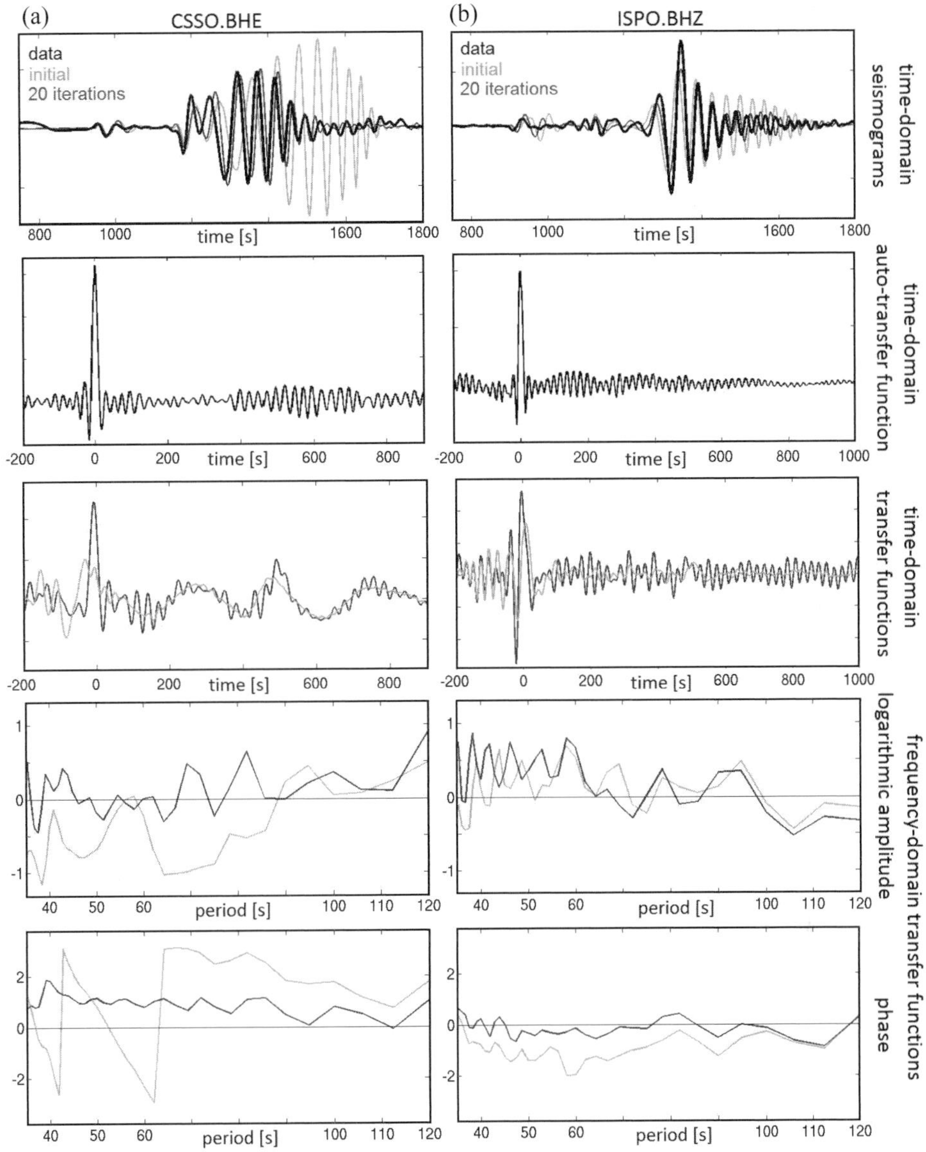

Figure 5.13 Representation of the transfer function between observed and synthetic seismograms in the time and frequency domains. In the upper panel the observed trace is shown in black, the seismogram for the initial model in light grey, and the seismogram for the model obtained after 20 iterations of the waveform inversion scheme in mid grey. In the middle panels the corresponding auto-transfer functions of the data and the actual transfer functions from the observed traces are shown with the same grey tones. In the bottom panels we show the response in the frequency domain with the logarithmic amplitude and phase spectra and the transfer functions. (a) EW component at station CSSO; (b) vertical component at station ISPO.

Illustration of time and frequency domain transfer functions

We demonstrate the nature of transfer functions in the time- and frequency-domains (Figure 5.13) for an event between Svalbard and Greenland (Ms 6.5, depth 13 km) recorded at the seismic stations CSSO and ISPO along a similar azimuth. We show results from a waveform inversion for structure beneath Europe (Fichtner & Trampert 2011a,b), illustrating synthetic seismograms for the initial model, S20RTS by Ritsema et al. (1999), and from the 3-D model achieved after 20 iterations of full waveform inversion. These seismograms are compared with the reference trace of the observations.

We show two examples of the time- and frequency-domain transfer functions. In Figure 5.13 we show the EW component at CSSO and the vertical component at ISPO, to illustrate the situation for both Love and Rayleigh waves accompanied by S-wave multiples. In the upper panel for each station we show the observed seismic traces, together with the simulated seismograms for the initial model and for the final model obtained after 20 iterations of the full waveform inversion. In the panels below we show the auto-transfer functions of the data, as well as the actual transfer functions between the simulations and the observations in the time domain. In the bottom group of two panels we show the logarithmic amplitude and phase characteristics of the transfer functions in the frequency domain.

At CSSO (Figure 5.13a) the initial model produces a rather late high-amplitude surface wave arrival, but by the 20th iteration of the inversion the time and amplitude alignment is much improved. The initial transfer function has distributed amplitude with long-period oscillations and little concentration near time zero. After the waveform inversion, although there are still long-period contributions, there is now a concentrated peak at the time origin, as desired for a successful fit.

For the vertical component at ISPO (Figure 5.13b) we again see a dramatic improvement between the initial model and the 20th iteration of the waveform inversion with a considerable sharpening of the transfer function. For this station the phase correspondence at the end of the inversion is good, but there are problems with the match to the amplitude spectrum. This situation is reflected in the presence of both long-period and short-period components in the transfer functions and a bi-modal character to the main peak in the time-domain transfer function.

Weighted transfer functions

When using the long-period part of the seismic signal, there is a likelihood that there will be overlap and interference between phases. One approach is to extract a set of localised windows and look at the measure of seismogram fit through a sum of segments, with a subsidiary transfer function for each. However, it is more in keeping with the original concept to use a common weighting function for both comparator and reference seismograms. Thus we can construct

$$v_r(t) = w(t)u_r(t), \qquad v_c(t) = w(t)u_c(t), \tag{5.3.19}$$

where the weighting function $w(t)$ is designed to modify the amplitude of the trace to enhance specific phase features. We can then introduce the weighted transfer operator $\mathfrak{T}^w_{cr}$ such that

$$v_c(t) = \mathfrak{T}^w_{cr} v_r(t), \tag{5.3.20}$$

and once again our objective is to seek to minimise the difference between $\mathfrak{T}^w_{cr}$ and the identity operator $\mathfrak{I}$. By choosing a function $w(t)$ that isolates, or enhances, separate phase windows we can work with the transfer operator even where attention is oriented towards specific parts of the seismogram rather than the whole trace. As with the original form we can extract and work with envelope and phase information in the frequency or frequency–time domain.

When working with higher-frequency traces where timing discrepancies may be important in determining the level of seismogram fit, it may be preferable to allow for a time shift in a segmented operator. In this case we introduce a modified transfer function $\mathcal{T}^{w\prime}_{cr}$ so that

$$v_c(t) = \mathfrak{T}^{w\prime}_{cr} v_r(t - \tau_w), \tag{5.3.21}$$

where the time shift τ_w is to be adjusted to bring the transfer operator $\mathfrak{T}^{w\prime}_{cr}$ towards the identity operator $\mathfrak{I}$.

Filter sequences

The transfer operator formalism provides a natural way to incorporate multiple temporal scales of fit via a hierarchy of filters. If a linear filter is applied to (5.1.9),

$$\mathcal{F}u_c = \mathcal{F}\mathfrak{T}_{cr} u_r = \mathfrak{T}_{cr}\left[\mathcal{F}u_r\right], \tag{5.3.22}$$

since simple frequency-domain filters will be commutative when applied to a single component. Consider then a filter set $\{\mathcal{F}_j\}$ progressing from, e.g., a lower-frequency to a higher-frequency perspective with increasing index j. We can, e.g., examine the set of filtered transfer operators $\{\mathcal{F}_j\mathfrak{T}_{cr}\}$ via the level of match between the synthetics and the observations, and refine the model for generating the synthetics by working from lower-frequency to higher-frequency content. This is the strategy adopted by Pratt (1999) in the inversion of reflection seismic data using a set of isolated frequencies.

An alternative approach to filtering is that provided by the isolation filters of Gee & Jordan (1992), designed to extract aspects of the seismogram tuned to specific properties by cross-correlation with both comparator and reference seismograms. The subsequent extraction of *secondary variables* via a sequence of narrow-band Gaussian filters can be used in a similar way to optimise the transfer properties of specific aspects of the seismograms, rather than the whole trace.

Yoshizawa & Kennett (2002a) have used a seismogram misfit criterion related to the analytic transfer function (5.3.18) with a set of n_f frequency filters applied to both observed and synthetic seismograms. Their misfit function is then a sum over the contributions from each of the set of filters in terms of the filtered seismograms $\mathcal{F}_i u(t)$ and their envelopes $E\mathcal{F}_i u(t)$ as follows,

$$\mathcal{P}_f = \sum_{i=1}^{n_f} \int dt \Big\{ |\mathcal{F}_i u_c(t) - \mathcal{F}_i u_r(t)|^p + \gamma_i |E\mathcal{F}_i u_c(t) - E\mathcal{F}_i u_r(t)|^p \Big\}, \tag{5.3.23}$$

where γ_i is the weighting factor for the ith filtered envelope. Yoshizawa & Kennett (2002a) use an L_3 norm, i.e., $p = 3$, that is very sensitive to discrepancies in the waveforms or envelopes. They note that the envelope fit is helpful for stabilizing the waveform fit by avoiding phase-cycle skips that may occur when only the first term is employed. For their inversion for one-dimensional velocity models they use a constant value $\gamma_i = 1.5$. Larger or smaller γ_i tended to diminish the waveform fit for their waveform inversions.

Multi-taper spectral estimates

We have adopted a direct definition of the transfer function so that the frequency domain representation of the complex transfer function $\bar{T}_{cr}(\omega)$ is by spectral division with appropriate regularisation. An alternative approach is employed by Laske & Masters (1996) and Zhou et al. (2004), who adopt a least-squares procedure to define their style of transfer function for a multi-taper spectral estimate.

When windowing is applied with a set of orthogonal tapers $h_j(t)$ the effect in the frequency domain is a set of convolutions; defining

$$\bar{u}^c_j(\omega) = \bar{u}_c(\omega) \star \bar{h}_j(\omega), \quad \bar{u}^r_j(\omega) = \bar{u}_r(\omega) \star \bar{h}_j(\omega), \tag{5.3.24}$$

the least-squares form of the transfer function $\check{T}_{cr}(\omega)$ is to be determined such that $\sum_j [\bar{u}^c_j(\omega) - \check{T}_{cr}(\omega)\bar{u}^c_j(\omega)]^2$ is minimised. Thus,

$$\check{T}_{cr}(\omega) = \frac{\sum_j \bar{u}^c_j(\omega)\bar{u}^{r*}_j(\omega)}{\sum_j \bar{u}^r_j(\omega)\bar{u}^{r*}_j(\omega)}, \tag{5.3.25}$$

and the deviation of this style of transfer function from the desired identity, $\delta\check{T}_{cr}(\omega)$, can be represented in terms of the difference between the projections of the comparator and reference traces on the multi-taper basis,

$$\delta\check{T}_{cr}(\omega) = \frac{\sum_j \left(\bar{u}^c_j(\omega) - \bar{u}^r_j(\omega)\right) \bar{u}^{r*}_j(\omega)}{\sum_j \bar{u}^r_j(\omega)\bar{u}^{r*}_j(\omega)}, \tag{5.3.26}$$

from which specific derivative forms can be calculated when the trace difference is small.

6

Correlations and Interferometry

The exploitation of correlations between seismograms has opened up new ways of studying the Earth. The correlation wavefield emphasises different aspects of seismic wave propagation than are revealed by direct excitation by a source. The process of correlation enhances similarities in the traces that are separated in time, and so enhances features that individually have small amplitude but collectively become significant.

The process of correlation can be interpreted as exploiting constructive interference between portions of seismogram time series. The correlation wavefield associated with a pair of stations resembles aspects of a virtual source–receiver pair, so that the extracted component can be interpreted in terms of subsurface structure. By suitable choice of the segments of seismograms to be correlated such *seismic interferometry* can be used to enhance specific features of the response.

In the context of ambient noise, primary attention is focussed on waves with near-horizontal propagation, such as microseisms. In consequence, surface-wave contributions emerge most readily from stacked cross-correlations between stations. Where a broader range of slowness is present in the ambient field, body waves can be extracted in favourable circumstances.

When correlation is applied to the coda of seismic events, the later portions of the wavetrain are dominated by small slownesses, so that complex reflected phases are emphasised. Correlations of different portions of seismograms at the stations can be exploited to direct attention to particular propagation phenomena.

6.1 Correlation of Seismic Signals

The nature of the correlation wavefield obtained from the correlation of seismograms at different stations can be understood by examining the case of a radial Earth model. We start by developing the theory for the correlation of seismic waveforms at a pair of seismic stations from a single event, and then examine the way in which the correlation wavefield emerges as a multiplicity of sources are considered.

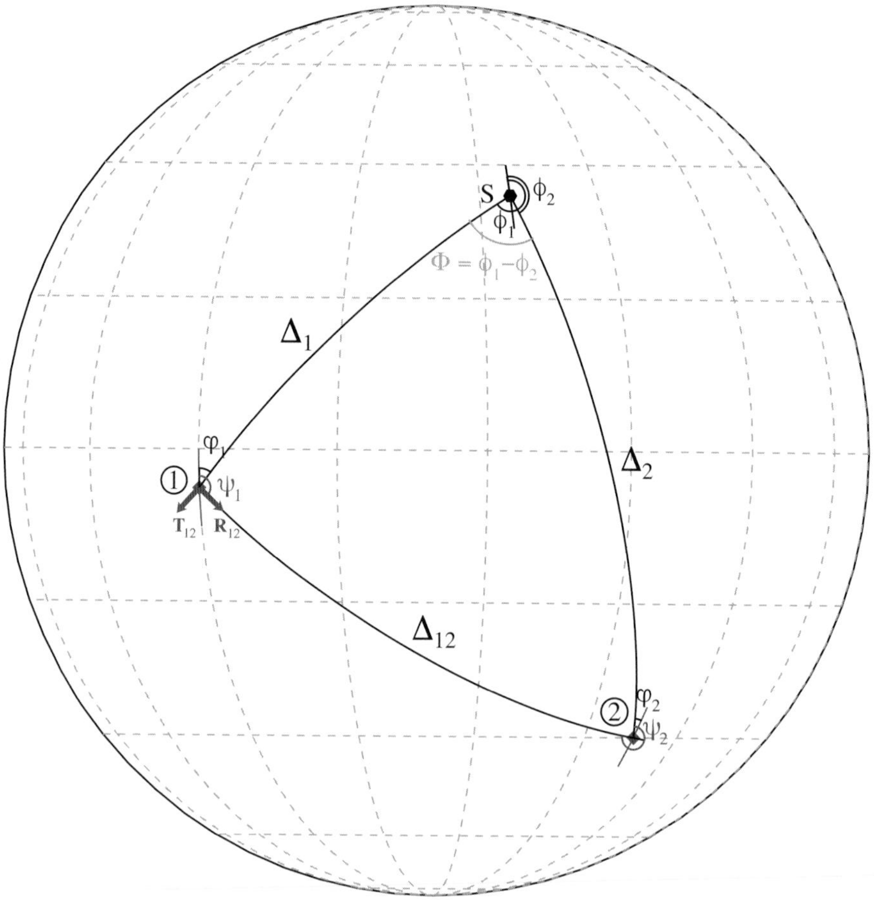

Figure 6.1 Configuration of a seismic source recorded at two stations at different epicentral distances Δ_1, Δ_2 and azimuths ϕ_1, ϕ_2, in relation to the great-circle path between the two stations.

A single source

We start with the situation with a single source and examine seismic-wave propagation on a radially stratified sphere, with stations at locations $\mathbf{x}_1$ and $\mathbf{x}_2$ separated by inter-station distance Δ_{12}. We will construct the cross-correlation of the seismograms produced by the source at the two stations.

We introduce a source S with epicentral distances Δ_1 to station 1, and Δ_2 to station 2 (Figure 6.1). For simplicity we will consider the source to be sufficiently far from the two stations that we can use the asymptotic representation (2.1.27) in terms of travelling wave components in the frequency (ω) – slowness (p) domain. This means that we have a common phase-field contribution (2.1.28) for all components of motion. We introduce a source spectrum $M(\omega)$ and extract an azimuth term $s(\phi)$ that depends on the nature of the source and its size. For station 1 at azimuth ϕ_1 from the source, we therefore have the vector displacement

$$\mathbf{u}_0(\Delta_1, \phi_1, \omega) = \mathbf{E}_1 \omega^{\frac{1}{2}} \mathrm{M}(\omega) \mathrm{f}(\Delta_1) \mathrm{s}(\phi_1) \int_0^{p_0} \mathrm{d}p\, p^{\frac{1}{2}} \mathbf{G}(p, \omega) e^{i\omega p \Delta_1},$$
$$\mathbf{E}_1 = [\hat{\mathbf{e}}_z, i\hat{\mathbf{e}}_{r1}, -i\hat{\mathbf{e}}_{\phi 1}]^{\mathrm{T}}. \tag{6.1.1}$$

In (6.1.1) $\mathbf{G}(p, \omega)$ represents the response of the radially stratified Earth to excitation by the source. The geometrical spreading factor $\mathrm{f}(\Delta) = (1/\sin\Delta)^{\frac{1}{2}}$. The vector $\mathbf{E}_1$ expresses the orientation of the components in the vertical direction $\hat{\mathbf{e}}_z$, along the radial direction from the source $\hat{\mathbf{e}}_{r1}$ and tangential to the great-circle path $\hat{\mathbf{e}}_{\phi 1}$. The limit of the slowness integral in (6.1.1) needs to include the full response, so as in (2.2.1) we take $p_0 > 1.25\beta_0^{-1}$, where β_0 is the *S* wavespeed at the surface. A comparable development to (6.1.1) can be made for the displacement $\mathbf{u}_0(\Delta_2, \phi_2, \omega)$ at station 2.

As noted in Chapter 2, the representation (6.1.1) is based on the travelling-wave forms for the Legendre functions, and is suitable for use away from the poles of the sphere ($\Delta = 0°$ and $180°$), but requires $\omega p \Delta_1 \gg 1$. In any passage through the poles, an extra phase increment of $\pi/2$ is acquired. This phase contribution to the seismograms can be included through the use of a KMAH index σ for the path to the station (Chapman, 2004), so that the additional phase is $\sigma\pi/2$.

Since we wish to consider seismogram components relative to the inter-station path between $\mathbf{x}_1$ and $\mathbf{x}_2$, we need to rotate the horizontal components at the two stations. For station 1 the rotation angle from the great circle from the source to the inter-station path is $\vartheta_1 = \psi_{12} - \varphi_1$, where φ_1 is the back azimuth to the source and ψ_{12} the azimuth to the great-circle path to station 2. Thus the components relative to the inter-station great circle at stations 1 are

$$R_{12} = R_1 \cos\vartheta_1 + T_1 \sin\vartheta_1, \quad T_{12} = -R_1 \sin\vartheta_1 + T_1 \cos\vartheta_1, \tag{6.1.2}$$

where R_1, T_1 are the components relative to the source path. Only if ϑ_1 is small, i.e., the source path and the inter-station path are approximately aligned, will the original *S*-wave polarisation be transferred under rotation.

The inter-station distance Δ_{12} on the sphere has a somewhat complex dependence on Δ_1, Δ_2 and the difference between the azimuths to the two stations from the source $\Phi = \phi_1 - \phi_2$,

$$\cos\Delta_{12} = \cos\Delta_1 \cos\Delta_2 + \sin\Delta_1 \sin\Delta_2 \cos\Phi,$$
$$= \cos(\Delta_1 - \Delta_2) + \sin\Delta_1 \sin\Delta_2(\cos\Phi - 1),$$
$$\Delta_{12} = |\Delta_1 - \Delta_2| - \delta_{12}. \tag{6.1.3}$$

Once again, only if the paths from the source to the stations and the inter-station path are approximately aligned will δ_{12} be small, and Δ_{12} be close to the difference in epicentral distances $|\Delta_1 - \Delta_2|$.

We now construct the frequency-domain representation of the cross-correlation of the two seismograms generated by the same source as

$$\mathcal{U}^{(12)}(\omega) = \mathcal{U}(\Delta_1, \phi_1; \Delta_2, \phi_2; \omega) = \mathbf{u}_0(\Delta_1, \phi_1, \omega)\mathbf{u}_0^{T*}(\Delta_2, \phi_2, \omega), \tag{6.1.4}$$

where * denotes the complex conjugate.

Using the representations of the form (6.1.1) for the displacement at each station we can construct the tensor cross-correlation $\mathcal{U}^{(12)}$ in the form of a product of slowness integrals,

$$\mathcal{U}^{(12)}(\omega) = \mathbf{E}_1\mathbf{E}_2^T\, \omega|M(\omega)|^2\, f(\Delta_1)f(\Delta_2)\, s(\phi_1)s^*(\phi_2) \cdot \int_0^{p_0} dp\, p^{\frac{1}{2}}\mathbf{G}(p,\omega)e^{i\omega p\Delta_1} \int_0^{p_0} dq\, q^{\frac{1}{2}}\mathbf{G}^{T*}(q,\omega)e^{-i\omega q\Delta_2}, \tag{6.1.5}$$

$$= \mathbf{E}_1\mathbf{E}_2^T\, \omega|M(\omega)|^2\, f(\Delta_1)f(\Delta_2)\, s(\phi_1)s^*(\phi_2) \int_0^{p_0} dp\, p^{\frac{1}{2}} \int_0^{p_0} dq\, q^{\frac{1}{2}} \cdot \mathbf{G}(p,\omega)\mathbf{G}^{T*}(q,\omega)e^{i\omega(p\Delta_1 - q\Delta_2)}. \tag{6.1.6}$$

We would now like to express the results in terms of the inter-station configuration using components oriented along the great circle between the stations as in (6.1.2), and the inter-station separation Δ_{12}. To reduce the complexity we focus on the contribution from the cross-correlation of the vertical components $\mathcal{U}_{ZZ}^{(12)}(\omega)$. We introduce the slowness difference $\zeta = p - q$, so that $q = p - \zeta$. We can then represent the correlation field $\mathcal{U}_{ZZ}^{(12)}(\omega)$ between the two stations from the same source as

$$\mathcal{U}_{ZZ}^{(12)}(\omega) = \omega|M(\omega)|^2\, f(\Delta_1)f(\Delta_2)\, s(\phi_1)s^*(\phi_2) \int_0^{p_0} dp\, p^{\frac{1}{2}} e^{i\omega p\Delta_{12}} \cdot \int_{-p_0}^{p_0} d\zeta\, (p-\zeta)^{\frac{1}{2}} G_Z(p,\omega)G_Z^*(p-\zeta,\omega)e^{i\omega(p\delta_{12}+\zeta\Delta_2)}, \tag{6.1.7}$$

where δ_{12} was introduced in (6.1.3). We are thus able to extract a phase component relating to propagation between the stations, modulated by a second slowness integral.

Multiple sources

Now consider summing the correlation result (6.1.7) for seismograms at stations 1 and 2 over many sources, with a broad span of azimuths ϕ_1, ϕ_2 and epicentral distances Δ_1 and Δ_2. The inter-station distance Δ_{12} will remain fixed.

For simplicity we assume a common source spectrum $M(\omega)$ for all sources. The summed correlation field, in the frequency domain, for station separation Δ_{12} is then

$$\left\langle \mathcal{U}_{ZZ}^{(12)}(\Delta_{12},\omega)\right\rangle = \sum_{\{\Delta_1,\phi_1\}}\sum_{\{\Delta_2,\phi_2\}} \mathcal{U}_{ZZ}^{(12)}(\omega), \text{ subject to } \Delta_{12}, \psi_{12}\ \textit{constant}. \tag{6.1.8}$$

From (6.1.7) the dependence of the summed correlation field $\langle \mathcal{U}_{ZZ}^{(12)}(\omega)\rangle$ on the properties of the paths from a source to the two stations enters through:

(1) source orientation effects through $s(\phi_1)s^*(\phi_2)$;
(2) the product of the distance dependencies $f(\Delta_1)f(\Delta_2)$;
(3) a phase contribution in the integral over ζ.

We now conduct the summation over all the sources to the two stations, so that the distance between the stations and the orientation of the path are maintained constant. We first consider the summation over the trajectories to the second station $\{\Delta_2, \phi_2\}$ applied to the differential slowness integral:

$$I_2 = \sum_{\{\Delta_2,\phi_2\}} F(\Delta, \phi_1; \Delta_2, \phi_2) \int_{-p_0}^{p_0} d\zeta\, (p-\zeta)^{\frac{1}{2}} G_Z(p,\omega) G_Z^*(p-\zeta,\omega) e^{i\omega\zeta\Delta_2}, \tag{6.1.9}$$

where $F(\Delta, \phi_1; \Delta_2, \phi_2) = f(\Delta_1)f(\Delta_2)s(\phi_1)s^*(\phi_2)$ allows for the geometric spreading and source radiation terms. In the integral over differential slowness, *constructive interference* will occur in the neighbourhood of $\zeta = 0$, since then the phase term in (6.1.9) is nearly constant and independent of the specific value of Δ_2. Thus

$$I_2 \sim \left[\sum_{\{\Delta_2,\phi_2\}} F(\Delta, \phi_1; \Delta_2, \phi_2) \right] \omega^{-\frac{1}{2}} p^{\frac{1}{2}} G_Z(p,\omega) G_Z^*(p,\omega), \tag{6.1.10}$$

The constructive interference condition is equivalent to the stationary phase treatment employed by Snieder (2004) in discussing surface waves.

With the second, constrained, summation over the trajectories to the first station $\{\Delta_1, \phi_1\}$ we also encounter a phase term depending on the extent that the difference in epicentral distances to the two stations deviates from the inter-station distance,

$$\Psi(\omega) = \sum_{\{\Delta_1,\phi_1\}} e^{i\omega p \delta_{12}}. \tag{6.1.11}$$

The contribution $\Psi(\omega)$ in (6.1.11) will have significant variation with source position unless $\delta_{12} \sim 0$. Thus we have a second *constructive interference* condition that emphasises contributions from sources such that the paths from the sources to the two stations are approximately aligned with the inter-station path. This isolates two regions of significant sources for the correlation process: the first aligns with the trajectory from station 1 to station 2; the second with the path traversed in the opposite direction (Figure 6.2). These source zones extend to depth, and so can be effective for both body waves and surface waves. Each constructive interference contributes an additional phase increment of $\pi/4$.

The two constructive interference conditions require that the slowness contributions to the summed correlation field $\langle \mathcal{U}_{ZZ}^{(12)}(\omega) \rangle$ are such that the two stations share the same surface slowness p, i.e., an extended version of Snell's law, and the dominant contributions come from nearly hyperbolic areas with foci at the two stations (Figure 6.2).

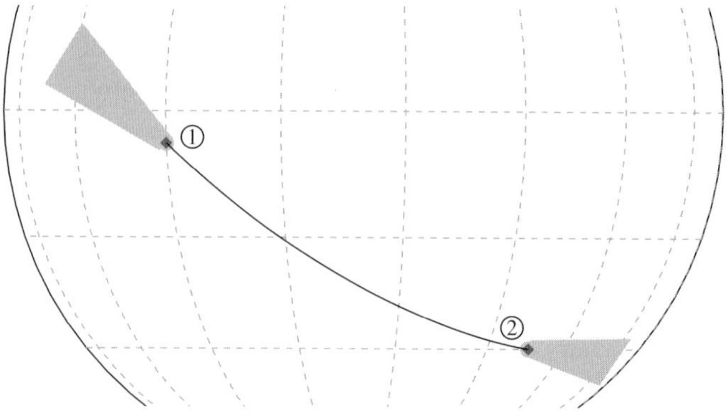

Figure 6.2 Schematic representation of the two source zones where constructive interference will occur in the construction of the correlation wavefield between two stations

Hence we have a summed correlation field for the two specified stations

$$\left\langle \mathcal{U}_{ZZ}^{(12)}(\Delta_{12},\omega)\right\rangle \sim i|M(\omega)|^2\mathcal{F}_{12} \int_0^{p_0} dp\, p^{\frac{1}{2}}\, G_Z(p,\omega)G_Z^*(p,\omega)[e^{i\omega p\Delta_{12}}+e^{-i\omega p\Delta_{12}}], \tag{6.1.12}$$

which is equivalent to that for a virtual source–receiver pair at the two station locations. The correlation wavefield thus directly carries information on the sub-surface structure along the inter-station path.

The integral representation (6.1.12) of the summed correlation field $\langle \mathcal{U}_{ZZ}^{(12)}(\omega)\rangle$ is in a similar form to that for the original seismogram (6.1.1). Now we have a source term $|M(\omega)|^2$ so that there is a stronger dependence on the source spectrum $M(\omega)$, and also the spectrum of the medium response $G_Z(p,\omega)G_Z^*(p,\omega)$ replaces $G_Z(p,\omega)$. This means that we can readily use specific representations for the response of the radially stratified model to look at the behaviour of the correlation wavefield.

The amplitude contribution from the various sources,

$$\mathcal{F}_{12} = \sum_{\{\Delta_1,\phi_1\}}\sum_{\{\Delta_2,\phi_2\}} F(\Delta,\phi_1;\Delta_2,\phi_2), \tag{6.1.13}$$

will not have a simple relation to the path between stations 1 and 2, but will accumulate as more sources are added.

The representation of the summed correlation field for stations 1 and 2 $\langle \mathcal{U}^{(12)}(\omega)\rangle$ for the remaining components of the tensor field takes a similar form to (6.1.12). We need to consider contributions with the same slowness p travelling closely aligned with the inter-station path. This means that the radial (RR) and transverse (TT) components can be directly interpreted in terms of the expected *S*-wave polarisations – *SV* waves on the RR, RZ, ZR, and ZZ components and

SH waves on TT. In a purely radial Earth model there would be no contribution to TZ, ZT, but heterogeneity effects will infill.

Although the principal contributions to the correlation wavefield come from restricted source areas, as in Figure 6.2, we need a wide range of source azimuths arriving at each station to ensure that *destructive interference* removes the contributions that do not have an association with the inter-station path.

We have invoked the presence of multiple sources, so where are these to be found? Direct sources of seismic energy come from earthquakes, with a reasonable distribution across the globe relative to the increasing numbers of high-quality seismic stations. As discussed in Section 3.2.2, deviations from assumed structure lead to the presence of equivalent scattering sources that will be distributed zones of heterogeneity. Such scattering sources will be present at depth as well as at the surface, and, although individually small, when taken in total can be significant. The other major class of sources is associated with the ambient noise field; this includes microseisms from storms at sea or coastal interactions. Noise fields also interact with scatterers to add complexity to the seismic environment. The nature of the correlation wavefield differs with the segments of data being correlated, and their slowness content.

Auto-correlation

A special case of the stacked correlation field is when we consider just a single station, so that $\Delta_1 = \Delta_2$ and $\Delta_{12} = 0$. In this case we recover a summed auto-correlation, e.g., for the vertical ZZ component:

$$\left\langle \mathcal{U}^{0}_{ZZ}(0,\omega) \right\rangle \sim i|M(\omega)|^2 \mathcal{F}_0 e^{-i\omega t} \int_{-p_0}^{p_0} dp\, p^{\frac{1}{2}} G_Z(p,\omega) G^*_Z(p,\omega). \qquad (6.1.14)$$

Contributions come from a full range of slowness, not just from $p = 0$, but reinforcement by the stacking process is most likely for steeply travelling phases. Such behaviour with a range of incident slownesses has been observed by Poli et al. (2017) in auto-correlations of late coda from the very deep Mw 8.3 Sea of Okhotsk earthquake in 2013 across an array of stations in the central United States.

6.1.1 Modal Contributions

As in Section 2.2 we can extract both body-wave and surface-wave contributions from the correlated wavefield (6.1.12), using a suitable contour path in slowness space (as in Figure 2.2).

In (6.1.12) we have contributions from waves travelling the distance interval from point 1 to point 2, and in the opposite direction, as both stations act as apparent sources. We will concentrate attention on the outgoing waves propagating away from $\mathbf{x}_1$, and look at the scenario where we have extracted shallow propagation in a modest suite of modal contributions, including the fundamental mode and a few higher modes.

If we take the vertical component of the inter-station correlation we extract N Rayleigh wave modes, and then

$$\left\langle \mathcal{U}^{(12)}_{ZZ}(\Delta_{12},\omega)\right\rangle_{\text{modes}} \sim \mathcal{F}_{12}\, i|M(\omega)|^2 \cdot \sum_{j=0}^{N} p_j(\omega)\left[\operatorname*{Res}_{\omega,p=p_j(\omega)} G_Z(p,\omega)G_Z^*(p,\omega)\right] e^{i\omega p_j(\omega)\Delta_{12}}. \tag{6.1.15}$$

The trajectories of the pole locations for the various modes in slowness $p_m(\omega)$ as a function of frequency will be the same as in the regular wavefield from a source (2.5.2). Thus the group and phase slowness will match those of regular modes, and the propagation of energy will proceed with the group slowness, to give the same position of peak energy in time.

A major difference from the normal wavefield is that there is a second-order pole at the mode points, rather than a simple pole singularity. This means that the residue for the jth mode depends on the derivative of $G_Z(p,\omega)G_Z^*(p,\omega)$ with respect to slowness rather than on just the value of $G_Z(p,\omega)$ evaluated at the pole $p_j(\omega)$. The structure of this residue contribution will have the form

$$\operatorname*{Res}_{\omega,p=p_j(\omega)} G_Z(p,\omega)G_Z^*(p,\omega) = 2\left(\frac{g_j}{2\omega I_j}\right)^2 [\mathbf{w}_j^{eT}(0)S_T\mathbf{w}_j^{e}(0)]\frac{\partial}{\partial p}[\mathbf{w}_j^{eT}(0)S_T\mathbf{w}_j^{e}(0)], \tag{6.1.16}$$

in terms of the eigenvector $\mathbf{w}_j^e$ at the surface. The nature of S_T will depend on the specific distribution of sources in the construction of the summed cross-correlations. The term $g_j/2\omega I_j$ is squared compared with the regular situation, the eigenvector contribution also appears squared, with the distinction between 'source' and 'receiver' contributions removed. The resemblance to the normal wavefield will be strongest for the fundamental Rayleigh mode with the simplest eigenfunction.

6.1.2 Body Waves

For the stacked cross-correlation field (6.1.12) the principal contribution to the body-wave field will come from the line integral along the real slowness axis, up to the hinge point p_c where the contour is turned into the upper half-plane to pick up the modal poles (Figure 2.2). The leaking mode effects from the branch D_+ will only be significant for very small epicentral distances. We can therefore concentrate on the real-axis integral, and consider again outgoing waves propagating away from station 1:

$$\left\langle \mathcal{U}^{(12)}(\omega)\right\rangle \sim \mathcal{M}(\omega)\int_0^{p_c} dp\, p\, \mathbf{G}(p,\omega)\mathbf{G}^{T*}(p,\omega)e^{i\omega p\Delta_{12}}, \tag{6.1.17}$$

where

$$\mathcal{M}(\omega) = \mathcal{F}_{12}\, i|M(\omega)|^2. \tag{6.1.18}$$

A suitable form for the slowness–frequency response $\mathbf{G}(p, \omega)$ is to work with a *generalised ray* expansion (Müller, 1970; Gilbert & Helmberger, 1972). As noted by Kennett & Pham (2018a,b), this representation of the seismic response in the stacked correlation result allows attention to be directly focussed on the interaction of seismic phases.

We build the generalised ray expansion by considering the composition of all the possible paths through the segments of the Earth model, taking into account the time accumulation along the path and the product of all the reflection and transmission processes encountered in passage to the receiver (§ SWI:14.7). If we work in terms of the major discontinuities of a radial Earth model we can set up a direct correspondence between the standard seismological phase notation, such as *PKIKP*, and the legs encountered in the generalised ray.

The generalised ray expansion expresses the medium response in the form

$$\mathbf{G}(p, \omega) = \sum_{I} \mathbf{g}_I(p) e^{i\omega\tau_I(p)}. \tag{6.1.19}$$

$\tau_I(p)$ includes the delay-time contributions for propagation on all the different legs encountered along the Ith path. The amplitude term $\mathbf{g}_I(p)$ is composed of the product of all reflection and transmission coefficients at interfaces, modulated by the slowness dependence of the source radiation and the displacement response at the receiver.

The spectral product in the slowness integral for the stacked cross-correlation (6.1.17) then takes the form

$$\mathbf{G}(p, \omega)\mathbf{G}^{T*}(p, \omega) = \sum_{I} \sum_{J} \mathbf{g}_I(p)\mathbf{g}_J^{T*}(p) e^{i\omega[\tau_I(p) - \tau_J(p)]}, \tag{6.1.20}$$

with the result that the correlation process will extract the difference in the delay-time contributions between pairs of generalised rays.

Consider a pair of generalised rays I and J: their contribution to the slowness integral

$$\left\langle \mathcal{U}^{(12)}(\omega) \right\rangle_{IJ} \sim \mathcal{M}(\omega) \int_0^{p_c} dp\, p^{\frac{1}{2}}\, \mathbf{g}_I(p)\mathbf{g}_J^{T*}(p)\, e^{i\{\omega[\tau_I(p) - \tau_J(p)] + p\Delta_{12}\}}. \tag{6.1.21}$$

The slowness integral is then simply the contribution from an arrival at range Δ_{12} with a differential propagation contribution of $\omega[\tau_I(p) - \tau_J(p)]$.

We can therefore expect that all discrete phases appearing in the stacked correlation results will arise from a suite of contributions whose apparent travel times will correspond to the difference of the times of actual seismic phases arriving at the original station locations Δ_1 and Δ_2, with the same surface slowness p. A slight complication is that we may be comparing seismic phases with different numbers of circuits of the globe or interaction with other caustics that introduce additional phase effects, which will produce some distortion of the apparent waveform of a differential phase.

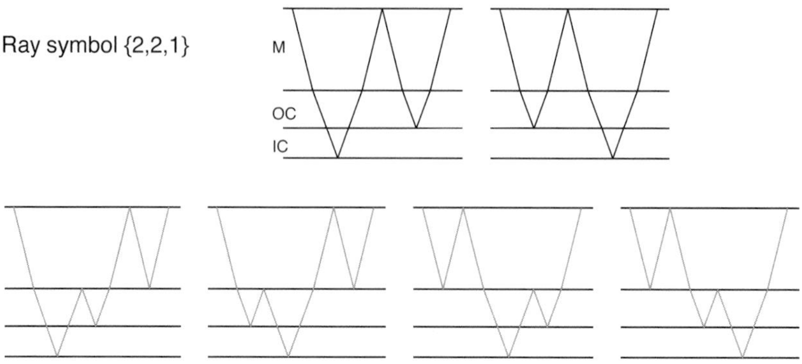

Figure 6.3 Illustration of *kinematic* and *dynamic* groups for a specific generalised ray described by code {2, 2, 1} in a three-layer model. All the illustrated paths will have the same travel time, but separate into two groups with common amplitude behaviour.

The generalised ray representation of the stacked correlation field (6.1.21) implies that all correlation phases emerge from differences. Yet, as we shall see in Chapters 7 and 8 the correlation wavefield can display apparently standard seismic phases. To extract a feature with the travel time properties of a regular seismic phase from a phase difference, we require that both the Ith and Jth generalised rays include the same suite of legs, say *Y*, but that one of them has an additional contribution corresponding to the legs in the regular phase, e.g., *SKS*. Subtraction of *Y* from *YSKS* or *SKSY* leaves the *SKS* contribution isolated.

A similar situation prevails for non-standard arrivals. In each case they are built from a situation where some element of propagation is in common between the different generalised rays. For example, the suite of seismic phases *YPKP* and *YScS*, arriving at stations with a common slowness, yield a correlation phase with the differential time corresponding to *PKP*−*ScS*. In a similar way, the combination *YcS*−*YcP*, with just a difference in the final leg from the core-mantle boundary to the surface, produces *cS*−*cP*, which is equivalent to a virtual source at the core-mantle boundary (Pham et al., 2018).

Amplitude considerations

We can describe the general behaviour of propagation in an Earth model by working with a three-layer configuration of mantle, outer core, and inner core, which mirrors seismological notation. This enables us to describe the character of generalised rays in terms of a ray symbol derived from the nature of the propagation legs in each layer. All rays with the same number of legs of a particular wave type in each layer will have the same passage time $\tau_I(p)$ for a specific slowness. All such configurations are *kinematic* analogues, and within this set there will be groups with the same common product of reflection and transmission coefficients $g_I(p)$ – these are *dynamic* analogues. We illustrate the behaviour in

Figure 6.3 for a simple configuration with six elements in the kinematic group, divided into two dynamic subgroups of two and four with the same amplitude behaviour. In the Appendix we summarise the properties of the analogues and how the number of elements can be evaluated for a given ray symbol. An important result (6.A.5) is that the sum over generalised rays can be reduced to an outer sum over the different phase terms associated with the various kinematic groups, with an inner sum over their constituent dynamic groups.

Suppose now we consider a seismic phase *Y*, without conversions, described by n_{Y1} legs in the top layer, n_{Y2} legs in the middle layer, and n_{Y3} legs in the bottom layer so the ray symbol is $\{n_{Y1}, n_{Y2}, n_{Y3}\}$. From (6.A.2), the number of elements in the kinematic group is given by

$$N_Y = \frac{(n_{Y1} + n_{Y2} - 1)!}{n_{Y2}!\,(n_{Y1} - 1)!} \cdot \frac{(n_{Y2} + n_{Y3} - 1)!}{n_{Y3}!\,(n_{Y2} - 1)!}. \tag{6.1.22}$$

If then *Y* is combined with another phase *X* to produce *YX* we now have a new symbol $\{n_{Y1} + n_{X1}, n_{Y2} + n_{X2}, n_{Y3} + n_{X3}\}$, and the number of elements in the combined phase N_{YX} can be obtained by analogy with (6.1.22).

When we create a difference phase *X* by correlating *YX* with *Y*, all kinematic analogues of *Y* can combine with *X* and so contribute. There is also the possibility that *X* itself has a multiplicity of manifestations. The number of possibilities for extracting phase *X* from the combination *YX* is then N_{YX}/N_Y and grows rapidly with the number of legs in phase *X*. Thus, for example, consider combining core reflections with the phase *I3*≡*PKIKPPKIKPPKIKP* that has multiple surface reflections. We can produce a phase with the timing of *I3* in 100 ways, this increases to 200 for *cI3*≡*ScSI3* and 350 for *cI3c*. The most important dynamic analogues will be those with the fewest reflections, but the large number of the remainder will help to compensate for their reduced individual contributions. If we add an extra double leg in both layers 1 and 2, the number of kinematic equivalents rises to 700, and more complex phases can have a very large number of possibilities.

6.2 The Nature of the Correlation Wavefield

Our analysis of the correlation field for a stratified Earth has shown how the stacking of the cross-correlation of the seismograms at two stations over the contributions from many well-distributed sources can produce a coherent result that corresponds to a virtual source–receiver pair at the two sites.

The nature of the correlation wavefield will depend on the slowness content of the segments of seismograms being cross-correlated. Where sources are shallow, such as those due to atmospheric and oceanic disturbances or human activity, correlation of horizontally travelling waves will enhance the contribution from surface waves, particularly the fundamental mode (Chapter 7). For deeper sources as in the coda of large earthquakes, body waves with much smaller slowness will be significant and a complex of arrivals materialises with exotic combinations of phases that do not appear in the regular wavefield from a source (Chapter 8).

In no case do we recover directly the response of a simple source at site 1 recorded at site 2 (or the reverse), which would be the true *Green's function*. There are fundamental differences:

(1) the cross-correlation process extracts the square of the frequency spectrum of the sources;

(2) the medium response involves $\mathbf{G}(p, \omega)\mathbf{G}^*(p, \omega)$ rather than just $\mathbf{G}(p, \omega)$ as in the direct response; thus there will be a different emphasis on aspects of wave propagation;

(3) all apparent seismic arrivals arise by differencing phase contributions, and so it is possible to obtain features that have no counterpart in the regular wavefield.

The size of the correlation field contribution between a pair of stations will depend on the disposition of the available sources and the particular mechanisms of the sources.

Nevertheless, contained within the general correlation field are elements which have the same time behaviour as the regular field, though the amplitude will be different. This means that cross-correlation stacking can be used to enhance information about the Earth between two station locations by exploiting a multitude of weak sources, rather than having to rely on the fortuitous placement of major earthquake sources.

The stacked cross-correlation results have often been termed 'empirical Green's functions' (e.g., Ritzwoller & Feng, 2019), which provides a convenient shorthand for the extracted results for the virtual source–receiver pair between the stations.

To achieve the extraction of the desired propagation component between the stations, there needs to be a full azimuthal distribution of sources, even though constructive interference is concentrated along the great circle passing through the sites. An irregular distribution of sources will produce a markedly stronger contributions in one direction than the other, and potentially distortion of the results because of incomplete destructive interference.

Our analysis has focussed on the effects of averaging over sources, so that the correlation wavefield has the strongest resemblance to the conventional wavefield. However, strong and sustained sources that lie outside the stationary phase zone can introduce classes of 'spurious' arrivals that do not correspond to expectations (Snieder et al., 2008; Li et al., 2020).

Modal representation

In the derivation above we have made use of the representation of the wavefield on a sphere in terms of a continuous integral over slowness and travelling waves. Tanimoto (2008) has made an alternative development for the stacked cross-correlation between two stations directly in terms of normal modes. He employs a distribution of sources across the globe, with a vertical force applied near the surface to simulate atmospheric and oceanic effects, such as the Longuet-Higgins (1950) mechanism for the generation of microseisms that produces pressure through nonlinear interaction of ocean waves. Rather than having constructive interference between travelling waves, this approach integrates the Legendre function product $P_l(\cos\Delta_1)P_l(\cos\Delta_2)$ over all the sources to extract a representation in terms of $P_l(\cos\Delta_{12})$. Figure 2 of Tanimoto (2008) demonstrates that the interference of all the Legendre functions, associated with the distributed sources, produces a stationary phase region along the great circle passing through the receivers, as we have found above.

The normal mode results demonstrate that the summed correlation field between the stations depends on the square of the source spectrum, and also the square of the eigenfunctions of the modes, in direct correspondence with (6.1.12). The normal mode derivation also makes it explicit that the time dependence of the stacked cross-correlation contains one extra integration over time compared to the regular response to a vertical force source. Tanimoto (2008) also points out that the presence of the squared eigenfunctions tends to place more emphasis on higher-frequency energy and the fundamental modes since the sources are shallow. The results for moment-tensor sources are more complex than those for a suite of vertical forces. As we have already noted the stacked correlations depend on the disposition of the sources and the orientation of their mechanisms, so that an analytic result is not available.

Representations in terms of Green's functions

In Section 3.2.1 we have introduced a relation between the Green's tensor and its complex conjugate in terms of a surface integral involving the correlation of Green's tensor displacements and associated tractions over a surface ∂V surrounding the points of interest. This result (3.2.12) derived from the representation theorem for a perfectly elastic medium

$$G_{ij}(\mathbf{x}_1,\mathbf{x}_2,\omega) - G^*_{ij}(\mathbf{x}_1,\mathbf{x}_2,\omega) = \int_{\partial V} d^2\xi\, \left[G^*_{iq}(\mathbf{x}_2,\xi,\omega)H_{qj}(\mathbf{x}_1,\xi,\omega) - G_{iq}(\mathbf{x}_1,\xi,\omega)H^*_{qj}(\mathbf{x}_2,\xi,\omega)\right], \tag{6.2.1}$$

involves reciprocal propagation between $\mathbf{x}_1$ and $\mathbf{x}_2$ and cross-correlation of contributions from the Green's tensor G and its traction elements H on the enveloping surface ∂V. The complex conjugate implies time reversal on back transformation to the time domain.

The structure of (6.2.1) has resemblances to the stacked cross-correlation result (6.1.12), but is not directly equivalent because the direct correlation of seismograms does not extract the Green's function. Many attempts have been made to try to convert (6.2.1) into a simpler form that could be considered to have a direct relation to the actual process used in stacked cross-correlation of seismograms (e.g., Wapenaar & Fokkema, 2006). None are wholly satisfactory, and it is therefore appropriate to work directly with the correlation field, as is explored further in the next section.

Kimman & Trampert (2010) demonstrate how simplifications of the surface integral in (6.2.1) can lead to the introduction of cross-terms of significant amplitude between higher modes of surface waves. Such cross-terms vanish in the full implementation (6.2.1). Only the fundamental mode can be reasonably well represented with simpler forms, such as those discussed by Halliday & Curtis (2008). For an isolated surface-wave mode ν,

$$G^{\nu}_{ij}(\mathbf{x_1},\mathbf{x_2},\omega) - G^{\nu*}_{ij}(\mathbf{x}_1,\mathbf{x}_2,\omega) \approx \qquad (6.2.2)$$
$$-2i\omega U^{\nu}(\omega)\int_{\partial V} d^2\xi.\mathbf{n}\,\rho G^{\nu}_{ip}(\mathbf{x_1},\mathbf{x_2},\omega)G^{\nu*}_{jp}(\mathbf{x}_2,\xi,\omega),$$

where $U^{\nu}(\omega)$ is the group velocity associated with the mode (Love or Rayleigh). In (6.2.2) the cross-correlation works directly on the modal components of the Green's tensor $\mathbf{G}^{\nu}$ on a distant boundary ∂V.

6.3 Generalised Interferometry

In the previous sections we have seen that the cross-correlation of the seismic wavefield can provide information on Earth structure. With a sufficiently even coverage of sources, we recover expressions whose structure is reminiscent of Green's functions between the stations being correlated. Indeed, the time characteristics can be well represented, but the presence of squared amplitude terms precludes direct correspondence.

Here we consider the extraction of information on wavefield sources and Earth structure directly from inter-station correlations building on the work of Fichtner et al. (2017). We explore the situation when the distribution of noise sources is arbitrarily heterogeneous, or where earthquakes dominate, with the aim of understanding the nature of the correlation wavefield from practical observations, and the way it can be exploited.

We seek to achieve a number of different objectives:

(i) Provide a unified theory for interferometry by correlation that accounts for heterogeneous noise sources, linear and nonlinear processing, and the presence of transient sources such as earthquakes.

(ii) Enable the exploitation of seemingly unphysical correlation waveforms that result from heterogeneous noise sources and that are not present in the actual inter-station Green's function.

(iii) Bridge the gap between inter-station correlations of ambient noise (noise interferometry) and earthquake recordings (e.g., the two-station method for surface waves).

(iv) Function for any type of medium, including a realistic elastic Earth with attenuation and anisotropy.

We will show that through the process of constructing stacked cross-correlations between the seismograms at different stations, we extract an effective propagation process and effective sources rather than the actual wave physics and sources. Nevertheless, these results can be recognised to contain the information we seek.

6.3.1 Effective theory for modelling inter-station correlations

We first establish a representation for inter-station correlation without processing, which can be directly compared to similar constructs from synthetic seismograms calculated including attenuation. We then turn attention to the modifications introduced by data processing, and how these can be accommodated in the theoretical scheme.

Raw correlations

We consider three-component seismograms recorded at stations at locations $\mathbf{x}_i$, $\mathbf{x}_k$ and construct tensor cross-correlations by averaging over time. We take a suite of time intervals $[t_n, t_{n+1}]$, $n = 1, \dots, N$ for which the Fourier transforms of the seismograms are calculated and use the superscript $^{[n]}$ for the results for the nth time interval. To avoid contamination of the Fourier spectrum by the finiteness of the time window, we assume that $(t_{n+1} - t_n)^{-1}$ is much smaller than the minimum frequency of interest.

From the correlation of recordings at positions $\mathbf{x}_i$ and $\mathbf{x}_k$, we compute the observed interferogram, or correlation function, for time interval n

$$\mathcal{I}^{[n]}_{\text{obs}}(\mathbf{x}_i, \mathbf{x}_k, \omega) = \mathbf{u}^{[n]}_{\text{obs}}(\mathbf{x}_i, \omega)\mathbf{u}^{[n]T*}_{\text{obs}}(\mathbf{x}_k, \omega), \tag{6.3.1}$$

We will use the expressions *interferogram* and *correlation* interchangeably, noting that the former term is more general, comprising more than correlation.

Now taking the average of the interferograms between stations i and k for the intervals $n = 1, \dots, N$, we obtain the ensemble interferogram for this station pair:

$$\mathcal{I}_{\text{obs}}(\mathbf{x}_i, \mathbf{x}_k, \omega) = \frac{1}{N}\sum_{n=1}^{N} \mathbf{u}^{[n]}_{\text{obs}}(\mathbf{x}_i, \omega)\mathbf{u}^{[n]T*}_{\text{obs}}(\mathbf{x}_k, \omega). \tag{6.3.2}$$

Synthetic interferograms

We assume that we are able to model the seismograms at the recording locations using seismic wave theory, with allowance for attenuation. Thus we have the equation of motion (3.2.1),

$$\partial_j(c_{ijkl}\partial_k u_l) + \rho\omega^2 u_i = -F_i, \tag{6.3.3}$$

for a source $F(\mathbf{x})$. We can then make use of the representation theorem from Section 3.2.1, in the frequency domain, to cast the response to the source in integral form over a volume V:

$$\mathbf{u}^{[n]}(\mathbf{x}_i) = \int_V d^3\xi\, G(\mathbf{m}; \mathbf{x}_i, \xi) F^{[n]}(\xi). \tag{6.3.4}$$

In this way we connect sources $F^{[n]}(\xi)$ to a synthetic wavefield $u^{[n]}(\mathbf{x}_i)$ via the Green's tensor $G(\mathbf{m}; \mathbf{x}_i, \mathbf{x})$, which in turn depends on a structural model $\mathbf{m} = (\rho, c_{ijkl})$. The sources $F^{[n]}(\xi)$ may comprise any combination of quasi-random, deterministic, point-localised, and distributed sources. Thus we can consider sources of ambient noise, earthquakes and explosions, and the effects of scattering.

For each time interval we can construct a tensor interferogram as

$$\begin{aligned}\mathcal{I}^{[n]}(\mathbf{x}_i, \mathbf{x}_k, \omega) &= \mathbf{u}^{[n]}(\mathbf{x}_i, \omega)\mathbf{u}^{[n]T*}(\mathbf{x}_k, \omega) \\ &= \int_V d^3\boldsymbol{\xi} \int_V d^3\boldsymbol{\eta}\, \mathsf{G}(\mathbf{x}_i, \boldsymbol{\xi}, \omega)\mathsf{F}^{[n]}(\boldsymbol{\xi}, \omega)[\mathsf{G}(\mathbf{x}_k, \boldsymbol{\eta}, \omega)\mathsf{F}^{[n]}(\boldsymbol{\eta}, \omega)]^{T*}.\end{aligned} \tag{6.3.5}$$

Now grouping together the source terms and using Green's tensor reciprocity we have

$$\mathcal{I}^{[n]}(\mathbf{x}_i, \mathbf{x}_k, \omega) = \int_V d^3\boldsymbol{\xi} \int_V d^3\boldsymbol{\eta}\, \mathsf{G}(\mathbf{x}_i, \boldsymbol{\xi}, \omega)[\mathsf{F}^{[n]}(\boldsymbol{\xi}, \omega)\mathsf{F}^{[n]T*}(\boldsymbol{\eta}, \omega)]\mathsf{G}^{T*}(\boldsymbol{\eta}, \mathbf{x}_k, \omega). \tag{6.3.6}$$

We now introduce the power-spectral density (psd) tensor for the nth time interval

$$\mathsf{S}^{[n]}(\boldsymbol{\xi}, \boldsymbol{\eta}, \omega) = \mathsf{F}^{[n]}(\boldsymbol{\xi}, \omega)\mathsf{F}^{[n]T*}(\boldsymbol{\eta}, \omega). \tag{6.3.7}$$

The diagonal elements of $\mathsf{S}^{[n]}$ describe the correlation of force systems acting in the same direction in the nth time interval; whereas the off-diagonal elements represent the correlation of forces in different directions. Force models are well developed for vertical-component noise, as in the case of microseisms, but the nature of horizontal-component excitation of seismic noise is still being elucidated (see, e.g., Fukao et al., 2010; Saito, 2010).

The time-averaged psd tensor for the source distribution is then:

$$\mathsf{S}(\boldsymbol{\xi}, \boldsymbol{\eta}, \omega) = \frac{1}{N}\sum_{n=1}^{N} \mathsf{S}^{[n]}(\boldsymbol{\xi}, \boldsymbol{\eta}, \omega). \tag{6.3.8}$$

Combining (6.3.6) and (6.3.7), we construct an ensemble synthetic interferogram tensor by summation over multiple time intervals:

$$\begin{aligned}\mathcal{I}(\mathbf{x}_i, \mathbf{x}_k, \omega) &= \frac{1}{N}\sum_{n=1}^{N} \mathcal{I}^{[n]}(\mathbf{x}_i, \mathbf{x}_k, \omega) = \frac{1}{N}\sum_{n=1}^{N} \mathbf{u}^{[n]}(\mathbf{x}_i, \omega)\mathbf{u}^{[n]*}(\mathbf{x}_k, \omega) \\ &= \frac{1}{N}\sum_{n=1}^{N} \int_V d^3\boldsymbol{\xi} \int_V d^3\boldsymbol{\eta}\, \mathsf{G}(\mathbf{x}_i, \boldsymbol{\xi}, \omega)\mathsf{G}^*(\boldsymbol{\eta}, \mathbf{x}_k)\mathsf{S}^{[n]}(\boldsymbol{\xi}, \boldsymbol{\eta}, \omega).\end{aligned} \tag{6.3.9}$$

We can therefore obtain an equation for the ensemble interferogram in the form of a representation theorem

$$\underbrace{\mathcal{I}(\mathbf{x}_i, \mathbf{x}_k, \omega)}_{\text{interf. wavefield}} = \int_V d^3\boldsymbol{\xi}\, \mathsf{G}(\mathbf{x}_i, \boldsymbol{\xi}, \omega) \underbrace{\left[\int_V d^3\boldsymbol{\eta}\, \mathsf{S}(\boldsymbol{\xi}, \boldsymbol{\eta}, \omega)\mathsf{G}^*(\boldsymbol{\eta}, \mathbf{x}_k, \omega)\right]}_{\text{source of interf. wavefield}}. \tag{6.3.10}$$

Choosing the first station position $\mathbf{x}_i$ as a free variable, we can interpret $\mathcal{I}(\mathbf{x}_i, \mathbf{x}_k, \omega)$ as a tensorial interferometric wavefield that is driven by the source $\int d^3\boldsymbol{\eta}\, \mathsf{S}(\boldsymbol{\xi}, \boldsymbol{\eta}, \omega)\mathsf{G}^*(\boldsymbol{\eta}, \mathbf{x}_k, \omega)$. With an exchange of the order of integration we get a comparable result in terms of the second station position. The ensemble interferogram thus satisfies the seismic wave equation (6.3.3).

The calculated interferometric tensor $\mathcal{I}(\mathbf{x}_i, \mathbf{x}_k, \omega)$ can be compared directly to the time averaged results for the same station pair constructed from the observations $\mathcal{I}_{obs}(\mathbf{x}_i, \mathbf{x}_k, \omega)$ as in (6.3.2).

If the force systems at the locations $\boldsymbol{\xi}$ and $\boldsymbol{\eta}$ can be assumed to be uncorrelated when averaged over time, so that

$$S(\boldsymbol{\xi}, \boldsymbol{\eta}, \omega) = S(\boldsymbol{\xi}, \omega)\delta(\boldsymbol{\xi} - \boldsymbol{\eta}), \tag{6.3.11}$$

then equation (6.3.10) for the ensemble tensor interferogram reduces to

$$\underbrace{\mathcal{I}(\mathbf{x}_i, \mathbf{x}_k, \omega)}_{\text{interf. wavefield}} = \int_V d^3\boldsymbol{\xi}\, G(\mathbf{x}_i, \boldsymbol{\xi}, \omega) \underbrace{[S(\boldsymbol{\xi}, \omega) G^*(\boldsymbol{\xi}, \mathbf{x}_k, \omega)]}_{\text{source of interf. wavefield}}. \tag{6.3.12}$$

In these circumstances the source of the interferometric wavefield can be computed without requiring a spatial integral (e.g., Woodard, 1997; Tromp et al., 2010).

Most applications focus on the vertical–vertical (Z–Z) component of the interferometric tensor, but useful signals can be found on other components, Section 7.3, and can be used to investigate seismic anisotropy and its temporal variations (e.g., Roux, 2009).

The merit of this effective wavefield theory for interferograms is that only the averaged power-spectral density (psd) of the sources is required to compute the synthetic interferogram for a station pair. This obviates the need for long-duration realisations of seismic wavefields that would be computationally prohibitive for complex 3-D media. In principle we should be able to find a set of material parameters (ρ, c_{ijkl}) such that the synthetic ensemble interferogram $\mathcal{I}(\mathbf{x}_i, \mathbf{x}_k, \omega)$ matches the raw observed ensemble interferogram $\mathcal{I}_{obs}(\mathbf{x}_i, \mathbf{x}_k, \omega)$ to within the observational uncertainties

$$\mathcal{I}_{obs}(\mathbf{x}_i, \mathbf{x}_k, \omega) = \mathcal{I}(\mathbf{x}_i, \mathbf{x}_k, \omega) + \mathcal{E}(\mathbf{x}_i, \mathbf{x}_k, \omega). \tag{6.3.13}$$

The residual term $\mathcal{E}(\mathbf{x}_i, \mathbf{x}_k, \omega)$ is path specific.

Influence of processing

Commonly some class of data processing is applied to each time interval with the aim of enhancing the coherent part of the cross-correlation signal or removing large-amplitude transients, for instance, from earthquakes (e.g., Bensen et al., 2007). As a result of the effects of processing the time series, the interferogram is modified from its raw state. For each time interval, the *processed interferogram* $\tilde{\mathcal{I}}^{[n]}$ can then be related to the raw interferogram through a frequency-dependent transfer tensor $\mathfrak{T}_{ik}^{[n]}(\omega)$,

$$\tilde{\mathcal{I}}_{obs}^{[n]}(\mathbf{x}_i, \mathbf{x}_k, \omega) = \mathfrak{T}_{ik}^{[n]}(\omega)\mathcal{I}_{obs}^{[n]}(\mathbf{x}_i, \mathbf{x}_k, \omega). \tag{6.3.14}$$

Some aspects of processing will be specific to the station pair $\{i, k\}$ and so independent of the time interval, but much will depend on the nature of the seismograms being correlated in a particular window. We therefore factorise the transfer tensor as

$$\mathfrak{T}_{ik}^{[n]}(\omega) = \mathfrak{g}_{ik}(\omega)\mathfrak{f}^{[n]}(\omega) + \mathfrak{e}_{ik}^{[n]}(\omega), \tag{6.3.15}$$

where $\mathfrak{f}^{[n]}(\omega)$ is a scalar source correction, $\mathfrak{g}_{ik}(\omega)$ is a tensor propagation correction, and $\mathfrak{e}_{ik}^{[n]}(\omega)$ is the tensor factorisation residual. In the following section, we demonstrate how the factorisation (6.3.15) can be constructed such that the factorisation residual $\mathfrak{e}_{ik}^{[n]}(\omega)$ is minimised.

We can apply the transfer tensor $\mathfrak{T}_{ik}^{[n]}$ to our synthetic interferogram to introduce the effects of processing. The processed interferometric wavefield for the nth time interval is then

$$\begin{aligned}
&\tilde{\mathcal{I}}^{[n]}(\mathbf{x}_i, \mathbf{x}_k, \omega) = \\
&\quad \int_V d^3\boldsymbol{\xi}\, \mathfrak{g}_{ik}(\omega) G(\mathbf{x}_i, \boldsymbol{\xi}, \omega) \left[\int_V d^3\boldsymbol{\eta}\, \mathfrak{f}^{[n]}(\omega) S^{[n]}(\boldsymbol{\xi}, \boldsymbol{\eta}, \omega) G^*(\boldsymbol{\eta}, \mathbf{x}_k, \omega)\right] \\
&\quad + \int_V d^3\boldsymbol{\xi}\, G(\mathbf{x}_i, \boldsymbol{\xi}, \omega) \left[\int_V d^3\boldsymbol{\eta}\, \mathfrak{e}_{ik}^{[n]}(\omega) S^{[n]}(\boldsymbol{\xi}, \boldsymbol{\eta}, \omega) G^*(\boldsymbol{\eta}, \mathbf{x}_k, \omega)\right].
\end{aligned} \tag{6.3.16}$$

We can identify an *effective* Green's tensor $\mathfrak{g}_{ik}(\omega)G(\mathbf{x}_i, \boldsymbol{\xi}, \omega)$ for the path $\mathbf{x}_k \to \mathbf{x}_i$ that is independent of the time interval n. Further, when we take the ensemble average over all the time intervals as in (6.3.9) we can recognise an *effective source psd*:

$$\widehat{S}(\boldsymbol{\xi}, \boldsymbol{\eta}, \omega) = \frac{1}{N} \sum_{n=1}^{N} \mathfrak{f}^{[n]}(\omega) S^{[n]}(\boldsymbol{\xi}, \boldsymbol{\eta}, \omega). \tag{6.3.17}$$

As a result we can express the processed interferometric wavefield in the form

$$\begin{aligned}
\underbrace{\tilde{\mathcal{I}}(\mathbf{x}_i, \mathbf{x}_k, \omega)}_{\text{proc. interf. wavefield}} &= \int_V d^3\boldsymbol{\xi}\, \underbrace{\mathfrak{g}_{ik}(\omega) G(\mathbf{x}_i, \boldsymbol{\xi}, \omega)}_{\text{eff. Green function}} \underbrace{\left[\int_V d^3\boldsymbol{\eta}\, \widehat{S}(\boldsymbol{\xi}, \boldsymbol{\eta}, \omega) G^*(\boldsymbol{\eta}, \mathbf{x}_k, \omega)\right]}_{\text{effective source}} \\
&\quad + \mathbf{E}(\mathbf{x}_i, \mathbf{x}_k, \omega),
\end{aligned} \tag{6.3.18}$$

with the modelling residual

$$\mathbf{E}(\mathbf{x}_i, \mathbf{x}_k, \omega) = \int_V d^3\boldsymbol{\xi} \int_V d^3\boldsymbol{\eta}\, G(\mathbf{x}_i, \boldsymbol{\xi}, \omega) \left[\frac{1}{N} \sum_{n=1}^{N} \mathfrak{e}_{ik}^{[n]}(\omega) \mathbf{S}^{[n]}(\boldsymbol{\xi}, \boldsymbol{\eta}, \omega)\right] G^*(\boldsymbol{\eta}, \mathbf{x}_k, \omega). \tag{6.3.19}$$

In this way we can express the processed ensemble interferogram in terms of an *effective* interferometric wavefield and the residual contribution (6.3.19):

$$\underbrace{\tilde{\mathcal{I}}(\mathbf{x}_i, \mathbf{x}_k, \omega)}_{\text{proc. interf. wavefield}} = \underbrace{\widehat{\mathcal{I}}(\mathbf{x}_i, \mathbf{x}_k, \omega)}_{\text{effective interf. wavefield}} + \mathbf{E}(\mathbf{x}_i, \mathbf{x}_k, \omega). \tag{6.3.20}$$

The effective interferometric field propagates through an effective medium described by the effective Green's tensor $\mathfrak{g}_{ik}(\omega)G(\mathbf{x}_i, \boldsymbol{\xi}, \omega)$. Excitation is provided by a source determined by the effective source psd $\widehat{\mathbf{S}}(\boldsymbol{\xi}, \boldsymbol{\eta}, \omega)$. This treatment of the construction of the effective wavefield and its relation to the distribution of seismic sources provides the basis for full waveform inversion of interferograms (Sager et al., 2018, 2020) as discussed in Chapter 18.

Paitz et al. (2019) have shown how the treatment of the effective wavefield we have discussed can be extended to rotation and strain. This means that coorelation results using new classes of seismic sensor, such as rotational ground motion and distributed acoustic sensing technology, can be treated in a unified approach.

Unphysical components of processed interferograms

The modelling residual $\mathbf{E}(\mathbf{x}_i, \mathbf{x}_k, \omega)$ defined in (6.3.19) can be interpreted as a wavefield excited by the spatially distributed source

$$F_{ik}^{res}(\boldsymbol{\xi}) = \int_V d^3\boldsymbol{\eta} \left[\frac{1}{N} \sum_{n=1}^{N} \mathfrak{e}_{ik}^{[n]}(\omega) \mathbf{S}^{[n]}(\boldsymbol{\xi}, \boldsymbol{\eta}, \omega) \right] G^*(\boldsymbol{\eta}, \mathbf{x}_k, \omega). \tag{6.3.21}$$

However, in contrast to the source of a raw (unprocessed) interferogram, the source $F_{ik}^{res}(\boldsymbol{\xi})$ depends on the receiver pair $\mathbf{x}_i, \mathbf{x}_k$. Thus, the residual for each receiver pair is potentially excited by a different source, which makes $\mathbf{E}(\mathbf{x}_i, \mathbf{x}_k, \omega)$ an unphysical quantity that we wish to minimise. To do so, without prior knowledge of the temporal variability of the source psd $\mathbf{S}^{[n]}$, we first rewrite the term in square brackets in (6.3.21) as

$$\frac{1}{N} \sum_{n=1}^{N} \mathfrak{e}_{ik}^{[n]}(\omega) \mathbf{S}^{[n]}(\boldsymbol{\xi}, \boldsymbol{\eta}, \omega) = \tag{6.3.22}$$

$$\frac{1}{N} \mathbf{S}(\boldsymbol{\xi}, \boldsymbol{\eta}, \omega) \sum_{n=1}^{N} \mathfrak{e}_{ik}^{[n]}(\omega) + \frac{1}{N} \sum_{n=1}^{N} \mathfrak{e}_{ik}^{[n]}(\omega) \left[\mathbf{S}^{[n]}(\boldsymbol{\xi}, \boldsymbol{\eta}, \omega) - \mathbf{S}(\boldsymbol{\xi}, \boldsymbol{\eta}, \omega) \right].$$

We can force the first term on the right hand side of (6.3.22) to vanish by requesting that the time-averaged factorisation residual be zero:

$$\sum_{n=1}^{N} \mathfrak{e}_{ik}^{[n]}(\omega) = 0. \tag{6.3.23}$$

The remaining term is then controlled by the temporal variations of the source psd relative to its time average, that is, $\mathbf{S}^{[n]}(\boldsymbol{\xi}, \boldsymbol{\eta}, \omega) - \mathbf{S}(\boldsymbol{\xi}, \boldsymbol{\eta}, \omega)$. Thus, in the special case of stationary sources, the modelling residual vanishes.

To ensure uniqueness of the transfer function factorisation (6.3.15), we require, in addition to (6.3.23), that the source correctors $\mathfrak{f}^{[n]}(\omega)$ average to 1:

$$\frac{1}{N} \sum_{n=1}^{N} \mathfrak{f}^{[n]}(\omega) = 1. \tag{6.3.24}$$

Averaging the factorisation (6.3.15) over all time intervals n yields an equation for the tensorial propagation corrector,

$$\mathfrak{T}_{ik}(\omega) = \frac{1}{N}\sum_{n=1}^{N} \mathfrak{T}_{ik}^{[n]}(\omega) = \mathfrak{g}_{ik}(\omega). \tag{6.3.25}$$

To minimise the residual $\mathfrak{e}_{ik}^{[n]}(\omega)$ in a least-squares sense, we force the derivative of

$$\sum_{i,k=1}^{N} |\mathfrak{e}_{ik}^{[n]}(\omega)|^2 = \sum_{i,k=1}^{N} |\mathfrak{T}_{ik}^{[n]}(\omega) - \mathfrak{g}_{ik}(\omega)\mathfrak{f}^{[n]}(\omega)|^2 \tag{6.3.26}$$

with respect to $\mathfrak{f}^{[n]}(\omega)$ to zero for each n. This yields

$$\mathfrak{f}^{[n]}(\omega) = \frac{\mathrm{Re}\sum_{i,k=1}^{N} \mathfrak{T}_{ik}^{[n]}(\omega) : \mathfrak{g}_{ik}^{*}(\omega)}{\sum_{i,k=1}^{N} |\mathfrak{e}_{ik}^{[n]}(\omega)|^2}, \tag{6.3.27}$$

where : denotes the component-wise scalar product. The imaginary part of the source correctors $\mathfrak{f}^{[n]}(\omega)$ is unconstrained by the minimisation procedure. The physically most meaningful choice is to require $\mathrm{Im}\,\mathfrak{f}^{[n]}(\omega) = 0$, thereby ensuring that the original psd and the effective source psd have identical phase.

6.3.2 Illustrative Examples

We have developed the theoretical basis for generalised interferometry for a 3-D elastic medium, but to illustrate the way that the approach works we will use a 2-D acoustic scenario. The Green's function for this situation is given by

$$G(\mathbf{x}_i, \mathbf{x}_k) = -i\frac{1}{4\rho c^2}\sqrt{\frac{2v}{\pi\omega r}}\, \exp\left(-i\frac{\omega}{c}r\right) \exp\left[-\frac{\omega r}{2cQ}\right] \exp\left(i\tfrac{\pi}{4}\right), \tag{6.3.28}$$

with source–receiver distance $r = |\mathbf{x}_k - \mathbf{x}_i|$, and quality factor Q, the real part of the bulk modulus κ, and the phase velocity $c = \sqrt{\kappa/\rho}$.

We illustrate the construction of the interferometric wavefield $\mathcal{I}(\mathbf{x}_i, \mathbf{x}_k)$, using the scalar equivalent to (6.3.12), in Figure 6.4. The medium is homogeneous with $c = 3$ km/s, density $\rho = 3000$ kg/m^3, and $Q = 1000$. We display results for an array of receiver locations and two patches of non-zero source psd.

The heterogeneitiy of the source psd leads to asymmetric correlations for the various station pairs, and wave packets arriving earlier than expected for a direct wave travelling between the receiver pairs. The correlation wavefield for this configuration is two-sided, with energy at negative times travelling towards the reference station (*anti-causal*) and at positive times travelling away from the reference station (*causal*). Fichtner et al. (2017) show the significant effects that arise from varying the level of attenuation.

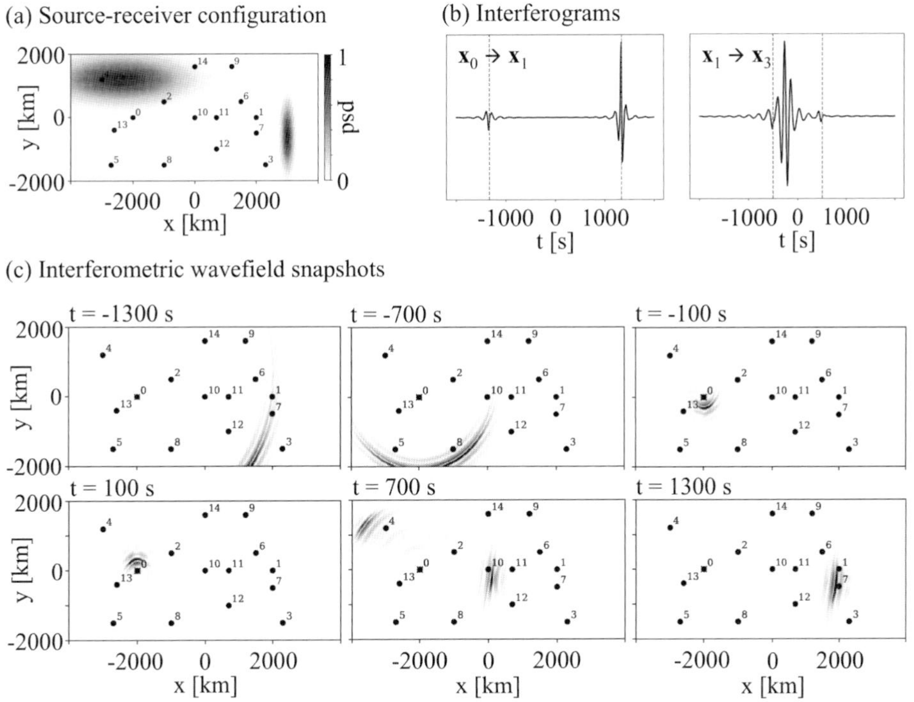

Figure 6.4 Interferometric wavefield in a 2-D homogeneous medium. (a) Normalised source psd distribution. Receiver positions are marked by numbered black dots. (b) Interferograms $\mathcal{I}(\mathbf{x}_0, \mathbf{x}_1)$ and $\mathcal{I}(\mathbf{x}_1, \mathbf{x}_3)$. Vertical dashed lines indicate the arrival times of direct waves travelling from one of the receivers to the other. (c) Snapshots of the interferometric wavefield, starting at $t = -1300$ s (anti-causal branch of the interferogram) and moving towards $t = 1300$ s (causal branch of the interferogram).

Effects of noise processing

To investigate the effects of commonly applied processing schemes, we compute artificial ambient noise recordings at the stations shown in Figure 6.4. For this, we represent the noise source distribution in Figure 6.4(a) by a large number of discrete points, each radiating 5000 different realisations of random signals of 8000 s duration. Thus, the total duration of the artificial noise time series is 5000×8000 s (~463 days).

The first case we consider is processing with one-bit normalisation, where the samples of the raw time series are set to 1 for positive and to -1 for negative amplitudes. This approach is designed to suppress the influence of large-amplitude bursts, such as from earthquake signals. Figure 6.5(a) shows the raw and the processed interferograms in the time domain for station pair $(\mathbf{x}_{i=0}, \mathbf{x}_{k=1})$ in Figure 6.4. Since earthquakes were not included in the computation of artificial data, raw and processed interferograms are very similar. The propagation corrector $\mathfrak{g}_{ik}(\omega)$ in Figure 6.5(b) is the average of the transfer functions over all time windows, as demonstrated in (6.3.25). Throughout the considered frequency band, from around 0.005 Hz to 0.035 Hz, the propagation corrector is nearly flat, with only a comparatively small imaginary part. Therefore, the original Green's function and the effective Green's function are almost identical, up to a scaling factor. Similarly, the processed and effective interferograms in Figure 6.5(c) are nearly indistinguishable. Consequently, the modelling residual, i.e., the unphysical component introduced by one-bit normalisation, is very small. Similar properties are found for all other receiver pairs. The behaviour for one-bit normalisation conforms to theoretical predictions of vanishing phase shifts between raw and processed correlations in the case of stationary random sources (van Vleck & Middleton, 1966).

In addition to one-bit normalisation, which is applied routinely, a large variety of processing schemes have been developed including spectral whitening, time- and frequency-dependent amplitude normalisations, and clipping of recordings with amplitudes above a certain threshold. In an attempt to equalise the contributions from an irregular source distribution many workers have averaged the causal and anti-causal branches of the correlation function for station pairs. This averaging, summarised in Figure 6.6, produces a nearly flat propagation corrector, and therefore has very little effect on the phase of the Green's function. Waves in the original and effective media are almost in phase for all frequencies. However, the processed and effective interferograms in Figure 6.6(c) differ significantly, meaning that the unphysical modelling residual is large. The processed correlation is symmetric by design, but the effective correlation is not. As a result there is no combination of effective medium and effective source for which the processed correlation can be represented as a proper correlation function. Averaging causal and anti-causal branches combines dominant contributions at different times, which would require two or more waves travelling at different velocities instead of a single wave for a true acoustic medium.

An alternative approach to suppressing the imprint of the source is to exploit the phase equivalence of correlation and spectral division. Saygin & Kennett (2012) have used the transfer function between station pairs rather than correlation. Spectral division (5.1.12) removes the common temporal source spectrum from the two station terms to give a broader spectral response than plain cross-correlation. To avoid problems with spectral holes, the spectral division needs to be regularised with, e.g., a water-level technique or the addition of a small bias term to the denominator. In terms of the Fourier transforms of the recorded seismic displacement field at stations $\mathbf{x}_i$ and $\mathbf{x}_k$, $u(\mathbf{x}_i, \omega)$ and $u(\mathbf{x}_k, \omega)$, the regularised frequency-domain transfer function,

$$\Phi = \frac{u(\mathbf{x}_i, \omega) u^*(\mathbf{x}_k, \omega)}{\varphi(\omega)}, \tag{6.3.29}$$

has the same phase as for the cross-correlation. The denominator in (6.3.29)

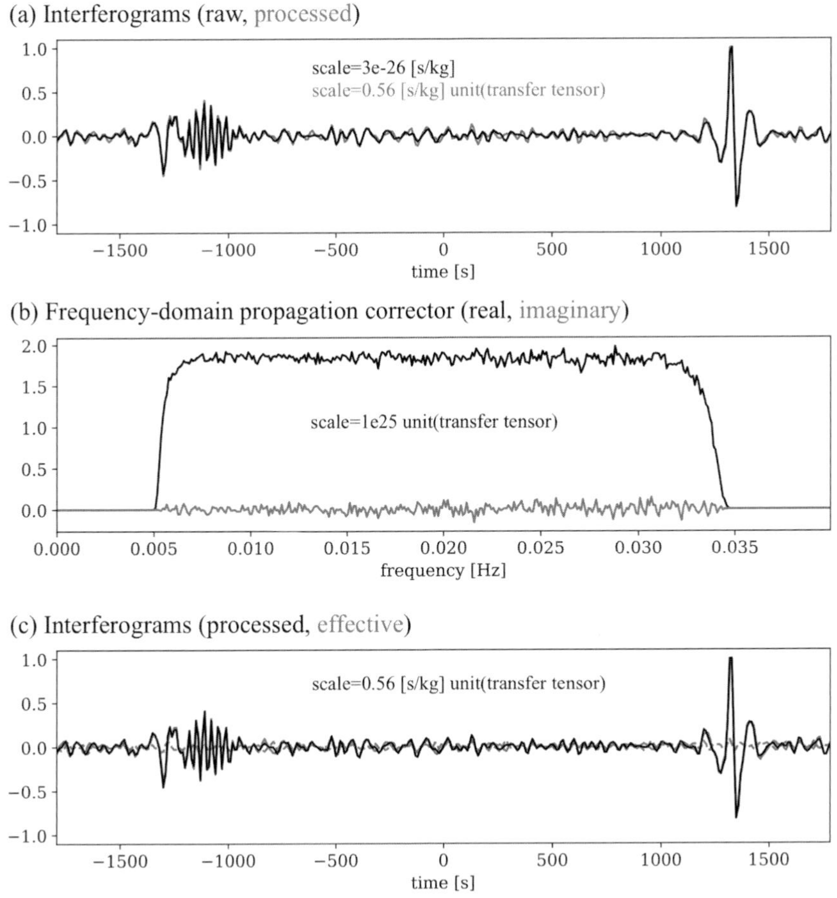

Figure 6.5 Effects of one-bit normalisation for station pair $(\mathbf{x}_{i=0}, \mathbf{x}_{k=1})$ in Figure 6.4. (a) Raw interferogram $\mathcal{I}(\mathbf{x}_i, \mathbf{x}_k, t)$ (black) and the corresponding processed interferogram $\widetilde{\mathcal{I}}(\mathbf{x}_i, \mathbf{x}_k, t)$, resulting from one-bit normalisation of the individual recordings (grey). (b) Real (black) and imaginary (grey) parts of the propagation corrector $\mathfrak{g}_{ik}(\omega)$. (c) Processed interferogram $\widetilde{\mathcal{I}}(\mathbf{x}_i, \mathbf{x}_k, t)$ (black) and effective interferogram $\widehat{\mathcal{I}}(\mathbf{x}_i, \mathbf{x}_k, t)$ (grey). The difference between the two, i.e., the modelling residual $E(\mathbf{x}_i, \mathbf{x}_k, t)$, is plotted as grey dashed curve. Note that some scaling is relative to the unit of the transfer tensor.

is regularized by an adjustable water-level parameter w_c. We take $\varphi(\omega) = \max[\varphi_1, \varphi_2]$ with

$$\varphi_1 = u(\mathbf{x}_k, \omega)u^*(\mathbf{x}_k, \omega), \qquad \varphi_2 = w_c \max[u(\mathbf{x}_k, \omega)u^*(\mathbf{x}_k, \omega)]. \tag{6.3.30}$$

The presence of the water-level term sets a floor to the minimum value of the denominator, and so suppresses large spectral excursions in the transfer function. Results using the transfer-function approach are illustrated in Figure 6.7. The division by $u(\mathbf{x}_k, \omega)u^*(\mathbf{x}_k, \omega)$ has an effect similar to spectral whitening, thereby compensating for the amplitude factor $1/\sqrt{\omega}$ in the acoustic Green's function

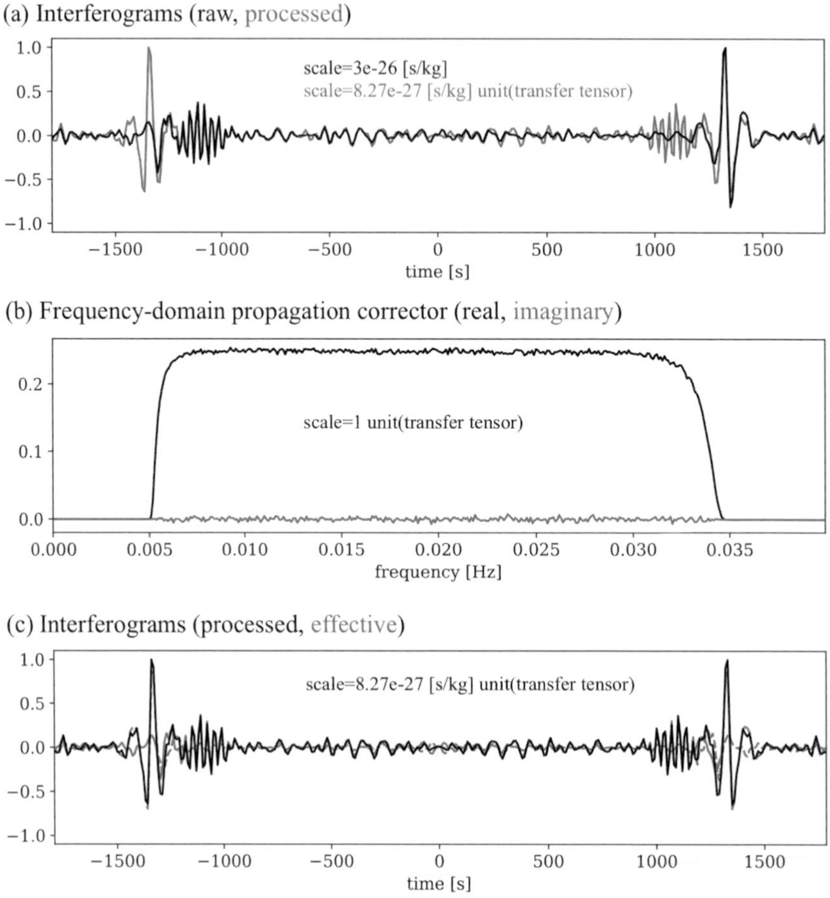

Figure 6.6 Effects of processing as in Figure 6.5 but for the averaging of the causal and anti-causal correlation branch. In panel (c), the difference between processed (black) and effective (grey) interferograms, i.e., the unphysical modelling residual, is large compared to the interferograms themselves.

(6.3.28). As a consequence, higher frequencies are enhanced in the processed and effective interferograms. This effect is also reflected in the propagation corrector in Figure 6.7(b), which has relatively larger amplitudes at higher frequencies.

As we can see from Figures 6.5 to 6.7, the character of effective propagation depends strongly on the nature of processing, and often mimics a heterogeneous medium even starting from a simple homogeneous state. The only case where one may hope – also on a theoretical basis – that the forward modelling residual can be regarded as negligible for practical purposes is one-bit normalisation. The distortions introduced by processing mean that it is necessary to be careful in the way that information is extracted from processed inter-station correlograms. Cross-correlation time shifts (Section 5.1.1), and time–frequency phase misfits (Section 5.3.2) may be used with caution for all processing schemes we have

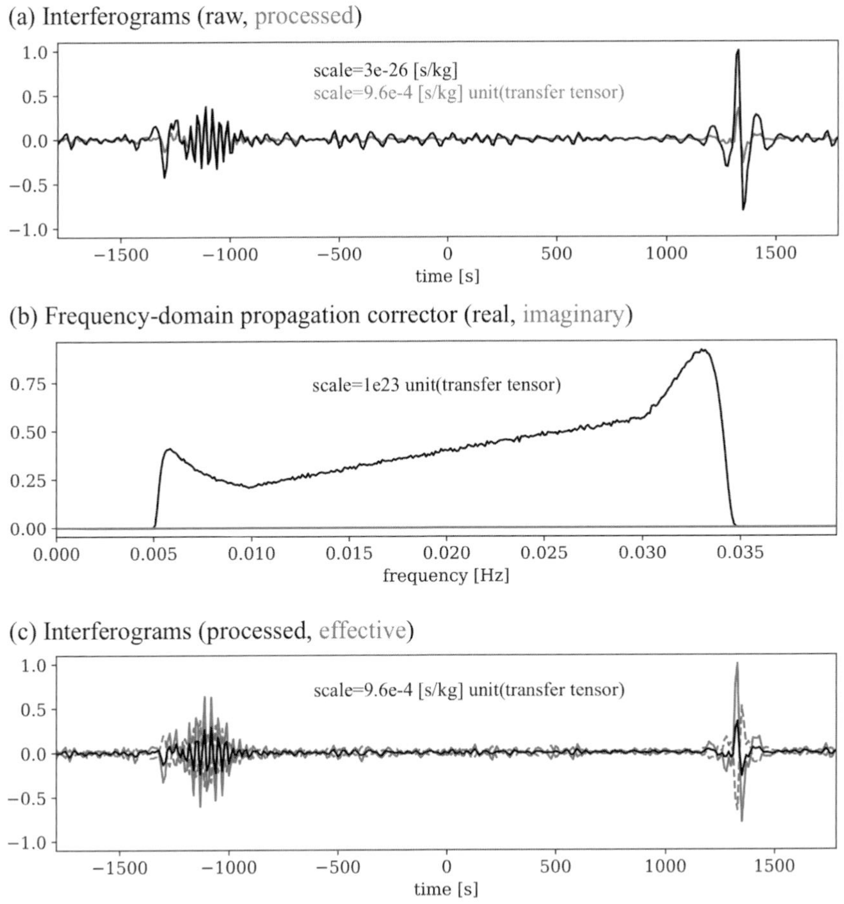

Figure 6.7 Effects of processing as in Figure 6.5 but for the transfer-function approach.

considered. However, it is not generally possible to extract reliable amplitude information.

Combining random noise sources with earthquakes

To concentrate attention on the ambient noise field, numerous processing schemes have been developed to automatically suppress large-amplitude transient signals such as earthquakes. Since we have made no restrictive assumptions on the nature of the wavefield sources $\mathsf{F}^{[n]}(\mathbf{x})$ in (6.3.3), the style of analysis above can be extended to scenarios where earthquakes are present.

For this, we repeat the calculation of inter-station correlation functions from the previous section, the only difference being the addition of 10 artificial earthquakes of variable magnitude to the source distribution. These additional events are indicated by black stars in Figure 6.8, and are modelled as delta functions in space and time. The introduction of these earthquake sources modifies the ensemble correlation between station pairs (Figure 6.9a). Now the behaviour is dominated

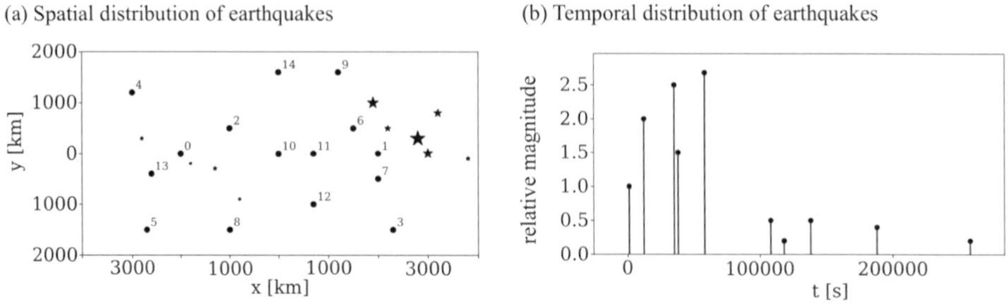

Figure 6.8 Spatial (a) and temporal (b) distribution of artificial earthquakes (marked by stars) used for the ensemble calculation of correlation functions.

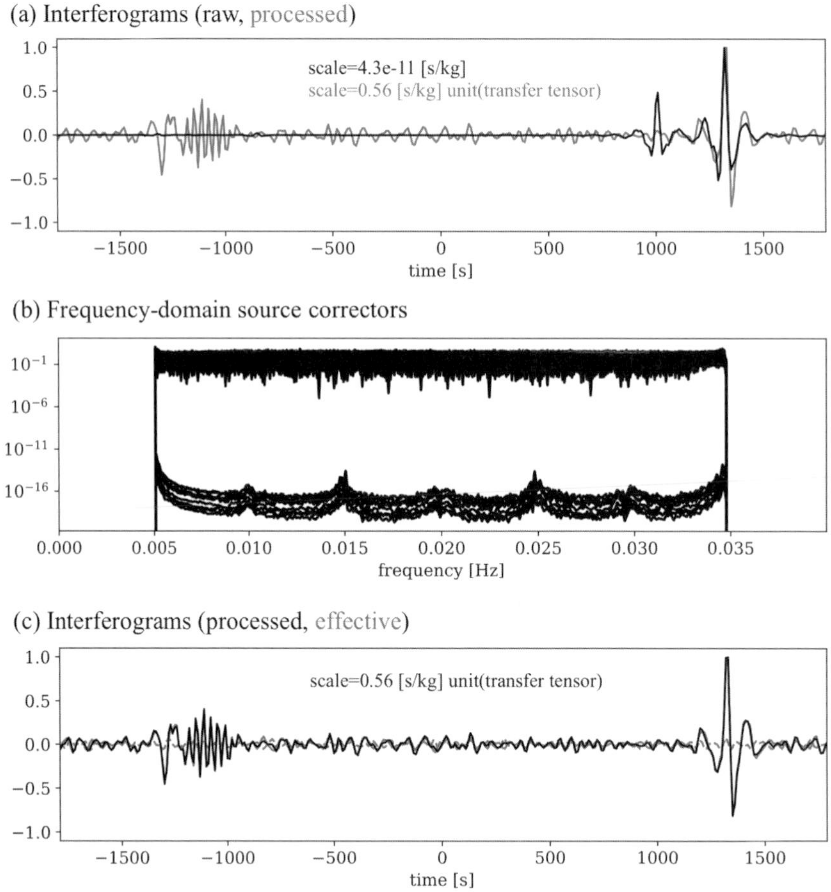

Figure 6.9 Effects of one-bit normalisation for station pair $(\mathbf{x}_{i=0}, \mathbf{x}_{k=1})$ in Figure 6.4, including the artificial earthquakes from Figure 6.8. (a) Raw interferogram $\mathcal{I}(\mathbf{x}_i, \mathbf{x}_k, t)$ (black) and the corresponding processed interferogram $\widetilde{\mathcal{I}}(\mathbf{x}_i, \mathbf{x}_k, t)$, resulting from one-bit normalisation of the individual recordings (grey). (b) Frequency-domain source correctors $\mathfrak{f}^{[n]}(\omega)$ for all 5000 time windows. (c) Processed interferogram $\widetilde{\mathcal{I}}(\mathbf{x}_i, \mathbf{x}_k, t)$ (black) and effective interferogram $\widehat{\mathcal{I}}(\mathbf{x}_i, \mathbf{x}_k, t)$ (grey). The difference between the two, i.e., the modelling residual $\mathrm{E}(\mathbf{x}_i, \mathbf{x}_k, t)$, is plotted as a grey dashed curve.

by the earthquakes and so there is a significant difference from the result in Figure 6.5 where earthquakes are absent and only noise sources were used.

With processing we can attempt to remove large-amplitude transients. We illustrate the effect of using one-bit normalisations in Figure 6.9(b). The one-bit processing primarily affects the source correctors $\mathfrak{f}^{[n]}(\omega)$. While 4990 of the source correctors have values close to 1 for most frequencies, the 10 source correctors that correspond to time windows with earthquakes are orders of magnitude smaller. Thus, the source psd for these 10 windows, $S^{[n]}$, are strongly downweighted by their source correctors, and so the effective source, $\widehat{S}$, is practically free of earthquakes. The interferograms after processing (Figure 6.9c) show strong suppression of the event contribution in the long-term ensemble results with little modelling residual. Other processing schemes to remove large-amplitude transients, such as normalisation by the maximum amplitude or rms clipping, function in a similar way.

In summary, the generalised interferometry scheme provides a way of exploiting a wide range of scenarios with both ambient and transient sources in a consistent manner. It provides the basis for effective modelling of inter-station interferograms without the need for too complex calculations. This opens the way for inversion of the interferometric field for properties of both sources and structure, as will be developed further in Chapter 19.

Appendix: Amplitude Effects for Generalised Rays

6.A Kinematic and Dynamic Analogues

An individual generalised ray path within a multi-layered medium can be represented by a code indicating the nature of the ray. The layer number and wave type are indicated by assigning an ordered pair of integers to each layer $\{C_j, i_j\}$, where i_j is the layer number and C_j indicates the wave type in that layer ($C_j = 1$, *P* waves; $C_j = 2$, *S* waves). For n ray segments there will be n ordered pairs $\{C_j, i_j\}$. Rays which do not include conversions of wave type are completely described by a *ray symbol* composed of the layer indices $\{i_j\}$. The properties of such ray codes have been extensively studied by eastern European seismologists, and a convenient summary of results and algorithms is presented by Hron (1972).

When we consider a class of rays with permutations of the same ray codes we may divide these into:

kinematic groups, for which the phase term τ_I will be the same; and

dynamic groups, which form subclasses of the kinematic groups and have the same products of interface coefficients $g_I(p)$.

For generalised rays for which all legs are in a single wave type it is possible to enumerate all the possible kinematic and dynamic groups. For a surface source and receiver, there will be an even number of ray segments. For a ray without conversions, the time characteristics τ_I can be described by the set

$$\{n_1, n_2, \ldots\ldots, n_J\}, \quad J \geq 2, \tag{6.A.1}$$

where n_j is half the number of segments in the jth layer since each downgoing leg is matched by an upgoing. All ray numbers of a kinematic group will share the same code (6.A.1) and the number of members can be determined by combinatorial considerations

(Hron, 1972). The number of rays in a kinematic group is given by a product of combinatorial terms

$$N_k(n_1, n_2, ..., n_J) = \prod_{j=1}^{J-1} \frac{(n_j + n_{j+1} - 1)!}{n_{j+1}!(n_j - 1)!}. \tag{6.A.2}$$

The members of a dynamic group can be specified by the numbers of reflections from interfaces. We define m_j as the number of reflections from the jth interface when the ray is in the jth layer. The set of $2j - 1$ integers

$$\{n_1, n_2, ..., n_j; m_1, m_2, ..., m_{j-1}\}, \tag{6.A.3}$$

then completely describes the amplitude function $g_I(p)$ for all the members of the same dynamic group, since $m_j \equiv n_j$. The number of members in each dynamic group is

$$N_{dk}(n_1, ..., n_j; m_1, ..., m_{j-1}) = \prod_{j=1}^{j-1} \left[\frac{n_j!}{m_j!(n_j - m_j)!}\right] \left[\frac{(n_{j+1} - 1)!}{(n_j - m_j - 1)!(n_j - n_{j+1} + m_j)!}\right]. \tag{6.A.4}$$

When we consider the properties of the sum of generalised rays we only need to take one ray from each dynamic group and can then use the multiplicity factor N_{dk} to account for all the other rays.

We may thus organise a generalised ray sum such as to exploit the benefits of the kinematic and dynamic groupings by writing the slowness integral as

$$\int dp\, p \sum_k \left\{\sum_d N_{dk} g_d(p) g_d^*(p)\right\} e^{i\omega[\tau_k(p) + p\Delta]}. \tag{6.A.5}$$

The frequency dependent portions are then the same for each kinematic group k, and the inner sum over dynamic groups accounts for different reflection processes with the same phase delays.

The concepts of dynamic and kinematic groups are just as useful for rays with converted legs, but the combinatorial mathematics becomes very difficult for more than a single conversion leg (Hron, 1972; Vered & BenMenahem, 1974).

7

Correlations and Ambient Noise

In this chapter we examine the way in which information on the structure of the Earth can be extracted from the ambient seismic wavefield by exploiting its correlation properties. We first consider the nature of this ambient field, which was formerly regarded as so much unfortunate "noise" that got in the way of recording earthquake signals. Indeed in the days of photographic recording separate seismometer systems were developed to look at the seismic response above and below the main microseismic peak. The advent of broadband sensors and digital recording means that it is now possible to record the seismic wavefield over a broad frequency range, and then retrospectively extract different aspects of the behaviour.

We then consider the ways in which the noise field can be exploited by long-term stacking of the auto-correlations of records at pairs of stations, and the classes of data processing that are needed for successful results. Most work exploiting the ambient wavefield concentrates on fundamental mode surface waves, and such methods have allowed a considerable extension of seismic analysis that is not dependent on the specific location of earthquakes. In certain circumstances it is also possible to extract body-wave signals.

7.1 Nature of Ambient Noise Field

The surface of the Earth is active across a wide sweep of frequencies (Figure 7.1). At very low frequencies (up to 20 mHz), *seismic hum* is induced by atmospheric disturbances and oceanic gravity waves (Tanimoto & Um, 1999; Webb, 2008). Such signals show stationary stochastic properties on timescales of several hours and can be analysed in terms of continuous excitation of the free oscillations of the Earth. The amplitude of seismic hum tends to increase with decreasing frequency (Figure 7.1). At higher frequencies we encounter the larger-amplitude microseismic peaks. The primary microseisms occupy the band from 0.02Hz to 0.1 Hz, and the secondary microseisms extend up to around 1 Hz. Above 0.5 Hz much of the ambient seismic disturbance is associated with human activities.

The level of noise is normally high on oceanic islands, and this defines the locus

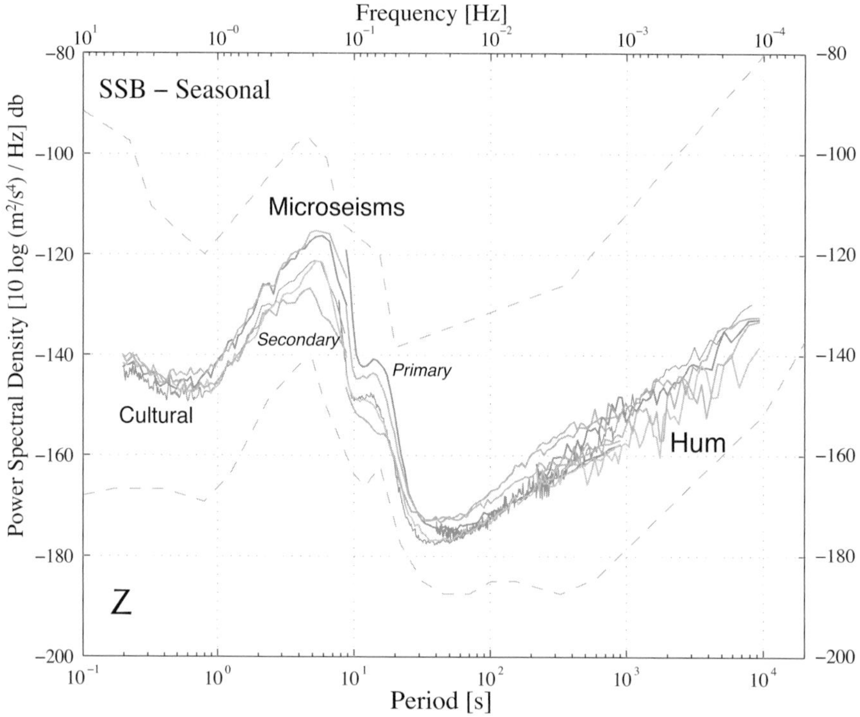

Figure 7.1 Seasonal variation of vertical-component noise at station SSB in France, showing a typical continental pattern as a function of frequency, compared with representative high and low noise models indicated by dashed lines (Peterson, 1993). The various classes of noise sources are indicated. [Noise data courtesy of GEOSCOPE.]

of the high noise model shown in Figure 7.1. In contrast, the lowest noise levels tend to be reached in continental interiors away from human settlement.

The excitation mechanism of the primary microseisms comes from the direct coupling of ocean swell and the sea floor, associated with topography. This coupling is strongest in the transitional zone between the deep ocean and shallow waters on the continental shelf, and so tends to be concentrated around coastlines. The amplitude of excitation is roughly proportional to the ocean wave height. The much larger secondary microseisms arise from the interaction of strong ocean wave systems (Longuet-Higgins, 1950; Hasselman, 1963) that give rise to locally stationary disturbances equivalent to a vertical force system proportional to the square of ocean wave height, enhanced by seismic resonance (Hudson & Douglas, 1975). Most of the secondary microseisms are produced in deep-water situations, but some coastal sources can occur when storm systems interact near land or where reflections from the coastline impinge on ocean waves continuing to propagate towards the shore.

Further contributions to the seismic noise spectrum come from microbaroms generated by atmospheric disturbances. The frequency distribution is similar to the microseisms, but amplitudes are generally much lower. However, strong effects can be produced when intense depressions cross continental areas, which can provide effective sources in the immediate vicinity of seismic stations.

Much of the daily variation in noise activity comes from the residuum of earthquake signals. The late coda of shallow earthquakes consists of multiply scattered fundamental mode surface waves whose randomised propagation directions are helpful for correlation studies. Both seismic hum and such surface-wave coda have relatively large horizontal slowness and can be exploited in global and regional studies. Mixed in with the strong surface waves are much weaker body waves with smaller horizontal slowness, and these can only be extracted by analysing the noise field over long time periods (Boué et al., 2013; Nishida, 2013). Deep earthquakes produce much less in the way of surface waves, and for many hours after the larger events multiply reflected phases traverse the Earth with steep paths, so that there is a change in the character of the correlation field (Boué et al., 2014; Poli et al., 2017; Pham et al., 2018).

Although microseisms are most evident in the form of surface waves, body-wave contributions also exist. The horizontal interactions associated with the primary microseisms produce *P* waves that can travel to regional distances. Whereas, the vertical excitation of the secondary microseisms produces steep travelling *P* waves that can be detected at teleseismic distances.

7.2 Exploitation of Ambient Noise Correlation

The central goal of interferometry exploiting ambient noise is the retrieval of coherent signal from the nearly random variations of the continuous ground motion at the Earth's surface. The success of such procedures based on the correlation of the noise field at multiple pairs of stations has had many applications, ranging from the imaging of noise sources, to seismic tomography and time-lapse monitoring.

A major motivation for the exploitation of correlations for structural studies comes from the observation that the cross-correlation of a diffuse wavefield recorded at a pair of receivers approximates an empirical Green's function, as has been demonstrated in laboratory experiments (Malcolm et al., 2004). A variety of theoretical justifications for Green's function recovery have been proposed for models with variable degrees of approximation to the true physics of wave propagation in the complex Earth. The various approaches depend on different but related assumptions such as wavefield equipartitioning or a homogeneous distribution of noise sources. Failure to meet these conditions for seismic observations may lead to biases in travel times, amplitudes, or waveforms in general, thereby limiting the accuracy of the method.

For homogeneous closed systems, with a finite volume, Lobkis & Weaver (2001) demonstrated that the inter-station correlation of a wavefield with equipartitioned modes is proportional to the inter-station Green's function. Equipartitioning requires that all propagation modes are equally strong and statistically uncorrelated. Unfortunately, although such mode arguments are theoretically appealing, they have limited applicability to the Earth because the dominant variation of structure with depth introduces complexity in eigenfunction shape, so that the square of an eigenfunction generated by correlation differs from the eigenfunction itself (see Section 6.1.1). In addition, heterogeneously distributed sources – in the form of noise sources, earthquakes, or scatterers – prevent the ambient field from being equipartitioned across the complete seismically observable frequency band (see, e.g., Nishida, 2013).

As we have seen in (6.2.1) we can represent the combination of the Green's functions between two stations (acting as virtual sources and receivers) in terms of a surface integral involving the correlation of displacement and traction terms, which is directly derived from representation theorems (Section 3.2.1). The derivation depends on perfect elasticity, i.e., the absence of attenuation. Considerable efforts have been made to recast the representation theorem results in a form that is more directly related to correlations of displacements (e.g., Wapenaar 2004; Wapenaar & Fokkema, 2006). The various approximations depend on a homogeneous distribution of uncorrelated sources on a closed surface, and commonly on a rather smooth Earth structure. Examination of the integral terms shows that the coherent part emerges from very subtle interference between the contributions from the surrounding surface, since the propagation effects from nearby segments on the surface are almost identical. Only in a stationary phase region nearly aligned with the path between the pair of stations is there constructive interference (Snieder, 2004; cf. Section 6.1), the rest of the surface coverage is needed to force destructive interference away from the path.

The frequency-dependent arrival times of fundamental mode surface waves extracted from inter-station correlations are empirically found to be rather robust, but other components of the wavefield are less well recovered. As we have seen in Section 6.1 the nature of the correlation field depends strongly on the horizontal slowness content of the noise field and its azimuthal distribution. Inadequate azimuthal coverage can contribute to biased estimates of travel times and, particularly amplitude, errors. Excitation of higher mode surface waves tends to be patchy, with strong excitation occurring at limited locations along coastlines (e.g., Gal et al, 2017). Body-wave arrivals tend to be rather weak, unless there is strong heterogeneity in the neighbourhood of the stationary phase zones that induces surface wave to body wave scattering. If there are residual contributions from earthquake activity, a variety of seismic phases can be found in the correlograms that have no counterpart in the direct wavefield; such arrivals have often been termed 'spurious', but actually are a direct product of regular propagation (Section 8.1).

Data processing

Many different processing schemes have been developed to produce robust results from long-term stacks of inter-station correlograms. Major efforts have been made to reduce the 'pollution' from earthquake signals and to attempt to compensate for the presence of spatially heterogeneous, non-stationary sources. For many parts of the world there is strong seasonality in the distribution of noise sources, and so the best results are obtained from long durations of continuous seismic records, for which stacking of correlations for day segments (or similar) can even out the pattern of sources and produce a more homogeneous effective noise field.

In Section 6.3.2 we have introduced a number of processing schemes in the context of their influence on an artificial noise field. We considered one-bit normalisation, the averaging of causal and anti-causal correlation branches, spectral whitening, the use of transfer functions rather than correlations, and phase-weighted stacks. Many other approaches have been tried and can have merits in particular circumstances. Bensen et al. (2007) provide a practical guide to the extraction of surface-wave dispersion from correlograms, particularly for Rayleigh waves. Further discussion of pre- and post-processing is provided by Ritzwoller & Feng (2019), who present a range of processing procedures and a useful discussion of the reliability of the estimate of the empirical Green's function between the stations being correlated.

Much processing is directed at minimising the effect of earthquakes and persistent sources of noise, as well as achieving the best possible azimuthal content in the ambient noise field. When attention is concentrated on fundamental mode surface waves, the size of earlier arrivals can be used as a measure of the quality of the recovered signal since it should be suppressed by effective processing. The span of time used for correlation should be long enough to allow capture of multiple samples of the propagation phenomena being studied, so that each interval can make a contribution to the final stack. With modest time intervals, segments contaminated by large events can be dropped. Otherwise, time-dependent weighting needs to be employed to reduce the influence of such segments. One-bit normalisation removes such effects on a continuous basis, but tends to need longer duration of continuous records to produce a stable result. Such normalisation also does not interact well with efforts to broaden the recovered frequency spectrum. An alternative mode of suppression is to use running absolute-mean weighting (Ritzwoller & Feng, 2019) so that the weight $W(t_n)$ for the nth time sample is given by

$$W(t_n) = \frac{1}{N+1} \sum_{j=n-N/2}^{n+N/2} |u(t_j)|. \tag{7.2.1}$$

The time interval used for this running mean (N samples) should be significantly longer than the maximum period employed.

As we have noted above, the correlation process produces the square of the original spectrum of the noise, and thus emphasises the microseismic peak. Spectral whitening procedures can be employed to broaden the available bandwidth before correlation, but need to be used with caution. The transfer-function approach introduced by Saygin & Kennett (2010, 2012) also achieves spectral broadening by cancelling out the noise spectrum contributions at the two stations.

As experience with ambient noise analysis has grown it has become clear that optimal processing depends on the nature of a specific data set and its location relative to noise sources, as well as the wave type that is to be extracted. In most applications, a quantitative measure for the quality of a processing sequence is difficult to obtain because there is no specific reference, since the propagation characteristics between all receiver pairs are not well known. In consequence, the choices made in processing have a significant subjective component, and tend to be guided by the desire to obtain stable measurements, with the maximisation of some measure of signal-to-noise ratio or cross-correlation symmetry. Differences in processing can lead to significant differences in the correlation functions (Bensen et al., 2007). These differences leave an unavoidable imprint on the sensitivity to Earth structure and noise sources (e.g., Fichtner, 2014).

7.3 Extraction of Surface Waves

Surface waves are the most easily retrieved components of the seismic field from noise correlation, because of their large amplitude compared to other components of the wavefield when excited by shallow sources. Information can readily be obtained on the paths between specific seismic stations, without the need for alignment with the great circle to a specific source as in the conventional two-station surface-wave analysis (Section 5.1.3). With a network of N three-component stations, a suite of $\frac{1}{2}\mathrm{N}(\mathrm{N}-1)$ inter-station paths can be built. With a suitable network geometry, this dense set of paths can provide a good basis for tomography exploiting the dispersion characteristics of fundamental mode surface waves. Surface-wave tomography exploiting ambient noise with networks of many sizes has been applied worldwide at scales ranging from shallow layers for engineering purposes to lithospheric imaging.

The use of the ambient noise field means that large sets of local measurements can be produced, which can achieve exceptional resolution with a significantly shorter span of data than would be needed for direct use of earthquake sources. Even where earthquake results are available, ambient noise correlation can provide an extension of dispersion information to shorter periods.

In Figure 7.2 we illustrate the TASMAL deployment in Australia from 2003 to 2005, with 20 portable broadband stations in a configuration designed to bracket the Tasman Line that represents the transition from Precambrian to Phanerozoic crust. The 190 possible inter-station paths provide good coverage of the area and with three-component recording, in principle, nine cross-components can be extracted via correlation. This means that in addition to vertical–vertical (Z–Z)

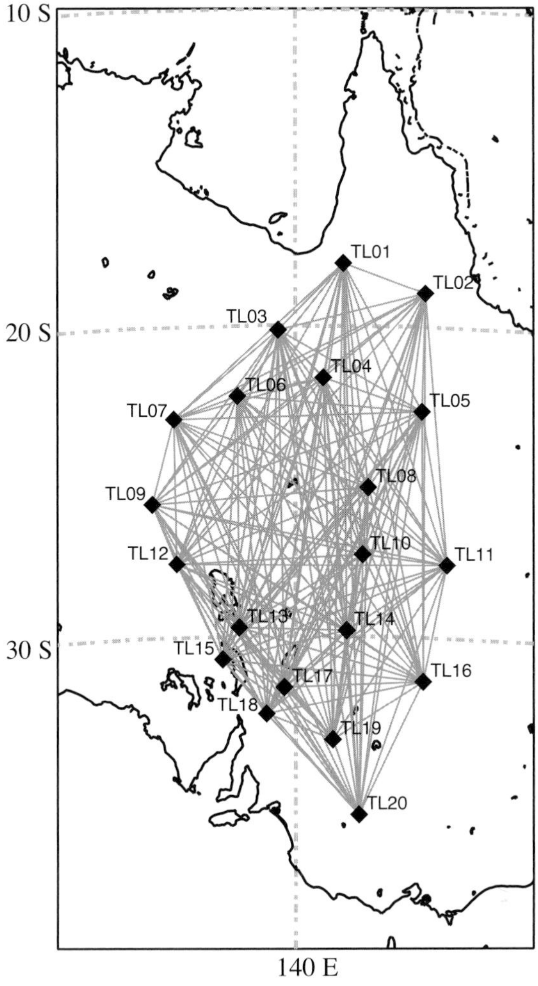

Figure 7.2 Configuration of TASMAL deployment in Australia, and the network of interstation paths.

correlations dominated by fundamental mode Rayleigh waves, the horizontal components can also be exploited.

In Figures 7.3 and 7.4 we show record sections of stacked correlograms for the horizontal components between all the station pairs in the TASMAL deployment as a function of inter-station separation. These results illustrate some of the practical difficulties that can be encountered in exploiting ambient noise.

The most pronounced features in the record sections are the fundamental mode surface waves that are much stronger for positive than negative time. This asymmetry is the result of stronger noise sources to the east, e.g., in the Tasman Sea, compared with the west. The group of stations span a region with notable contrasts in crustal structure, and the paths between stations cross this region

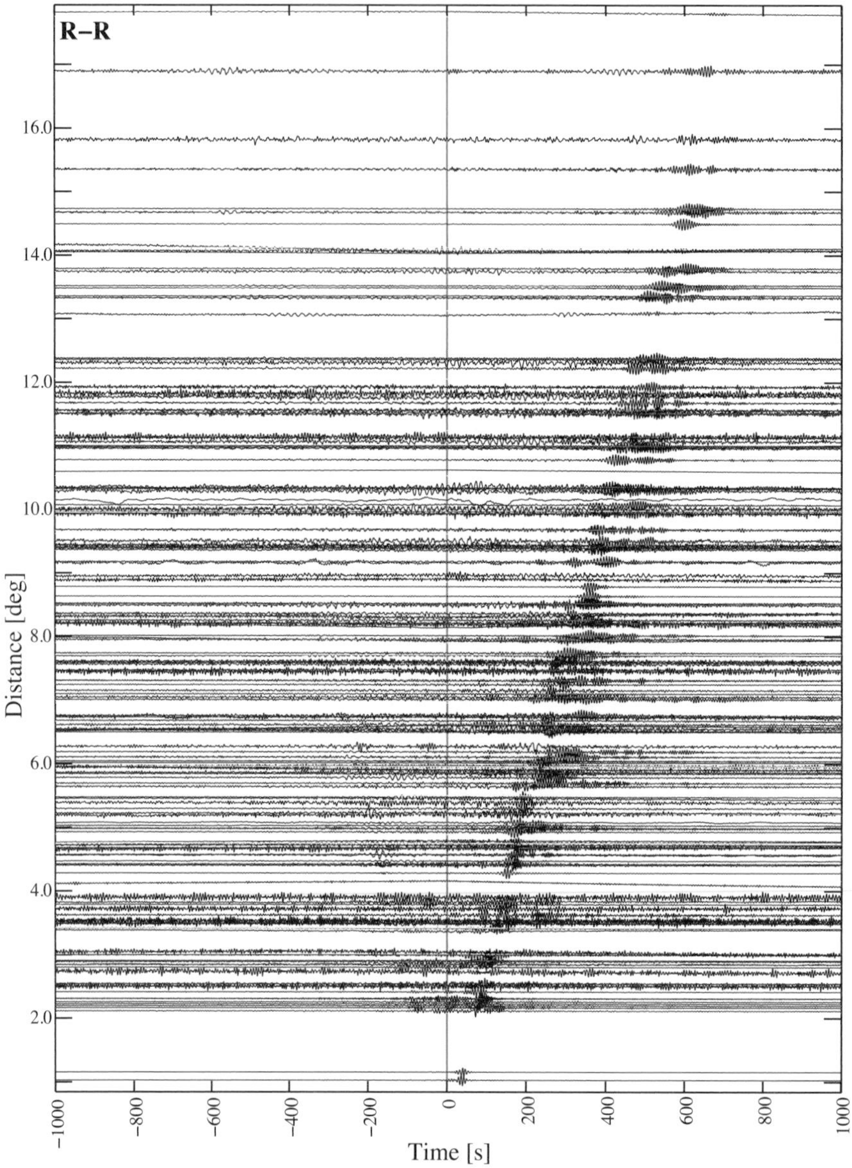

Figure 7.3 Stacked correlation results from all stations in the TASMAL deployment from radial-component seismograms. Positive times correspond to propagation from the east to the west. [Courtesy of E. Saygin.]

in many directions. In consequence there is not a simple moveout for the surface-wave traces with distance, and indeed there are suggestions of multiple arrivals. The differences in the travel times of the surface waves as a function of frequency can be exploited in a tomographic inversion to map out contrasts in structure. The correlation of radial component pairs (R–R) shows clear Rayleigh

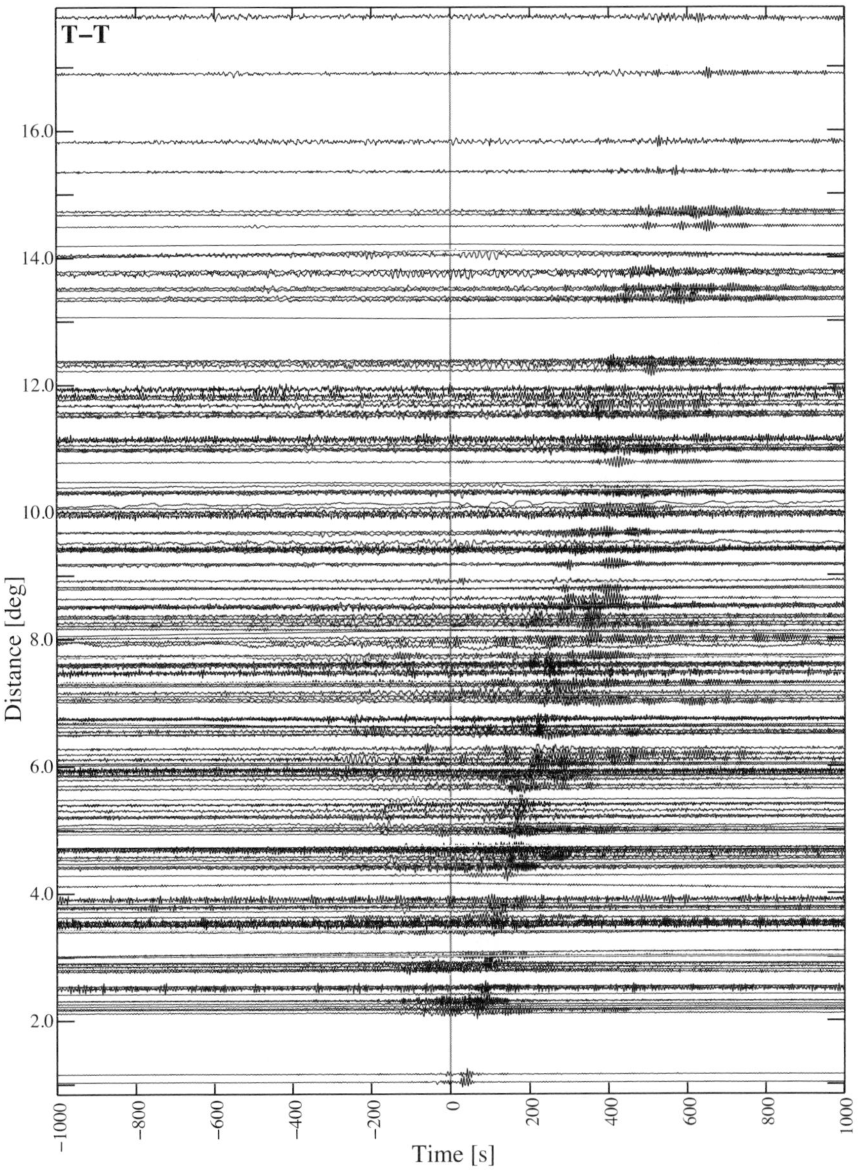

Figure 7.4 Stacked correlation results from all stations in the TASMAL deployment from tangential-component seismograms. Positive times correspond to propagation from the east to the west. [Courtesy of E. Saygin.]

waves. Whereas, the tangential component pairs (T–T) show earlier arriving Love wave energy, but with some later influence from Rayleigh waves. This apparent contamination is due to the strong lateral heterogeneity in the region traversed by the set of paths.

The record sections in Figures 7.3 and 7.4 were generated by rotating the three-component seismograms at the two stations to lie along and perpendicular to the great circle between the station pair, and then constructing the stacked cross-correlograms. An alternative is to construct the full nine-component tensor of cross-component correlograms from the original three-component seismograms followed by tensor rotation (Shapiro, 2019). The tensorial approach works well for limited separation between stations. When when the stations are far apart, the varying azimuth of the great circle should be taken into account.

From the stacked correlograms, phase and group speeds can be measured as a function of frequency to characterise the surface-wave dispersion along the path between the stations. As in Section 5.1.3 such analysis can be achieved by a combination of frequency-domain and time-domain filtering to isolate specific portions of the response. For the correlograms we are looking at the equivalent of a single station dispersion from a seismic event, since the two stations act as a virtual source–receiver pair.

Using a sequence of filter bands with centre frequency ω_0, the group travel time $t_g(\omega_0)$ can be measured from the maximum of the envelope of the filtered trace in each band (e.g., Levshin & Ritzwoller, 2001). The group speed is then $U(\omega_0) = t_g(\omega_0)/\Delta_{12}$, where Δ_{12} is the distance between the stations being correlated. The phase of the correlation function depends on the phase speed $c(\omega_0)$,

$$\phi(t, \omega_0) = \omega_o[c(\omega_0)\Delta_{12} - t] + \phi_0 + 2n\pi, \tag{7.3.1}$$

where n is an integer. The conventional assumption for the initial phase ϕ_0 is $\pi/4$, based on the asymptotic properties of the travelling waves. However, depending on the actual character of the noise sources there might be some perturbation to this value. From (7.3.1) the phase velocity $c(\omega_0)$ can be determined once the appropriate value of n is resolved. Commonly this involves the use of reference dispersion curves and iteration (see, e.g. Lin et al., 2008).

As in the approach described in Section 5.1.3 this process is easier for the lowest frequencies, and once n is found it is valid for all frequencies. The frequency dispersion of the apparent group velocity between the pair of stations is somewhat easier to estimate than the corresponding phase dispersion, but is liable to suffer from more scatter (Ritzwoller & Feng, 2019).

With dispersion results for a network of paths, the interpretation in terms of structure is normally based on ray approximations (e.g., Woodhouse, 1974; Wang & Dahlen, 1995) with the ray path controlled by the phase speed distribution. The first step is to construct phase or group wavespeed maps at a suite of frequencies by combining the information from the network of inter-station paths. This nonlinear tomographic inversion requires iterative updating of the paths, particularly when paths extend beyond 10° station separation.

Where data from temporary networks are used in ambient noise studies, areal coverage can be extended by correlation with permanent stations that are in operation over the period spanned by different experiments. For example, Saygin & Kennett (2012) combine results from more than 250 portable stations across

Australia with stations from the Australian and global networks. They used passage times extracted from dispersion analysis of the group wavespeed in a tomographic inversion spanning the whole Australian continent with a cellular model representation at each frequency with a $1.4^\circ \times 1.4^\circ$ grid. Saygin & Kennett (2012) used the inversion approach of Rawlinson & Sambridge (2004) with wavefronts iteratively tracked through the model with the Fast Marching method. They worked directly with group slowness, and point out that comparable results to the use of phase slowness will be obtained if the gradients of the two slownesses have similar geographic patterns. The results of Young et al. (2011) show such a correspondence for a study of southeastern Australia. Further, the results of the Australia-wide inversion using group slowness match well with known geological features, such as the patterns of sedimentary basins that tie in well with the distribution of lowered velocities at shorter periods.

For the moderate frequencies usually employed in ambient noise studies, the zone of structural influence is concentrated around the propagation path. In a ray-based approach an effective procedure is to allow a spatial extent of approximately 1/6 of the first Fresnel Zone for the appropriate period (Yoshizawa & Kennett, 2002b).

As noted in Section 3.5.1, Woodhouse (1974) has demonstrated that for slowly varying media the local dispersion corresponds to the structure directly beneath a point. However, in zones of rapid change this correspondence can not be assured (Yanovskaya, 1984). Normally it is possible to use the set of dispersion maps at different frequencies and to extract a local dispersion relation for a specified point. The local dispersion can then be inverted to extract a 1-D velocity distribution associated with the point, normally in terms of the shear wavespeed $\beta(z)$ as a function of depth z. Saygin & Kennett (2012) use a five-layer representation of crustal structure and undertake a fully nonlinear inversion using the Neighbourhood algorithm (Sambridge, 1999; Wathelet, 2008) that allows an exploration of the model space and hence a measure of the model uncertainty (cf. §SWII:28.2.2). A final 3-D shear wavespeed model can then be constructed by interpolating from the point wavespeed estimates.

Ambient noise tomography has provided results on scales from the global (e.g., Nishida et al., 2009) to the local, but most emphasis has been on lithospheric structure with exploitation of the microseism band. Such ambient noise tomographic results have yielded spectacular images of crustal structure, particularly where relatively dense networks of stations are available, such as the US Array exploited by Shen & Ritzwoller (2016).

Even where the conditions for inversion to local structure are not satisfied, the distribution of group speed as a function of frequency can provide useful geological controls. Saygin et al. (2013) exploit a temporary network of stations in northern Australia that extends from the Proterozoic craton of the Mt. Isa block into the deep sedimentary basin to the east with a very strong contrast in the top few kilometres. Group velocity dispersion behaviour is reasonable for north–south paths that stay in the individual geological domains. But east–west-oriented paths that cross the geological boundary have dispersion characteristics that cannot be represented by a 1-D shear wavespeed model. Nevertheless, the group wavespeed results can be exploited to gain significant information on structure at depth.

7.3.1 Local Properties of the Correlation Field

In the same way as the local wavefield can be expressed as a superposition of plane waves via Fourier analysis, a similar approach can be applied to the cross-correlation field in the neighbourhood of a specific location. We can envisage cross-correlations between the noise field at a reference location and a continuous distribution of stations. Consider the normalised correlation of vertical components $\mathcal{C}_{ZZ}$, then in terms of cartesian coordinates (x, y):

$$\mathcal{C}_{ZZ}(x, y, \omega) = \int dk_x \int dk_y\, \bar{\mathcal{C}}_{ZZ}(k_x, k_y, \omega) \exp[i(k_x x + k_y y)], \tag{7.3.2}$$

or in polar coordinates (X, θ) :

$$\mathcal{C}_{ZZ}(X, \theta, \omega) = \sum_{m=-\infty}^{\infty} \int dk\, \tilde{\mathcal{C}}_{ZZ}(k, m, \omega) J_m(kX) \exp(im\theta), \tag{7.3.3}$$

where $J_m(kX)$ is the Bessel function of the first kind.

If the local field were to be composed of waves with a single propagation velocity $c(\omega)$ with a uniform distribution in azimuth then

$$\mathcal{C}_{ZZ}(X, \theta, \omega) = J_0(k_c X), \tag{7.3.4}$$

with the wavenumber $k_c = \omega/c(\omega)$. This relation lies at the core of the SPAC method introduced by Aki (1957). The original SPAC approach used spatial auto-correlation with the assumption that the local microtremor field is stationary in space and time. As the method has developed, averaging of the coherency across an array of stations over a period of time has been used to try to secure uniformity in azimuthal coverage.

Asten (2006) provides a useful summary of applications in which surface wave dispersion is extracted by fitting the coherency distribution as a function of station separation X at a suite of frequencies, with suitably scaled Bessel functions, that depend on the phase-velocity distribution $c(\omega)$. Asten suggests that a suitable description is ‘spatially averaged coherence’, which is compatible with the original SPAC acronym, but better reflects modern practice.

A wide range of array designs have been used in local studies, frequently with configurations dictated by local logistics, particularly in engineering applications. The potential recovery of a single plane wave for a particular geometry can be assessed by the array response function $S(\Delta\mathbf{s}, \omega)$ (§ SWII:23.1) for the deviations

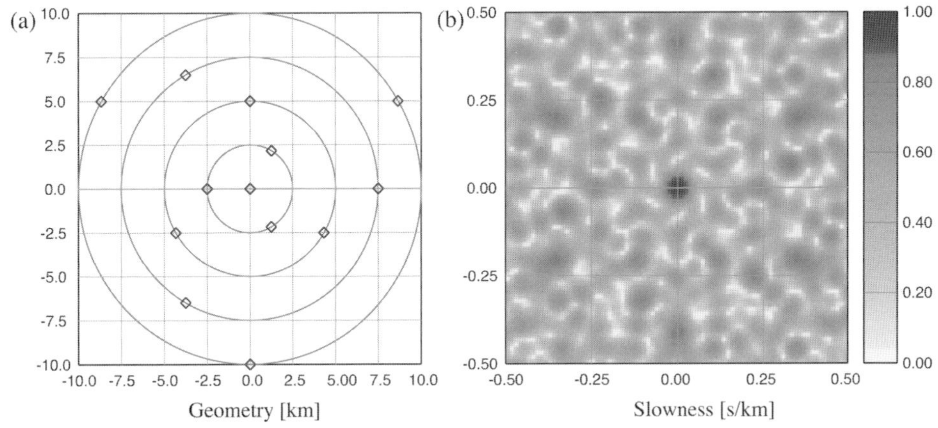

Figure 7.5 (a) Configuration of a 13-element spiral arm array suitable for local correlation studies, and (b) the array response in slowness space.

$\Delta\mathbf{s}$ in slowness from the target value. For stations at locations $\{\mathbf{x}_j\}$ the array response is

$$S(\Delta\mathbf{s}, \omega) = \sum_{j-1}^{N} \exp\{i\omega[\Delta\mathbf{s}.\mathbf{x}_j]\}\,. \tag{7.3.5}$$

An effective design with a modest number of stations is to use a spiral-arm array (Kennett et al., 2015a), which can be thought of as a thinned set of circles (Figure 7.5). This design provides a high number of station pairs with a wide range of directions and inter-station distances. The array response is highly concentrated considering the limited number of elements. The design is readily scaled, and insensitive to moderate deviations from the strict geometric pattern.

Exploitation of the local coherence has been used in many different ways. A number of authors have explored the close relation between the SPAC approach and the conventional extraction of inter-station surface waves by correlation of ambient noise (e.g., Yokoi & Margaryan, 2008; Tsai & Moschetti, 2009). Ekström et al. (2009) have made a direct application of (7.3.4) to dispersion estimation, making a fit to the zero crossings of the Bessel function for stacked cross-correlations using groups of stations from USArray. An alternative approach has been used by Wang et al. (2019) to extract an estimate of $\tilde{C}_{ZZ}(k, 0, \omega)$ by working with just the radial dependence as a function of inter-station separation X. They project the results into a phase-velocity–frequency (c–ω) presentation, in which modal dispersion is represented directly. With the use of stations extending several hundred kilometres from the reference point they demonstrate the capacity to extract the dispersion of higher mode surface waves, as well as the fundamental mode. This approach requires the assumption of a local 1-D structure extending across the entire domain of stations, but lateral heterogeneity will tend to broaden the dispersion branches and increase the error in dispersion estimates.

7.3.2 Correlations of Correlogram Coda

One of the earliest efforts to extract an empirical Green's function from observed seismograms was made by Campillo & Paul (2003) who worked with the cross-correlation of earthquake coda. The concept has been extended further by Stehly et al. (2008) who have computed the correlation of the coda of cross-correlations. This approach has been termed C3, in contrast to traditional cross-correlations of ambient noise (C1).

The coda wavefield is diffuse and this property carries over into the stacked correlograms, improving the scenario for extraction of inter-station propagation. The C3 approach is built on the assumption that the long-term stacking of correlation functions produces stable coda, with little time variation. With C3, the scenario for signal extraction depends more on the distribution of stations and less on the original noise distribution. This means that the C3 function can be arranged to provide greater symmetry between the two directions of propagation than in simple ambient noise cross-correlation.

The results of C3 correlation for Rayleigh wave recovery tends to be more band-limited than with C1, and longer time spans may be needed to extract effective empirical Green's functions. Nevertheless, Zhang & Yang (2013) suggest that C3 is more suitable for the extraction of attenuation information than C1.

A major merit of the C3 approach is that it can be used even when the pair of stations being used are not operating at the same time, provided that there are other stations whose operational windows overlap those for the two stations. Ma & Beroza (2012) have demonstrated the feasibility of the asynchronous use of C3 with stations from southern California.

7.3.3 Cross-Correlation of Correlograms

An alternative approach to the exploitation of stations operating at different times has been developed by Chen & Saygin (2020). They use the direct correlation of cross-correlation functions between networks to provide estimates of empirical Green's functions. This C2 approach exploits the energetic surface waves in the correlograms, rather than the weaker coda.

The first step is to extract stacked C1 cross-correlations between pairs of stations $(\mathbf{x}_i, \mathbf{x}_j)$, so that the two stations can be regarded as virtual sources contributing to the causal and anti-causal branches:

$$C_1(\mathbf{x}_i, \mathbf{x}_j, t) = u(\mathbf{x}_i, t) \otimes u(\mathbf{x}_j, t), \tag{7.3.6}$$

where $\otimes$ represents the stacked cross-correlation operation.

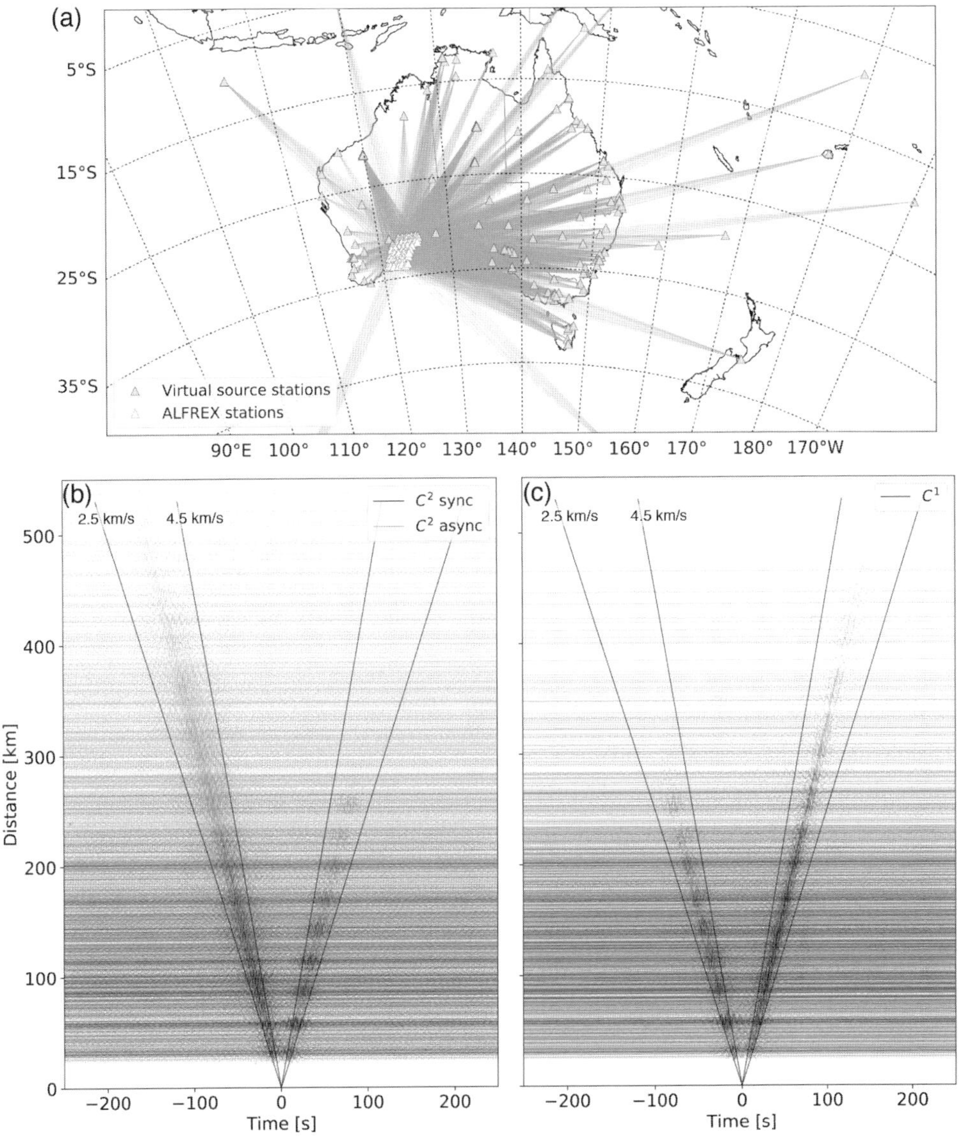

Figure 7.6 Empirical Green's functions (EGFs) for station pairs in the ALFREX deployments of portable seismometers in southwestern Australia. (a) The ray paths between virtual sources (permanent stations) and the ALFREX stations used in calculating the C2 functions. (b) The EGFs between synchronous station pairs using the C2 procedure in black and asynchronous pairs in grey. The waveforms are normalised to unity and bandpass filtered (0.05–0.2 Hz). The guidelines mark moveout velocities of 2.5 km/s and 4.5 km/s, corresponding to the range of group speeds for surface waves in western Australia (Saygin & Kennett, 2012). (c) The EGFs extracted using the C1 approach using synchronous stations. [Courtesy of Y. Chen & E. Saygin.]

The second step extracts C2 empirical Green's functions by stacked cross-correlation of the correlograms corresponding to different virtual sources:

$$G_2(\mathbf{x}_i, \mathbf{x}_j, t) \approx \frac{1}{N_s} \sum_{k=1} N_s C_1(\mathbf{x}_A, \mathbf{s}_k, t) \otimes C_1(\mathbf{x}_B, \mathbf{s}_k, t), \tag{7.3.7}$$

where $\mathbf{x}_A, \mathbf{x}_B$ represent temporary stations that have been correlated with the same set of longer-term stations $\{\mathbf{s}_k\}$.

The stations $\mathbf{x}_A$ and $\mathbf{x}_B$ need not operate at the same time, provided that both can be correlated with the same station set, e.g., from a permanent backbone seismic network. The set of longer-term stations $\{\mathbf{s}_k\}$ can be regarded as a fixed network of virtual sources, and make most contribution when they lie in the stationary phase zone for the path between A and B (as in Figure 6.2).

Chen & Saygin (2020) suggest that a minimum of three months of continuous data are required for the C1 correlations, and that suppression of earthquake contributions is not necessary, since they help to enhance the surface-wave component. The causal and anti-causal parts of the C1 functions are treated separately for the C2 stacked correlation. All contributions form the virtual sources are stacked to produce the C2 estimates of the empirical Green's functions between $\mathbf{x}_A$ and $\mathbf{x}_B$.

Chen & Saygin (2020) have demonstrated the effectiveness of the C2 procedure with application to temporary networks in southern Australia, using the permanent networks in the region to produce a broad distribution of virtual sources. Figure 7.6 shows results from the ALFREX experiment in Western Australia (Sippl et al., 2017a,b), which was conducted over 2 years with sub-arrays of 27 stations in each year linked to a fixed group of 13 stations. Thus, both within and outside the network there are combinations of both synchronous and asynchronous stations. The C2 results link well with the direct C1 estimates across the ALFREX array. The addition of results from stations that were not operating at the same time extends the range of inter-station path configurations, and hence gives enhanced recovery of local structure by combining C1 and C2 results, particularly in the frequency band of the secondary microseisms. The dispersion characteristics from C1 and C2 analysis for common station pairs match well, with differences below the uncertainty level for the C1 dispersion estimates despite the lower signal-to-noise ratio for the C2 results. Chen & Saygin (2020) also demonstrate the successful application of the C2 approach to spatially separated arrays operating at different times.

The key to the extraction of high-quality C2 empirical Green's functions is a good distribution of virtual sources (i.e., long-term stations) relative to the area under investigation; as well as good-quality C1 results. As previously discussed (Section 6.1), a good azimuthal distribution of sources relative to the path between stations is required to secure constructive interferences along the path and suppression of secondary effects. Because the location of the virtual sources are known, the generalised interferometry approach of Fichtner et al. (2017), discussed in Section 6.3 can be employed to compensate for bias introduced in the C2 results from the virtual source distribution.

7.4 Body Waves in the Ambient Field

Although the ambient noise field readily yields fundamental mode surface waves with suitable processing, it has proved more difficult to extract body waves. Most noise sources lie in the surface zone, and so there are limited possibilities within the stationary phase zones aligned with the great-circle path between stations to excite any specific body-wave phase. Scattering effects are also strongest in the shallower parts of the lithosphere, and thus the distribution of secondary sources is also unfavourable for teleseismic paths. In consequence, excitation of body-wave components by ambient noise is weak and considerable stacking is commonly needed to render them visible. Such stacking can be associated with the exploitation of long durations of continuous records (e.g., Nishida, 2013), or with the use of seismic arrays (e.g., Poli et al., 2012a). Indeed, for correlations between well separated locations, there can be considerable advantage in stacking over nearby stations at both ends (*double-beam forming*, Nakata & Nishida, 2019). Despite the difficulties of the relative weakness of body-wave contributions to the ambient noise field, inter-station correlation studies exploiting noise have been carried out on all scales from the local to the global, as discussed below.

In addition to the use of stacked cross-correlation between different stations, auto-correlation at a single station has been used to emphasise steeply travelling waves (e.g., Tibuleac & von Seggern, 2012; Poli et al., 2017). For auto-correlation, the virtual source and receiver are coincident, but the response still involves a range of slownesses (6.1.14), with concentration near vertical incidence. In the presence of strong local scattering, there is a possibility of contamination of the auto-correlogram traces. Arrivals can appear as if they have propagated from depth, at the time for propagation to and back from the scatterer. Such effects can be distinguished by their inconsistent slowness if a local array analysis can be undertaken.

7.4.1 Studies of Local Structure

Local heterogeneity can be favourable for the extraction of body waves, since it induces scattering between surface waves and body waves. For example, in an early study, Roux et al. (2005) worked on data from the neighbourhood of the San Andreas fault near Parkfield, California. With binned stacking they were able to track P-wave refractions from 2 km to 10 km station separation. Nakata et al. (2017) used seismic noise recorded at a dense array in Long Beach, California. They were able to extract coherent diving body waves and use these to image subsurface structure using travel-time tomography.

Clayton (2020) has used stacks of near-offset traces from virtual source gathers of ambient noise, mostly from oceanic microseisms, to create an approximation to local auto-correlation for arrays in California. To enhance reflections from depth, the survey-wide average auto-correlation is subtracted from each trace. The net result is a zero-offset reflection image of the subsurface generated by ambient noise correlations that allow investigation of the middle and lower crust, zones for which traditional seismic reflection methods have penetration problems. The basement interface beneath the San Bernardino Basin in southern California has been imaged with both reflected *P* and reflected *S* waves, extracted from an array of 92 nodes at approximately 300 m spacing, to reveal an apparent fault in the basement.

At an exploration scale, Draganov et al. (2007) were able to retrieve reflected body waves with a recording duration less than 24 h. The recordings were carried out in a desert region with 16 active three-component geophones at 50 m spacing along a line, with multiple stacking of cross-correlograms to produce virtual shot gathers. The likely origin of the seismic noise is from local microtremors. More recently, Zhou & Paulssen (2017) analysed seismic noise data from boreholes at the Groningen gas reservoir in the Netherlands, and were able to extract both *P* and *S* waves.

Shirzad & Shomali (2015) used a 35 km aperture seismic array around Tehran, and demonstrated the extraction of both *P* and *S* body waves from the ambient noise field, in addition to surface waves. Following data preparation using one-bit normalisation, they employed a stacking procedure targeted at the expected arrival times of the seismic phases to enhance the body-wave contributions.

In an area where the Moho is known to be sharp, Tibuleac & von Seggern (2012) exploited the auto-correlation of ambient noise to look for near vertical reflections on a coincident virtual source-receiver configuration. They employed stations in the USArray in the western Great Basin and the Sierra Nevada, and constructed stacked auto-correlations for all three components of continuous ground motion from 2005 to 2008. Tibuleac & von Seggern (2012) worked in the 0.5–1.0 Hz band with automatic gain control applied to enhance the response at later times. Relatively consistent arrivals were found across the Nevada region near 10 s and 17 s reflection time. The earlier time corresponds to near-vertical *PmP* reflections from the Moho and the later, somewhat more variable arrival, to *SmS*. The estimated Moho depths are in good accord with other evidence for the region, and demonstrate the possibility of exploiting noise directly for crustal mapping.

7.4.2 Regional Studies

Where the Moho is sharp, the amplitude of reflected *S* waves can be large particularly near critical reflection for *SmS*. Indeed *SmS* and its multiples play a major role in determining the character of the seismic wavefield at near-regional distances (cf. § SWII:19.1.1). The high amplitudes of these reflected phases helps their recovery from the ambient noise field. Zhan et al. (2010) demonstrated the extraction of *SmS* exploiting networks of stations in South Africa and Canada, in each case using a dense seismic array in association with nearby temporary stations. Their study used spectral whitening to enhance the higher-frequency components and so favour the recognition of body waves. For the stations near

Kimberley in South Africa, Zhan et al. (2010) were able to demonstrate a good tie between the ambient noise results in the 0.5–1 Hz frequency band and array recordings of a local earthquake at slightly greater range. *SmS* could be tracked for inter-station distances of 130–270 km, with the multiple *SmSSmS* most evident around 270–280 km separation.

Similar results have been obtained by Poli et al. (2012a) exploiting the LAPNET experiment with 42 broadband stations across northern Finland, with inter-station separation ranging from 50km to 600 km. In the 0.5–1 Hz frequency band they were able to track *SmS* and its multiples across the full span of distances from the stacked cross-correlation of vertical component records (Z–Z). In addition, the correlation of radial components (R–R) reveals *PmP* and its multiples. The polarisations may appear surprising, but are consistent with horizontal excitation of the ambient noise field, e.g., near the Atlantic coastline of Norway.

The full suite of vertical correlograms from LAPNET were subsequently used to look for reflections from the upper-mantle transition zone (Poli et al., 2012b). With stacking of all suitable station pairs along the predicted trajectory for reflections from the 410 km and 660 km discontinuities, distinct reflection packets *P410P*, *P660P* were extracted. The timing and amplitude behaviour suggest minor regional deviations from the AK135 global model.

Feng et al. (2017) have exploited using correlations of ambient noise at a dense seismic array in the eastern part of the North China craton. They project their results onto two nearly orthogonal profiles. To enhance the *P410P*, *P660P* arrivals they employ distance bins 200 km long incremented by 100 km, with phase-weighted stacking within each bin to enhance the signal-to-noise ratio. The results indicate lateral variations in both the depth and sharpness of the 410 km and 660 km discontinuities, and are reasonably consistent with teleseismic receiver functions results exploiting conversions.

7.4.3 Global Body Wave Propagation

By cross-correlating vertical-component records between stations in LAPNET and the broadband network F-net in Japan, Poli et al. (2015) were able to identify *P*-wave reflections from the core-mantle boundary with stacking over both networks. The reflection points for the *PcP* phase sample a region beneath Siberia. In addition to *PcP*, there are indications of *PdP* reflections from the top of the D″ region at a depth around 2530 km. The results are consistent with earthquake data for a similar path that samples the southern part of the expected reflection zone.

A similar approach has been used by Xia et al. (2016) exploiting over 1000 stations in regional networks in China and 310 portable stations in the western part of South America, using a 4-year span of data. This configuration gives inter-station distances from 145° to 180°. The amplitude of the body-wave phases is enhanced by stacking the results in 0.5° distance bins. For the frequency band 0.02–0.2 Hz, just the use of ambient noise reveals the triplication of *PKP* with all three branches visible (DF, BC and AB).

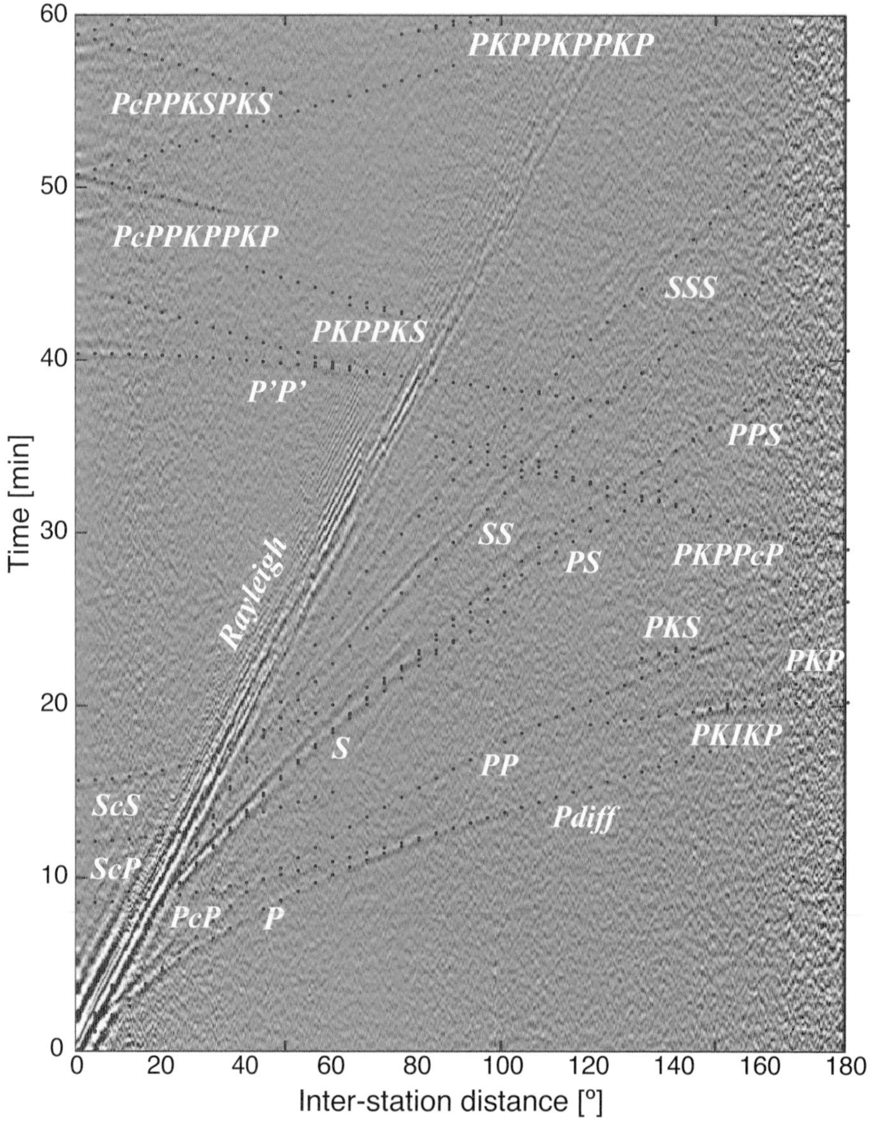

Figure 7.7 Stacked vertical-component cross-correlograms for global stations during 2008, excluding the 30 days with significant earthquake activity. The record section corresponds closely to the expectation for a radial Earth model. [Courtesy of P. Boué.]

At the full global scale, Nishida (2013) has exploited 7 years of continuous broadband records for a global distribution of stations, for the frequency range with strong seismic hum (5–40 mHz). His stacked cross-correlations between stations exclude all intervals with noticeable earthquake activity. Nishida has produced record sections for vertical, radial, and transverse correlations that generally reproduce the major features of the direct wavefield for a shallow source, though

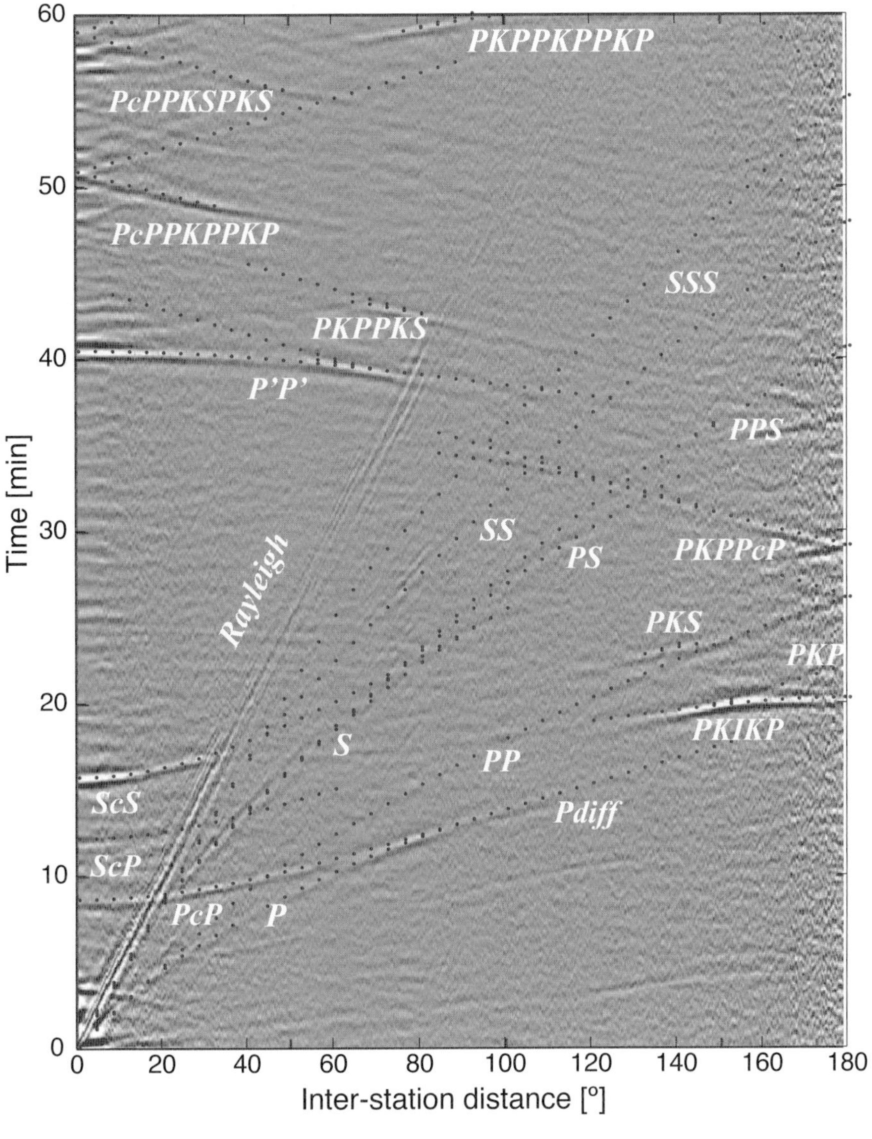

Figure 7.8 Stacked vertical-component cross-correlograms for global stations during 2008, for just the 30 days with significant earthquake activity. The record section now shows a preponderance of reflected phases with steep paths through the Earth. [Courtesy of P. Boué.]

some features associated with long-period *P* waves are more prominent than might be expected at the global scale. Nishida reports strong *PL* and shear-coupled *PL* phases. Such phases are normally fairly large for continental paths and can be enhanced by noise excitation mechanisms such as the shear traction associated with the generation of hum.

Body-wave phases can also be extracted from the ambient field at higher frequencies. Boué et al. (2013, 2014) have exploited one year of data for 2008 at over 300 stations across the globe, and present stacked correlograms for the vertical component with 0.1° bins in the frequency band 0.025–0.1 Hz (Figure 7.7). They also demonstrate the presence of body waves at even higher frequencies (0.1–0.5 Hz). Boué et al. (2014) point out that many of the prominent features in a record section constructed with continuous data come from days with significant earthquake activity. When such days are suppressed, as in Figure 7.7, the character of the global correlation wavefield is similar to that extracted by Nishida (2013).

If just the days with strong earthquake occurrence are used, the character of the record section changes markedly (Figure 7.8). Complex core phases become much stronger, and readily visible. Also a number of features emerge that have no counterpart in the direct wavefield from a shallow source, such as strong apparent *ScS* waves on the vertical component at short distances. These days with strong seismic activity are dominated by coda energy from the earthquakes that is large enough to suppress much of the influence of the ambient noise field.

Lin et al. (2013) have exploited the continuous records from the USArray and the New Zealand national seismic network and demonstrate the extraction of body-wave phases reflected off the outer core (*ScS*), and twice refracted through the inner core ($PKIKP^2$). As in the work of Boué et al. (2014), they come to the conclusion that much of these signals originate from distant earthquakes and emerge due to array interferometry. Undoubtedly, there is a small component of correlation phases corresponding to deep propagation in the ambient noise field, but this is enhanced in the presence of earthquakes.

As discussed in the next chapter, by concentrating attention on the correlation of earthquake coda we can enhance correlation phases corresponding to steep paths within the Earth that sample the deep interior.

8

Coda Correlations

In the previous chapter we have noted how the character of the correlation wavefield is modified on days when significant earthquakes have occurred. Here we look explicitly at the exploitation of the coda of earthquakes to provide a means of examining the properties of the deep Earth. The early parts of the wavefield generated by an earthquake include the large horizontal slownesses of surface waves and the moderate slownesses of body waves propagating in the mantle. As time progresses, the response of the Earth is dominated by steeply travelling waves with small horizontal slowness. Although individually small in amplitude, in combination such waves can reveal complex propagation processes. As discussed in Section 6.1.2, stacked inter-station cross-correlograms extract difference phases. Many of these combinations have travel times that correspond to regular phases in direct excitation by a source. However, other apparent phases emerge that have no simple correspondence to the direct wavefield. Although often termed 'spurious phases', such features are a direct consequence of the correlation process.

The preponderance of steeply travelling waves in the correlation of the late coda of earthquakes makes the use of the correlation wavefield valuable for examining the properties of the deep Earth. The patterns of the correlation field depend on the material properties at depth. Thus, comparisons of observed and synthetic correlation fields can place constraints of structure in the deep Earth.

The properties of the coda of teleseismic arrivals can also be exploited using seismometer arrays to characterise local structure. Such correlation results depend on the presence of body waves generated by scattering near the receivers.

8.1 Constructing Earth's Correlation Wavefield

In Section 6.1 we have shown how with many sources the inter-station correlation field converges on a scenario where the arrivals at the pair of stations have the same horizontal slowness, with properties directly related to the inter-station separation.

Pham et al. (2018) have demonstrated the construction of the correlation field for the coda of large earthquakes exploiting all events larger than Mw 6.8 in the period 2010 to 2016 (169 events). Records from 165 seismic stations in the Global Seismographic Network were used for correlation (Figure 8.1). The

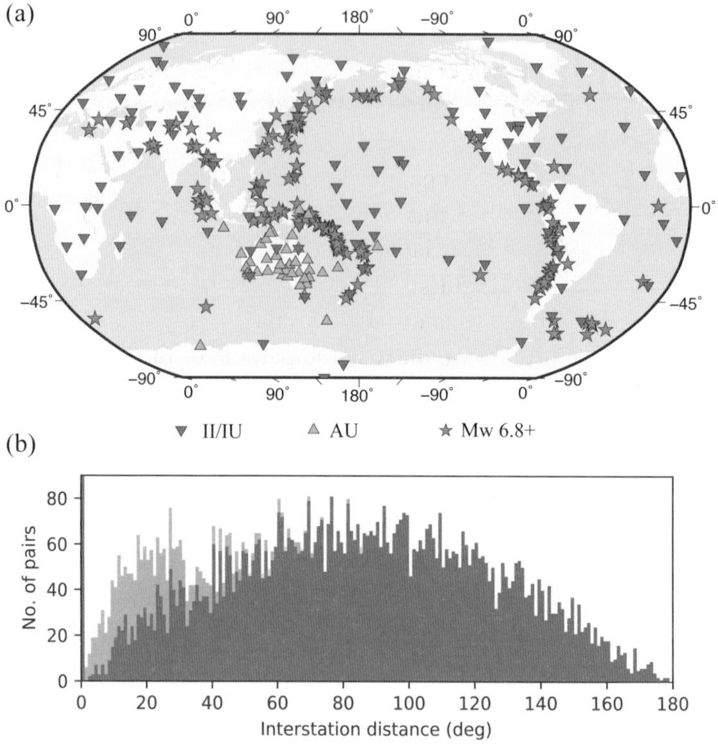

Figure 8.1 (a) Map of stations and earthquakes used for the construction of the correlation wavefield. Stations are used from three networks: the Australian National Seismograph Network (network code AU), and the networks IU and II from the Global Seismographic Networks administrated through the Incorporated Research Institutions for Seismology (IRIS). Mw 6.8+ earthquakes for the time interval 2010–2016, from the Global CMT catalogue, are marked by stars. (b) Histogram of inter-station distances using 1° bins. The combined IU and II global networks are shown in dark grey and the AU network in light grey.

vertical-component waveforms for the time window from 3 hours to 10 hours after event initiation were cross-correlated and filtered in a intermediate period band (15–50 s; 20–67 mHz), and then stacked into 1° bins in inter-station separation. Almost 2 hours of correlation lapse time is displayed in Figure 8.2(a). A wide variety of correlation phases can be recognised including features with many internal reverberations through the whole Earth (e.g., *I4*, *I5*).

For comparison with the observed correlation field we show simulations using the same geographical source and station distribution as the observations, for sources with a dominant period of 10 s. Ten-hour-long synthetic seismograms were calculated by numerical simulation in a spherical Earth, using the axisymmetric spectral element code AxiSEM (Nissen-Meyer et al., 2014). The AK135 reference model (Kennett et al., 1995) was used for isotropic seismic wavespeeds and density as a function of radius, together with the elastic attenuation model from Montagner

& Kennett (1996). For simplicity, explosive sources were placed at a fixed depth of 50 km at each source position. The synthetic correlation field, with the same processing as the observations, is displayed in Figures 8.2(b), 8.3(a), and shows a very strong similarity to the correlation field constructed from the observed seismograms.

Thus in the 15–50 s period range, the features of the correlation wavefield are dominated by interactions of seismic waves with the Earth's radial structure. The deviations in timing introduced by lateral heterogeneity are typically much smaller than half the shortest period and so do not disrupt the correlation stacks. The correlation wavefield differs substantially from the direct response of the Earth to a surface source. In Figure 8.3 we make a comparison of the first two hours of the correlation wavefield and the direct wavefield from the same numerical simulation for a radially stratified sphere. We see that some aspects of the wavefield are in common, e.g., *PKP (I1)* and its multiples *I2* and *I3* but there is little general correspondence in the appearance of the two aspects of the seismic wavefield.

The construction of the correlation wavefields was carried out in exactly the same way for both real and synthetic waveforms (Pham et al., 2018). Temporal and spectral normalisation was employed for all input seismograms. Firstly, running-absolute mean normalisation (Bensen et al., 2007) was used to suppress surface waves which are dominant in the period range 15–50 s (Lin et al., 2008). Then, a spectral whitening operation (Pham & Tkalčić, 2017) was employed to balance the contribution of individual frequencies in the spectral domain. The appearance of the correlation field is influenced quite strongly by the nature of the normalisation applied during the practical construction of the correlation field. Here the fully normalised case with both temporal and spectral normalisation was used, since the correlation phases are then recovered with the greatest clarity. The data processing operations have the effect of modifying the results away from simple cross-correlations, but for the intermediate frequency band used, distortion of the timing should be small (Hanasoge & Branicki, 2013; Fichtner, 2014; Fichtner et al., 2017).

The biggest contributions to the direct wavefield in Figure 8.3(b) are the large surface-wave arrivals, which decay steadily over multiple circuits of the globe and so have rather weak energy in the coda window (3–10 hours) used for correlation. There is just a hint of such surface-wave energy in the correlations for the observations (Figure 8.2a), and slightly more in the synthetics (Figure 8.2b, 8.3a). The vertical component of the direct wavefield (Figure 8.3b) also shows strong multiple *S* phases, but these too have quite large slowness and make almost no contribution to the correlation wavefield. The travel times of a large number of the phases that can be identified in the correlation results correspond to conventional seismic body-wave phases, but there are also a number of prominent correlation features in Figures 8.2(a) and 8.2(b) that have no counterpart in the direct wavefield in Figure 8.3(b).

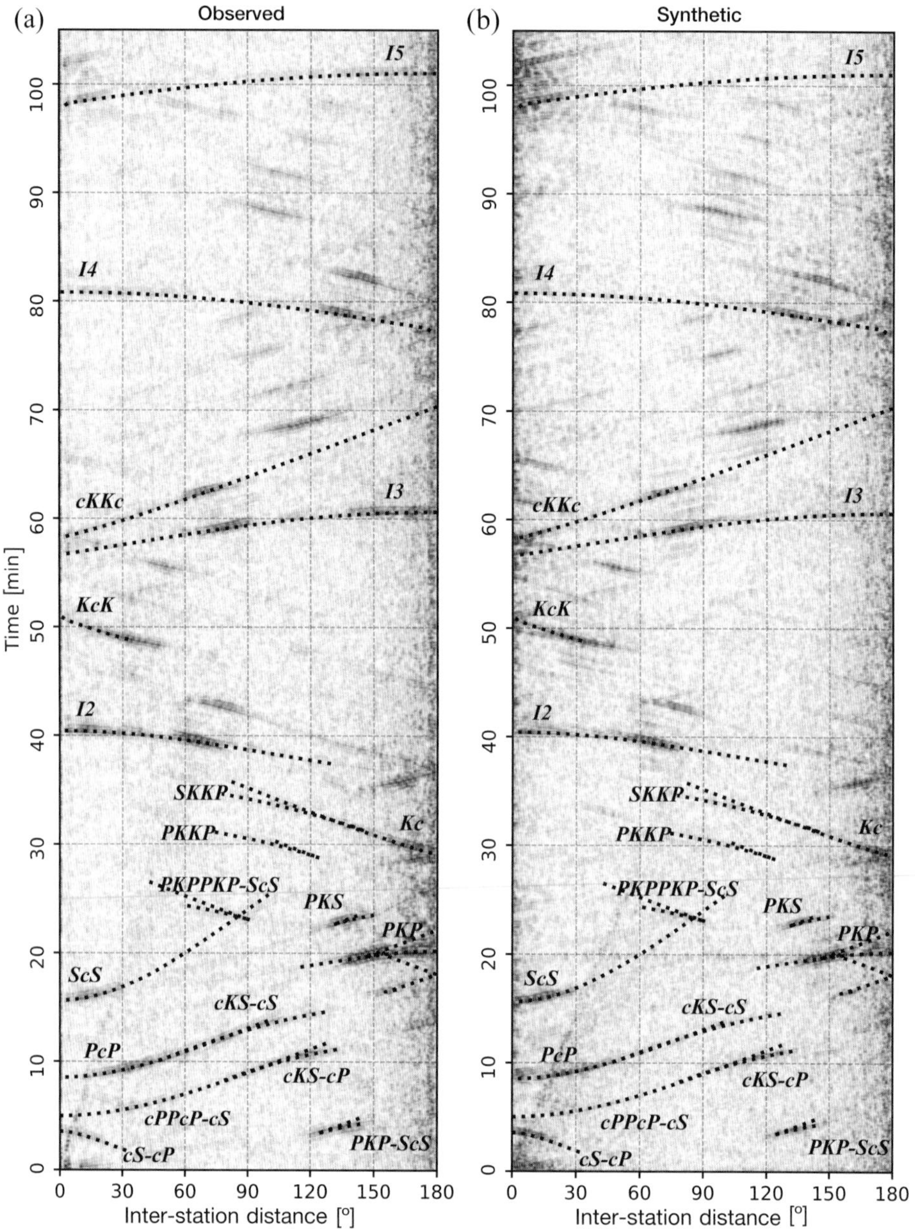

Figure 8.2 The correlation wavefield constructed from the vertical component of the late coda of seismic events. (a) Observed and (b) synthetic cross-correlograms for binned inter-station distance with bin size of 1°. Cross-correlograms are pre-filtered using a 15–50 s period band. Travel-time curves with corresponding phase names are predicted using the reference model AK135 (Kennett et al., 1995), using a compressed notation (see text).

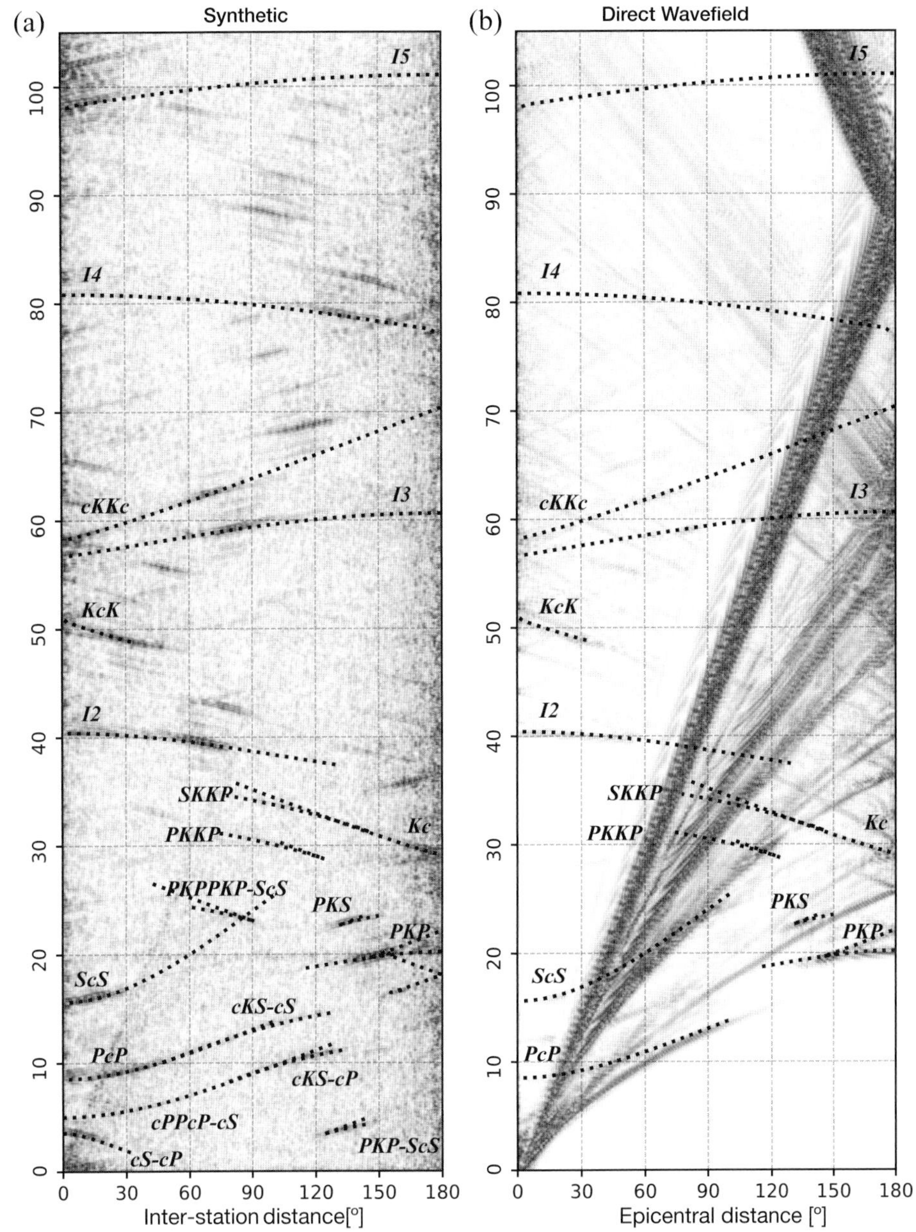

Figure 8.3 The correlation wavefield constructed from the vertical component of the late coda of seismic events. (a) Synthetic cross-correlograms for binned interstation distance with bin size of 1°. Cross-correlograms are pre-filtered using a 15–50 s period band. (b) The regular seismic wavefield for a surface source. Travel-time curves with corresponding phase names are predicted using the reference model AK135 (Kennett et al., 1995), using a compressed notation (see text).

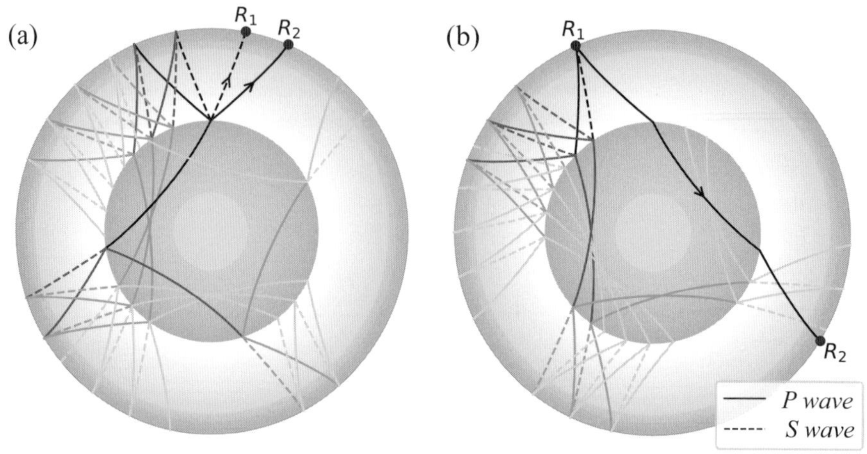

Figure 8.4 Geometrical ray paths contributing to two correlation phases between receivers at locations R_1 and R_2: (a) *cS-cP* and (b) *PKP*. All legs including *P* and *S* legs from the mantle and the *K* leg from the outer core (*P* wave) have the same ray parameter. *P*-wave segments are shown with solid lines and *S*-wave segments with dashed lines. The darker paths represent rays that contribute more directly to the correlation phases. [Courtesy of T.-S. Pham.]

In particular, correlation phases appear at lapse times shorter than the *P*-wave propagation time for that distance, and also with unlikely negative move-out so that the lapse time decreases with the separation of the stations.

In the synthetic seismograms, all *S* waves are generated by conversion because we have used explosive sources. The attenuation of *S* is significantly greater than for *P* in the mantle so that phases with only a few *S* legs will be favoured. This means that the late coda is much richer in phases with dominantly *P*-wave propagation, notably those with multiple near-vertical reverberations. As a result, the *S* waves from the earthquake sources used to create the observed correlation field in Figure 8.2(a) are largely lost, and hence there is a good match to the synthetics with the simpler explosive sources.

As discussed in Section 6.1.2 we can use a generalised ray representation of the body-wave field in correlation to demonstrate that all the apparent phases arise from the difference of propagation elements in the direct wavefield. Thus, even those features that have the travel times of regular phases are due to phase interactions.

In the portion of the late coda of seismic events that is being correlated to produce the results in Figure 8.2, much of the seismic energy will be composed of steeply travelling reverberations. In consequence, a very complex suite of difference phases can arise that are not observed in the direct earthquake wavefield. The standard phase name convention used in seismology rapidly becomes unwieldy for large numbers of propagation legs. We have therefore used in Figure 8.2 the simplified notation convention introduced by Pham et al. (2018), in which only the parts of the phase name are retained that indicate the significant subdivisions of the Earth traversed by the phase. Thus *PcP* is represented as *c*, *PKPPcP* as *Kc*. We abbreviate *PKIKP* as *I*, so that *PKIKPPKIKPPKIKP* becomes *I3*. Other prominent phases in the correlation field (Figures 8.2b, 8.3a) include *Kc*, *KcK*, *cKKc*, *I2*, *I4*, and *I5*.

A similar notation has been used for correlation phases as for the regular wavefield. Yet, the origin of features with travel times appropriate to a particular suite of propagation legs comes from the correlation of more complex arrivals, with these phase legs in common. Apparently 'spurious' correlation phases are those where the combinations do not correspond directly to regular propagation paths.

Pham et al. (2018) have analysed the genesis of the phase *cS-cP* with negative moveout. This correlation phase corresponds to arrivals that have the same slowness at the two stations and just a difference in wave type in propagation from the core–mantle boundary to the two stations. The *cS-cP* correlation phase can be regarded as arising from a virtual source at the core-mantle boundary, and the stationary phase condition forces common slowness. The difference property leads to the negative moveout in time as a function of epicentral distance that is evident in Figure 8.2(a),(b). The group of seismic arrivals that can contribute to *cS-cP* for a common station separation are illustrated in Figure 8.4(a). Contributions from paths with high numbers of legs will be attenuated as they reverberate through the Earth, and so the main contributions arise from paths with a relatively small number of propagation legs. In Figure 8.4(b) we also show the way in which the apparently regular phase *PKP* is created from the interaction of phase pairs that differ only in the presence of legs that correspond to *PKP*.

Snieder & Sens-Schönfelder (2015) show that the stationary phase zone is larger in the neighbourhood of a caustic, and that this will enhance the recovery of signals, particularly those associated with the inner core. Indeed in Figure 8.2(a),(b) we see strong contributions near the caustics of core-related correlation phases.

For the correlation of the late coda, the dominant generalised rays will be those with many near-vertical legs, so that the portions of regular phases that can couple into steep propagation will be favoured. This is why we see *PcP*, *ScS* restricted to a limited distance span and ceasing once their slowness is too high to link to the core phases. Thus in the correlation fields shown in Figures 8.2(a),(b) the core reflection *ScS* stops at around 27° and *PcP* around 67° once their slowness exceeds 4.45 s/°. In a similar way the combination *cS-cP* is restricted to inter-station separations of less than 32°.

In the intermediate frequency band (20–67 mHz) used to construct Figure 8.2, the stacking procedure across the many stations and sources is tolerant to small time shifts due to 3-D heterogeneity within the Earth. As a result, constructive interference emphasises the dependence on radial structure, and so gives a good correspondence between the stacked correlation from the observations (Figure 8.2a) and synthetic seismograms (Figure 8.2b).

8.2 Understanding the Nature of the Correlation Field

Distinct arrivals emerge from the correlation field when there are many ways in which combinations of seismic phases can arise with the appropriate differences in propagation legs. Thus we require a situation where there are a large number of kinematic analogues (Section 6.1.2), which generate the same composite delay time $\tau_C(p)$. Yet the phase combinations may well have such propagation legs in different order, and hence different physical significance. The average of many different seismic phase combinations appears in the stacked correlation pulse.

Because the amplitude terms are squared, the slowness dependence for a correlation phase is rather different from the regular wavefield. This means that the pattern of distance dependence for correlation phases is also modified from that of the direct wavefield, but this will not be very obvious for steeply travelling phases. Since very large numbers of kinematic analogues are possible for many classes of reverberations, features such as *I4* and *I5*, with multiple passages through the whole Earth, become visible in the correlation wavefield (Figure 8.2).

Although strong *ScS* waves at small epicentral distances are to be expected for *SH* where the core–mantle boundary is a perfect reflector, the presence of strong vertical component *ScS* has been regarded as anomalous (e.g., Boué et al., 2014). Once again, this correlation phase is extracted from the differences of steeply travelling waves with strong vertical-component excitation, and these properties are passed to the correlation result.

8.2.1 Emergence of Complex Phases

In Section 6.1.2 we have used a generalised ray procedure in which we build propagation paths with equivalent time and amplitude response by combining propagation legs in different parts of the Earth model, with allowance for reflection and transmission at interfaces. We have shown that the number of possibilities for creating a combination of propagation legs increases rapidly as the number of legs itself increases. The multiplicity of ways in which phase combinations can occur explains the occurrence of exotic features such as *cI3c* in the correlation field, that are never seen in the direct wavefield. In the complex of low-amplitude contributions in the coda, with enormous numbers of steep reverberations, there are many ways to build contributions. The amplitudes are largest near the caustics associated with the correlating phases, and so the arrivals appear at the slowness and inter-station separation dictated by the caustics.

For any particular window used for correlation, the clarity of the correlation arrivals depends on how many ways are available to extract the same time difference at a particular station separation. Thus, the kinematic and dynamic analogues are modulated by the specific time window employed. For example, if we use 2-hour windows five core multiples of the type *In* will fall within the

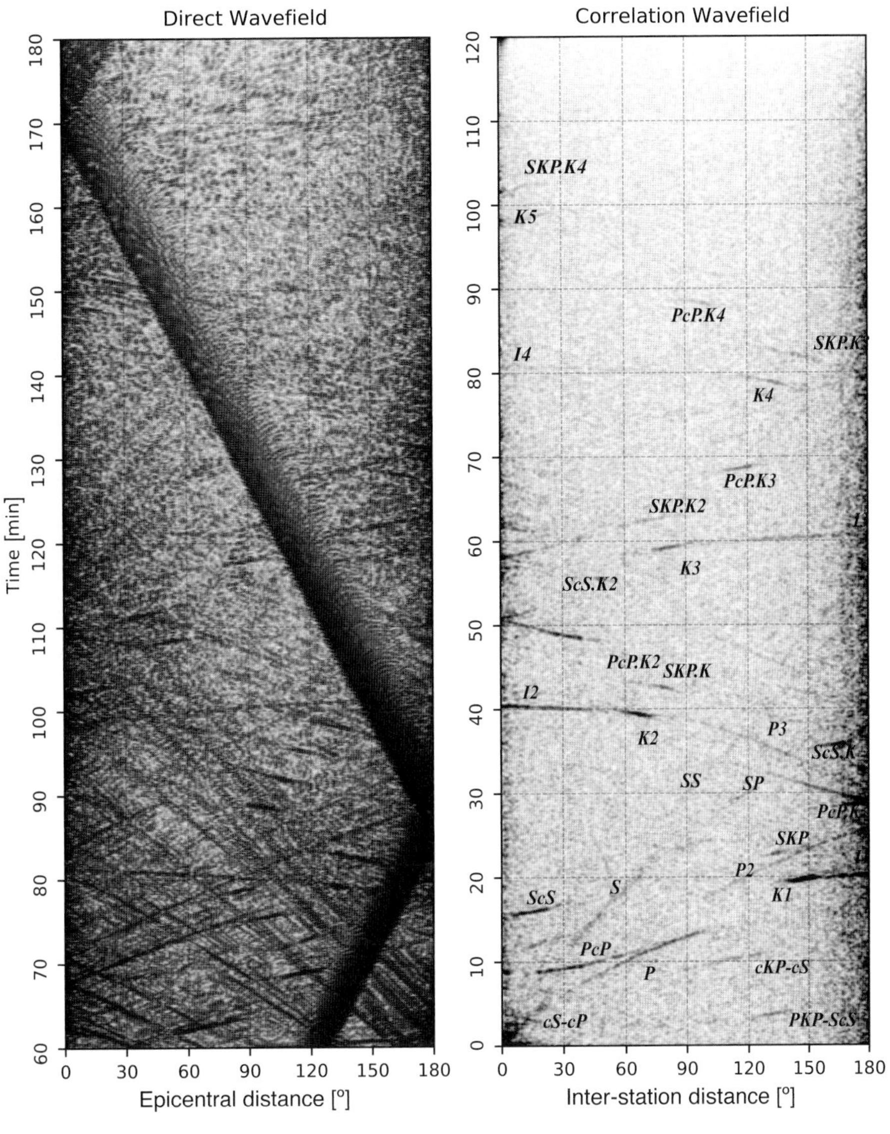

Figure 8.5 The correlation wavefield for the period of 1–3 hours after seismic events. (a) Portion of the wavefield being correlated. (b) Cross-correlograms with bin size of 1° in inter-station distance, and filtering with a pass band from 15–50 s.

window. This means that the number of options for extracting phases such as *I2*, *I3* are limited, and similar for successive time windows. Whereas, in the window from 3–10 hours after the event employed to create Figure 8.2 we include the regular phases *I9* to *I30*. This means that even from just the direct *PKIKP* multiples, there will be a very large number of ways of producing a correlation phase such as *I5*.

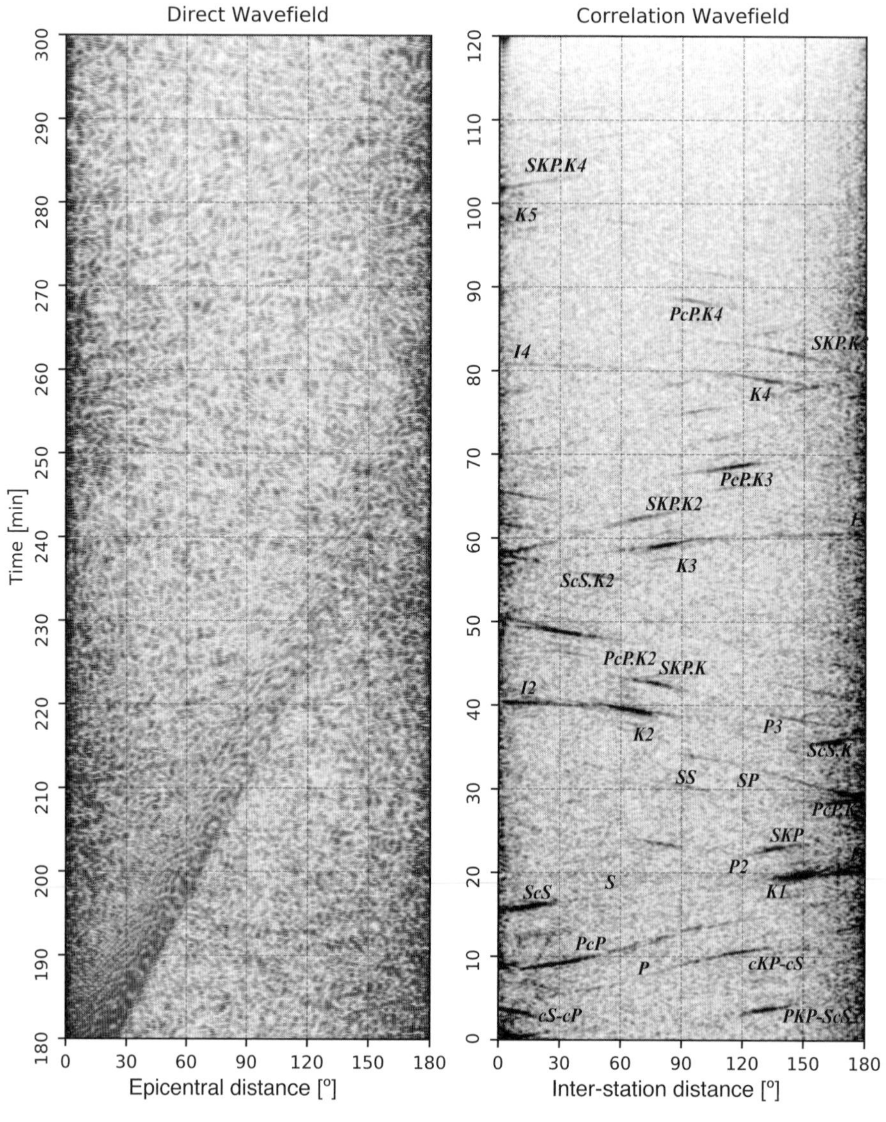

Figure 8.6 The correlation wavefield for the period of 3–5 hours after seismic events. (a) Portion of the wavefield being correlated. (b) Cross-correlograms with bin size of 1° in inter-station distance, and filtering with a pass band from 15–50 s.

Kennett & Pham (2018b) have looked at the evolution of the correlation field as a function of time along the coda, using a sequence of 2-hour windows. It takes some time before a clear correlation field is established, but after 1 hour from event initiation a set of weak phases emerge that have the time behaviour expected for the regular wavefield (Figure 8.5). This weak 'regular' field is accompanied by many phases that have no counterpart in direct source excitation produced by the interaction of seismic phases with common propagation legs, whose time–distance behaviour is controlled by the differences in accumulated phase. The weakness of the 'regular' phases relates directly to the limited number of options that have the appropriate differential behaviour. This situation is illustrated in Figure 8.5, with correlograms constructed from synthetics for 1–3 hours after event initiation using the AK135 model as in Figure 8.2. Figure 8.5 shows both the segment of the seismic wavefield used for correlation and the resulting correlograms.

The patterns of correlation phases are closely linked to the dominant ranges of horizontal slownesses present in the time window for which the correlations are performed. The apparently regular phases fade with time as the larger slowness components of the wavefield are exhausted. Then, distinct arrivals in the correlation field arise when there are many ways in which combinations of seismic phases have the same difference in propagation legs. There are many more such possibilities for steeply travelling waves in the late coda, so that a relatively stable correlation field develops. This stable field corresponds well to the long-duration correlation results displayed in Figure 8.2. In Figure 8.6 we show the way that the more complex correlation phases emerge from the later part of the event coda, using the next 2-hour segment of the coda. As in Figure 8.5, both the direct wavefield and the resulting correlation field are displayed. Apart from the evident surface waves, this segment of the wavefield shows little in the way of obvious phases. Yet, the correlation processes exploits the many different minor phases to build a complex pattern of coherent correlation phases.

8.2.2 Steeply Travelling Normal Modes

Since we have established in (6.1.10) that the principal contribution to the correlation field will be such that the slowness is the same at each of the pair of stations being correlated, it is appropriate to use the organisation of the Earth's normal modes in terms of spherical slowness $\wp = \mathrm{p}\mathrm{r}_e$, where r_e is the radius of the Earth, as introduced in Section 2.1.3. We present in Figure 8.7 a portion of the modal spectrum as a function of slowness at an enlarged scale compared to Figure 2.1.

As discussed in the Appendix to this chapter, there are three main groups of asymptotic modes corresponding to steeply travelling waves in the Earth: *PKIKP*-equivalent modes, the inner-core modes that involve propagation in the core, and *ScS*-equivalent modes linked to reflection from the core–mantle boundary.

In Figure 8.7 we can recognise *ScS*-equivalent modes by their arcuate moveout in frequency across the entire slowness span. Because of their dominant *S*-wave propagation, such modes have stronger attenuation and tend to decay moderately quickly. In contrast, the *PKIKP*-equivalent modes display the least attenuation, and so persist for a long time in the coda (cf. Poli et al., 2017).

For constructive interference in the correlation wavefield, groups of modes with common slowness need to have consistent properties over a range of frequencies

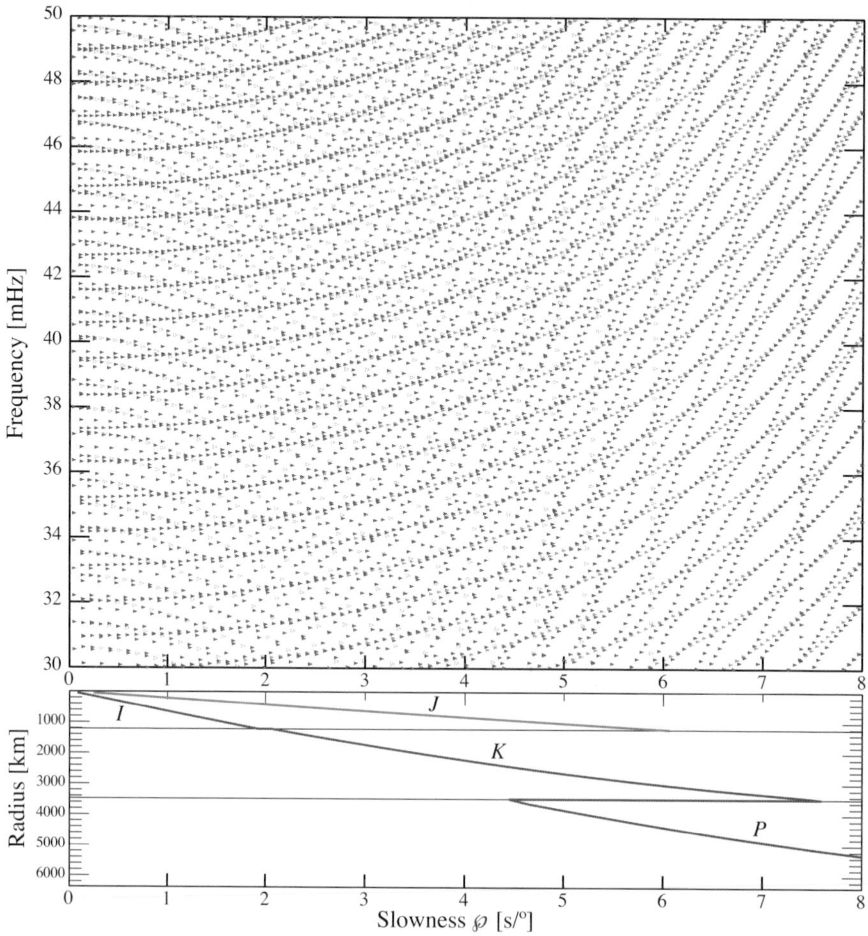

Figure 8.7 Normal modes of the Earth as a function of slowness ℘, for the frequency range 30–50 mHz contained in the construction of Figures 8.2, 8.3 8.5, and 8.6 compared with a representation of the Earth model AK135 in the slowness domain. Mantle *S* waves lie outside the plot interval for slowness. Both spheroidal and toroidal modes are displayed, with grey for the toroidal modes and open symbols for the toroidal inner-core modes.

(Nolet & Kennett, 1978). The behaviour will be controlled by the differences in eigenfrequency between modes. Asympotically, the eigenfrequencies have a strong dependence on the phase delays for passage through the mantle and core (Appendix 8.A; Dahlen & Tromp, 1998). These eigenfrequency differences again emphasise features of the wavefield with time relations that are the difference between those of regular seismic phases.

In the process of correlation and stacking, combinations of *PKIKP*- and *ScS*-like elements can be coupled for small slowness, but will only be significant in the earlier parts of the earthquake coda. Such coupling leads to apparent *ScS* arrivals on the vertical component for short epicentral distances as pointed out by Poli et al.

(2017). The enhanced attenuation in *ScS* modes compared with *P*-wave dominated modes means that multiple *ScS*-type phases will be suppressed, as is indeed seen in Figures 8.2(a),(b).

8.3 Exploitation of Coda Correlations in Deep Earth Studies

A major limitation on many studies of the deep Earth is dependence on the locations of earthquakes. In many situations only a few combinations of suitable sources and receivers can be found so that geographic coverage is limited. Correlation methods offer the possibility of reducing dependence on the fortuitous placement of earthquakes. The coda of signals from earthquakes has complex character with a strong component of scattering. Stacked cross-correlations between stations can exploit the higher amplitudes in the coda compared with ambient noise to extract different parts of the wavefield. An early application was by Campillo & Paul (2003), who recovered surface wave-signals from the correlation of earthquake coda.

We have seen in Section 7.4 that it is possible to recover body waves at the global scale from ambient noise, but that stronger signals emerge from the presence of the coda of earthquakes. Sens-Schönfelder et al (2015) have made an empirical study of the character of the coda of a major earthquake, using records from the USArray from the Mw 8.3 earthquake at 605 km depth beneath the Sea of Okhotsk on 2013 May 24. This is the largest deep earthquake yet recorded, and so provides an opportunity to examine the body waves in the coda without undue influence from surface waves. At the highest frequencies, around 0.5 Hz, multiply reflected *PKIKP* travelling along great-circle paths stay above the noise for around 5000 s. Whereas, in the 0.1 Hz band in this time interval, the wavefield is almost isotropic in all slowness ranges. This distribution arises from strong scattering of *S* energy in the heterogeneous lithosphere, particularly the crust. The early part of the coda can therefore be used to retrieve *P* and *S* waves with inter-station correlations at regional distances, as in the work of Tonegawa et al. (2009).

The attenuation of shallowly travelling *S* waves means that *P* waves travelling deeply into the Earth tend to dominate after 5000 s. Such waves have only weak interaction with shallow heterogeneity. As the lapse time into the coda gets larger the slowness of the body-wave arrivals is observed to steadily decrease, indicating that the dominant propagation is at steep angles – as is expected for a dominantly radially stratified Earth. The interval from 10 000 s to 30 000 s contains a rich collection of small slowness arrivals, as we have also seen from the global correlation in Figure 8.1. This part of the coda is very suitable for studies of wave propagation penetrating deep into the Earth's interior.

8.3.1 Targeted Seismic Phases

The coda interval from 10 000 to 30 000 s has been used in a number of studies, and is similar to that employed in the construction of Figure 8.2. Lin et al. (2013) have exploited cross-correlations between nearly-antipodal seismic stations

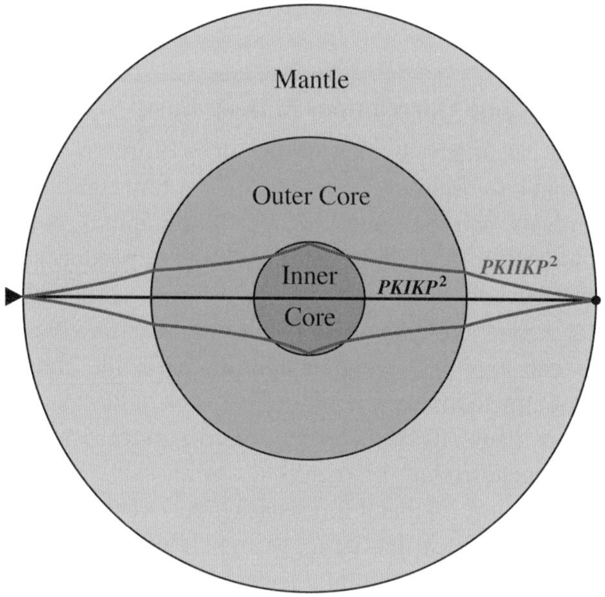

Figure 8.8 Ray paths for the *PKIKP*^{2} and *PKIIKP*^{2} phases for antipodal propagation.

to extract *PP*, *PKPPcP*, and *SKSP* with benefit from the focussing with nearly 180° separation. *PKIKP* has a smaller amplitude in the correlations since there is no focussing effect.

In addition to the use of ambient noise, Xia et al. (2016) have exploited earthquake coda for cross-correlation between stations in China and South America to examine the triplications of *PKP*. They find differences with the frequency band employed, the AB branch can be detected in the 0.05–0.2 Hz band and DF in the 0.02–0.067 Hz band, but not AB. Only a few of their earthquake sources lie close to the great-circle zone as needed for constructive interference (Figure 6.2), and this may help to explain the differences.

Wang et al. (2015) have used stacked auto-correlations from the coda of large events (Mw $\geq$ 7) at a number of networks of stations across the globe to investigate near-antipodal propagation. They concentrate on the phases *PKIKP*^{2} and *PKIIKP*^{2}, whose paths proceed from a station to reflection on the opposite side of the Earth with return to the same station. For this phase pair, the main difference in ray path lies in the core. *PKIKP*^{2} propagates directly through the center of the Earth. *PKIIKP*^{2} is reflected at the underside of the core–mantle boundary, and so samples the outer part of the inner core. There is also some deviation in paths in the outer core (Figure 8.8). Near-antipodal focussing helps to raise the amplitude of *PKIIKP* (Rial & Cormier, 1980), which is otherwise weak because of the small reflection coefficient at the inner-core boundary. Wang et al. (2015) have used the frequency band from 0.02-0.067 Hz, and need to stack over all stations in their arrays to get adequate signal strength at the expected arrival times of the phases. They see variations in the differential times between the two phases across the globe, which they attribute to inner-core anisotropy.

Huang et al. (2015) employ all available stations from USArray with stacked auto-correlation and cross-correlation between nearby stations. Their target is again the phase pair $PKIKP^2$ and $PKIIKP^2$. They have used the coda window from 20 000 to 40 000 s for 143 events with Mw ≥ 7 with correlations for those stations that were operational at the time of each event. They use the 0.02–0.067 Hz frequency band, and work with vertical-component correlations. To enhance the correlated signal, slant-stacking is applied to all the correlation functions in a 300 km radius around each specified location, so that a stack equivalent to tens of thousands of sources is produced. The arrival times of the core phases are then picked from the full suite of localised stacks using the adaptive stacking procedure described in Section 5.1.2. The residual patterns for the two phases show a dominantly east–west variation with earlier arrivals in the east. The patterns for $PKIKP^2$ and $PKIIKP^2$ are displaced from each other by about 10° with the neutral line for $PKIKP^2$ time anomalies further west. The difference between the arrival times ($PKIKP^2$-$PKIIKP^2$) produces a prominent north–south trending anomaly of delayed arrivals across the middle of the continental USA, with an average value of 2.5 s.

Huang et al. (2015) suggest that the dominant effects on the differential travel times for the two core phases come from the inner core. However, there may be some influence from lowermost mantle structure in the antipodal zone on the far side of the Earth (beneath the Indian Ocean for USArray), particularly on $PKIKP^2$. They propose that the observed differential time anomaly is linked to differences in inner-core properties that are approximately hemispherical (see, e.g., Waszek & Deuss, 2011, Lythgoe et al., 2014).

8.3.2 Exploitation of the Global Correlation Field

As we have seen in Figure 8.2, the observed global correlation wavefield can be well represented by simulations for the standard Earth model AK135, with a diversity of phases that provide sampling of the deep Earth. The interference conditions that give rise to the presence of the correlation phases are quite sensitive to the details of Earth structure. Thus, if the velocity model is modified, there can be distinct shifts in the pattern of arrivals. This sensitivity of the global correlation wavefield to Earth structure has been used by Tkalčić & Pham (2018) to provide new constraints on the shear wavespeed in the inner core.

The main evidence for the rigidity of the inner core comes from normal mode studies. Although a number of studies have attempted to find phases with a shear leg in the inner core, such as *PKJKP*, none has been conclusive. Indeed Shearer

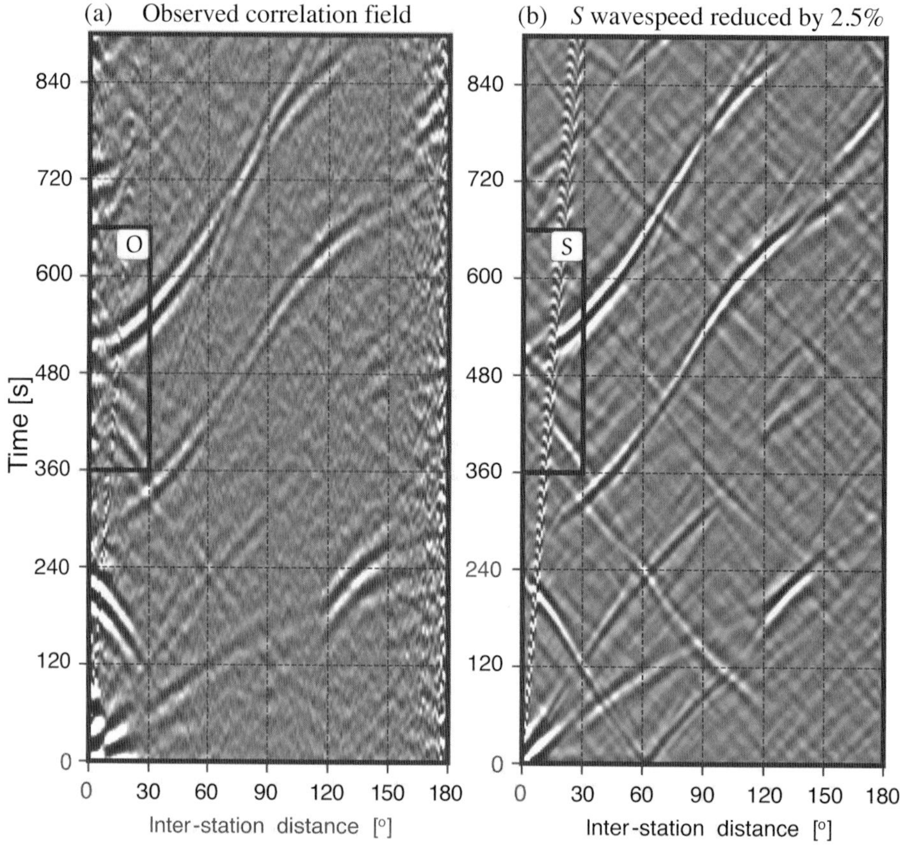

Figure 8.9 Comparison of observed and synthetic correlograms in relation to the *I2-PKJKP* cusp. (a) The observed global correlogram; and (b) the best-fit simulated global correlogram for a central period of 23.1 s, with a reduction of shear wavespeed from PREM by 2.5%. [Courtesy of H. Tkalčić and T.-S. Pham.]

et al. (2011) have shown that direct observations of *PKJKP* waves are unlikely for periods greater than 10 s.

Rather than make a direct search, Tkalčić & Pham (2018) have looked for features of the correlation wavefield that are sensitive to a shear propagation leg *J* in the inner core. They have identified three correlation phases *PKIKPPKIKP–PKJKP* (*I2–PKJKP*), *PKIKP–PKJKP*, and *PKiKP–PKJKP* that can be recognised in synthetic correlation fields and show such sensitivity. The combination *I2–PKJKP* occurs for small angular distance and so gets contributions from all azimuths, which enhances its visibility. The correlation phases *PKIKP–PKJKP* and *PKiKP–PKJKP* are weaker because there is no such focussing effect for the 120° to 150° angular distance in which they emerge.

Tkalčić & Pham (2018) have made systematic searches for the properties of the inner core that best match the observed correlation field (Figure 8.9). They

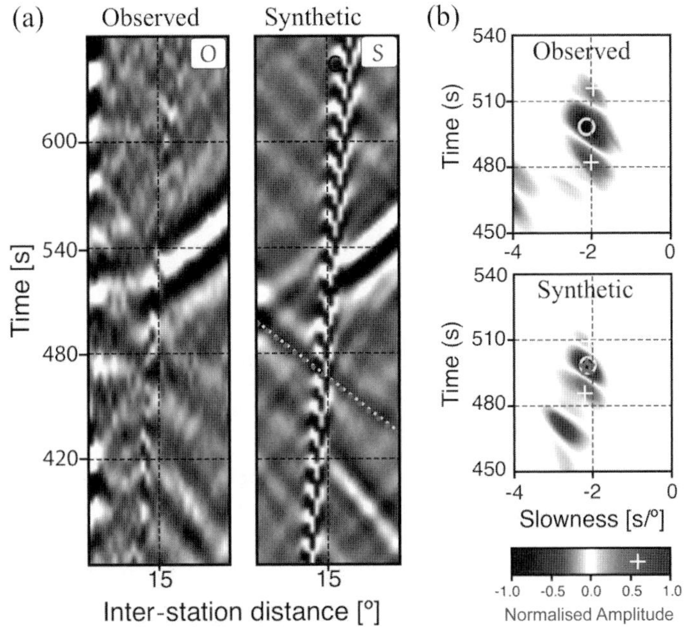

Figure 8.10 Comparison of observed and synthetic correlograms in relation to the *I2–PKJKP* cusp. (a) Enlargements of windows focussed on the *I2–PKJKP* cusp from Figure 8.9 (b) Observed and simulated slant stacks. [Courtesy of H. Tkalčić and T.-S. Pham.]

exploit visual comparisons of the cusp positions for the *I2–PKJKP* correlation phase, and of the relative location of negative peaks in slant stacks produced from their observed and synthesised correlation wavefields (Figure 8.10).

The best fit is achieved with a shear-wavespeed reduction of $2.5 \pm 0.5\%$ relative to the reference model PREM. This converts to a shear wavespeed of 3.42 ± 0.02 km/s and shear modulus of 149.0 ± 1.6 GPa near the inner-core boundary and 3.58 $\pm$ 0.02 km/s and 167.4 ± 1.6 GPa at the centre of the Earth center. The reduced values imply a rather soft inner core, and explain the absence of *PKJKP* waves in the direct seismic wavefield produced by earthquake sources.

8.4 Local Structure from Teleseismic Coda

We have so far focussed on the use of earthquake coda for global studies, but the local properties of the coda of teleseisms can be exploited with the aid of seismometer arrays. Incident teleseismic *P* and *S* waves can be scattered by heterogeneity near the receivers, and produce a combination of body-wave energy and local surface waves. Using examples from a number of seismic arrays across the globe, Saygin & Kennett (2019) have demonstrated that it is possible to exploit the cross-correlograms of coda segments at the suite of station pairs to extract the refracted and reflected body waves associated with the local structure. They have shown examples for a number of medium-aperture seismic arrays, across the globe,

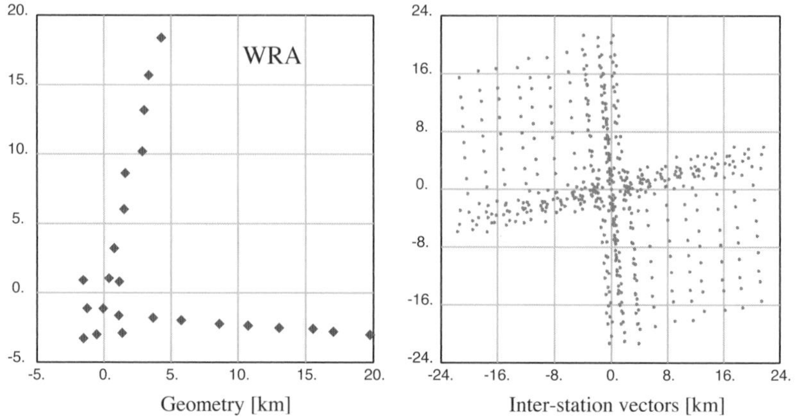

Figure 8.11 Configuration of the Warramunga array, Northern Territory, Australia (WRA), and the pattern of inter-station vectors.

in Australia, Alaska, Scotland, and Kazakhstan. The Warramunga array (WRA) in northern Australia is illustrated in Figure 8.11, together with the complex pattern of inter-station vectors used when cross-correlations are made between pairs of stations.

The approach taken by Saygin & Kennett (2019) exploits stacking over records from large numbers of teleseisms in the epicentral distance range 40–50° from the seismic arrays. At these ranges the interaction of the seismic waves with oceanic and continental structure yields a long duration of complex coda. This coda energy is then scattered in the vicinity of the arrays. The scattered waves can be exploited via correlation to raise the local body-wave components above the noise level.

The early part part of the coda, immediately following the *P* arrival, is dominated by local reverberations and conversions that can be exploited in receiver function or auto-correlation studies (see Section 9.1). After about 120 s, scattering in the lithosphere beneath and around the stations becomes dominant. *P* to *P* scattering dominates the vertical-component records ahead of the *S* phase. Scattering extends beyond *S*; efficient scattering from *S* to *P* appears on the vertical component.

A suite of 100 s time windows in the coda are used to construct the cross-correlations between the elements of the array. The correlations for each station pair are stacked over all the different events and then binned to enhance the clarity of the body-wave arrivals, since this suppresses the influence of lateral heterogeneity across the array. In Figure 8.12 we show the evolution of the extracted body waves at the Warramunga array with analysis windows at increasing times after the onset of *P*. In this case over 5 000 teleseismic events with magnitudes between 5.5 and 7.5 in the period 2000–2017 were used. The event-stacked correlograms were further stacked into bins 0.5 km wide in inter-station separation, out to 26 km. In the later part of each section shown

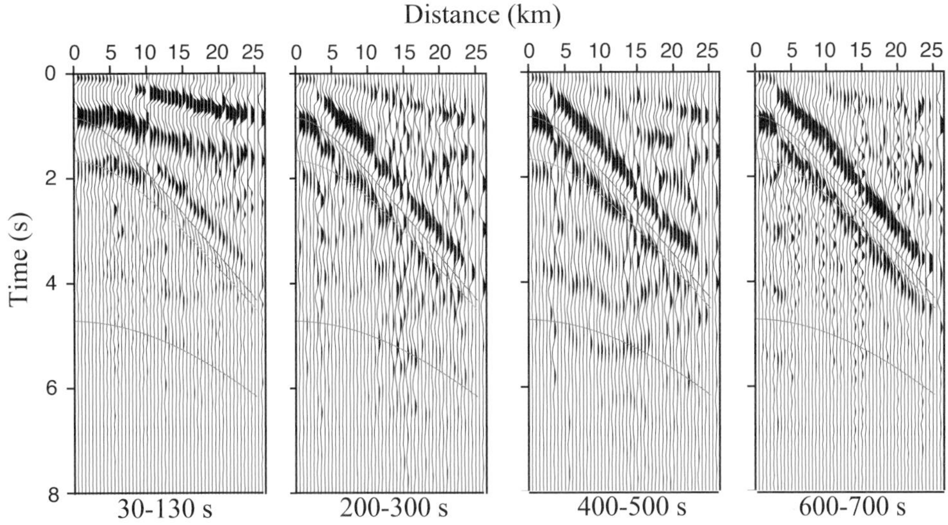

Figure 8.12 Stacked cross-correlograms at WRA using 100 s portions of earthquake coda for events in the distance range 40-50°, with superimposed travel time curves for local reflection structure. The waveforms are filtered with a bandpass between 1 and 5 Hz, and normalised to unity.

in Figure 8.12, we see a coherent arrival starting at around 1 s with a clear moveout across the full suite of traces, and also weak arrivals at around 4.5 s. For the correlation intervals 400–500 s and 600–700 s, even though the incoming teleseismic wavefield comprises planar *S* waves, the sections continue to show *P* waves with similar moveout.

Reflection and refraction experiments in the general vicinity of the Warramunga array provide some control on the local structure, though also indicate the presence of local variability, particularly at depth. Superimposed on the record sections in Figure 8.12 we show the reflection trajectories for *P* waves for a simple 1-D model for the neighbourhood. The first two reflections tend to merge to give travel times corresponding to the main observed branch; the deep reflection is in general agreement with the weak arrivals around 4.5 s.

Appendix: Asymptotic Results for Normal Modes with Small Slowness

8.A Relation of Mode Frequencies to Structure

For relatively high frequencies, the frequencies of the normal modes corresponding to spherical slowness $\wp$ are determined by the poles of the reverberation operator

$$[\mathbf{I} - \mathbf{R}_{\mathsf{F}}(\wp)\mathbf{R}_{\mathsf{D}}^{\mathsf{E}}(\wp, \omega)]^{-1}, \tag{8.A.1}$$

where $\mathbf{R}_{\mathsf{F}}(\wp)$ is the reflection matrix at the free surface, and $\mathbf{R}_{\mathsf{D}}^{\mathsf{E}}(\wp, \omega)$ is the reflection back from all the structure beneath the surface down to the centre of the Earth.

We can build an expression for $\mathbf{R}_{\mathsf{D}}^{\mathsf{E}}$ in terms of contributions from the major zones of Earth structure and the interfaces between them. We will adopt the compressed notation introduced in § SWII:26.1, in which we designate elements attached to the mantle, outer core, and inner core as M, K, I, the interface matrices for the inner-core boundary with i, and the core–mantle boundary with c.

We start from the inner core and build $\mathbf{R}_D^E$ by adding the contributions from successive shells using the addition formulae for reflection matrices (2.3.2). We designate the reflection matrix for the inner core, down to the centre of the Earth, as $\mathbf{R}_D^I$. Now moving just above the inner-core boundary, the reflection from beneath takes the form

$$\mathbf{R}_D(i-) = \mathbf{R}_D^i + \mathbf{T}_U^i[\mathbf{I} - \mathbf{R}_D^I\mathbf{R}_U^i]^{-1}\mathbf{R}_D^I\mathbf{T}_D^i. \tag{8.A.2}$$

We recall that the outer core is a fluid so only *P* waves can propagate in this zone. Adding in the outer core, the reflection response from just below the core–mantle boundary is

$$\mathbf{R}_D(c+) = \mathbf{R}_D^K + \mathbf{T}_U^K[\mathbf{I} - \mathbf{R}_D(i-)\mathbf{R}_U^K]^{-1}\mathbf{R}_D(i-)\mathbf{T}_D^K. \tag{8.A.3}$$

Including the contrasts at the core–mantle boundary, the response just above is then

$$\mathbf{R}_D(c-) = \mathbf{R}_D^c + \mathbf{T}_U^c[\mathbf{I} - \mathbf{R}_D(c+)\mathbf{R}_U^c]^{-1}\mathbf{R}_D(c+)\mathbf{T}_D^c. \tag{8.A.4}$$

The final stage is to include the contribution from the mantle so that we obtain

$$\mathbf{R}_D^E = \mathbf{R}_D^M + \mathbf{T}_U^M[\mathbf{I} - \mathbf{R}_D(c-)\mathbf{R}_U^M]^{-1}\mathbf{R}_D(c-)\mathbf{T}_D^M. \tag{8.A.5}$$

Thus we have a nested sequence of reverberations all coupled to the free surface through (8.A.1). The dominant contribution to the phase terms will come with propagation through the major elements of the model M, K and I in transmission, or direct reflection. At steep incidence, only discontinuities will be significant for reflection, so that contributions from the main zones will be dominated by reflections from the major interfaces at the core–mantle and inner-core boundaries. The small slowness range lies well away from any critical angles, and so the interface matrices remain simple.

We recall from (2.3.18) that

$$(\mathbf{I} - \mathbf{A} - \mathbf{B})^{-1} = (\mathbf{I} - \mathbf{A})^{-1} + (\mathbf{I} - \mathbf{A})^{-1}\mathbf{B}(\mathbf{I} - \mathbf{A} - \mathbf{B})^{-1}, \tag{8.A.6}$$

and so when we break up a reverberation operator we modulate the residual response with further pre- and post- reverberation terms. This means that the dispersion relation (8.A.1) can be broken up to some extent in terms of subsidiary dispersion terms.

For *SH* waves, the inner-core boundary and the core–mantle boundary act as perfect reflectors. In consequence the dispersion relation for toroidal modes factors into two terms:
(i) inner core modes governed by $[I - R_D^I R_U^i]_{HH}^{-1}$, and
(ii) *ScS*-equivalent modes represented through $[I - R_F \bar{R}_D^M]_{HH}^{-1}$,
where $\bar{R}_D^M \approx R_D^M + T_U^M R_D^c T_D^M$ with a dominant effect from the reflection from the core–mantle boundary.

The underside reflection for *SH* waves at the inner-core boundary will be nearly perfect, and allowing for the phase shift associated with the turning point for the shear leg in the inner core

$$[\mathbf{R}_D^I]_{HH}(\wp, \omega) \sim \exp\{i\omega\tau_I(\wp) + \pi/2\}, \tag{8.A.7}$$

where $\tau_I(\wp)$ represents the phase accumulated in the passage through the inner core,

$$\tau_I(\wp) = 2\int_0^{r_i} dr \left[\frac{1}{\beta^2} - \frac{\wp^2}{r^2}\right]^{\frac{1}{2}} = 2\int_0^{r_i} dr\, q_\beta(\wp). \tag{8.A.8}$$

The dispersion condition for the inner-core modes reduces to

$$\omega_n\tau_I(\wp) \approx 2\pi(n + \tfrac{5}{4}), \tag{8.A.9}$$

where n is the conventional radial mode index ($n = 0, 1, 2, \dots$). For *ScS* modes reflected directly back from the core–mantle boundary there is no additional phase shift, so

$$[\bar{R}_D^M]_{HH}(\wp, \omega) \sim \exp\{i\omega\tau_M(\wp) + [R_D^M]_{HH}(\wp, \omega)\}, \tag{8.A.10}$$

where $\tau_M(\wp)$ is the accumulated phase in *SH* wave propagation between the surface and the core–mantle boundary,

$$\tau_I(\wp) = 2\int_{r_c}^{r_e} dr\, q_\beta(\wp). \tag{8.A.11}$$

Thus the dispersion curves for the *ScS* modes are controlled by

$$\omega_n \tau_M(\wp) \approx 2\pi(n+1) - \psi_M(\wp, \omega), \tag{8.A.12}$$

where n is again the radial mode number. The presence of the term $\psi_M(\wp, \omega)$ associated with reflection from mantle discontinuities introduces modulation in the separation of mode branches, known as the *solotone effect* (see Lapwood & Usami, 1981). The difference between the eigenfrequencies for different mode branches from (8.A.9) and (8.A.12) depend directly on the accumulated phase in the inner core and mantle.

For the coupled *P–SV* wave system, the *SV* waves will have some interaction with *P* waves in the core. Nevertheless, for small slownesses the core–mantle boundary remains a near-perfect reflector for *SV* waves. This means there is a direct analogue of the inner-core modes through $[I - R_D^I R_U^i]_{SS}^{-1}$. On the mantle side, the modes have to interact with the free surface and thus one modulating factor on the dispersion will be $[I - R_F \bar{R}_D^M]_{SS}^{-1}$, which represents the *ScS* equivalent reverberations. The *P* waves can penetrate right through the Earth, and we can extract the most significant part of the reflection and transmission response through the core–mantle and inner-core boundaries as

$$\bar{\mathbf{R}}_D^{KI} \approx \mathbf{T}_U^c[\mathbf{T}_U^K(\mathbf{R}_D^i + \mathbf{T}_U^i\{\mathbf{R}_D^I\}\mathbf{T}_D^i)\mathbf{T}_D^K]\mathbf{T}_D^c. \tag{8.A.13}$$

The dominant part of the *PKIKP* equivalent reverberations then come from the term $[\mathbf{I} - \mathbf{R}_F\mathbf{T}_U^M\bar{\mathbf{R}}_D^{KI}\mathbf{T}_D^M]_{PP}^{-1}$. We can construct asymptotic approximations for the reflection and transmission contributions in a similar way to (8.A.7) and (8.A.10), but with the extraction of *P*-wave propagation legs in mantle, outer core, and inner core. The dominant part of the dispersion of the *PKIKP*-type modes will be controlled by the accumulated phase for *P* waves from the surface down to the turning point in the inner core.

The presence of the additional contributions we have identified in (8.A.1)–(8.A.6) means that the full dispersion relation for the eigenmodes introduces coupling between the different styles of propagation, particularly around slownesses where multiple reverberation terms are close to singular.

Dahlen and Tromp (1998) present a single equation for asymptotic mode dispersion that brings together all the propagation components we have identified, with full coupling. However, the resulting expression with over twenty terms is too complex to reveal the structure associated with the different shells of the Earth without making further approximations, such as $\wp \approx 0$.

9

Correlations in Receiver Studies

We have seen in Chapter 6 that with a suitable distribution of sources the stacked cross-correlation of ground motion between two stations corresponds to a situation with a virtual source–receiver pair. In the limit where the two stations are coincident, the cross-correlation reduces to the auto-correlation. For energy arriving from distant sources, the auto-correlation can be related directly to the reflectivity beneath the receiver location building on the results we have developed in § 5.2.1. For ambient noise, we have the possibility of local scattering from around the receiver location whose temporal behaviour cannot be directly separated from reflections from depth. Nevertheless if we exploit the auto-correlation of continuous records at a receiver, the dominant coherent component corresponds to reflection return from depth.

The use of auto-correlation procedures opens up new opportunities for structural studies that do not depend on conversion between wave types. With care, good results can be obtained for the delineation of shallow boundaries such as ice–rock interfaces. The Moho can also be tracked as the base of crustal reflections. With a broad distribution of sources, imaging techniques can be applied to migrated auto-correlogram records to delineate complex structure.

In exploration seismology, the properties of correlation have been exploited to create specific virtual sources from recordings of the regular reflection field, and thereby illuminate structures in different ways. This approach relies on the difference properties of the correlation field that we have discussed in the context of whole Earth studies in Chapter 8.

9.1 Correlations and Receiver Response

In Section 5.2 we have shown how the auto-correlation of the surface response from teleseismic events could extract the reflection behaviour associated with the incoming slowness, modulated by the free surface. The result depends on perfect elasticity, but should be a good approximation unless seismic attenuation is particularly strong.

We now look at the way that this result is modified in a 3-D varying medium and examine the properties of the reflected field.

9.1.1 General Development

In a 3-D medium we can use the correspondence principle between reflection and transmission operators and results for the frequency–slowness domain (Section 3.3.2) to produce a comparable expression to (5.2.16) for the auto-correlation of the surface response. We again consider a situation with waves from distant sources impinging on a level J beneath the structure of interest. We can represent the auto-correlation of the surface displacement in the frequency domain as

$$[\mathbf{w}_0(\mathbf{x},\omega)]^*[\mathbf{w}_0(\mathbf{x},\omega)]^T = \mathsf{Z}^*(\omega)\mathsf{C}_R^*(\omega)\mathsf{C}_R^T(\omega)\mathsf{Z}^T(\omega)|S_I(\omega)|^2, \tag{9.1.1}$$

where the operator Z describes propagation effects before arrival at the level J. The transmission operator C_R from J to surface displacement is related to the shallow structure by

$$\mathsf{C}_R = \mathsf{W}_F\mathsf{T}_U^{FJ} = \mathsf{W}_F[\mathsf{I} - \mathsf{R}_D^{0J}\mathsf{R}_F]^{-1}\mathsf{T}_U^{0J}. \tag{9.1.2}$$

The operator W_F produces surface displacement from an upcoming wave, and R_F is the surface reflection operator. On auto-correlation of $\mathbf{T}_U^{FJ}$ we have the operator relation

$$[\mathsf{T}_U^{FJ}(\omega)]^*[\mathsf{T}_U^{FJ}(\omega)]^T = \mathsf{I} + \mathsf{R}_F\,[\mathsf{R}_D^{FJ}(\omega)]^* + [\mathsf{R}_D^{FJ}(\omega)]^T\mathsf{R}_F^T, \tag{9.1.3}$$

where R_D^{FJ} is the reflection operator including all free-surface reflections. As a result, the tensor auto-correlation of the surface displacement (9.1.1) is determined by the sum of the causal and anti-causal parts of the reflectivity including free-surface multiples. As noted by Galetti & Curtis (2012), this result provides a direct correspondence of receiver results with seismic interferometry. Further, as discussed in Section 3.3, the reflection and transmission operators are non-local, and so (9.1.1) implies interaction with the neighbourhood of the receiver, over a range of slownesses.

Wapenaar et al. (2004) have taken an alternative approach to describing this situation, working with propagation invariants, equivalent to those in Section 3.1, expressed in terms of upgoing and downgoing waves. For two station locations $\mathbf{x}_A$ and $\mathbf{x}_B$, the reflection response between these locations $\mathbf{R}_F(\mathbf{x}_A,\mathbf{x}_B,\omega)$, including free-surface effects, can be related to an integral of the correlation of upward transmission terms from the surface J. Thus

$$\mathbf{R}_F(\mathbf{x}_A)\mathbf{R}_D^{FJ}(\mathbf{x}_A,\mathbf{x}_B,\omega) - [\mathbf{R}_D^{FJ}(\mathbf{x}_A,\mathbf{x}_B,\omega)]^*\mathbf{R}_F(\mathbf{x}_A)^{T*} =$$
$$\int_J d\xi\,[\mathbf{T}_U^{FJ}(\mathbf{x}_A,\xi,\omega)]^*[\mathbf{T}_U^{FJ}(\xi,\mathbf{x}_B,\omega)]^T - \mathbf{I}\delta(\mathbf{x}_{h,A} - \mathbf{x}_{h,b}), \tag{9.1.4}$$

where, as in Section 3.3, $\mathbf{x}_h$ designates horizontal position. In the limit as $\mathbf{x}_A$ fuses with $\mathbf{x}_B$ we still have an integral over the lower surface J.

The arrival of waves from a distant earthquake at the level J will illuminate this surface in the general neighbourhood of an observation point $\mathbf{x}_A$. The narrow bundle of slowness around the geometric arrival will be coherent over a substantial distance and provide apparent sources along the surface J, not just immediately below the surface station. With many different earthquakes arriving with varying slownesses, the composite result can provide a good characterisation of the subsurface reflectivity.

Both of these treatments make it clear that the local reflection response is assembled from elements that sample well away from the receiver position (cf. Figures 9.5, 9.7). Stacking with compensation for the moveout for different events will produce an apparent vertical profile that is actually an averaged version of the nearby structure. Indeed, at later times contributions can come from virtual surface sources significantly displaced from the recording locations.

Although the auto-correlation of the surface response produces an estimate of the subsurface reflectivity including free-surface multiples, this is generally not a major problem unless there is a very strong contrast in the near surface. The reflection coefficients encountered at depth rarely exceed 0.05, and so the squared term in a surface multiple makes very little contribution. In the case of an ice–bedrock interface, as in Section 9.2.2, surface multiples are visible and their timing can help to constrain ice thickness.

9.1.2 Behaviour with Slowness

Even for a simple stratified model beneath a station there is considerable variation in the reflection response as a function of slowness, as imposed by the epicentral distance of events.

We illustrate the behaviour of the reflection response with slowness using synthetic calculations for a 1-D lithospheric model taken from the study by Kennett & Furumura (2016) on the nature of lithospheric heterogeneity across Australia. As discussed in Section 13.3 the resulting multi-scale model includes strong crustal variations and variations in the lithospheric mantle with horizontal correlation length of several kilometres and vertical correlation length around 0.5 km. These fine-scale features are needed to explain the long durations and large amplitudes of the *P* and *S* coda recorded by stations in northern Australia from events in Australia and from the Indonesian subduction zone.

In Figure 9.1 we show the reflection response for the 1-D model, including free-surface effects, for both incident *P* and *S* waves. We show the full span of slowness appropriate to teleseismic arrivals. We have constructed this reflection response by auto-correlation of the transmitted waves, incident at 290 km depth. The strong peak near zero time is reduced by tapering. The bands of slowness associated with the main seismic phases are indicated by shading behind the traces. Since the traces are represented in terms of reflection transit time τ the duration of the *S* wavetrain is longer. The reflection responses are built up by complex interference effects between the arrivals from the impedance variations and their internal multiples. The longer delays for *S* waves give somewhat more extended pulses and the apparent frequency is further reduced by the inclusion of attenuation in the calculations.

Across the full suite of teleseismic slowness there is little moveout for crustal

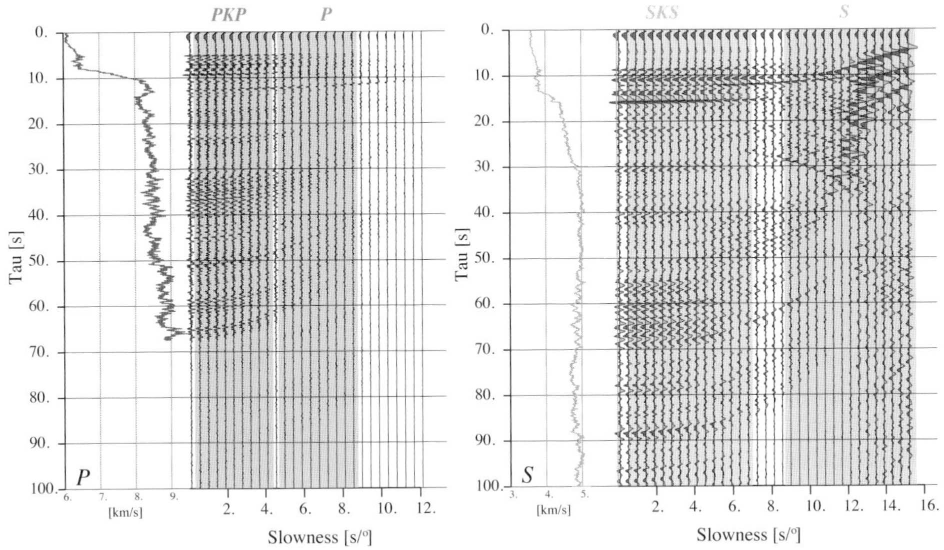

Figure 9.1 Simulation of reflection response including free-surface reflections constructed using the auto-correlation of the transmission response for incident *P* and *S* waves across the full suite of teleseismic slowness in the frequency band 0.5–3 Hz. The slowness regimes associated with the major seismic phase regimes are indicated by blocks of colour. The synthetics are calculated for a 1-D model extracted from a multi-scale heterogeneity model for Australia (Kennett & Furumura, 2016) at the cratonic location 21°S, 119°E. The *P* and *S* wavespeeds are plotted as a function of vertical reflection time τ to the left of each suite of reflection traces.

reflections, so that Moho reflections can be expected to stack well. However, for reflections from greater depth there is more change in reflection timing as a function of slowness.

For incident *P* waves, even for waves returned from 200 km depth (approximately 50 s two-way time) the change in moveout with slowness is slow but not entirely negligible. For very distant events, *PKP* phases show very little moveout and so can be expected to stack coherently without any corrections; this corresponds to the stationary phase condition employed by Ruigrok and Wapenaar (2012) for a study in Tibet. For most parts of the world, there is limited seismicity in near-antipodal configurations and so teleseismic *P* waves become important. Stacking is still possible provided that appropriate moveout corrections are made for each slowness value, as in (5.2.18).

Even without moveout corrections, teleseismic *P* waves are effective in rendering reflections from the Moho and later reflectivity to about 100 km depth (30 s). But, better results will be obtained for the mantle lithosphere when moveout corrections are applied to the auto-correlograms for individual events based on their arrival slowness before stacking. The build-up in the size of the moveout corrections with increasing depth means that it may become difficult to extract reflections from discontinuities deeper than 300 km even when using distant *PKP*.

For *S* waves, the best results for *S* reflectivity are to be obtained with *SKS*, provided there are events of sufficient size to overcome the noise. Far distant *S* can also give good results, but the moveout corrections are quite significant and need to be applied. A complication for incident *S* from closer events is that their slownesses couple to *P* waves in the crust, so that the apparent reflectivity includes *S*–*P* coupling. The effects are clearly seen in Figure 9.1 by the complex band of arrivals for slowness greater than 12 s/° for *S* waves. The coupling phenomena is the same as gives rise to the shear-coupled *PL* phase following the onset of teleseismic *S*. In certain circumstances such arrivals can be exploited directly to extract a virtual *PmP* arrival as discussed in Section 9.2.3.

9.2 Receiver-Based Studies

A variety of approaches have been used to exploit the stacked auto-correlation of seismograms at a single station. With the use of continuous data, both ambient noise and earthquake contributions help build the apparent reflectivity. Whereas, use of teleseismic arrivals alone provides explicit control on the slowness of the incident wave, and hence stacking can be made with well-defined moveout corrections. With stacking over many earthquakes, complications from the effect of source-side structure will be suppressed since these are incoherent between events.

9.2.1 Exploitation of Continuous Data

For the auto-corrrelation of records of ambient noise at a single station, we have coincident virtual source and receiver. The stacked correlation response as a function of time can include both energy that travels near vertical and is reflected beneath the station, and side scattering that returns to the station with a timing equivalent to a return from depth. Fortunately, if continuous data are employed without removal of earthquake effects, there is reinforcement of near-vertical propagation by the contributions from distant earthquakes. By working with just the vertical component of surface displacement attention is concentrated on *P* waves and thus we extract an estimate of *P* reflectivity. Because seismic energy arrives from events of many sizes at a wide range of distances, stacking over time will encompass a span of slowness and so induce some blurring of the apparent reflection response at later times (Gorbatov et al., 2013).

Ito et al. (2012) demonstrated the potential of the autocorrelation approach with a study exploiting 300 days of continuous records from stations in northern Japan. They used surface stations from the Tohoku University network and Hi-net stations in 100 m boreholes. The auto-correlograms showed little temporal variation and produced very distinct results with a strong association with the structure beneath the stations. A common feature at around 10 s lag time could be associated with *P*-wave reflections from the continental Moho. Later arrivals from 20–50 s lag time

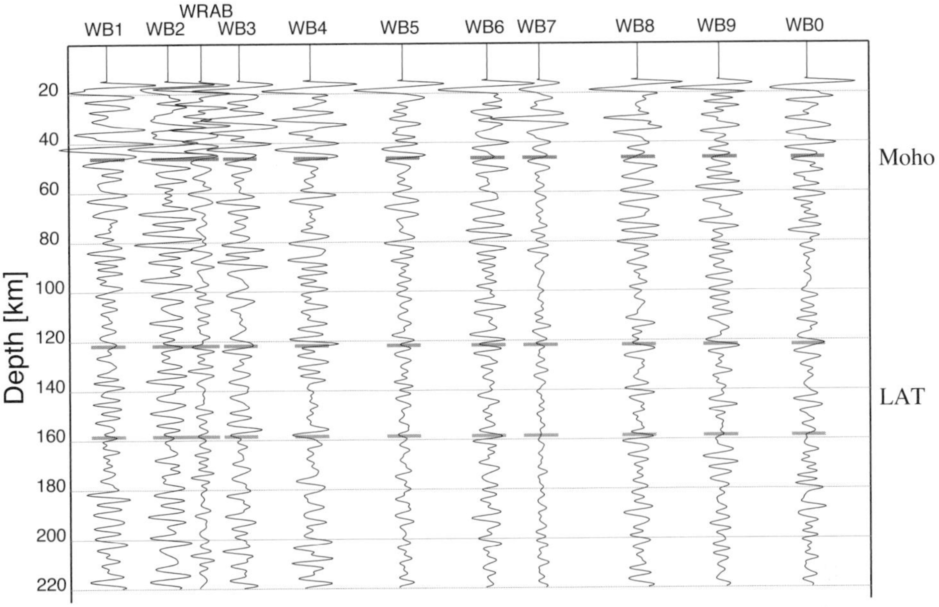

Figure 9.2 Spatial coherence of the stacked auto-correlograms for the stations on the northern (blue) arm of the Warramunga array (WRA) in northern Australia over a span of 25 km. The Moho depth is marked from the model of Salmon et al. (2013), and the shallower and deeper bounds on the lithosphere–asthenosphere transition (LAT) from Yoshizawa (2014) are indicated on each trace.

displayed strong geographic variations that could be linked directly to structure at the subducting Pacific plate and in the mantle wedge above.

Gorbatov et al. (2013) carried out stacked auto-correlations for 223 broadband stations, both permanent and temporary, across Australia. They employed running windows 6 hours long on continuous data from the vertical component, and then stacked the resulting correlograms. To minimise the effect of the dominant pulse at zero time lag, the first 5 s were zeroed, and a bandpass filter from 1.5 - 4 Hz was employed to suppress the effect of sidelobes. The use of frequencies above 0.3 Hz also tends to enhance the earthquake content of the continuous seismic records, and so provide more illumination from depth. The results showed strong *P* reflectivity in the lower crust, with some extension into the mantle. In locations where the Moho is sharp, the reflection from the crust–mantle boundary can be readily identified in the auto-correlogram traces. In many parts of Australia, the crust–mantle transitional is gradational and marked more by the end of a consistent style of crustal reflectivity than a dominant reflector.

This work was subsequently extended to more than 1200 stations across the continent including both short-period and broadband stations. Kennett et al. (2015b) used the full suite of auto-correlogram results at more than 750 stations across southeastern Australia to extract 180 Moho estimates from spatial stacks

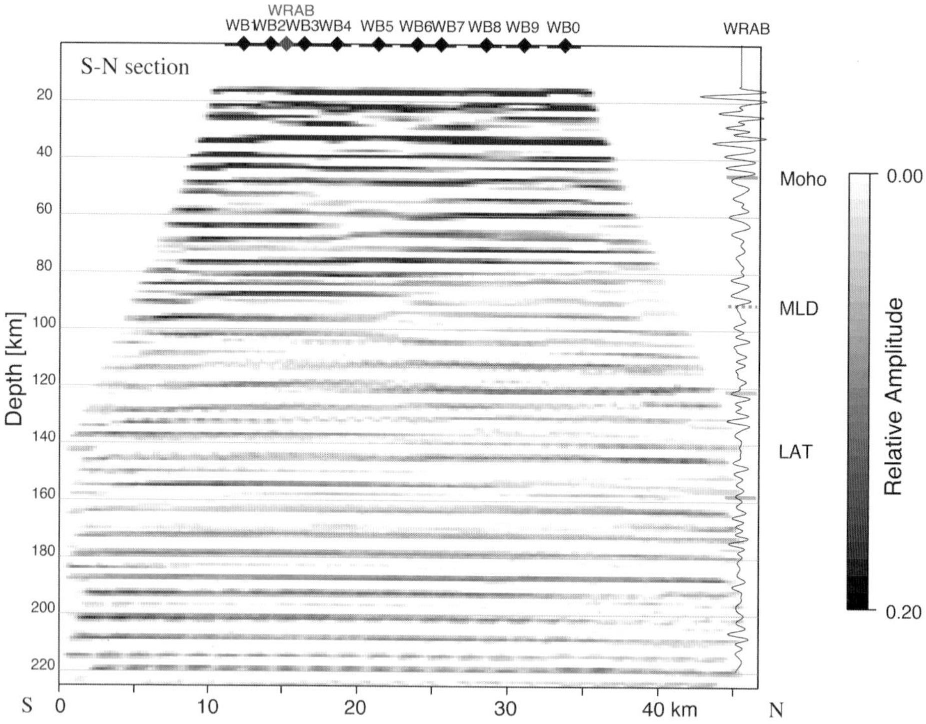

Figure 9.3 Migrated station auto-correlation results for the broadband stations of the Warramunga array (WRA) and the station WRAB projected onto profiles along the array arms. The MLD marker on the WRAB trace indicates the mid-lithospheric discontinuity depth inferred by Ford et al. (2010) from *Sp* receiver functions.

of the correlograms. Stations around the designated location are combined using Gaussian weighting with half width 0.5° to produce stacked *P* wave reflectivity. Picks of the base of crustal reflectivity were made with the aid of the previous Moho model, based on sparser data, and knowledge of the variation in the character of the crust-mantle boundary across the region.

The auto-correlogram results provide indications of noticeable reflectivity well beneath the base of the crust in the frequency band from 0.5– 4 Hz, which have been exploited by Kennett (2015) to study the character of the lithosphere and asthenosphere across a wide range of geological environments across Australia. The character of the apparent *P*-wave reflectivity varies with crustal age, in particular the Proterozoic shows stronger indications of scattering than Archean or Phanerozoic terrains. Nevertheless, where dense sampling is available the apparent reflectivity shows strong spatial coherence as illustrated in Figure 9.2 for the north-trending arm of the Warramunga array in northern Australia (see Figure 8.11 for the array configuration).

As demonstrated by Ito et al. (2012), spatial variations in apparent reflectivity can be imaged by depth migration of the auto-correlation results. They project their migrated auto-correlation results onto six corridors across northern Japan, approximately perpendicular to the plate boundary of the subducting Pacific plate. The images reveal the top of the subducting plate, and demonstrate features in the mantle wedge above that correlate well with regional seismic tomography.

A similar migration procedure has been applied in Figure 9.3 to the apparent reflectivity traces from the Warramunga array displayed in Figure 9.2. The stacked auto-correlograms at each station are built up from contributions from a range of slowness from beneath the station, though they appear as a single trace equivalent to a zero-offset seismic reflection record. In the migration process allowance is made for the range of sampling broadening with depth beneath each station by mapping each station trace onto a suite of trajectories for a range of slownesses, with conversion of time to depth. The depth migration was performed with ray tracing in the AK135 velocity model with the crust thickened to 45 km to reflect the local situation. All the migrated contributions from the full set of stations are then stacked, using cells 0.5 km wide and 1 km deep, to provide an image of the reflection structure beneath the array. The results are best constrained over the central span of the stations where many station contributions overlap. The quality of the results diminishes at the edges, because only one or two stations are available.

In Figure 9.3 the amplitudes are enhanced so that weak reflections at depth can still be recognised. The crustal reflection response shows some variability across the northern arm of the array. Distinctive reflections can be readily tracked to a depth around 90 km. The base of these reflections corresponds closely to the estimate of the depth to the mid-lithosphere discontinuity (MLD) obtained by Ford et al. (2010) from the *Sp* receiver function at much lower frequencies. The migration of the auto-correlation results provides one of the most direct ways of imaging heterogeneity in the lithosphere. The horizontal scale of variation of reflectivity is of the order of 10–20 km, and much shorter in the vertical direction, which is consistent with other strands of information on cratonic heterogeneity in Australia (Section 13.3).

9.2.2 Extraction of the Reflection Field from Teleseisms

The coda and reverberations associated with the arrival of major seismic phases such as *P*, *PKP*, *S*, and *SKS* have been extensively analysed using receiver function techniques exploiting conversions between wave types. Just the same portions of the seismogram can be used for auto-correlation to produce reflectivity estimates for *P* and *S* (Section 5.2.1).

The direct use of auto-correlation on teleseismic arrivals suffers from the strong influence of the spike at zero time lag, and so to extract reflectivity for short reflection times additional processing is needed. Pham & Tkalčić (2017, 2018) have demonstrated how such techniques can be used for shallow reflectors, such as the ice–rock interface in Antarctica. They use spectral whitening to reinforce the high-frequency end of the spectrum to compensate for the decay of earthquake

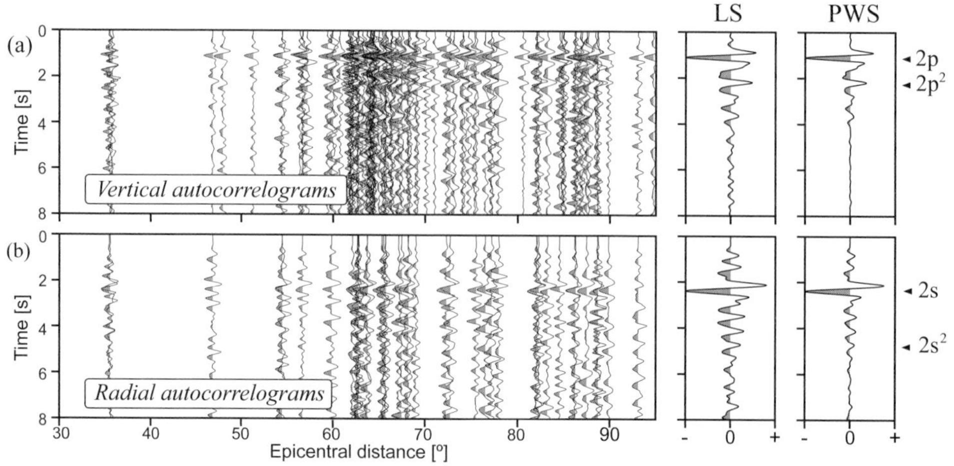

Figure 9.4 Example of auto-correlation of teleseismic arrivals at station BYRD in Antarctica (80.02°S, 119.47°E) as a function of epicentral distance: (a) vertical auto-correlograms, (b) radial auto-correlograms. Each set is accompanied with a linear stack (LS) and a phase-weighted stack (PWS) with the timing of major arrivals indicated. The traces are displayed after application of a taper function and a bandpass Butterworth filter with pass band 1–5 Hz. Spectral whitening was employed with width, ΔW, of 1.0 for the vertical and 0.5 for the radial auto-correlograms. [Courtesy of T.-S. Pham.]

source spectra and attenuation on the propagation path.

For P-wave coda with spectrum $\{s_n(\omega)\}$, the whitened spectrum $\{\hat{s}_n(\omega)\}$ can be constructed by smoothing around each frequency sample (Tauzin et al., 2019):

$$\hat{s}_n = s_n/w_n, \quad \text{with} \quad w_n = \frac{1}{2N+1}\sum_{j=n-N}^{n+N} |s_j|. \tag{9.2.1}$$

The phase spectrum is maintained, but the amplitude spectrum is modified by the average around the specific value controlled by the length of the averaging interval $\Delta W = 2N\Delta f$, where Δf is the discrete sampling interval in frequency. The whitened spectrum $\{|\hat{s}_n|^2\}$ is converted into the modified auto-correlogram by inverse Fourier transformation. This time series is then tapered with a cosine function to suppress the inherent zero-lag peak, and bandpass filtered. For the ice–bedrock system, Pham & Tkalčić (2018) use the frequency band 1–5 Hz.

Tauzin et al. (2019) demonstrate that the use of spectral whitening with correlation has the property of providing a better projection of P-wave energy onto the vertical and radial component than the conventional deconvolution procedures used in receiver function processing. Further improvement can be expected from the use of the LQT rotation or PSH transformation (Section 5.2) to emphasise P waves and isolate conversions. Figure 9.4 presents P and S reflectivity results for the station BYRD in west Antarctica, that lies on an ice sheet approximately 2.25-km thick. The left-hand panels show the vertical and radial autocorrelations of the *P*-wave coda for individual events as a function of epicentral distance. Negative pulses around 1 s (marked as 2p) and near 2 s (marked as 2s) can be seen on most auto-correlograms and arise from reflections from the ice–rock interface beneath the station. There is little moveout with epicentral distance and so the signals are markedly enhanced by direct stacking. The use of phase-weighted stacking with order 1 (Schimmel & Paulssen, 1997) provides better signal strength for the reflectors than a linear stack, at the expense of some mild signal distortion. As a result, the timing of the *P* and *S* reflections from the bedrock can be well determined. In Figure 9.4 we also see surface multiples ($2p^2$, $2s^2$), and their timing helps to confirm the interpretation of ice thickness.

Ruigrok & Wapenaar (2012) have exploited two dense lines of portable broadband seismometers from the Hi-CLIMB experiment in Tibet (2002–2005), to produce migrated reflection images. Each line has around 60 stations with spacing of 5 km in the line parallel to the Himalayan front and 10 km in the line that extends almost perpendicularly into Tibet. The resulting coverage is sufficient to image structure beneath the lines. *PKP* phases with small slowness have been extracted for events in South America and the eastern Pacific Ocean, avoiding epicentral distance ranges with interaction between the branches. The small slownesses approximate closely to Claerbout's (1968) concept of vertical propagation, and direct stacking can be made without moveout correction.

For the Hi-CliMB data, each segment of vertical-component seismogram has the instrument response removed by deconvolution before autocorrelation. This is followed by bandpass filtering (0.04–0.8 Hz) and spectral balancing of a similar style to (9.2.1). At each station the contributions from all the sources are then summed. To minimise the influence of the coherent zero-lag peak, a stack over all stations is used to estimate the net shallow time response. A scaled version is subtracted from each trace, and this process largely suppresses the artefact, with only a minor influence on the apparent reflectivity. From this point standard reflection processing techniques are employed. Predictive deconvolution is used to enhance primary reflection at the expense of surface multiples (Verschuur, 2006). The traces are then migrated using post-stack Kirchhoff migration with a local velocity model derived from seismic tomography down to 75 km and the AK135 structure beneath. Depth conversion is made with the same hybrid model. This study demonstrates that successful reflection imaging can be achieved for complex structures using low-frequency earthquake waves, building on procedures developed for conventional reflection work.

The Himalayan region is fortunate to have a range of seismicity at large epicentral distances, so that *PKP* phases can be directly exploited. For many regions there will be insufficient events at such distances beyond 135° to make an effective analysis, but substantially more events in the standard teleseismic range (30–90°). For the teleseismic regime, auto-correlation procedures can still

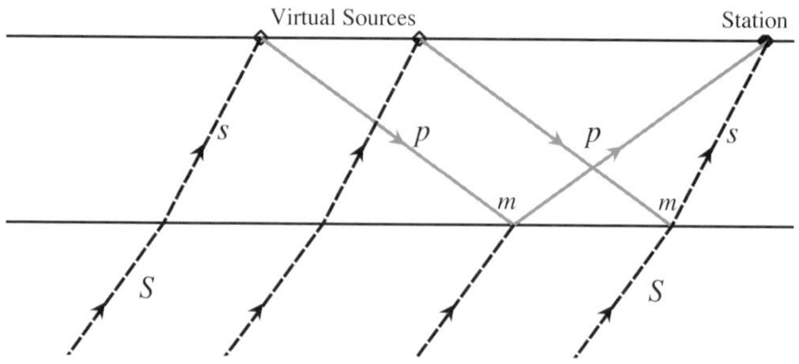

Figure 9.5 Ray diagram of the main phase contributions exploited in virtual seismic sounding from incident near-teleseismic *S* waves.

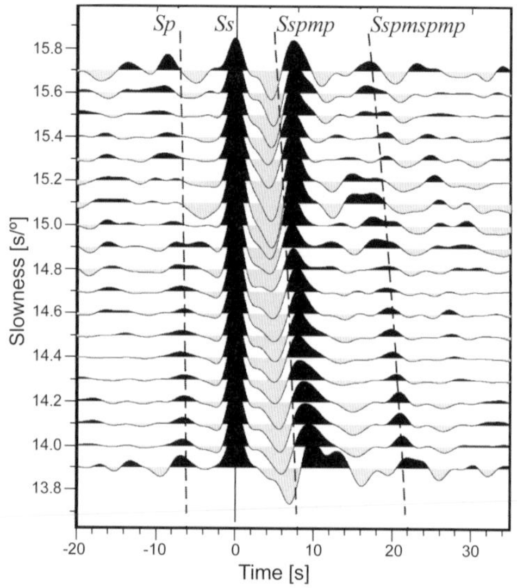

Figure 9.6 Teleseismic *S* waveforms at station WRAB, binned by slowness, and aligned on direct *S*. The *Sspmp* arrivals show clear moveout with slowness (epicentral distance 30–50°), and are accompanied by precursory *Sp* phase and later reverberations *Sspmspmp*. The dashed lines are the predicted arrival times for the local crust. [Courtesy of D. Thompson.]

be exploited but significant moveout corrections must be applied to bring all events onto a common footing, as can be seen from Figure 9.1.

9.2.3 Virtual Seismic Sounding

For earthquakes in the distance range from 30–50° incident *S* waves at the base of the crust have suitable slownesses to couple directly into crustal *P* waves. Such waves can travel for substantial distances in the crust and interfere to produce low-frequency shear-coupled *PL* wavetrains.

Conversion from incident *S* waves occurs dominantly at the free surface. The converted *P* waves can be then be reflected at the Moho, beyond the critical angle, to arrive at a receiver with significant amplitude. The ray paths are indicated in Figure 9.5. The relative timing of direct *Ss* waves compared to the conversion *Sspmp* is equivalent to that for a *PmP* arrival from a virtual source at the conversion point. Subsequent reverberations in the crust build the *PL* train.

This equivalence has been exploited in *virtual seismic sounding* introduced by Tseng et al. (2009) for a study in Tibet, using the frequency band from 0.05– 0.5 Hz. The use of lower frequencies means that the virtual *pmp* arrival is insensitive to local small-scale heterogeneity, including sediments, and still reflects well from a transitional Moho. The timing of *pmp* will represent the average Moho, because the post-critical reflection does not notice the details of the reflector. In consequence, such Moho estimates are likely to be shallower than those based on the base of the transition to the mantle, e.g., from seismic reflection work. The separation of the *pmp* contribution from later reverberations is most direct for thick crust, but with care the approach can be applied in many situations.

The original application of this concept used direct *S* waves from deep earthquakes, to achieve a simple pulse form impinging on the crust, and minimal impact of attenuation in the mantle. Convenient deep events are often not available in the required distance range from an observation point, and then shallower, more complex, events will need to be employed. With such a broader distribution of source depths, effects from source-side structure may get mapped into apparent arrivals near the receiver. To minimise such source-side contamination, Yu et al. (2013) have used a transfer function approach. From the vertical and radial components they make a rotation to the *VWT* frame (Section 5.2.1) to construct an estimate of the incident *S* waveform from the *W*-component. This wavetrain is then used to deconvolve both the vertical and radial components of the observations. In this way the source waveform and near source effects are removed. Further since the *W*-component contains *S*-type signals due to scattering near the receiver, *S* multiples such as *Sspms* and related reverberations are suppressed. Yu et al. (2013) advocate the use of an empirical rotation angle to the *VW* coordinates determined from the nature of the particle motion.

In Figure 9.6 we show an example of the application of the virtual seismic sounding approach to the station WRAB at the Warramunga array in Australia (Thompson et al., 2019). The traces are organised by slowness, so that closer epicentral distances lie at the top, and are aligned on the direct *S* arrival. The deconvolution procedure of Yu et al. (2013) was used to reduce source-side effects, and later reverberations. The Moho beneath WRAB shows gradational character (cf. Figure 9.3) but in the 0.05–0.5 Hz band employed produces strong post-critical reflections. Ahead of direct *S* we see the precursory *Sp* arrivals exploited in *S*-wave receiver function studies.

9.3 Imaging with Auto-correlation

Where stations are relatively closely spaced, a very effective procedure for the exploitation of receiver functions is common conversion point (CCP) stacking

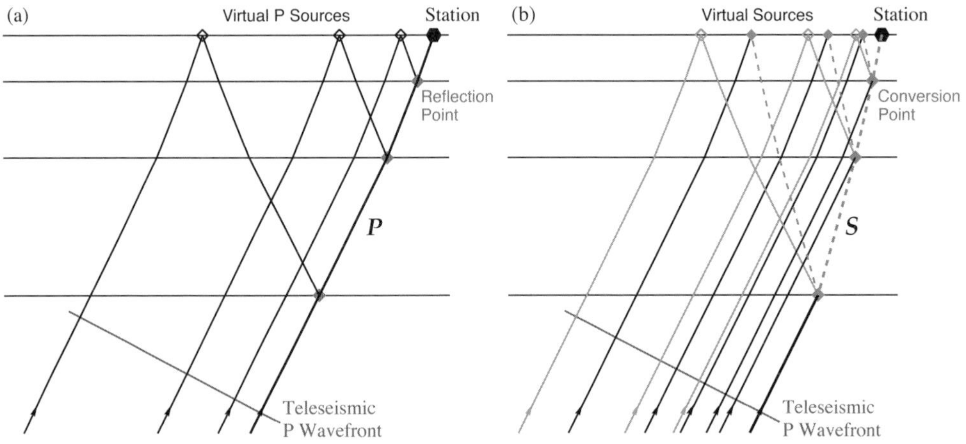

Figure 9.7 Comparison of near-receiver imaging using direct and converted waves. (a) Reflections from the surface for transmitted *P* waves provide virtual sources for reflection from depth. The trajectory of the reflection points follow the *P* ray path. (b) Common conversion point stacking of receiver functions tracks the *S* ray path to the station, with virtual sources at the surface for both *P* and *S* waves associated with surface reflection.

(e.g., Wittlinger et al., 2004). For incident *P* waves, the *Ps* receiver functions from multiple events are plotted along the *S* ray path to the station as a function of depth, which can be regarded as a simple form of migration. The results are are then spatially stacked into discrete cells to provide continuous coverage of the subsurface. The sampling of the converted waves in transmission lies in a relatively narrow cone beneath each station. Thus, unless stations are tightly spaced, there are gaps in near-surface coverage. Nevertheless, such CCP stacks can provide highly informative images of structure as, e.g., the work of Kawakatsu & Wadata (2007) that tracks the dehydrating serpentinite layer on the top of the subducting Pacific plate beneath central Japan, with clear rendering of the continental Moho and mantle wedge.

Similar CCP stacking has been employed with *Sp* receiver functions, for which the span of sampling is wider because the converted *P* waves travel at a larger angle to the vertical than *S*. The frequencies used for *Sp* receiver functions are much lower than for *Ps*, in part due to the stronger teleseismic attenuation for *S* waves than *P* waves. As a result, images from *Sp* studies are less detailed, but have been employed for full lithospheric studies, e.g., the work of Hansen et al. (2015) using the full set of data from the USArray to study the base of the lithosphere and mid-lithospheric discontinuities.

For incident *P* waves, the common conversion point approach can be extended to include waves reflected from the free surface as well as direct transmission (Figure 9.7b). All the modes of propagation link to the same points on the *S* raypath to the station where *P* is converted in transmission. The contribution from reflected waves can be regarded as arising from virtual sources at the surface

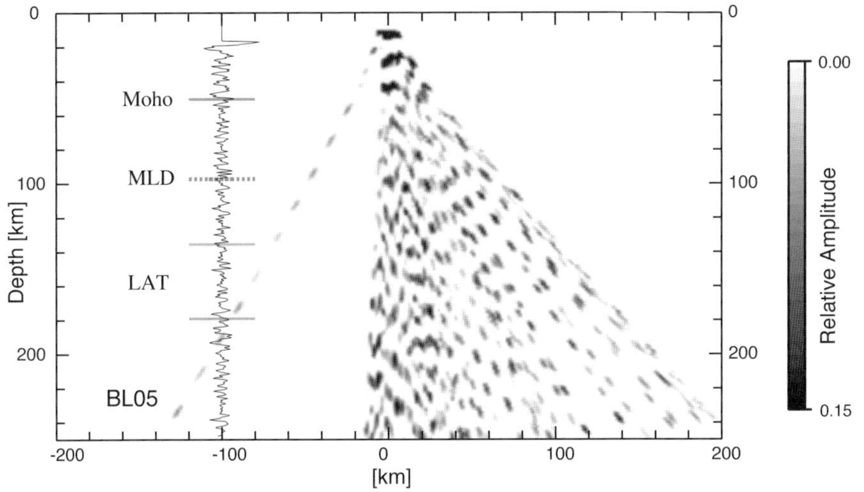

Figure 9.8 Mapping of the auto-correlation of teleseismic events recorded at station BL05 onto a south–north profile. Each event trace is mapped from time to depth following the ray path for its incident slowness in the AK135 model, modified to include a 45 km thick crust. At the left is displayed the stacked auto-correlation of continuous data with depth conversion using the same model. The depth of the Moho and the lithosphere–asthenosphere transition (Yoshizawa, 2014) are shown, together with an estimate of the location of a mid-lithospheric discontinuity.

that radiate *P* waves for *P–P–S* interactions and *S* waves for *P–S–S* paths. These additional contributions contain extra information that can be exploited to obtain improved images, but also sample near-surface zones away from the immediate vicinity of the stations. Tauzin et al. (2016) recommend the use of phase-weighted stacking (Schimmel & Paulssen, 1997) to build a multi-mode image that minimises artefacts.

In a similar way we can exploit the reflected waves from virtual surface source in the auto-correlograms from incident teleseismic *P* waves. Now we conduct common reflection point stacking (CRP), since the reflection points for all interfaces lie along the *P* ray to the station (Figure 9.7a). As depth increases, the virtual sources lie further from the station and so variations in local structure can get mapped into apparent reflectivity. For each station, we map the auto-correlation traces onto the *P* raypath and then stack all the contributions at each spatial location. As in CCP stacking, it is often convenient to project the results onto a single profile, rather than work directly with 3-D variation.

In Figure 9.8 we show the application of the common reflection point approach to station BL05 from the BILBY experiment in central Australia. This nearly north–south deployment of portable broadband stations links to the permanent Alice Springs (ASAR) and Warramunga (WRA) seismic arrays (Figure 9.9). The trend of geological structures is perpendicular to the line.

Figure 9.8 shows the individual auto-correlograms for incident *P* waves plotted

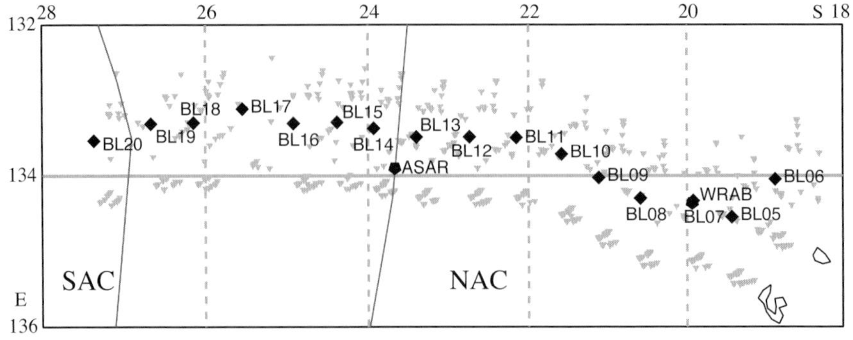

Figure 9.9 Piercing points at 100 km depth for teleseismic *P* waves used in the construction of the reflection point images shown in Figure 9.8. The station locations and craton boundaries are indicated, along with the line of section at 134°E. SAC – South Australian Craton; NAC – North Australian Craton.

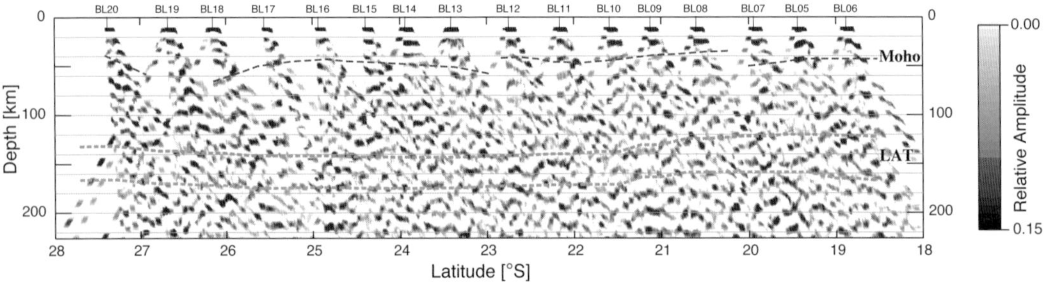

Figure 9.10 Common reflection point stacking from teleseismic arrivals at the BILBY stations projected onto a profile along 134°E. The section is shown with a slight vertical exaggeration. The Moho surface from Sippl (2016) and the shallower and deeper bounds on the lithosphere–asthenosphere transition (LAT) from Yoshizawa (2014) are indicated with dashed lines.

along the ray paths in the AK135 model, modified to include a thicker 45 km crust. The slowness is determined by the epicentral distance of the event. We also display the stacked auto-correlation from continuous data (vertical component) for the station BL05 with depth conversion, and indications of major features such as the Moho and the location of the lithosphere–asthenosphere transition from Yoshizawa (2014). Most of the events lie to the east and north of the station, so the projection onto a south–north profile shows strong asymmetry. The Moho is evident in the group of teleseismic auto-correlograms, and there is a possible link to a mid-lithospheric discontinuity. As in the results from continuous data, there is no specific signature of the lithosphere–asthenosphere transition. Strong interfaces at depth can produce contamination of CRP results with converted *S* waves (Tauzin et al., 2019), but at BL05 and other BILBY stations such effects will be weak.

The full configuration of the BILBY deployment is shown in Figure 9.9, together with the piercing points for the incident teleseismic *P* waves at 100 km, which

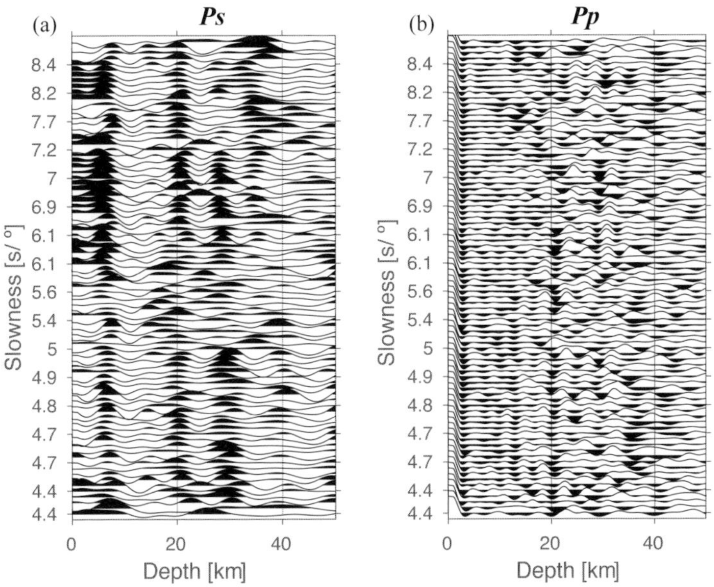

Figure 9.11 Station ME25 of the FAME Mendocino experiment: (a) depth converted radial receiver functions (*Ps*); (b) auto-correlation of vertical component (*Pp*). Traces are ordered by slowness. [Courtesy of B. Tauzin.]

provide an indication of the region which is sampled in the construction of the CRP stack in Figure 9.10 (as in Kennett & Sippl, 2018). With the combination of teleseismic auto-correlograms from all the BL stations projected onto a south-north profile along 134°E, we produce the equivalent of a low-fold reflection section through the entire lithosphere.

The Moho lies at the base of crustal reflectivity and agrees well with the results of Sippl (2016) from CCP stacking. There are a number of intriguing features at depth, including the south-dipping reflector visible from 26–23°S that has a possible link to just below the Moho at the northern end of the line. A similar feature may be discerned in the CCP stacks of Sippl (2016). This dipping feature may represent a former detachment surface associated with multiple stages of compression and extension in central Australia (Kennett & Sippl, 2018).

Tauzin et al. (2019) have made a detailed comparison of CRP stacking and CCP stacking, exploiting all conversion modes, for the FAME Mendocino array in northern California. In Figure 9.11 we show the results from auto-correlation of the vertical component (*Pp*) at station ME25, compared with the *Ps* receiver function constructed by deconvolution of the radial component by the vertical. Both traces are converted from time to depth. The higher frequency content of the auto-correlogram *Pp* is immediately apparent, and offers the possibility of higher-frequency imaging.

Projected profiles using data from the whole FAME Mendocino array are displayed in Figure 9.12. The upper panel shows the results from multimode CCP

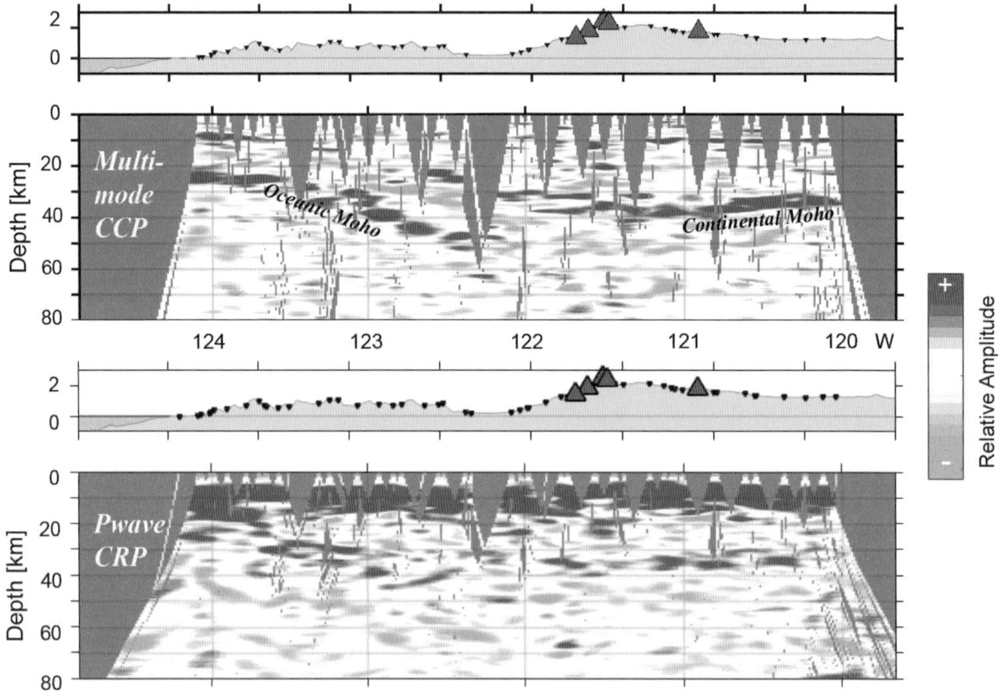

Figure 9.12 Seismic imaging from the dense broadband seismic network of the FAME Mendocino experiment. Upper panel: multimode CCP stacking of radial receiver functions. Lower panel: CRP stacking using vertical-component auto-correlograms. [Courtesy of B. Tauzin.]

analysis, and the lower the use of CRP from vertical-component auto-correlograms (Tauzin et al., 2019). The stacks are projected onto a common profile, and show good correspondence in the major features, such as the continental Moho and the oceanic Moho associated with the subducting Juan de Fuca plate. Both images have been constructed using the same filtering and lateral smoothing, and so the frequency differences evident in Figure 9.11 are suppressed. The wider sampling around individual stations with incident *P* waves is very clear, and as a result the CRP stacks, in principle, provide better definition of the near surface. However, as noted above, lateral heterogeneity in the surface zone can be mapped into apparent structure.

9.4 Correlations in the Reflection Field

Applications of correlation procedures in exploration seismology have largely focussed on exploiting the time-subtractive aspect of correlation to transform data recorded in one configuration so that it can be exploited in a different way. Schuster (2009) provides a wide range of examples of such interferometry, linking surface and down-hole observations to provide improved illumination of structure at depth.

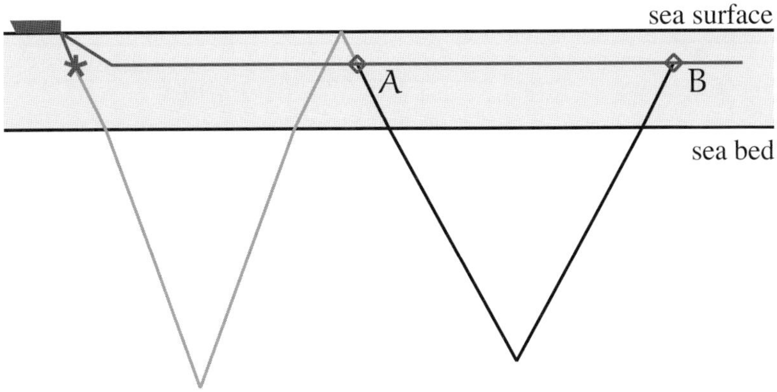

Figure 9.13 Creation of virtual sources by correlation: correlation of multiples in a marine streamer to extract the reflection response R_{AB} for the outer part of the cable, indicated as a heavy line.

By cross-correlating portions of the wavefield that have propagation elements in common, this common component is suppressed and the difference field emphasised. The situation is similar to our discussion of the emergence of regular seismic phases in correlation on a global scale (Section 6.1.2), and the conversion of the transmission response for teleseismic arrivals to reflections.

As a simple example we consider a marine reflection experiment in a scenario where surface multiples are important (Figure 9.13). Seismic waves are produced by an airgun array and reflected from the surface, and then from the sea surface before return to depth. The multiple reflection will be recorded on many traces of the recording cable. If we can isolate the multiple as it passes a hydrophone at moderate offset (A) and correlate this with the same multiple recorded on the outer traces of the cable (B), then we can extract the reflection leg for a virtual source at A recorded at B as indicated schematically in Figure 9.13, with a heavier line for the extracted component. As we have seen in Section 6.1, quite stringent conditions are needed for the cross-correlation process between two stations to produce the equivalent of propagation between the two points. All exploration configurations have a limited width for the source and receiver arrays, which are smaller than required for the correlation process. Further, the azimuthal distribution of effective sources is limited, unless there is strong local scattering. As a result the correlation result will provide only a partial recovery of the response due to a source at A and receiver at B, with contamination by artefacts. Schuster (2009) discusses the range of mitigation measures that can be used to improve the results.

Bakulin & Calvert (2006) describe the concept of extracting a virtual source by exploiting down-hole recordings from surface sources (Figure 9.14), based on a patent application submitted in 2002. Their idea is to use down-hole information

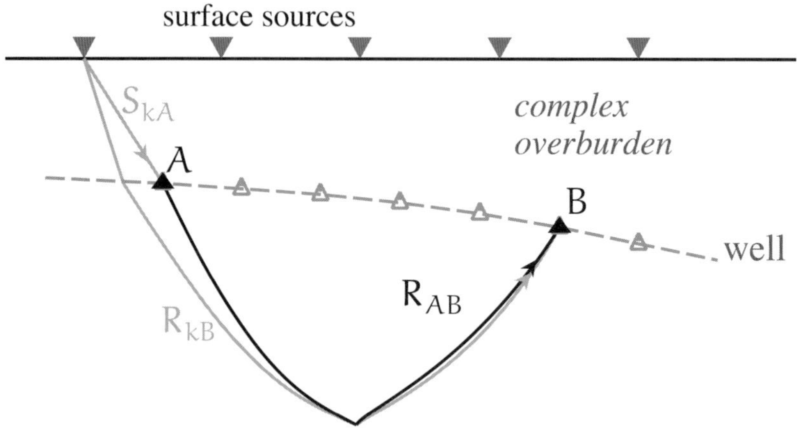

Figure 9.14 Creation of virtual sources by correlation: use of a down-hole geophone to create a virtual source at depth at A and reduce the influence of complex shallow structure by extracting R_{AB}.

to get below complex structure in the near surface. With a number of geophones deployed in a well with a strong horizontal deviation, the correlation of the downgoing field recorded at one site with the full reflected field at another provides a virtual source–receiver pair focussed on the structure below the well. In this situation the radiated energy from sources at or near the surface is recorded at A as S_{kA} and correlated with the reflection response (R_{kB}) at B, to leave an approximation to the reflection R_{AB} between the geophones from beneath the well string. Once again considerable care is needed to focus attention on the appropriate part of the correlated field, and Bakulin & Calvert (2006) use an extra step of source wavelet deconvolution to mitigate artefacts.

A further use of correlation methods in reflection work has been in the suppression of surface waves (e.g., Halliday et al., 2007). Such ground roll is strongly excited by vibrator sources and receiver groups are usually used to minimise the effect on the reflected field, but at some cost in resolution. As we have seen above, the strongest component of the correlation field for surface sources is that associated with fundamental mode surface waves. Hence, cross-correlation between surface receivers can be used to extract this surface-wave signal from the recorded wavefield without any specific knowledge of local structure. Then adaptive subtraction can be used to suppress the ground-roll component, with little damage to the body-wave field reflected from depth. Such an approach has particular value in complex experiments where individual node recording is employed, and so ground roll arrives undiluted.

Part III

Interaction of Seismic Waves with Heterogeneity

10

Deterministic and Stochastic Heterogeneity

The Earth shows significant variations in structure in three dimensions on many scales, though the dominant variation is with depth as a consequence of the effect of pressure. One of the early successes of seismology was the delineation of the main features of the radial variations in seismic wavespeeds. In consequence, the term *heterogeneity* is often used just to describe deviations from this radial structure.

As knowledge of Earth structure has improved, striking contrasts in seismic structure such as the presence of fast wavespeeds in subduction zones have emerged. In studying such features it is desirable to take account of the main contrasts, and then concentrate on deviations from this basic structure. Heterogeneity at the microscopic level is likely to manifest itself in the form of anisotropy (see Section 10.4), which subtly modifies the nature of wave propagation.

Many of the most readily extracted characteristics of the seismic wavefield depend on integral properties along a propagation path, as in the travel time and attenuation of a seismic phase. With sufficient observations it is possible to build up the broad features of seismic wavespeed structure, but inevitably fine detail is suppressed even though it is undoubtedly present in the Earth. Direct evidence for small-scale features comes from the nature of seismic-wave propagation at high frequencies, where complex trains of scattered waves can have considerable duration.

Because the fine scales of Earth structure cannot be directly probed, we have to resort to indirect representations. It is common to present stochastic models for heterogeneity in which material properties are described by a limited number of parameters. Except in unusual circumstances, such as associated with volcanic activity, the properties of the Earth do not vary on human time scales. Thus, for seismic waves we are not able to exploit the averaging methods developed in oceanographic and atmospheric studies that equate temporal results to ensemble properties of a stochastic system. Instead we must regard the stochastic forms for seismic heterogeneity as a convenient summary of behaviour. The nature of heterogeneity will be position dependent, with significant potential variations in depth, e.g., between crust and mantle.

We have therefore to understand the interaction of various styles of heterogeneity, and their influence on the wavefield. Advances in computational seismology provide valuable tools for simulating wave propagation through complex structures, but attention needs to be focussed on one, or a few, representative structures drawn from the statistical ensembles. Fortunately, much can be learnt about the nature of heterogeneity in a broad sense, but detail can be elusive. We can find structures that are compatible with observations, but should not assume that they are necessarily required for a suitable fit.

10.1 Heterogeneity in the Earth

A range of seismological techniques and observations have been used to extract information about 3-D variations in structure in the mantle of the Earth. Many of the properties of seismic waves reflect averages along the different pathways that energy has travelled. Our understanding of the properties of the whole globe comes from a mix of information from low-frequency energy sampling the whole Earth, and results from higher-frequency arrivals with more specific geographic coverage. The highest frequencies are used in understanding the complex structure of the outer skin of the Earth – the crust.

The splitting of the frequencies of the free oscillations of the Earth with respect to the angular order m is an important source of information for the lowest frequency modes that have sensitivity to density and seismic wavespeed. Splitting studies constrain the longest wavelengths of 3-D variation (e.g., Ishii & Tromp, 2001; Kuo & Romanowicz, 2002; Resovsky & Trampert, 2003).

At intermediate frequencies, information on 3-D structure has been developed through the fitting of portions of seismograms using a perturbation development based on normal mode theory. Surface waves can be described through the summation of simple mode branch contributions, but body waves need multiple branch contributions with coupling between the coefficients. Such waveform tomography (e.g., Dziewonski & Woodhouse, 1987; Megnin & Romanowicz, 2000; Masters et al., 2000; Panning & Romanowicz, 2006) can provide coverage of much of the mantle for shear wavespeed.

The arrivals of seismic phases are exploited in a variety of ways. High-frequency information can be derived from the compilations of arrival-time readings from seismic stations across the globe. Careful reprocessing of such catalogues, including relocation of events and association of arrivals, provides a major source of information for *P* and *S* waves and many later phases (see, e.g., Engdahl et al., 1998). Good results can be obtained for both *P* wavespeed (e.g., van der Hilst et al., 1997) and *S* wavespeed (e.g., Widiyantoro et al., 2000). The bulletin results for *S* are much noisier than for *P*, but the effects of structure are also larger.

At somewhat lower frequencies a substantial data set has been built up for both absolute and differential travel times through the use of cross-correlation methods, including the use of synthetic seismograms. The way in which such lower-frequency waves interact with structure is not fully described by ray theory. A number of schemes have been developed to allow for the effect of the wavelength of the propagating phase and so represent the zone of interaction around the propagation path for the frequency range under investigation. These extensions from ray theory are beneficial for the imaging of regions of lowered wavespeed, e.g., the work of by Montelli et al. (2003) imaging mantle plume structures. Nolet (2006) provides a detailed treatment of the use of finite-frequency techniques in seismic tomography.

With tomography exploiting the arrival times of seismic phases, relatively high resolution (200 km or better) can be achieved for about half the mantle (e.g., van der Hilst et al., 1997; Kennett et al., 1998; Bijwaard et al., 1998; Grand, 2002). A number of studies, e.g., Ritsema et al. (2011), use a wide range of different styles of information to try to achieve the maximum level of sampling of the Earth's mantle. However, such an approach combines multiple classes of information with very different frequency content and sensitivity to structure. The results are therefore somewhat dependent on the particular style of analysis, and can carry dependency on external factors, such as the attenuation structure used to correct for frequency dependence.

In the outer 400 km of the mantle, body-wave observations can be complemented by the exploitation of seismic surface waves that are strongly excited by shallower earthquakes. Images of 3-D structure can be built by combining information from many paths. Resolution of upper-mantle structure at about 200 km scale can be achieved in particular regions, e.g., Fishwick et al. (2005) for Australia. Across the globe, 500 km resolution is possible in most places (Debayle et al., 2005; Schaeffer & Lebedev, 2013).

With the development of full waveform inversion methods (see Chapter 17), direct calculations in fully 3-D models can be made and compared directly to observations. These methods require considerable computational resources, but avoid many of the limitations and approximations used in other styles of tomography. Three-dimensional models been extracted at many scales, from the regional (e.g., Fichtner et al., 2009) to the global (e.g., French & Romanowicz, 2014; Bozdağ et al., 2016). To keep computational effort in bounds, analysis has concentrated on lower frequencies. This frequency limit controls the resolution that can be achieved. The direct exploitation of waveforms means that amplitude information can be taken into account with allowance for focussing and defocussing by structure. Both *P*- and *S*-wavespeed structure can, in principle, be extracted. Though often simplifications to just *S*-wavespeed inversion are made to reduce the size of the model. Further, issues such as wavefront healing that can lead to misleading estimates of seismic travel times for moderate-frequency pulses (Malcolm & Trampert, 2011) are avoided completely. As a result, the full waveform methods have proved very effective in improving resolution of low-wavespeed features, such as those associated with plumes (e.g., French & Romanowicz, 2014).

Heterogeneity regimes

The highest levels of 3-D heterogeneity are found near the Earth's surface and near the core–mantle boundary. More subtle features appear in the mid-mantle, including relatively narrow zones of elevated wavespeed that are most likely associated with past subduction. In the uppermost mantle, the ancient cores of continents stand out with fast wavespeeds, while the mid-ocean ridge system and orogenic belts show slow wavespeeds. Below 400 km depth the high-wavespeed anomalies are mostly associated with subduction zones. In some regions the fast-wavespeed structures extend to around 1100 km depth, but in a few cases tabular features seem to extend to 2000 km or deeper. The base of the mantle shows long-wavelength regions of higher wavespeeds, most likely associated with past subduction, and two major regions of slow wavespeed beneath the central Pacific and southern Africa, which may represent sites of upwelling of hotter material.

The scale lengths of variation from the various types of tomography stay relatively constant throughout the mantle, though the amplitude varies (see Figure 10.2). Often models are depicted using the same size representation of the globe at all depths without allowance for the effect of sphericity, and the scales appear to get larger with depth. When the decrease in radius at greater depth is taken into account there is much greater concordance in scales of variation.

The complex structures at the base of the mantle provide a distorting lens for waves entering the core that can influence apparent structure. Nevertheless considerable progress has been made in elucidating the seismic velocity structure in the outer core, which appears to be close to adiabatic due to the rapid mixing in the convecting core. The solid inner core has enigmatic structures that have become progressively more complex with time as more and more detailed investigations have been made, which sample different parts of the inner core in detail (Tkalčić, 2015). There are indications of quasi-hemispherical differences in inner-core structure, as well as variations in radius.

The early success of seismic tomography came from the striking images of large-scale 3-D structure in the mantle and, later, of the details of the fast wavespeed features associated with subduction zones. The interpretation of such tomographic images is based on the patterns of variations of seismic wavespeed. Thermal processes can be expected to play a major role, but chemical heterogeneity could also be important, particularly in the regions with strong variability at the top and bottom of the mantle.

Results for a single wavespeed are not sufficient to indicate the nature of the observed anomalies. Recent developments in seismic imaging are therefore moving towards ways of extracting multiple images in which different aspects of the physical system are isolated. This may be from *P*- and *S*-wavespeed images

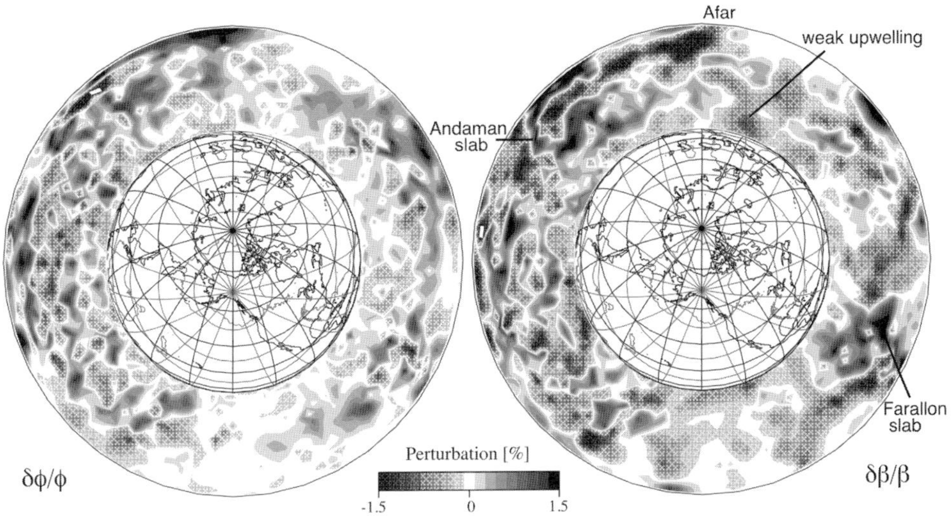

Figure 10.1 Cross-section through the variation of bulk-sound speed $\delta\phi/\phi$ and shear wavespeed $\delta\beta/\beta$ relative to the AK135 reference model, from the 3-D model of Kennett & Gorbatov (2004). The section crosses the mantle feature beneath South America associated with the Farallon slab, the Andaman slab, and the Afar hotspot.

(preferably from common data sources) or via the use of the bulk modulus, shear modulus, and density. Such multiple images of mantle structure encourage an interpretation in terms of processes and mineral physics parameters, since the relative variation of the different parameters adds additional information to the spatial patterns.

If both *P* and *S* information can be used, it is possible to gain a significant increase in understanding of the nature of heterogeneity, through the differences in dependencies on physical parameters. The *P* wavespeed α depends on both the bulk modulus κ and the shear modulus μ as

$$\alpha = [(\kappa + \tfrac{4}{3}\mu)/\rho]^{1/2}, \tag{10.1.1}$$

where ρ is the density. The dependence on the shear modulus μ and the bulk modulus κ can be isolated by working with the *S* wavespeed

$$\beta = [\mu/\rho]^{1/2}, \tag{10.1.2}$$

and the bulk-sound speed ϕ derived from both the *P* wavespeed α and the *S* wavespeed β,

$$\phi = [\alpha^2 - \tfrac{4}{3}\beta^2]^{1/2} = [\kappa/\rho]^{1/2}, \tag{10.1.3}$$

which just depends on the bulk modulus κ and density. This style of parameterisation has been employed in a number of studies (see Masters et al., 2000, for a comparison).

We show in Figure 10.1 an example of joint tomography exploiting both

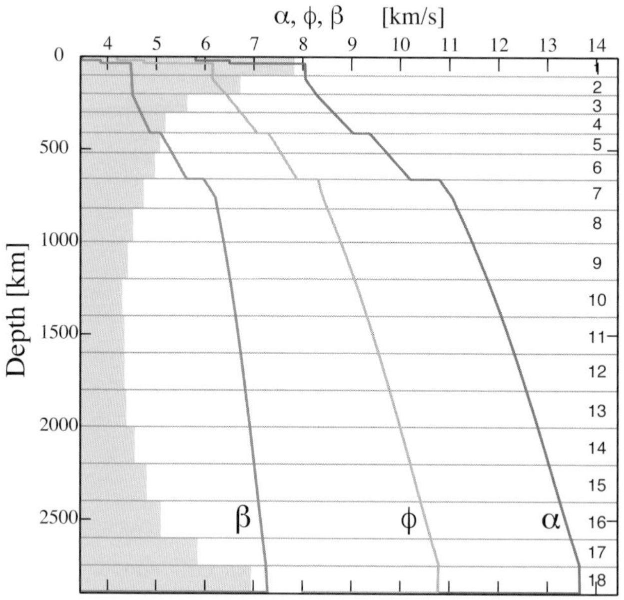

Figure 10.2 Variation of the *P* wavespeed α, *S* wavespeed β and bulk-sound speed ϕ with depth in the mantle for the reference model AK135, together with the 18 layer parametrisation used in the global tomography study illustrated in Figures 10.1,10.3. The grey tones indicate the relative variation of r.m.s. variation between the layers.

P and *S* travel times to extract models for the variation of both bulk-sound speed $\delta\phi/\phi$ and shear wavespeed $\delta\beta/\beta$. Figure 10.2 displays the *P*- and *S*-wavespeed distributions, and the corresponding bulk-sound speed profile for the radial reference model AK135 (Kennett et al., 1995), together with the 18-layer representation of structure that has been used in the construction and display of the tomographic image in Figure 10.1, and the correlation results in Figure 10.3. Figure 10.2 also shows the relative level of shear-wavespeed variations through the mantle by grey bars associated with each layer.

The study of Kennett & Gorbatov (2004), illustrated in Figure 10.1, employed just those paths for which both *P* and *S* arrival times were available for the same source. This choice restricts coverage but ensures that the sampling by both wavetypes is as close as possible. This avoids the difficulties encountered by Antolik et al. (2003) where near full global coverage of the mantle was available for *S*, but a much more limited geographic coverage for *P*.

Figure 10.1 displays a cross-section through the global bulk-sound speed and shear wavespeed model that passes through a range of different mantle features. There is a strong concentration of variability in both wave types near the surface, but we also see the presence of notable fast-wavespeed zones penetrating into the mantle. This cross-section cuts through the northern part of South America and passes obliquely through the zone of high shear wavespeeds that has been

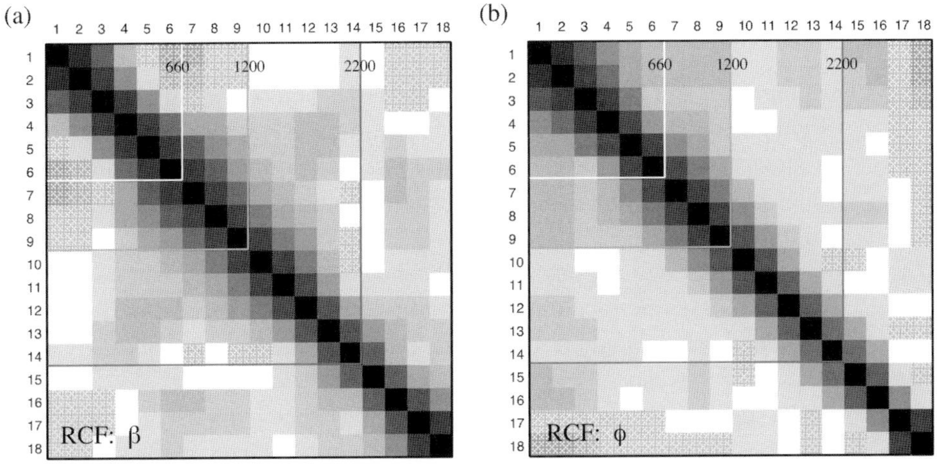

Figure 10.3 Radial correlation function between the variations in (a) S wavespeed ($\delta\beta/\beta$) and (b) bulk-sound speed ($\delta\phi/\phi$) for the joint tomographic model of Kennett & Gorbatov (2004). Positive correlation is indicated by grey tone, with deeper tones for stronger correlation; negative correlation is indicated by patterning.

associated with past Farallon plate subduction (e.g., Grand et al., 1997). The oblique cut produces a broad zone of higher wavespeeds. A more distinct subduction feature occurs on the other side of the globe, where the Andaman extension of the Indonesian subduction zone is cut directly. A narrow zone of faster wavespeeds descends to about 900 km depth. There is an apparent weak connection to a further pronounced shear-wave anomaly in the mid-mantle, which is likely to have been produced by successive phases of subduction at the northern margin of the Tethys Ocean (e.g., Van der Voo et al., 1999). Note that neither of the subduction-zone features have any significant expression in bulk-sound speed.

At the top of Figure 10.1, a prominent low-velocity zone for S reaches the surface at the Afar region. This zone appears to have a weak connection to a deep zone of lowered wavespeed beneath central Africa. Such an inclined link is consistent with a number of studies using different classes of seismological information (e.g., Ritsema et al., 1999). Possible upwelling from the core-mantle boundary can also be seen beneath the western Pacific. The blank zone in the eastern Pacific is a reflection of the limitations of the arrival-time data set. Structure cannot be imaged unless crossing ray paths traverse the region.

The use of common source–receiver pairs for both P and S has proved to be particularly effective for subduction-related features where the structures being imaged are faster than their surroundings. The presence of many sources in the upper-mantle subduction zones is very helpful. Further, for such faster-wavespeed anomalies the effects of wavefront healing are weak.

From the full set of patterns of relative variation of bulk-sound speed and shear wavespeed through the mantle, a number of different heterogeneity regimes in the mantle can be recognised:

Uppermost mantle: shield areas show strong fast shear-wave signature with a slight positive bulk-sound anomaly, the mid-ocean ridges are characterised by slow shear wavespeeds but slightly fast bulk-sound speed, and orogenic zones have strong slow shear-wavespeed features with little expression in bulk-sound speed.

Transition zone: subduction zones normally show a definite fast shear wavespeed which is sometimes accompanied by fast bulk-sound speeds; in regions of stagnant subduction there is a strong fast bulk-sound-speed anomaly with little shear-wavespeed expression.

Lower mantle (upper): in some parts subduction zones appear to penetrate into the lower mantle associated with significant fast shear-wavespeed anomalies but little variation in bulk-sound speed; this means that the 'remnant slabs' seen in high resolution *P*-wave imagery, e.g., Kárason and van der Hilst (2000), have their origin in shear wavespeed variations.

Lower mantle (mid): significant narrow zones of shear-wavespeed anomaly persist to around 1700 km depth and other features emerge around 1900 km, e.g., under Australia; away from these narrow zones both bulk-sound and shear-wavespeed variations are muted and tend to be weakly anti-correlated, suggesting mild thermal effects with $\delta\kappa/\kappa < \delta\rho/\rho < \delta\mu/\mu$.

Lowermost mantle: the bulk-sound and shear-wavespeed variations just above the core–mantle boundary are quite large but anti-correlated, which suggests the presence of chemical heterogeneity (Masters et al., 2000; Karato and Karki, 2001).

An alternative way to look at the properties of the Earth is through the correlations between the layers of the tomographic model. In Figure 10.3(a) we display the correlation matrix for shear wavespeed between the 18 layers. The amplitudes of the wavespeed variations are weighted by the sampling, so that most attention is directed to well-resolved features of the model. The diagonal entries represent the full correlation for a layer with itself. The bands of significant correlation extending away from the diagonal arise from similarities in the structure across a group of layers. The patterns of correlation show blocks of layers with a close relation in properties. To guide the interpretation of the correlation features, guidelines have been introduced at the 660 km discontinuity, 1200 km and 2200 km depth. The first two guidelines correspond well to changes in the correlation properties. The deepest change is less well defined, with a change in character commencing at about 1800 km depth and then becoming more prominent by 2200 km. The density of ray paths employed in the inversion diminished strongly near the core–mantle boundary, so correlations there are for a rather restricted global coverage.

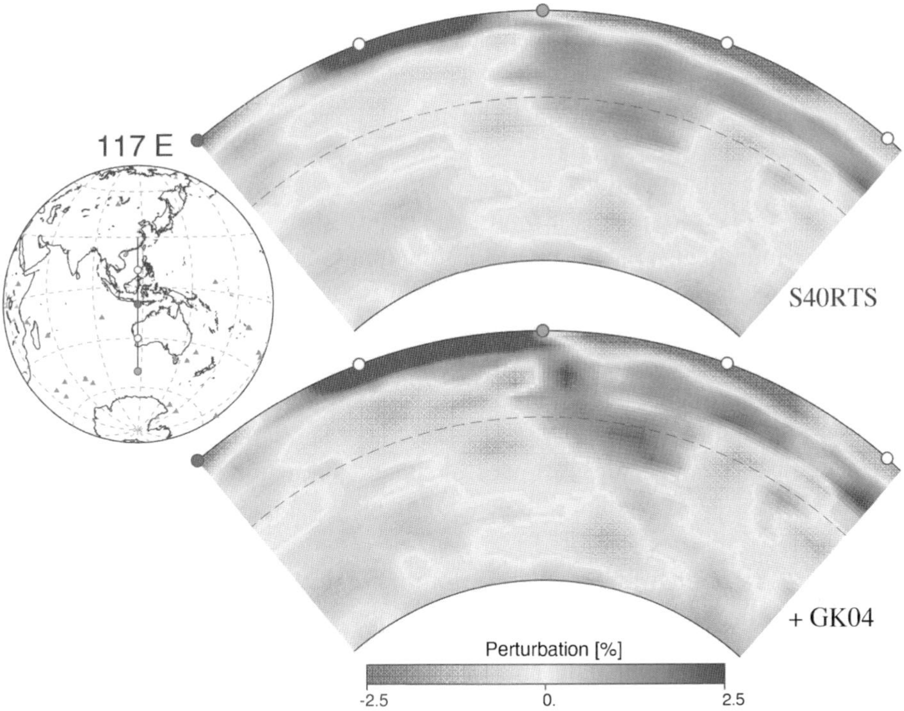

Figure 10.4 Cross-section at 117°E through the global tomography model for shear wavespeed S40RTS (Ritsema et al., 2011), and the augmented model including contributions from the model of Kennett & Gorbatov (2004) from travel-time tomography. The dashed line indicates 1000 km depth. *S*-wavespeed perturbations from model AK135 are displayed.

The comparable results for correlations of bulk-sound speed (Figure 10.3b) do not show the weak anti-correlation seen for shear wavespeed, e.g., layers 1, 2 with 5–9. Instead these are replaced by slight positive correlations. This leads to a greater contrast between the shallower and deeper parts of the mantle. Weak anti-correlation of shear wavespeed and bulk-sound speed is notable for individual layers from 660–1200 km depth and tends to extend to nearby layers as well.

10.2 Tomography and Beyond: Multi-Scale Structure

Most perceptions of seismic structure are based on seismic tomography. Yet such images are limited in their resolution by the nature of the information employed and the distribution of stations. By combining multiple sources of information we can build representations of structure that convey the full range of scales of variation in physical properties.

At the global scale, coverage of the entire mantle can be assembled from multiple sources of information derived from longer-period seismograms. Thus, Ritsema et al. (2011) developed the model S40RTS for shear wavespeed using a representation of the model in terms of spherical harmonics out to degree 40, exploiting both body-wave and surface-wave information. This model provides good coverage for the entire mantle, but displays somewhat lower variations in wavespeed than are indicated by more recent inversions using full waveform inversion techniques (e.g., French & Romanowicz, 2014). Although no data used were in common, S40RTS and the shear-wavespeed model GK04 of Kennett & Gorbatov (2004) from arrival-time tomography show strong similarity in the regions of common coverage. The travel-time tomography provides greater detail (2° resolution). A composite model with two parts S40RTS to one part GK04, gives an approximate balance in amplitude from each source. As can be seen from Figure 10.4, the addition of the travel-time results has the effect of enhancing the definition of structure. The cross-section is taken along the meridian 117°E and passes through Australia, the Indonesian region, and into eastern China. The composite image shows more distinct fast subduction features beneath Indonesia that link to apparently stagnant slab material in the zone from 900 km to 1100 km depth.

The primary source of broad-scale information on seismic structure in the lithosphere comes from surface-wave tomography exploiting the large amplitude contributions from earthquakes, or increasingly from ambient noise (cf. Section 7.3). The analysis is carried out in terms of near-horizontal propagation of modes, and the frequency limitations are based on ensuring modal independence. This range of frequencies also means that propagation paths are generally close to the great circle between source and receiver. For studies of the upper mantle, the frequency ranges employed extend up to around 0.03 Hz for the fundamental mode and a little higher ~0.05 Hz when higher modes are also analysed. With a suitable configuration of sources and stations, horizontal resolution of around 250 km and about 25 km in the vertical direction can be achieved (§ SWII:33.1). With a good distribution of stations, fundamental mode Rayleigh waves can be exploited to quite high frequencies (often around 0.5 Hz) with proportionately higher resolution. It is sometimes possible to link inter-station dispersion measurements from earthquakes directly to the corresponding ambient noise results (e.g., Soomro et al., 2016).

Studies exploiting body waves can frequently achieve higher resolution than for surface waves because resolution is largely limited by station spacing. Teleseismic arrivals can be correlated across broad networks (Section 5.1.2) to provide estimates of relative delay time that can be exploited with tomographic methods. For example, the FMTomo approach of Rawlinson & Urvoy (2006) uses the Fast Marching method to give rapid calculation of propagation paths and couples this with an iterative nonlinear inversion using a sub-space approach. With recalculation of the propagation paths at each step, rapid convergence can be achieved with good amplitude recovery. The propagation paths from teleseismic events have to arrive at the station location, so that even with broad azimuthal

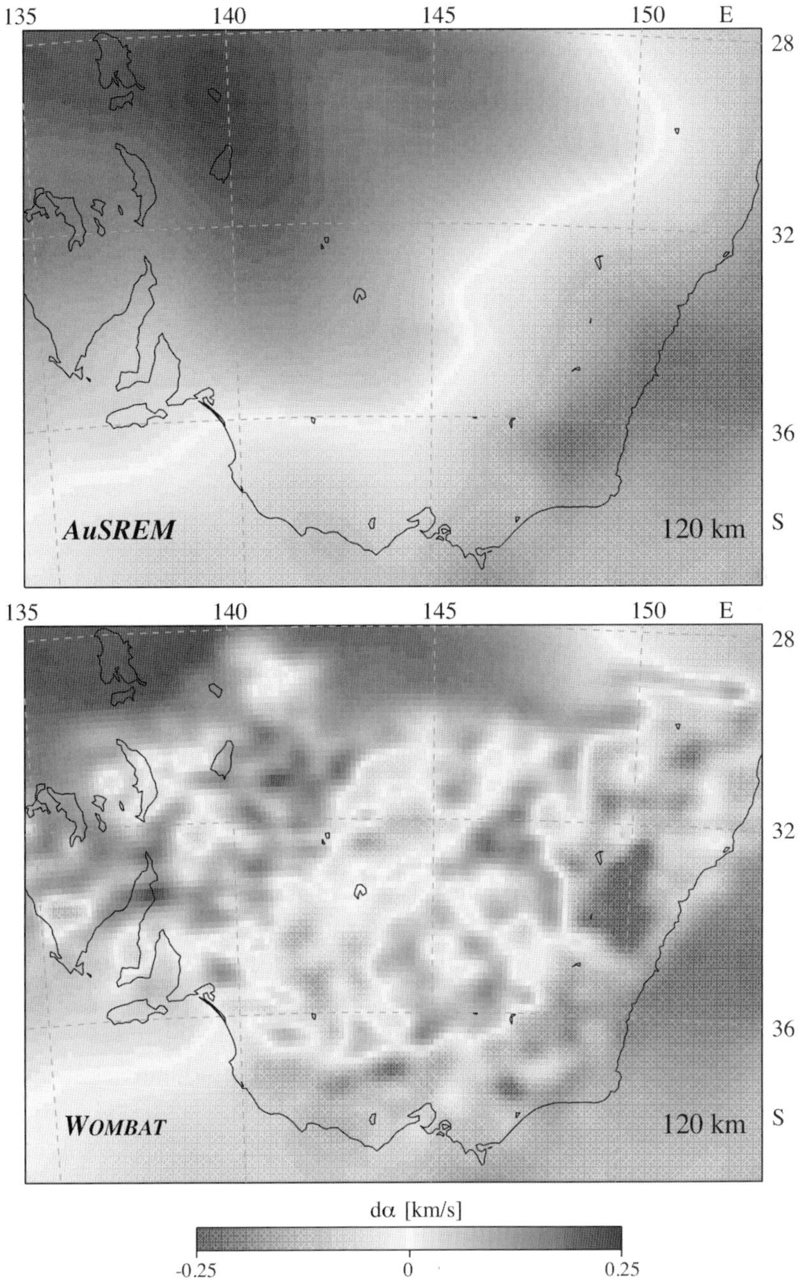

Figure 10.5 Tomographic results for southeastern Australia: above – the long-wavelength AuSREM model (Kennett et al., 2013) derived from surface-wave tomography, below – the results of delay time tomography exploiting the multiple deployments of portable instruments in the WOMBAT project (Rawlinson et al., 2015) displays much shorter scales of variation in all geological environments. [Courtesy of N. Rawlinson.]

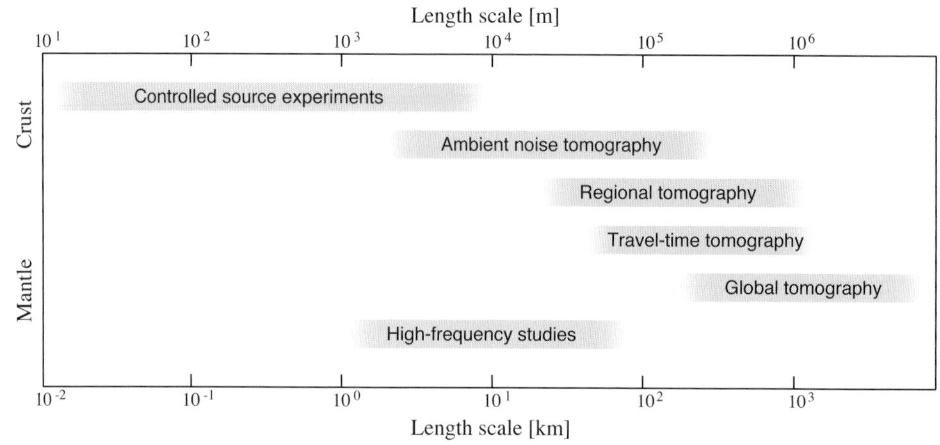

Figure 10.6 Schematic representation of the scales of heterogeneity that can be accessed using different seismological methods, including seismic tomography in a variety of forms. Studies exploiting high-frequency scattered waves provide information on a range of scales that are not readily addressed by other means.

distribution of sources paths cross at shallow angles just beneath the station. The result is inevitably some degree of vertical smearing with body waves alone, that typically will stretch structure by around 10 percent in the vertical direction. The gaps in coverage between stations at shallow depth can be filled in, to some extent, when local sources are available.

The results from such studies of regional structure indicate the presence of variations in seismic structure on scales smaller than those determined from surface-wave tomography. For southeastern Australia, a synthesis of results from a variety of deployments of portable seismometers in the WOMBAT project provides evidence for scales of variation down to at least 50 km horizontally across a broad range of geological environments: from cratonic to younger fold belts (e.g., Rawlinson et al., 2015). Similar results have been obtained in many other environments, and suggest the presence of medium-scale structure as an ubiquitous feature on the continents.

The contrast in the structures revealed by surface-wave tomography and detailed regional tomography using body waves is demonstrated in Figure 10.5. This figure presents a comparison of the mantle structure from the AuSREM model (Kennett et al., 2013) based on surface-wave tomography and the results of Rawlinson et al. (2015) from delay time tomography using the WOMBAT deployments. The mantle structures show medium-scale variability across the entire region superimposed on the larger-scale variations present in AuSREM. Ambient noise studies using the WOMBAT data reveal even shorter-scale variability in the crust (Young et al., 2011).

The scales of structure that can be imaged using tomographic methods by no means represent all the structural features that are present in the Earth (Figure

10.6). The complexity of the high-frequency coda of seismic events points to the presence of both organised and disorganised structural elements at smaller scales. Images from full-crustal reflection profiling (see Figure 4.11) show considerable complexity at multiple scales within the crust and a distinct change of character in uppermost mantle beneath. As we have seen in Section 9.3, similar styles of reflective structure are present in the mantle but on larger scales than in the crust. Such complex seismic structure is pervasive, yet except in favourable circumstances cannot be accessed directly.

Rather than provide a detailed specification of the spatial variation of seismic properties, we can invoke parametric representations that capture the character of observations. Such parametrisation can be readily produced by specifying the spatial power-spectral density of the heterogeneity, and multiple realisations can be achieved by specifying the phase. Such a stochastic approach can describe a wide range of circumstances by varying the parameters. In seismic studies we use the stochastic forms as a summary of fixed structure, using individual members of the ensemble as indicative of the structure.

10.3 Stochastic Representation of Heterogeneity

Stochastic representations have been extensively utilised in two different, but related, aspects of seismology. One class of usage arises from the description of seismic structure at scales finer than can be resolved with current patterns of observation, yet still leave an imprint on the seismic wavefield. Some idea of the character of likely structures can then be found by exploring suitable representations of heterogeneity described through spatial correlations. In these circumstances a stochastic representation in terms of a few parameters provides a way of generating suitable models for numerical simulation.

The other application is related through the complex nature of high-frequency seismic-wave propagation as evidenced in the coda of seismic phases and other scattering phenomena. The detailed nature of the medium is not resolvable, but useful concepts such as the rate of the decay of the coda can, under certain assumptions, be related to the correlation properties of the medium. An extensive treatment of such applications for body waves can be found in Sato, Fehler & Maeda (2012). Kennett (1990) developed a treatment for the attenuation of guided waves employing a coupled mode treatment of stochastic media (see Section 11.2.3). Flatté & Wu (1988) have used the correlation properties of the wavefield at the NORSAR array in terms of both phase and amplitude to propose a depth-dependent model for the heterogeneity in the vicinity of the array.

In seismology, the use of stochastic methods was adapted from the work of Chernov (1960) for acoustic waves, working in terms of the spatial correlation function of the heterogeneity. In the acoustic case, the temporal changes in the medium are slower than the propagation speed but still allow for considerable variation during a span of observation. However, the rate of change of Earth materials is many orders of magnitude slower than the time scales of elastic waves, and so the justification for using a spatial correlation function is based on characterising the properties of the medium via a summary description which ignores the details of heterogeneity. In effect a model is specified for the amplitude of the wavenumber spectrum and different members of the ensemble differ through their phase properties. Further, different patches of a single realisation of the stochastic medium will have different phase character.

10.3.1 Stochastic Models and Properties

Stochastic 'models' in seismology start from a parametric representation of the wavenumber spectrum of heterogeneity characterised by some class of correlation length, which are equivalent to a model of spatial auto-correlation of the heterogeneity. Different models in the ensemble are then realised through different phase representations initiated by varying random seeds.

A variety of different auto-correlation functions have been employed (Sato et al., 2012). The simplest is a Gaussian $N(r) = \epsilon^2 \exp[-r^2/a^2]$ in terms of a correlation length a and the amplitude of wavespeed deviation from the reference ϵ. The wavenumber spectrum is then also a Gaussian. The Gaussian model is easy to handle analytically, but produces rather smooth variation that lacks the short wavelengths found in the Earth. More complex correlation functions can rectify this deficiency.

The von Kármán distribution has the spatial auto-correlation function

$$N(r) = \frac{\epsilon^2}{2^{\nu-1}\Gamma(\nu)} \left(\frac{r}{a}\right)^{\nu} K_{\nu}\left(\frac{r}{a}\right), \tag{10.3.1}$$

where $\Gamma(\nu)$ is the gamma function and $K_\nu(x)$ is the modified Bessel function of the second kind of order ν. The nature of the spatial behaviour is characterised via the Hurst number ν and the correlation distance a. The case $\nu = 0.5$ is equivalent to an exponential correlation function. The corresponding power-spectral density for the von Kármán model is

$$P(k) = \frac{4\pi\epsilon^2 \nu a^2}{(1 + k^2a^2)^{\nu+1}}. \tag{10.3.2}$$

This wavenumber spectrum has a power-law dependence $(ka)^{-2(\nu+1)}$ for large wavenumbers, and is relatively rich in short-wavelength components compared with the Gaussian form. The von Kármán distribution is therefore very suitable for describing Earth heterogeneity.

The stochastic representations (10.3.1), (10.3.2) have isotropic properties and depend only on the distance between points. The assumption of isotropic correlation is commonly made, since it allows direct derivation of analytic results. Yet, we recognise the presence of a dominant gradient in material properties in the vertical direction through most of the crust and the mantle. The resulting quasi-stratification will be characterised by larger horizontal than vertical correlation distances. Stochastic treatments can be adapted to remove the assumption of isotropy by introducing directional variation in correlation length.

The wavenumber representation of the von Kármán distribution can be readily extended to situations with differing correlation lengths a_x in the horizontal and a_z in the vertical direction. For this stochastic model the probability density distribution in terms of horizontal wavenumber k_x and vertical wavenumber k_z takes the form

$$P(k_x, k_z) = \frac{4\pi\nu\epsilon^2 a_x a_z}{(1 + k_x^2 a_x^2 + k_z^2 a_z^2)^{\nu+1}}. \tag{10.3.3}$$

The representation in the spatial domain can then be obtained by inverse Fourier transformation over both k_x and k_z.

We illustrate the effects of deviations from isotropic correlation with a simple model of elastic wave propagation, using numerical simulation with a finite-difference technique. In Figure 10.7 we show snapshots of wave propagation in a sequence of models with increasing ratio of horizontal to vertical correlation length using the von Kármán model with a Hurst number of 0.4.

The 2-D model domain is square with sides 204.8 km long and a discretisation interval of 0.1 km. The background *P* wavespeed is 8.0 km/s, the *S* wavespeed is 4.8 km/s, and density is 3.4 Mg/m^3. Attenuation is included for both *P* and *S* waves, with Q_P set at 800, and Q_S at 400. These values are modulated by stochastic heterogeneities for both wavespeed and density generated using the representation (10.3.3). An explosive source is used in the corner of the model, generating just *P* waves with a peak frequency of 10 Hz. All *S* waves are produced by interactions with the heterogeneous medium.

We consider three situations with a common level of deviations ϵ from the uniform background of 2%. The first case has isotropic spatial correlation with $a_x = a_z = 0.5$ km. In the second the horizontal correlation length is enlarged by a factor of three, so that $a_x = 1.5$ km and the ratio $a_x/a_z = 3$. The third model has a further enlargement of the horizontal correlation length by a factor of three ($a_x = 4.5$ km) and the ratio of horizontal to vertical correlation length is 9:1.

Figure 10.7 displays snapshots of the ground-velocity distribution at 12 s and 19 s after source initiation. In Figure 10.8 we show the velocity waveforms as a function of angle from the vertical at a distance of 150 km from the source, as indicated in the left-hand frames of Figure 10.7.

For the case of isotropic correlation, with correlation length 0.5 km, the wavefronts show strong similarity at all take-off angles (Figure 10.7a). Modest local conversion from *P* to *S* helps fill the coda of *P*. but the strongest effects from heterogeneity involve *P* to *P* multiple scattering. The variation of the waveforms with azimuth (Figure 10.8a) shows some variety of behaviour with focussing and defocussing effects associated with specific paths through the heterogeneous medium. The effects lead to some minor fluctuations in onset time. The general character and duration of the *P* coda is similar at all angles.

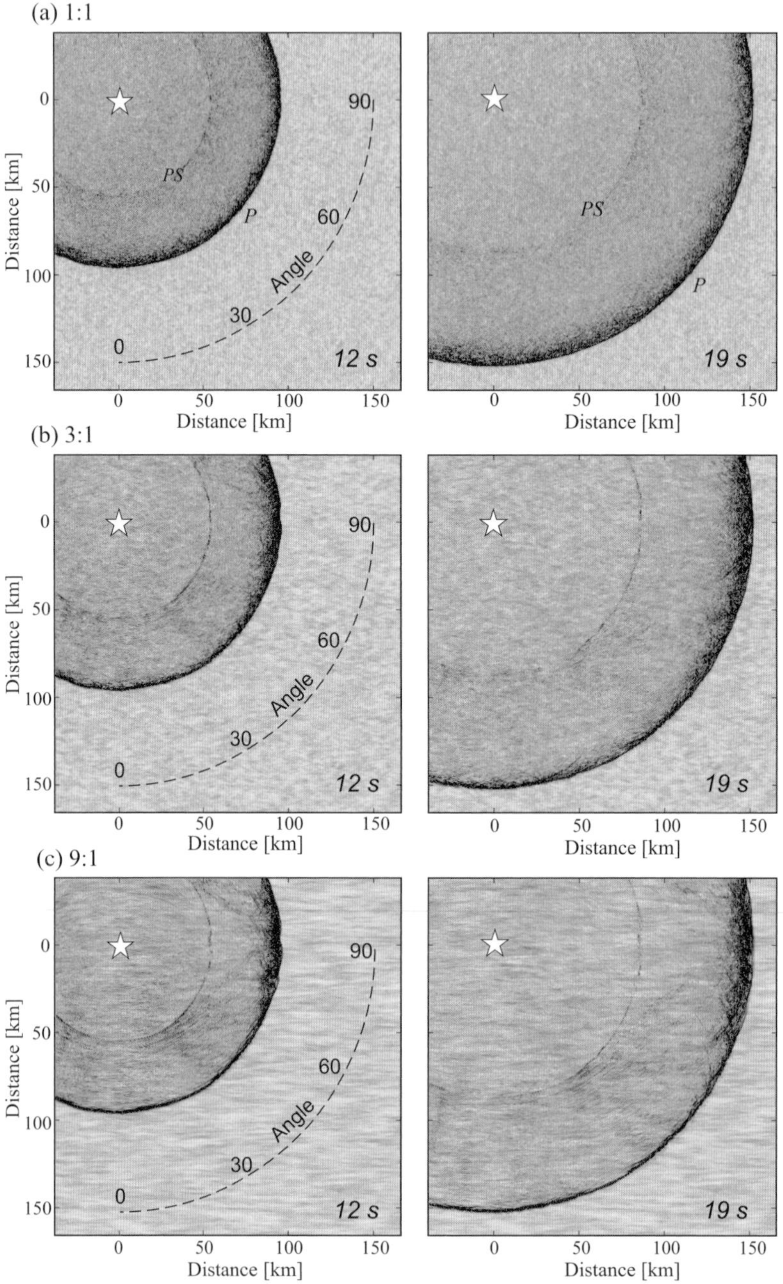

Figure 10.7 Snapshots of wave propagation through a sequence of heterogeneity models with increasing ratio of horizontal to vertical correlation length and a constant level of variation with ϵ set at 2%: (a) 1:1 ratio with correlation distance 0.5 km; (b) 3:1 ratio with horizontal correlation length 1.5 km and vertical correlation length 0.5 km; (c) 9:1 ration with horizontal correlation length 4.5 km and vertical correlation length 0.5 km. The wavespeed variations are plotted in the background. [Courtesy of T. Furumura.]

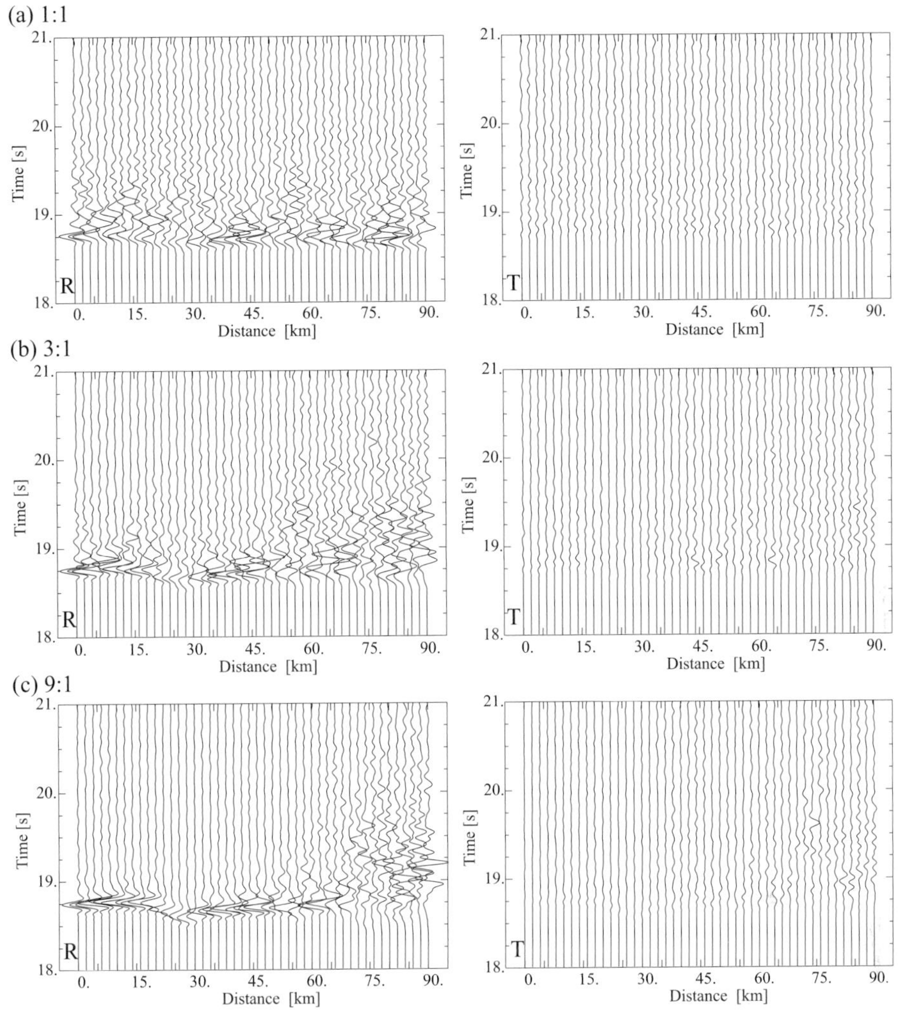

Figure 10.8 Snapshots of the wavefield as a function of angle of propagation from the z-axis for the three models illustrated in Figure 10.7 with increasing ratio of horizontal to vertical correlation length. Both the radial (R) and tangential (T) ground velocity are displayed. (a) 1:1 ratio; (b) 3:1 ratio; (c) 9:1 ratio. [Courtesy of T. Furumura.]

With the increase of the ratio between horizontal and vertical correlation length to 3:1, we start to see the development of more directional character in the wavefield. There is now a distinct contrast between propagation in the vertical direction, perpendicular to the grain of the structure, and that around the horizontal direction along the grain (Figure 10.7b). In the vertical direction (0°) the duration of the *P* coda is a little shorter than in the isotropic case. Whereas for angles of 60–90° from the vertical the length of the coda increases rapidly (Figure 10.8b). The onset of the *P* arrivals is less distinct and the position of maximum energy moves a little later in the wavetrain. The variations in onset time are stronger than for the isotropic case, with the largest changes at 30° to the vertical.

In the third case with the largest contrast between horizontal and vertical correlation length (9:1), directional effects are more pronounced. The *P* wavefront is relatively sharp in the vertical direction, but somewhat diffuse in the horizontal direction (Figure 10.7c). There are now noticeable variations in onset time out to 30° to the vertical (Figure 10.8c). The angular span with elongated coda is narrower than for the 3:1 ratio, and is largely confined to $\pm 20°$ from the horizontal. In near-horizontal propagation, the largest energy is displaced from the onset and arrives later than in the vertical direction. Tangential energy in this angular span is also a little larger than in the other cases.

These results show how organised heterogeneity can act to impose distinctive structure on the wavefield, with strong contrasts as a function of the direction of propagation relative to the dominant structure. The media with elongated heterogeneities channel energy and produce an apparently dispersed signal along the grain of the structure. As we shall see in Section 12.4 and Chapter 13, such stochastic waveguides can sustain guided energy for long distances to produce very complex seismograms.

10.3.2 Comparison of Spatial and Ensemble Averaging

From a stochastic representation such as (10.3.2), (10.3.3) different realisations can be achieved by using varying random number seeds to initiate the phase response, followed by Fourier transformation back to the spatial domain. In this way an ensemble of realisations with the same spatial correlation properties can be constructed.

The concept of ensemble averaging which is central to fully stochastic treatments of propagation characteristics comes from statistical mechanics, where observable quantities are related to ensemble averages under the assumption that the physical system samples the full range of possible states. This ergodic hypothesis can be applied directly to wave propagation through stochastic media when the system changes in time. For the Earth, however, we have essentially a fixed structure. Thus, different interaction with heterogeneity can only be achieved by looking at different propagation paths. Averages over a set of statistically independent samples are then to be achieved by spatial averaging within the single available system, i.e., the actual Earth.

Hudson (1982) has discussed the circumstances in which this equivalence might be justified. He suggests that spatial averaging over a single representation of an ensemble is suitable when the distance of observations from the heterogeneous region is much larger than the size of the region, which in turn is large compared to the correlation length. The total perturbation of the wavefield also needs to be small. The spatial average of observations crossing through a region several correlation lengths across can then be compared with the corresponding ensemble estimates. Such a comparison was made by Kennett (1990) for scattering attenuation of the regional phases *Lg* and *Sn* based on a treatment of stochastic energy loss through mode coupling.

When the observation point lies within the heterogeneous region, as is commonly the case, the theoretical treatment of Hudson (1982) is not applicable. We can, however, make use of numerical simulations with both multiple realisations of the stochastic medium and spatial sampling of a single realisation to examine the circumstances in which the ensemble and spatial averages are equivalent. Below, we employ a wavelet-based method (Hong & Kennett, 2002, 2003) that retains high accuracy for a strongly heterogeneous medium, and examine the properties of single wavetype propagation.

Wavenumber spectrum representations such as (10.3.2) have been employed directly in many studies of scattering attenuation for body waves based on the first-order Born approximation with semi-empirical corrections (e.g., Roth & Korn, 1993; Hong & Kennett, 2003). This approach requires a direct interpretation of the ensemble average in terms of observed amplitudes, and analytic results tie well to numerical simulations with low levels of heterogeneity.

Multiple scattering is important in the presence of strong heterogeneity but cannot readily be tackled directly with analytic techniques. As a result, simplifying assumptions have been employed without worrying about the detailed nature of the heterogeneity or the character of the scattering process. Aki & Chouet (1975) introduced the idea of homogenisation of the scattered field around a receiver in a pioneering study of coda development. Since that time a variety of ensemble methods have been developed that are quite successful in predicting the amplitude and form of the envelope of the coda (e.g., Sato et al., 2012).

With sufficient numerical accuracy it is possible to make a full allowance for multiple scattering for complex media using numerical simulation of the wave propagation process. Most numerical studies have looked at the behaviour of an individual realisation of the stochastic medium, and the statistical properties are examined by using different spatial sampling (e.g., Frenkel & Clayton, 1986). However, much then depends on the level of consistency between different realisations, which can be tested using Monte Carlo methods.

Modelling of Ensemble Results

We use a simple configuration for acoustic waves to examine the relation between spatial and ensemble averaging for receivers within a zone of heterogeneity. Multiple realisations of heterogeneous models, described by the same correlation function for the seismic wavespeed, are generated using different random seeds for the phase distribution. The properties of the wavefield are then compared with those obtained by spatial sampling within a single model. To examine the development of multiple scattering we use a single heterogeneous model using the isotropic von Kármán form, with Hurst number ν of 0.25, and varying levels of perturbation (1%, 3.3%, and 10%).

We consider a 40 km by 40 km domain, with a plane acoustic wave generated

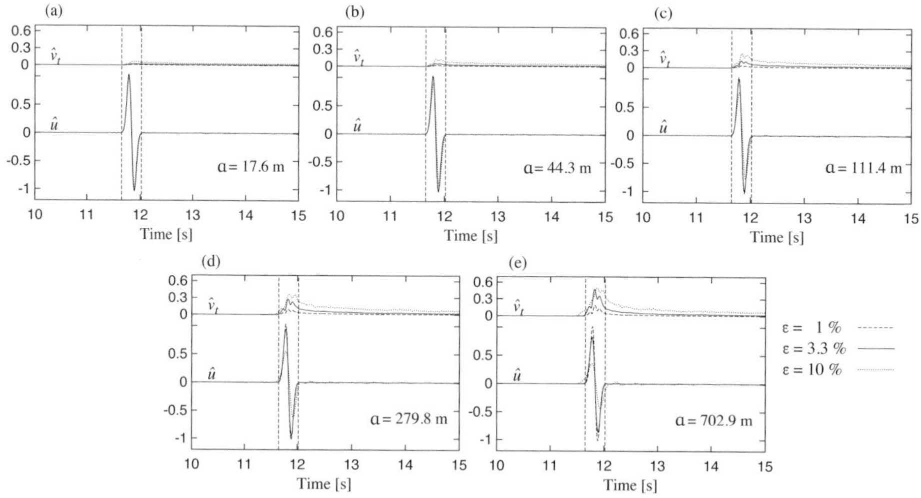

Figure 10.9 Mean time response $\widehat{u}(t)$ and waveform variance $\widehat{v}_t$ across the 128 receiver array for an incident plane acoustic wave on a von Kármán medium with $\nu = 0.25$ with varying correlation length a and perturbation level ε. [Courtesy of T.-K. Hong.]

across a line of sources at the top of the square. A set of 128 receivers are placed at 29.06 km away from the source line with receiver spacing of 312.5 m. The wavefield is calculated using the wavelet-based method of Hong & Kennett (2002, 2003), which is able to achieve high accuracy even in the presence of very strong heterogeneity. With this configuration, we can examine the change in wavefield characteristics with correlation distance a and perturbation level ϵ. We can also compare the properties of different realisations of the stochastic model.

Even for a 1% perturbation level there are slight indications of multiple scattering in the late coda, and these become more pronounced as the perturbation level increases (Figure 10.9). The 10% perturbation produces complex partially coherent features in the wavefront with a scale length of a few kilometres (about 10 times the correlation length).

The spatial average of the wavefield is generated by stacking across the N_s receivers to create the mean time response

$$\widehat{u}(t) = \frac{1}{N_s} \sum_{j=1}^{N_s} u(x_j, t), \tag{10.3.4}$$

where $u(x_j, t)$ is the time response at the jth receiver. A measure of the variance in the response across the receiver array is provided by

$$\widehat{v}_t(t) = \left[\frac{1}{N_s} \sum_{j=1}^{N_s} [u(x_j, t) - \widehat{u}(t)]^2 \right]^{1/2}. \tag{10.3.5}$$

Figure 10.9 shows the mean time responses and variances across the 128 receivers as a function of the correlation distance a and perturbation level ϵ for the heterogeneity. Averages over a more restricted set of receivers with greater separation give very similar results, although a little more irregularity is introduced at later times. As the correlation distance or the fractional fluctuation increases, the amplitude of the primary waves in the mean time response diminishes. The variance of the coda is more marked for large correlation distances.

The scale of the stochastic heterogeneity and the type of random medium control the level of scattering and the rate of attenuation. The variance $\widehat{v}_t$ scales approximately linearly with the perturbation level ϵ for small correlation lengths ($<$ 100 m) up to 10% variations. For larger correlation lengths, variations in $\widehat{v}_t(t)$ are concentrated near the primary wavefront and the tail of the coda maintains a nearly linear dependence on ϵ.

For multiple variants of the stochastic medium, the ensemble average at the jth receiver, $\langle u(x_j, t)\rangle$, can be evaluated as a sum across all the N_e realisations:

$$\langle u(x_j, t)\rangle = \frac{1}{N_e}\sum_{k=1}^{N_e} u_k(x_j, t), \tag{10.3.6}$$

where $u_k(x_j, t)$ is the contribution from the kth model, In a similar way we can generate an ensemble variance $\langle v_t(x_j, t)\rangle$ by analogy with (10.3.5).

We consider a von Kármán random medium (ν = 0.25, ϵ = 3.3%) with correlation length a of 279.8 m, so that for the dominant wavelength $k_d a = 2.31$. We examine realisations using 20 different random seeds. In Figure 10.10 we show the ensemble average response $\langle u(x_j, t)\rangle$ for five different receivers along the horizontal line (0.00, 7.81, 15.63, 23.44, 31.25 km). The dominant behaviour is very similar, but there are noticeable differences between the behaviour at different receivers. The variance across the ensemble shows a distinctive pattern, with only minor variations. The spatial variance $\langle \widehat{v}_t\rangle$ calculated from the five different ensemble averages, Figure 10.10(d), shows a comparable behaviour with time, indicating the general correspondence between the spatial and ensemble averaging. Nevertheless, there are considerable variations in the specific predictions of the ensemble average field at different positions within the heterogeneous medium.

In Figure 10.11, we compare the results from combining 20 spatial samples for a single realisation of the stochastic medium ($a = 279.8$m, $\epsilon = 3.3\%$). Each spatial set is composed of 20 receivers separated by 1875 m starting from the locations $x_0 = 0.0$, 312.5, 625.0, 937.5, 1250.0 m, so that there are no common receivers. We see a similar level of trace variations but no direct correspondence with the ensemble results. The spatial variance $\widehat{v}$ shows a similar rate of coda decay to the ensemble results (Figure 10.10c), but there are notable differences for the primary pulse.

Thus. although there is not a precise correspondence between the spatial and ensemble results, we can use the spatial results from a single realisation to provide

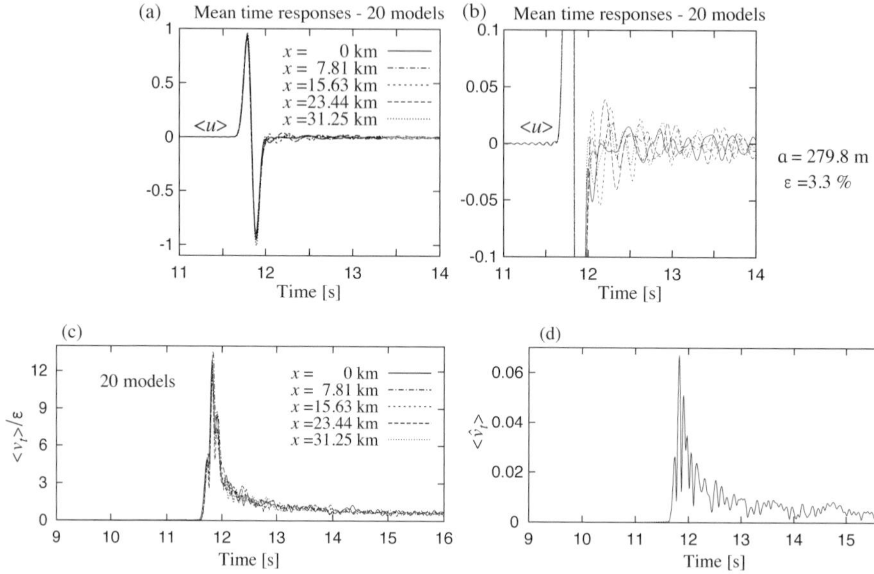

Figure 10.10 Ensemble averages over 20 realisations of a von Kármán medium with $\nu = 0.25$, illustrating the mean time response $\langle u \rangle$ at different locations x and the ensemble variability $\langle v \rangle$. The spatial variability is summarised through $\langle \widehat{v} \rangle$ in (d). [Courtesy of T.-K. Hong.]

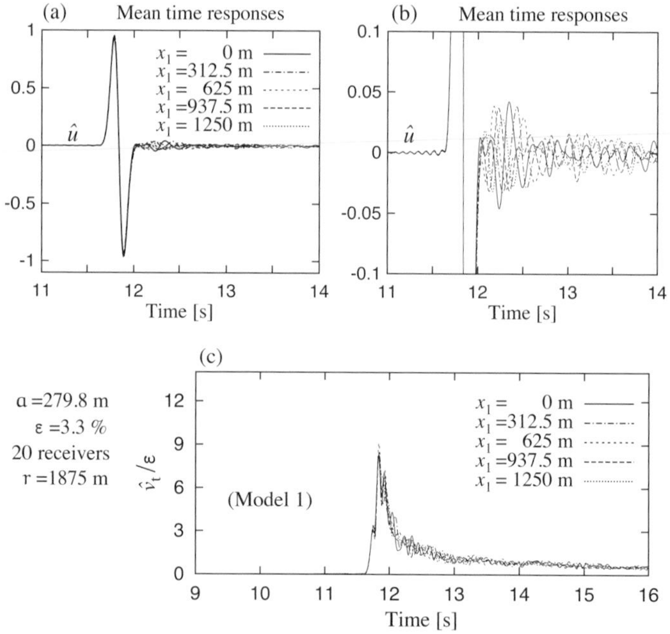

Figure 10.11 Spatial averages over 20 receivers spaced at 1875 m intervals for a single realisation of a von Kármán medium with $\nu = 0.25$, illustrating the mean time response $\widehat{u}$ at different locations x and the spatial variability $\widehat{v}$. [Courtesy of T.-K. Hong.]

a measure of the expected variance in the properties of the wavefield. However, the results reinforce the caution expressed by Hudson (1982) about the direct interpretation of ensemble averages. The results from any single realisation of the stochastic medium can be expected to capture much of the behaviour, but local spatial averages of numerical results are not equivalent to the ensemble result. It is not therefore surprising that studies of scattering attenuation based on different implementation of the stochastic media should arrive at different conclusions, especially with regard to the appropriate cut-off angle for the application of single scattering theory (see, e.g., Roth & Korn, 1993; Hong & Kennett, 2003). The level of correspondence between the observables from the numerical simulation and the ensemble average will depend on which part of the space of models has been sampled.

Multiple simulations of a stochastic medium are needed to gain the most meaningful results from numerical simulations. It would appear that about 20 such different media provides the potential for a stable ensemble average without excessive computational effort. Nevertheless, many more simulations would be required for confidence that adequate sampling has been achieved for the full ensemble.

We have a similar scenario when we consider any individual set of actual observations. The effective scattering Q from an ensemble averaging method should be comparable to the average of results from many observations taken over very different paths through the real medium (cf. Hong et al., 2005). There will be considerable variations in the way in which waves interact with the medium over different paths. Local averaging does not represent the same process as envisaged in the use of the ensemble.

The success of simple multiple scattering models of coda can be traced to the fact that this part of the seismogram contains contributions from a wide variety of different paths sampling the medium. The trend of the amplitude described by an envelope function can be well explained, but the detail of the seismogram will depend on the particular path and hence the actual deterministic structure.

10.4 Effective Media – a Macroscopic View of Heterogeneity

The presence of coda in the seismic wavefield, as illustrated in Figures 10.7 and 10.8, attests to the presence of fine-scale heterogeneities with length scales that are comparable to or smaller than the seismic wavelength, that is, certainly below 10 km. The length scales for the heterogeneity that is actually present in the Earth are orders of magnitude smaller than the scales that can be resolved by seismic tomography. This difference in scales naturally raises questions about the meaning of broad-scale heterogeneity, as in the global tomography in Figure 10.1, and their relation to the underlying smaller-scale structures. Establishing the relation between heterogeneity at different scales is the central theme of homogenisation or upscaling theories.

In a pioneering contribution, Backus (1962) considered the case of a finely layered medium, which he could study analytically. The fundamental result was that the propagation of low-frequency waves through the original finely layered medium and through a smooth *effective medium* are very similar, and this similarity increases with decreasing frequency. Furthermore, the finely layered medium and the smooth effective medium are related by a well-defined averaging procedure. There are two important consequences of this result:

(1) with the aid of the observation of waves with a defined maximum period, we cannot distinguish between media with fine sub-wavelength structure and an effective medium with smooth structure. Therefore, we typically bias seismic inversion towards the smooth option.

(2) the reason why the smooth option is reasonable is that it represents a form of average, and averages are quantities that we can interpret in a meaningful fashion.

Recently, upscaling results to extract a smooth model from complex structure have acquired additional relevance in the context of numerical wavefield simulations. The replacement of small-scale structure by a smooth equivalent medium allows a reduction in the number of grid points needed to represent that structure numerically. In this context, a number of 2-D and 3-D upscaling theories, as well as efficient computational implementations, have been developed (e.g., Capdeville & Marigo, 2013; Cupillard & Capdeville, 2018). In the following section, we will mostly follow the work of Fichtner & Hanasoge (2017) to provide a projection of structure onto the low-wavelength components of a Fourier basis.

10.4.1 The Upscaling Concept

To illustrate the concept, we first consider the simple scalar, frequency-domain wave equation

$$-\omega^2 \rho(x)\, u(x,\omega) - \partial_x[\mu(x)\, \partial_x u(x,\omega)] = f(x,\omega)\,, \tag{10.4.1}$$

which links the displacement u to an external forcing f, density ρ, and shear modulus μ. Our aim is to find a representation of the structure that gives equivalent results for longer periods, without the need to include fine detail.

We discretise the domain (x_1, x_N) with N subdivisions, as in Figure 10.12. In this discrete approximation we can express the wave equation (10.4.1) in terms of N-vectors $\bar{\mathbf{u}}$ for displacement and $\bar{\mathbf{f}}$ for forcing. Thus

$$-\omega^2 \mathbf{R}\bar{\mathbf{u}}(\omega) - \mathbf{DMD}\bar{\mathbf{u}}(\omega) = \bar{\mathbf{f}}(\omega)\,. \tag{10.4.2}$$

The finely sampled material properties ρ and μ are represented through the diagonal $N \times N$ matrices $\mathbf{R}$ and $\mathbf{M}$. The $N \times N$ matrix $\mathbf{D}$ implements the spatial derivative. The specific form chosen for $\mathbf{D}$ determines how far the continuous medium is approximated or represented by its discrete analogue. We here employ a second-order central finite-difference operator for $\mathbf{D}$, approximating the spatial

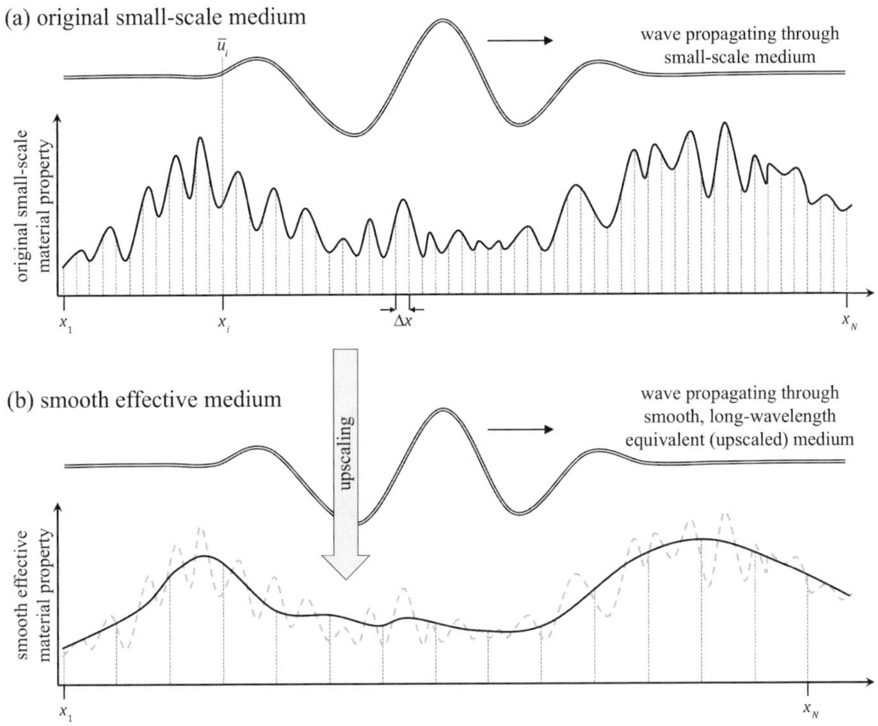

Figure 10.12 Summary of the discrete upscaling concept in 1-D. (a) Small-scale variations in material properties are represented on a large number N of grid points, $x_1, ..., x_N$, with equidistant spacing Δx. The wavefield at x_i is denoted by $\bar{u}_i$, which is one component of the N-dimensional displacement vector $\bar{\mathbf{u}}$. (b) For sufficiently low frequencies (long wavelengths) of the wave, the fine-scale medium from (a) can be represented by a smoother effective medium, sampled at only $K < N$ grid points. The wavefield in the original and the effective medium will be nearly identical, with similarity between the two increasing as frequency decreases. The mathematical procedure of translating the original (small-scale, finely sampled) medium into its effective (smoother, long-wavelength equivalent) medium is known as upscaling or homogenisation.

derivative at x_j in terms of the neighbouring points,

$$D_{ij}u_j = \frac{1}{2\Delta x}(u_{i+1} - u_{i-1})\,. \tag{10.4.3}$$

To simplify the structure, we seek an effective wave equation for the long-wavelength part of $\bar{\mathbf{u}}$ that

(i) has the same functional form as the discrete wave equation (10.4.2),

(ii) only depends on effective material parameters without small-scale detail, and

(iii) is sampled on $K < N$ grid points.

To achieve this end we transform the discrete wave equation (10.4.2) into the wavenumber domain using the discrete N-point Fourier transform, and separate behaviour depending on small and large wavenumbers. The discrete Fourier transform can be implemented as the action of the matrix F_N, whose components are defined by $\mathsf{F}_{N,kn} = N^{-1/2}e^{-2\pi ikn/N}$ for $k, n = 0, \ldots, N-1$. The inverse transform is provided by the action of F_N^{-1} with components $\mathsf{F}_{N,kn}^{-1} = N^{-1/2}e^{2\pi ikn/N}$. Following Fourier transformation, we have

$$-\omega^2 \mathsf{F}_N \mathbf{R}\bar{\mathbf{u}}(\omega) - \mathsf{F}_N \mathbf{DMD}\bar{\mathbf{u}}(\omega) = \bar{\mathbf{f}}(\omega)\,. \tag{10.4.4}$$

We can now use the partition of the $N \times N$ identity matrix $\mathbf{I}_N$ in terms of Fourier components $\mathbf{I}_N = \mathsf{F}_N^{-1}\mathsf{F}_N$, to rewrite (10.4.4) in the form

$$-\omega^2 \mathsf{F}_N \mathbf{R}\mathsf{F}_N^{-1}\mathsf{F}_N\bar{\mathbf{u}}(\omega) - \mathsf{F}_N\mathbf{D}\mathsf{F}_N^{-1}\mathsf{F}_N\mathbf{M}\mathsf{F}_N^{-1}\mathsf{F}_N\mathbf{D}\mathsf{F}_N^{-1}\mathsf{F}_N\bar{\mathbf{u}}(\omega) = \mathsf{F}_N\bar{\mathbf{f}}(\omega). \tag{10.4.5}$$

We now introduce the Fourier projections $\check{\mathbf{a}} = \mathsf{F}_N\mathbf{a}$ for a general vector $\mathbf{a}$, and $\check{\mathbf{A}} = \mathsf{F}_N\mathbf{A}\mathsf{F}_N^{-1}$ for a generic matrix $\mathbf{A}$. With these projections, the wavenumber-domain version of the discrete wave equation (10.4.5) reduces to

$$-\omega^2 \check{\mathbf{R}}\check{\mathbf{u}}(\omega) - \check{\mathbf{D}}\check{\mathbf{M}}\check{\mathbf{D}}\check{\mathbf{u}}(\omega) = \check{\mathbf{f}}(\omega)\,. \tag{10.4.6}$$

The projection of the derivative matrix $\check{\mathbf{D}}$ into the Fourier domain approximates the wavenumber representation of the spatial derivative ∂_x. In consequence $\check{\mathbf{D}}$ is nearly diagonal with components

$$\check{\mathsf{D}}_{jk} \approx i\kappa\,\delta_{jk} = \frac{2\pi ik}{N\,\Delta x}\,\delta_{jk}\,, \tag{10.4.7}$$

where κ denotes the continuous wavenumber distribution. The solution of the wavenumber-domain equation (10.4.6) is the N-vector $\check{\mathbf{u}}$ containing all wavenumber components of the discrete displacement field from 0 to $N-1$.

Intuitively, high-wavenumber components of the displacement will be more strongly affected by small-scale structure in the medium than the low-wavenumber components. We therefore look for a new equation involving only the low-wavenumber components of the wavefield that primarily depend on the low-wavenumber (smooth) structure of the medium.

We introduce a a threshold wavenumber $K < N-1$, and designate wavenumbers below this threshold as 'low' (L) and above as 'high' (H). We then make a sub-space partition of the Fourier space so that the $N \times N$ Fourier transform F_N is split into a $K \times N$ matrix $\mathbf{L}$ yielding the low wavenumbers from 0 to $K-1$, and an $(N-K) \times N$ matrix $\mathbf{H}$ yielding the high wavenumbers from K to $N-1$. Thus

$$\mathsf{F}_N = \begin{pmatrix} \mathbf{L} \\ \mathbf{H} \end{pmatrix}, \text{ with } (\mathbf{L}^{\ddagger}\,\mathbf{H}^{\ddagger}) \begin{pmatrix} \mathbf{L} \\ \mathbf{H} \end{pmatrix} = \mathbf{I}_N, \tag{10.4.8}$$

where ‡ denotes the Hermitian conjugate (complex conjugate of the transpose). We now use this new partition of the identity matrix and use it to repeat the steps that led from (10.4.2) to (10.4.5). We obtain

$$\begin{pmatrix}\mathsf{L}\\ \mathsf{H}\end{pmatrix}\bar{\mathsf{f}} = -\omega^2 \begin{pmatrix}\mathsf{L}\\ \mathsf{H}\end{pmatrix}\mathbf{R}\,(\mathsf{L}^\ddagger\,\mathsf{H}^\ddagger)\begin{pmatrix}\mathsf{L}\\ \mathsf{H}\end{pmatrix}\bar{\mathbf{u}} \\ -\begin{pmatrix}\mathsf{L}\\ \mathsf{H}\end{pmatrix}\mathbf{D}\,(\mathsf{L}^\ddagger\,\mathsf{H}^\ddagger)\begin{pmatrix}\mathsf{L}\\ \mathsf{H}\end{pmatrix}\mathbf{M}\,(\mathsf{L}^\ddagger\,\mathsf{H}^\ddagger)\begin{pmatrix}\mathsf{L}\\ \mathsf{H}\end{pmatrix}\mathbf{D}(\mathsf{L}^\ddagger\,\mathsf{H}^\ddagger)\begin{pmatrix}\mathsf{L}\\ \mathsf{H}\end{pmatrix}\bar{\mathbf{u}}\,. \tag{10.4.9}$$

We can simplify (10.4.9) by defining subspace projections onto low and high wavenumbers, $\check{\mathbf{a}}_{\mathsf{L}} = \mathsf{L}\mathbf{a}$ and $\check{\mathbf{a}}_{\mathsf{H}} = \mathsf{H}\mathbf{a}$ for a general vector $\mathbf{a}$, and $\check{\mathbf{A}}_{\mathsf{LL}} = \mathsf{L}\mathbf{A}\mathsf{L}^\ddagger$, $\check{\mathbf{A}}_{\mathsf{LH}} = \mathsf{L}\mathbf{A}\mathsf{H}^\ddagger$, $\check{\mathbf{A}}_{\mathsf{HL}} = \mathsf{H}\mathbf{A}\mathsf{L}^\ddagger$, and $\check{\mathbf{A}}_{\mathsf{HH}} = \mathsf{H}\mathbf{A}\mathsf{H}^\ddagger$ for a generic matrix $\mathbf{A}$. Now we use the approximation (10.4.7) to neglect the off-diagonal projections of the derivative matrix $\mathbf{D}$, and obtain a set of coupled equations between the low- and high-wavenumber components of the wavefield $\check{\mathbf{u}}_{\mathsf{L}}$, $\check{\mathbf{u}}_{\mathsf{H}}$ in the form:

$$\begin{pmatrix}\check{\mathsf{f}}_{\mathsf{L}}\\ \check{\mathsf{f}}_{\mathsf{H}}\end{pmatrix} = -\omega^2\begin{pmatrix}\check{\mathbf{R}}_{\mathsf{LL}} & \check{\mathbf{R}}_{\mathsf{LH}}\\ \check{\mathbf{R}}_{\mathsf{HL}} & \check{\mathbf{R}}_{\mathsf{HH}}\end{pmatrix}\begin{pmatrix}\check{\mathbf{u}}_{\mathsf{L}}\\ \check{\mathbf{u}}_{\mathsf{H}}\end{pmatrix} \\ -\begin{pmatrix}\check{\mathbf{D}}_{\mathsf{LL}}\check{\mathbf{M}}_{\mathsf{LL}}\check{\mathbf{D}}_{\mathsf{LL}} & \check{\mathbf{D}}_{\mathsf{LL}}\check{\mathbf{M}}_{\mathsf{LH}}\check{\mathbf{D}}_{\mathsf{HH}}\\ \check{\mathbf{D}}_{\mathsf{HH}}\check{\mathbf{M}}_{\mathsf{HL}}\check{\mathbf{D}}_{\mathsf{LL}} & \check{\mathbf{D}}_{\mathsf{HH}}\check{\mathbf{M}}_{\mathsf{HH}}\check{\mathbf{D}}_{\mathsf{HH}}\end{pmatrix}\begin{pmatrix}\check{\mathbf{u}}_{\mathsf{L}}\\ \check{\mathbf{u}}_{\mathsf{H}}\end{pmatrix}. \tag{10.4.10}$$

We now introduce the set of matrices

$$\begin{aligned} \mathbf{E}_{\mathrm{I}} &= [\omega^2\check{\mathbf{R}}_{\mathsf{LL}}+\check{\mathbf{D}}_{\mathsf{LL}}\check{\mathbf{M}}_{\mathsf{LL}}\check{\mathbf{D}}_{\mathsf{LL}}],\\ \mathbf{E}_{\mathrm{II}} &= [\omega^2\check{\mathbf{R}}_{\mathsf{LH}}+\check{\mathbf{D}}_{\mathsf{LL}}\check{\mathbf{M}}_{\mathsf{LH}}\check{\mathbf{D}}_{\mathsf{HH}}],\\ \mathbf{E}_{\mathrm{III}} &= [\omega^2\check{\mathbf{R}}_{\mathsf{HH}}+\check{\mathbf{D}}_{\mathsf{HH}}\check{\mathbf{M}}_{\mathsf{HH}}\check{\mathbf{D}}_{\mathsf{HH}}],\\ \mathbf{E}_{\mathrm{IV}} &= [\omega^2\check{\mathbf{R}}_{\mathsf{HL}}+\check{\mathbf{D}}_{\mathsf{HH}}\check{\mathbf{M}}_{\mathsf{HL}}\check{\mathbf{D}}_{\mathsf{LL}}], \end{aligned} \tag{10.4.11}$$

and then we can write the formal solution of (10.4.10) for the low-wavenumber component as

$$\check{\mathsf{f}}_{\mathrm{e}} = -[\mathbf{E}_{\mathrm{I}} + \mathbf{E}_{\mathrm{II}}\mathbf{E}_{\mathrm{III}}^{-1}\mathbf{E}_{\mathrm{IV}}]\,\check{\mathbf{u}}_{\mathsf{L}}. \tag{10.4.12}$$

Here the *effective source* $\check{\mathsf{f}}_{\mathrm{e}}$ is defined as

$$\check{\mathsf{f}}_{\mathrm{e}} = \check{\mathsf{f}}_{\mathsf{L}} - \mathbf{E}_{\mathrm{II}}\mathbf{E}_{\mathrm{III}}^{-1}\,\check{\mathsf{f}}_{\mathsf{H}}. \tag{10.4.13}$$

The process of solving (10.4.12) with the effective source (10.4.13) is equivalent to first solving (10.4.6) for the complete wavenumber-domain wavefield $\check{\mathbf{u}}$ and then separating the low wavenumbers.

Although the solution of (10.4.12) is the wavefield $\check{\mathbf{u}}_{\mathsf{L}}$, the equation (10.4.12) is not itself a wave equation. Thus the parameters cannot be directly interpreted in terms of effective medium properties. Furthermore, (10.4.12) still contains high-wavenumber projections of $\mathbf{R}$ and $\mathbf{M}$ that we wish to eliminate. Even so, (10.4.12) approximates a wave equation when ω is sufficiently small. When we assume that the cut off wavenumber K is significantly smaller than the maximum wavenumber N, and using (10.4.7), we find

$$||\check{\mathbf{D}}_{LL}||_1 \approx \frac{2\pi}{N\,\Delta x}\sum_{k=0}^{K-1} k \approx \frac{\pi}{N\,\Delta x}K^2\,,$$
$$||\check{\mathbf{D}}_{HH}||_1 \approx \frac{2\pi}{N\,\Delta x}\sum_{k=K}^{N-1} k \approx \frac{\pi}{N\,\Delta x}N^2\,. \tag{10.4.14}$$

This result implies that the high-wavenumber projection of the discrete derivative $\check{\mathbf{D}}_{HH}$ is numerically much larger than its low-wavenumber counterpart $\check{\mathbf{D}}_{LL}$. Hence, the frequency ω can be chosen such that terms involving ω^2 are insignificant in the terms II, III, and IV of (10.4.11) that involve $\check{\mathbf{D}}_{HH}$. Even lower frequencies would be needed to eliminate $\omega^2\check{\mathbf{R}}_{LL}$ from term I because $\check{\mathbf{D}}_{LL}\check{\mathbf{M}}_{LL}\check{\mathbf{D}}_{LL}$ may not be dominant when K is small.

We therefore assume we work with sufficiently small ω that we can omit ω-dependent terms in II, III, and IV; (10.4.12) then reduces to

$$\begin{aligned}\check{\mathbf{f}}_e = &-[\omega^2\check{\mathbf{R}}_{LL} + \check{\mathbf{D}}_{LL}\check{\mathbf{M}}_{LL}\check{\mathbf{D}}_{LL}]\check{\mathbf{u}}_L \\ &+ [\check{\mathbf{D}}_{LL}\check{\mathbf{M}}_{LH}\check{\mathbf{D}}_{HH}][\check{\mathbf{D}}_{HH}\check{\mathbf{M}}_{HH}\check{\mathbf{D}}_{HH}]^{-1}[\check{\mathbf{D}}_{HH}\check{\mathbf{M}}_{HL}\check{\mathbf{D}}_{LL}]\check{\mathbf{u}}_L\,.\end{aligned} \tag{10.4.15}$$

The quality of the approximation in the transition from (10.4.12) to (10.4.15) is frequency-dependent, and generally improves with decreasing frequency. We can rearrange (10.4.15) slightly to produce the simple form

$$-\omega^2\check{\mathbf{R}}_{LL}\check{\mathbf{u}}_L - \check{\mathbf{D}}_{LL}[\check{\mathbf{M}}_{LL} - \check{\mathbf{M}}_{LH}\check{\mathbf{M}}_{HH}^{-1}\check{\mathbf{M}}_{HL}]\check{\mathbf{D}}_{LL}\check{\mathbf{u}}_L = \check{\mathbf{f}}_e\,. \tag{10.4.16}$$

Defining effective, wavenumber-domain material properties

$$\check{\mathbf{R}}_e = \check{\mathbf{R}}_{LL}\,,\quad \check{\mathbf{M}}_e = \check{\mathbf{M}}_{LL} - \check{\mathbf{M}}_{LH}\check{\mathbf{M}}_{HH}^{-1}\check{\mathbf{M}}_{HL}\,,\quad \check{\mathbf{D}}_e = \check{\mathbf{D}}_{LL}\,, \tag{10.4.17}$$

we can write a *wavenumber-domain, discrete, effective wave equation*

$$-\omega^2\check{\mathbf{R}}_e\check{\mathbf{u}}_L - \check{\mathbf{D}}_e\check{\mathbf{M}}_e\check{\mathbf{D}}_e\check{\mathbf{u}}_L = \check{\mathbf{f}}_e\,. \tag{10.4.18}$$

The material properties in (10.4.18) are $K \times K$ matrices in the wavenumber domain. To return to the space domain, we insert $K \times K$ identity matrices $\mathbf{I}_K = \mathsf{F}_K\mathsf{F}_K^{-1}$ expressed in terms of K-point Fourier transforms into (10.4.17):

$$\begin{aligned}&\underbrace{\mathsf{F}_K\,\mathsf{F}_K^{-1}\check{\mathbf{f}}_e}_{\mathbf{f}_e} = \\ &-\omega^2\,\mathsf{F}_K\underbrace{\mathsf{F}_K^{-1}\check{\mathbf{R}}_e\mathsf{F}_K}_{\mathbf{R}_e}\underbrace{\mathsf{F}_K^{-1}\check{\mathbf{u}}_L}_{\bar{\mathbf{u}}_L} - \mathsf{F}_K\underbrace{\mathsf{F}_K^{-1}\check{\mathbf{D}}_e\mathsf{F}_K}_{\mathbf{D}_e}\underbrace{\mathsf{F}_K^{-1}\check{\mathbf{M}}_e\mathsf{F}_K}_{\mathbf{M}_e}\underbrace{\mathsf{F}_K^{-1}\check{\mathbf{D}}_e\mathsf{F}_K}_{\mathbf{D}_e}\underbrace{\mathsf{F}_K^{-1}\check{\mathbf{u}}_L}_{\bar{\mathbf{u}}_L}\,.\end{aligned} \tag{10.4.19}$$

The application of the inverse K-point Fourier transform F_K^{-1} to $\check{\mathbf{u}}_L$ has the effect of a spatial re-sampling that transforms the low-wavenumber part of the original N-dimensional displacement into a smaller, K-dimensional, low-wavenumber wavefield $\bar{\mathbf{u}}_L$. Similarly, $\mathbf{R}_e$ and $\mathbf{M}_e$ are smoother, effective material parameters.

Applying F_K^{-1} from the left to (10.4.19), finally gives the effective medium version of the space-domain wave equation

$$-\omega^2 \mathbf{R}_e \bar{u}_L - \mathbf{D}_e \mathbf{M}_e \mathbf{D}_e \bar{u}_L = f_e \,. \tag{10.4.20}$$

From (10.4.17) and (10.4.19) we see that upscaling the density distribution corresponds to the naive procedure of low-pass filtering ($\mathbf{R} \rightarrow \check{\mathbf{R}}_{LL} = \check{\mathbf{R}}_e$) followed by downsampling ($\check{\mathbf{R}}_e \rightarrow F_K^{-1} \check{\mathbf{R}}_e F_K = \mathbf{R}_e$). In contrast, the process of upscaling the elastic modulus, is significantly more involved, requiring the introduction of an additional correction term $\check{\mathbf{M}}_{LH} \check{\mathbf{M}}_{HH}^{-1} \check{\mathbf{M}}_{HL}$ in (10.4.17). This term is called the *corrector* in classical homogenisation theory (e.g., Capdeville & Marigo 2013).

Using the definitions of effective quantities from (10.4.19), we can study upscaling for the simplest case of a homogeneous medium. We choose the elastic parameters $\mathbf{M}$ to be the identity matrix $\mathbf{I}_N$. So, $\check{\mathbf{M}}_{LL} = \mathbf{L}\mathbf{M}\mathbf{L}^\ddagger = \mathbf{L}\mathbf{L}^\ddagger = \mathbf{I}_K$ and $\check{\mathbf{M}}_{LH} = \mathbf{L}\mathbf{M}\mathbf{H}^\ddagger = \mathbf{L}\mathbf{H}^\ddagger = 0$. Therefore, $\mathbf{M}_e = F_K^{-1} \mathbf{I}_K F_K = \mathbf{I}_K$. In a similar way, we find $\mathbf{R}_e = \mathbf{I}_K$ for an originally homogeneous density $\mathbf{R} = \mathbf{I}_N$. Thus, in agreement with our intuitive expectations, upscaling a homogeneous medium with N grid points produces a homogeneous medium with $K < N$ grid points.

10.4.2 Numerical Examples

To illustrate the upscaling procedure developed in the previous paragraphs, we consider numerical examples in 1-D and 2-D. They reveal a range of interesting phenomena related to long-wavelength propagation through small-scale media, such as non-local rheologies and apparent anisotropy.

Upscaling in 1-D

We begin in 1-D where (10.4.17) and (10.4.19) can be applied directly. Figure 10.13a shows a distribution of the elastic modulus μ along a line 3000 m long, discretised by N = 3000 grid points. The distribution features rapid oscillations, spikes, and other irregular variations. With the upscaling procedure, we produce a smoother, effective distribution, represented on K = 3000/4 = 750 grid points. The resulting effective elastic modulus matrix $\mathbf{M}_e$ is shown in Figure 10.13b.

A remarkable property of the 750×750 matrix $\mathbf{M}_e$ is the presence of non-zero off-diagonal elements, in stark contrast to the original and purely diagonal $\mathbf{M}$. The off-diagonal elements cause the effective rheology to be non-local. Stress at position x_i does not only depend on the strain at position x_i, as in standard elastic media with local rheology, but also on the strain at neighbouring positions $x_{j \neq i}$.

The off-diagonal elements in Figure 10.13(b) are small compared to the diagonal elements, despite the strong variations in μ. This means that, to a good approximation, the off-diagonal elements can be neglected. In addition to lowering the numerical cost of computing stresses from strains, this allows us to compare the original and the upscaled medium by visualising the diagonals of $\mathbf{M}$ and $\mathbf{M}_e$, and of $\mathbf{R}$ and $\mathbf{R}_e$. These diagonal components of the effective media are shown

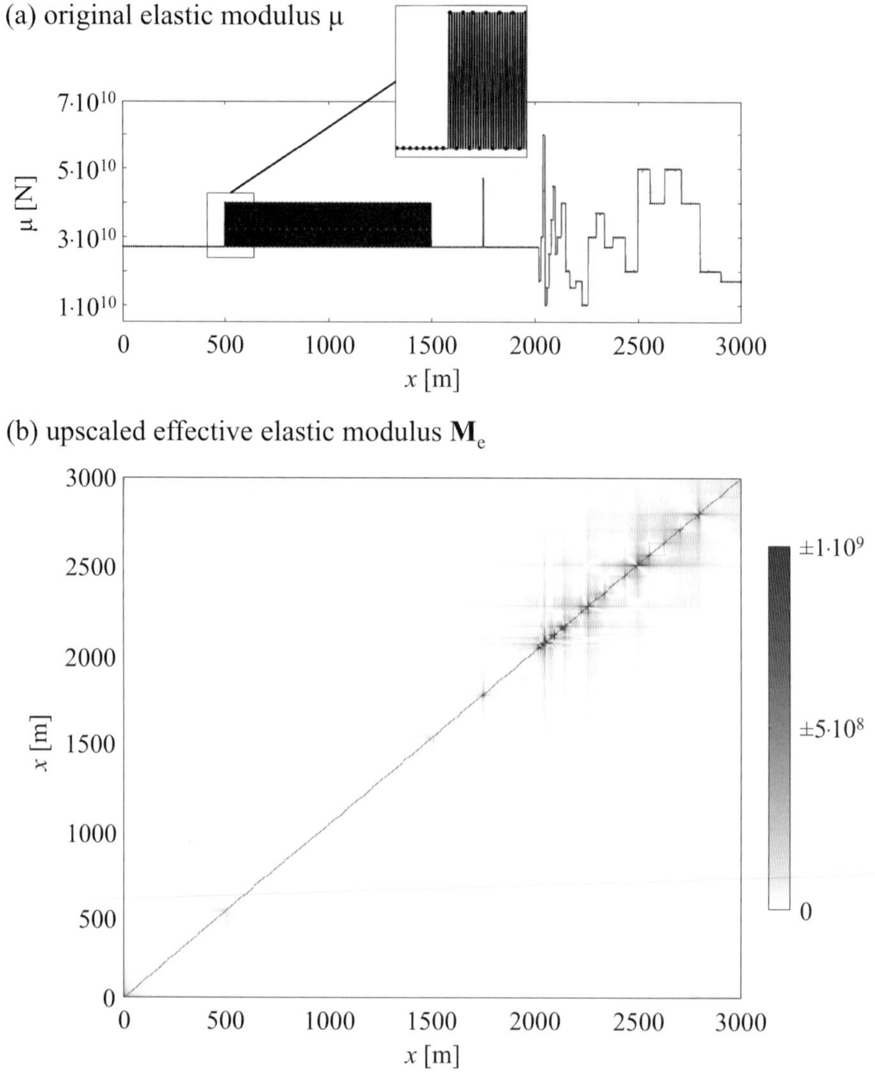

Figure 10.13 Upscaling of a 1-D elastic modulus μ. (a) The original, small-scale distribution of the elastic modulus is defined by N = 3000 equidistantly distributed points on a line 3000 m long. It contains very rapidly varying oscillations between 500 and 1500 m, an isolated spike at 1750 m, and some broader high-amplitude variations between 2000 and 3000 m. (b) In this example, the upscaling reduces the original 3000 grid points to K = 3000/4 = 750 grid points. The upscaled elastic modulus $\mathbf{M}_e$ is therefore a 750 × 750 matrix. In contrast to the original $\mathbf{M}$ from (10.4.2), the upscaled $\mathbf{M}_e$ has non-zero diagonal elements, which represent a non-local rheology.

in Figure 10.14. The upscaling process replaces the rapid oscillations in ρ and μ between 500 and 1500 m by a constant medium. While the constant ρ is exactly the average of the minimum and maximum oscillation amplitudes, the average μ is smaller than the average. This effect is due to the corrector $\check{\mathbf{M}}_{\mathrm{LH}}\check{\mathbf{M}}_{\mathrm{HH}}^{-1}\check{\mathbf{M}}_{\mathrm{HL}}$ in

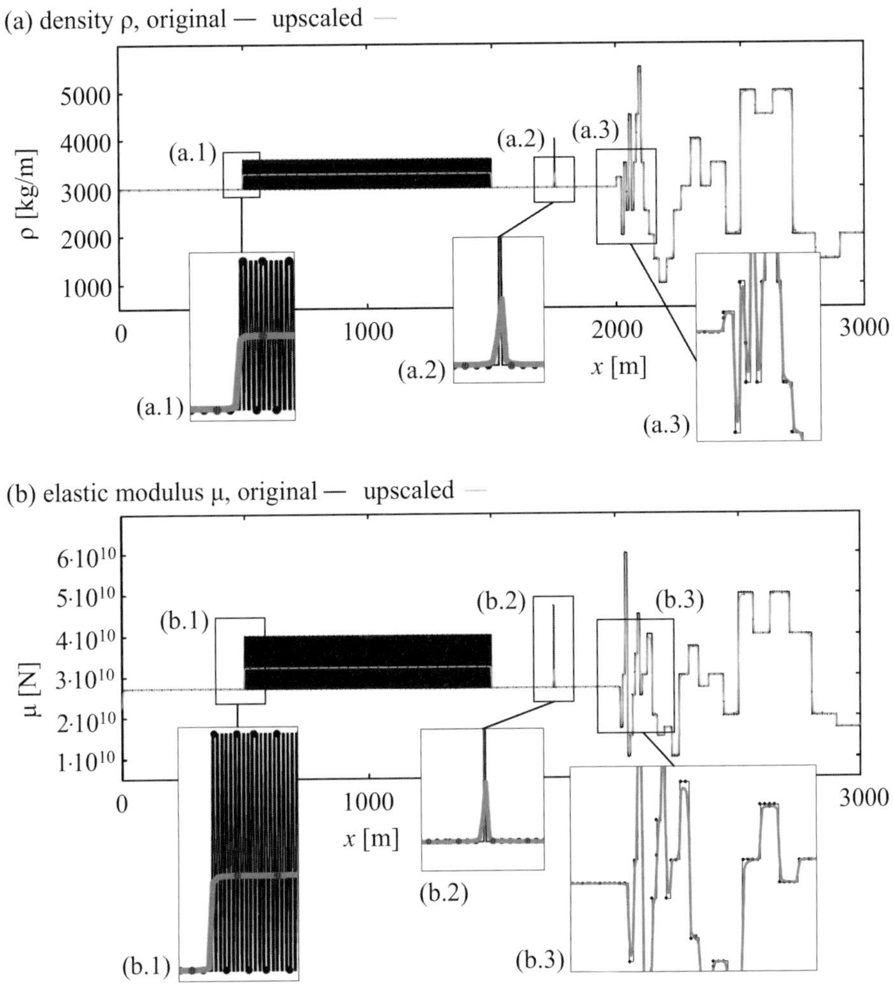

Figure 10.14 Upscaling of (a) density ρ and (b) elastic modulus μ. Original material properties, represented by 3000 grid points, are shown in black and upscaled properties on 750 grid points in grey.

(10.4.17) that transforms the simply downsampled $\check{\mathbf{M}}_{\mathrm{LL}}$ into the effective $\check{\mathbf{M}}_e$. As a consequence, naive downscaling of $\mathbf{M}$ would produce a smooth medium that is too fast.

In Figure 10.15 we consider pulse propagation through the effective medium compared to the results for the original complex structure. As expected, Figure 10.15 demonstrates that upscaling involves a low-frequency approximation, as already seen before in the transition from (10.4.12) to (10.4.15). At frequencies of 5 Hz and 20 Hz, the wavefields propagating through the original medium, with fine-scale variations, and through the upscaled smoother medium are similar to the extent that differences are hardly visible. However, at 40 Hz, the upscaling

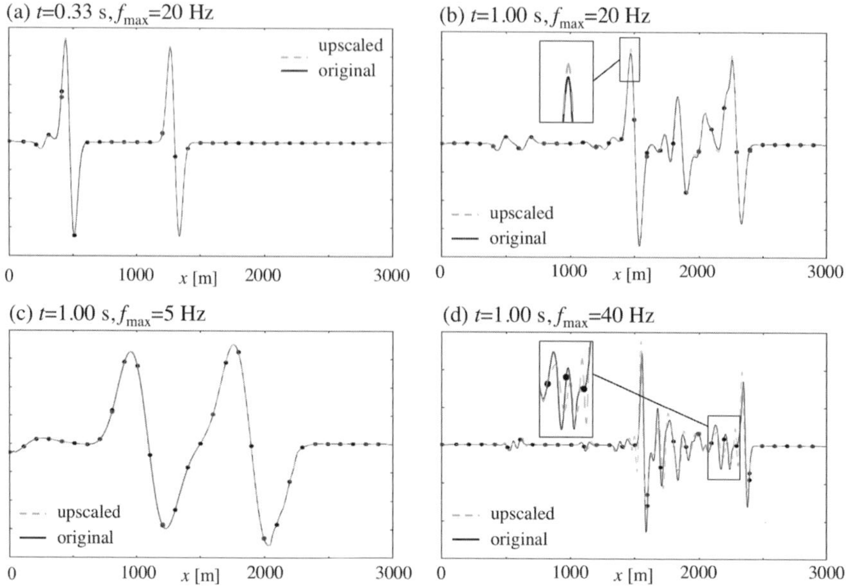

Figure 10.15 Wavefield snapshots at different times and maximum frequencies f_{max} for a source located at $x = 400$ m. For lower maximum frequencies – 20 Hz, cases (a) and (b), and 5 Hz case (c) – the differences between the wavefields propagating in the original medium (black) and the upscaled medium (grey) are so small as to be negligible. However, at a higher maximum frequency of 40 Hz, differences become clearly visible. This illustrates that the effective medium is long-wavelength equivalent, that is, equivalent to the original medium at sufficiently low frequencies.

procedure with a reduction of sampling by a factor of four begins to lose its validity, and clear differences appear.

Upscaling in 2-D

Though the upscaling concept in Section 10.4.1 has been illustrated with a 1-D medium, the approach can be translated straightforwardly to higher dimensions. The additional effort is mostly in the formulation of the discrete wave equation (10.4.2) and in the implementation of the discrete Fourier transform, which requires more bookkeeping when the dimension is higher than one.

In the following examples (Figure 10.16), we consider a plane-strain scenario where displacements and stresses are zero in the z-direction and non-zero only in the x- and y-direction. The corresponding stress–strain relationship is

$$\begin{pmatrix} \sigma_{11} \\ \sigma_{22} \\ \sigma_{12} \end{pmatrix} = \begin{pmatrix} c_{1111} & c_{1122} & c_{1112} \\ c_{1122} & c_{2222} & c_{2212} \\ c_{1112} & c_{2212} & c_{1212} \end{pmatrix} \begin{pmatrix} \varepsilon_{11} \\ \varepsilon_{22} \\ 2\varepsilon_{12} \end{pmatrix} . \tag{10.4.21}$$

Numerical examples of 2-D upscaling for this scenario are shown in Figure 10.16 for (a) a structured medium with lateral periodic striping and (b) a random medium with an isotropic correlation length scale of around 2 m. In both cases, the

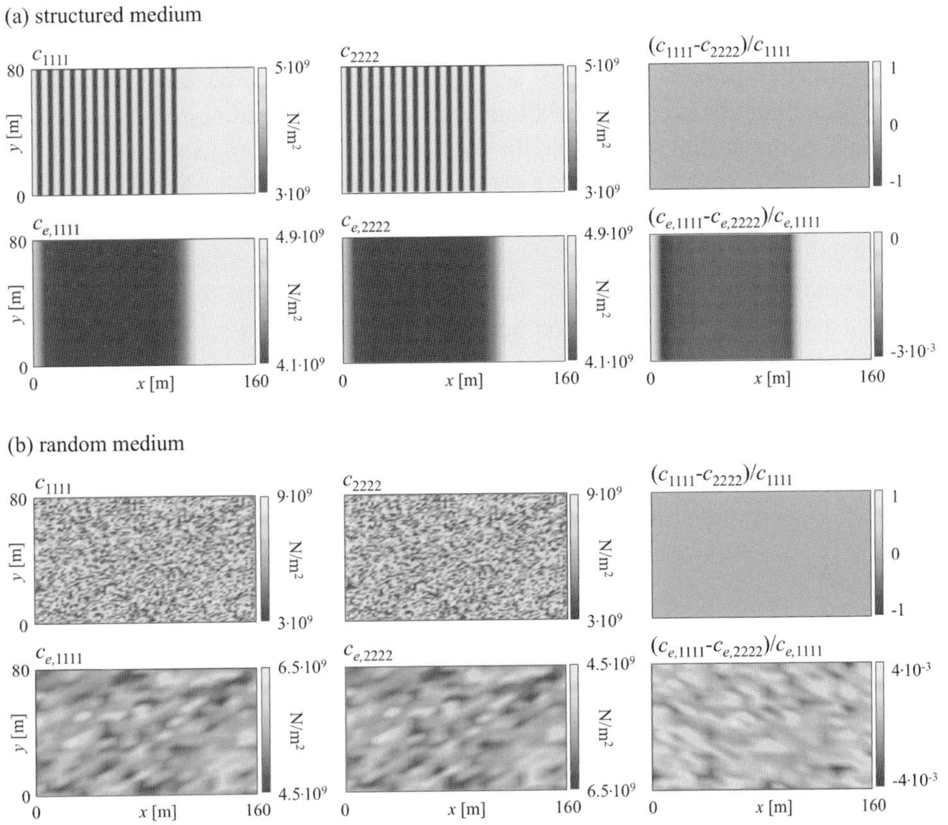

Figure 10.16 Upscaling in 2-D of a structured medium with lateral striping (a) and a random medium (b). The distributions of c_{1111}, c_{2222} and the anisotropy $(c_{1111} - c_{2222})/c_{1111}$ are shown on top. The corresponding effective distributions are shown below. The upscaling factor is $N/K = 4$, meaning that the upscaled medium is represented by four times less grid points in each of the two coordinate directions. While the original media are isotropic ($(c_{1111} - c_{2222})/c_{1111} = 0$), the upscaled media are effectively anisotropic ($(c_{e,1111} - c_{e,2222})/c_{e,1111} \neq 0$).

original medium is isotropic, so that $c_{1111} = c_{2222}$. In addition to producing a smoother medium represented by less grid points, upscaling induces apparent anisotropy, since $c_{e,1111} \neq c_{e,2222}$. This implies that we cannot distinguish if the low-frequency wave propagates through an isotropic medium with fine-scale heterogeneity or through a smooth effective medium that is anisotropic. This equivalent of long-wavelength anisotropy to small-scale anisotropy has profound implications for the interpretation of anisotropy in the Earth, because the origin of apparent anisotropy cannot be inferred on the basis of seismological data alone (Fichtner et al., 2013a).

This example demonstrates that a medium that has isotropic elasticity and uniform correlation length can apparently have anisotropic properties for the propagation of lower-frequency energy. As we have seen in Section 10.3.1 the introduction of differences in correlation length between the horizontal and vertical directions can have a significant effect on the character of the wavefield. Such aleatropic correlation, with variation according to direction, can establish a stochastic waveguide for high-frequency waves (Section 12.4) but will also have a significant effect on lower-frequency waves. With directional correlation, purely isotropic variations in medium properties will produce significant apparent anisotropy. If the aleotropic correlation is accompanied by anisotropic medium perturbations, which would be reasonable for some classes of mineral assemblages, then stronger anisotropy will result.

11

The Effects of Heterogeneity

As a consequence of variations in the physical properties of the Earth, the character of seismic wave propagation is distorted from that for a simple, radial structure. The broad-scale modifications of seismic structure are evident at low frequencies, and can be described with a limited numbers of parameters through, e.g., spherical harmonic analysis. Smaller- scale features become more important at higher frequencies and produce the complexity in observed seismograms.

Numerical simulation of the seismic wavefield can be achieved by a variety of methods for radially stratified structure. Spectral-element methods (e.g., Komatitsch & Tromp, 2001a,b) provide an effective means of calculation for 3-D structures on a global scale, but are ultimately limited by their assumption of locally smooth structure. For more local structure, spectral elements can also be employed but finite-difference schemes can also be very effective (e.g., Takemura et al., 2016).

Most approaches to the refinement of seismic structure build on the results from a previous model, incorporating perturbations to match additional data. The inclusion of fine-scale structure can also be achieved with a perturbation approach. In each case an important role is played by the *reference structure*, for which the propagation characteristics are regarded as known.

In this chapter we establish the theoretical foundations of the description of heterogeneity, for both body waves and modal fields. We then look at the way in which the varying nature of the distribution of heterogeneity in different parts of the Earth is manifest in the total effect on the seismic wavefield.

11.1 Reference Structures and Heterogeneity

We here explore how deviations from reference structures can be handled in a variety of ways, based on descriptions of perturbations to the wavefield. We build on the treatment of laterally varying media in Chapter 3, and look at approaches that can accommodate both smooth and rough structural features. We first consider approaches that consider the entire structure at once, and then a decomposition by layers that allows a more ready inclusion of different styles of heterogeneity with depth.

11.1.1 Deviations from a Reference Structure

We introduce a reference structure defined by variations in (complex) elastic moduli $c^o_{ijkl}(\mathbf{x})$, and density $\rho^o(\mathbf{x})$, as in (3.2.15). The equations of motion can then be written as

$$\mathcal{L}^o\{u_i\} = \partial_i(c^o_{ijkl}\partial_k u_l) + \rho^o\omega^2 u_j = -f_j, \tag{11.1.1}$$

with a solution via the representation theorem as

$$\mathbf{u}^o(\mathbf{x},\omega) = [\mathcal{L}^o]^{-1}\{\mathbf{u}\} = \int_V d^3\xi\, G^o(\mathbf{x},\xi,\omega)\mathbf{f}(\xi,\omega). \tag{11.1.2}$$

The integral over the volume with the Green's tensor for the reference structure acts as the inverse to the reference propagation operator $\mathcal{L}^o$,

$$[\mathcal{L}^o]^{-1}\cdot = \int_V d^3\xi\, G^o(\mathbf{x},\xi,\omega)\cdot \tag{11.1.3}$$

Even for a stratified medium, the construction of the reference Green's tensor throughout the region of interest is likely to require significant computation. For a 3-D reference model the situation becomes more complicated, but the operator form still provides useful insight into the character of the complex wavefield.

We now include the effects of additional three-dimensional structure through a heterogeneity operator $\mathcal{L}'$ so that the modified equation of motion becomes

$$\mathcal{L}^o_{\mathbf{x}}\{\mathbf{u}\} + \mathcal{L}'_{\mathbf{x}}\{\mathbf{u}\} = -\mathbf{f}(\mathbf{x}), \tag{11.1.4}$$

with a formal solution

$$\mathbf{u}(\mathbf{x}) = [\mathcal{L}^o + \mathcal{L}']^{-1}\mathbf{f}(\mathbf{x}), \tag{11.1.5}$$

in terms of the Green's tensor for the full heterogeneous medium.

At this point we have to expand the composite inverse operator to extract a solution in terms of the reference structure and the effect of heterogeneity. The usual procedure is to assume that the reference structure covers the significant features of the propagation and so $[\mathcal{L}^o]^{-1}$ is the most important component. We can write

$$[\mathcal{L}^o + \mathcal{L}']^{-1} = [\mathcal{L}^o[\mathcal{I} + [\mathcal{L}^o]^{-1}\mathcal{L}']^{-1} = [\mathcal{I} + [\mathcal{L}^o]^{-1}\mathcal{L}']^{-1}[\mathcal{L}^o]^{-1}, \tag{11.1.6}$$

where $\mathcal{I}$ is the identity operator. We now employ the operator identity

$$[\mathcal{I} - \mathcal{A}]^{-1} = \mathcal{I} + \mathcal{A}[\mathcal{I} - \mathcal{A}]^{-1}, \tag{11.1.7}$$

to extract a sequence of propagation contributions from the reference structure from (11.1.6).

$$\begin{aligned}[\mathcal{I} + [\mathcal{L}^o]^{-1}\mathcal{L}']^{-1}[\mathcal{L}^o]^{-1} &= [\mathcal{L}^o]^{-1} - [\mathcal{L}^o]^{-1}\mathcal{L}'[\mathcal{I} + [\mathcal{L}^o]^{-1}\mathcal{L}']^{-1}[\mathcal{L}^o]^{-1} \\ &= [\mathcal{L}^o]^{-1} - [\mathcal{L}^o]^{-1}\mathcal{L}'[\mathcal{L}^o]^{-1} \\ &\quad + [\mathcal{L}^o]^{-1}\mathcal{L}'[\mathcal{L}^o]^{-1}\mathcal{L}'[\mathcal{L}^o]^{-1} + \dots. \end{aligned} \tag{11.1.8}$$

The expansion of the composite inverse operator (11.1.8) corresponds to the standard Born series for the influence of heterogeneity. The contribution from the first-order term is most commonly used:

$$\mathbf{u} \approx \mathbf{u}^o + [\mathcal{L}^o]^{-1}\mathcal{L}'\mathbf{u}^o, \tag{11.1.9}$$

which has the explicit form

$$\mathbf{u}(\mathbf{x}, \omega) \approx \int_V d^3\boldsymbol{\xi}\, \mathrm{G}^o(\mathbf{x}, \boldsymbol{\xi}, \omega)\mathbf{f}(\boldsymbol{\xi}, \omega) + \int_V d^3\boldsymbol{\xi}\, \mathrm{G}^o(\mathbf{x}, \boldsymbol{\xi}, \omega)\, \mathcal{L}'_{\xi} \left[\int_V d^3\boldsymbol{\eta}\, \mathrm{G}^o(\boldsymbol{\xi}, \boldsymbol{\eta}, \omega)\mathbf{f}(\boldsymbol{\eta}, \omega) \right]. \tag{11.1.10}$$

This is the 'single scattering' result mentioned in Section 3.2.2. In (11.1.10) we see the combination of the unperturbed propagation with the second term that has propagation from the sources to the location of heterogeneity with the reference medium Green's tensor, the creation of a new effective source by the action of $\mathcal{L}'$, and then propagation in the reference medium to the location $\mathbf{x}$. Thus (11.1.10) can be viewed as the generation of a 'scattering' response at $\boldsymbol{\xi}$ with propagation through the reference medium from the sources to $\boldsymbol{\xi}$, and again from $\boldsymbol{\xi}$ to the observation point $\mathbf{x}$. We note that the perturbation term in (11.1.10) is simply added to the reference solution, and so this first-order Born approximation does not conserve energy.

This style of Born approximation, forms the foundation of many studies of body waves, but can also be used for surface waves. The mathematical development can be extended to provide useful expressions for the derivatives of the seismic wavefield with respect to variations in material properties exploiting the properties of adjoint fields as described in Section 11.1.3.

So far we have considered the first-order approximation. If the expansion for the inverse operator for the heterogeneous medium (11.1.8) is taken to second order

$$\mathbf{u} \approx \mathbf{u}^o + [\mathcal{L}^o]^{-1}\mathcal{L}'\mathbf{u}^o + [\mathcal{L}^o]^{-1}\mathcal{L}'[\mathcal{L}^o]^{-1}\mathcal{L}'\mathbf{u}^o, \tag{11.1.11}$$

so that

$$\mathcal{L}\mathbf{u} \approx \mathcal{L}^o\mathbf{u}^o + \mathcal{L}'\mathbf{u}^o + \mathcal{L}'[\mathcal{L}^o]^{-1}\mathcal{L}'\mathbf{u}^o, \tag{11.1.12}$$

which has a similar form to the 'mean-field' representation for stochastic media (Willis, 1981; Sato et al., 2012), with the operator effects of heterogeneity entering twice in the last term on the right-hand side.

Consider a model of variations $\mathfrak{L}'$ as a stationary random process such that the ensemble average $\langle \mathfrak{L}' \rangle = 0$. The equation for the mean field $\langle \mathrm{u} \rangle$, i.e., the displacement field after ensemble averaging, becomes

$$\mathcal{L}^o \langle \mathrm{u} \rangle = -\langle \mathfrak{L}'[\mathcal{L}^o]^{-1}\mathfrak{L}' \rangle \langle \mathrm{u} \rangle. \tag{11.1.13}$$

The mean field $\langle \mathrm{u} \rangle$ thus appears to propagate through a modified medium in which heterogeneity interaction occurs twice. The integral equation for the mean field $\langle \mathrm{u} \rangle$ can be solved in the frequency–wavenumber (ω–k) domain with explicit assumptions about the correlation properties of the random variations $\mathfrak{L}'$. The net effect is a stochastic dispersion of the mean wavefield, depending on the wavenumber convolution of the power spectra of the perturbations and the Green's function in the reference medium.

Energy transfer in such a random medium can be linked to the properties of the mean field. The net rate of energy transfer needs to be averaged over scales which are large compared to the correlation distance a and the wavelength of local fluctuations about the mean field λ_w, but small compared with the scale of variation in the amplitude of the mean field $\langle \mathrm{u} \rangle$. These concepts can be incorporated via multi-scale analysis and the local Born approximation to develop radiative transfer equations (Sato et al., 2012).

11.1.2 A Heterogeneity Series

The Born series provides a perturbation treatment of heterogeneity, with additional reference Green's tensor terms introduced at each level of the expansion. It is generally most useful when the heterogeneity is small, and so successive terms in the series diminish rapidly in size.

When heterogeneity is strong it is useful to look at the way in which the heterogeneity component interacts with the reference field. We consider a deterministic structural perturbation $\mathcal{L}'$, and make an alternative decomposition of the inverse operator $[\mathcal{L}^{\mathrm{o}} + \mathcal{L}']^{-1}$. Relative to the reference structure, the operator $\mathcal{L}'$ may well have negative apparent wavespeeds and so not be associated with a physical Green's tensor. We can compensate for such effects by introducing a shift $\mathcal{K}$ so that $\mathcal{L}' + \mathcal{K}$ represents a feasible medium, with an associated Green's tensor $\mathsf{G}'_{\mathcal{K}}$ so that the shifted operator is invertible. This shift operator will be simplest when $\mathcal{K}$ itself represents a simple structure, e.g., a stratified background medium.

We can express the original propagation operator for the heterogeneous medium $\mathcal{L}^{\mathrm{o}} + \mathcal{L}'$ in a form that incorporates this shift

$$\begin{aligned}\mathcal{L}^{\mathrm{o}} + \mathcal{L}' &= [\mathcal{L}^{\mathrm{o}} - \mathcal{K}] + \left[\mathcal{L}' + \mathcal{K}\right] \\ &= \mathcal{K}^{\mathrm{o}} + \mathcal{K}'.\end{aligned} \tag{11.1.14}$$

We now partition the composite inverse operator to emphasise the heterogeneity effects,

$$\begin{aligned}[\mathcal{L}^{\mathrm{o}} + \mathcal{L}']^{-1} = [\mathcal{K}^{\mathrm{o}} + \mathcal{K}']^{-1} &= [\mathcal{I} + [\mathcal{K}']^{-1}\mathcal{K}^{\mathrm{o}}]^{-1}[\mathcal{K}']^{-1}, \\ &= [\mathcal{K}']^{-1} - [\mathcal{K}']^{-1}[\mathcal{K}^{\mathrm{o}}][\mathcal{K}']^{-1} + \ldots \\ &= [\mathcal{K}']^{-1} - [\mathcal{K}']^{-1}[\mathcal{L}^{\mathrm{o}} - \mathcal{K}][\mathcal{K}']^{-1} + \ldots.\end{aligned} \tag{11.1.15}$$

This operator development offers a very different perspective on the effects of heterogeneity than the conventional Born series. The approximation for the wavefield to second order now has the form

$$\mathbf{u} \approx \mathbf{u}^{\mathrm{o}} + [\mathcal{K}']^{-1}\mathbf{u}^{\mathrm{o}} - [\mathcal{K}']^{-1}[\mathcal{L}^{\mathrm{o}} - \mathcal{K}][\mathcal{K}']^{-1}\mathbf{u}^{\mathrm{o}}. \tag{11.1.16}$$

The emphasis in (11.1.16) is on the effect of the heterogeneity through the shifted operator, and the original reference medium enters in the perturbation term. The Green's tensor $\mathsf{G}'_{\mathcal{K}}$ may well prove impractical to calculate, but nevertheless the split operator development shows how multiple interactions with the full effect heterogeneity modulate the propagation process of the reference field.

11.1.3 Variation of the Green's Function with Model Parameters

An alternative way of looking at the effect of changing the material properties of a medium is provided by using the adjoint method (see., e.g., Fichtner, 2011). This approach becomes important in practical methods for calculating derivatives of the wavefield, since the use of adjoint techniques dramatically reduces computational demands in inversion schemes (Chapter 16).

The frequency-domain Green's function $\mathsf{G}_j(\mathfrak{m}; \mathbf{x}, \xi)$ with vector components $\mathsf{G}_{ij}(\mathfrak{m}; \mathbf{x}, \xi)$ is the solution of the wave equation when the right-hand side is a point-localised force at position ξ acting in the j-direction, i.e.

$$\mathcal{L}^{o}_{\mathbf{x}}(\mathfrak{m})\mathsf{G}_j(\mathfrak{m}; \mathbf{x}, \xi) = \mathbf{f}_j\,\delta(\mathbf{x}-\xi)\,. \tag{11.1.17}$$

We have here emphasised the dependence of the Green's function on the material parameters $\mathfrak{m}$. The first variation of the ith-component of G_j with respect to the model parameters $\mathfrak{m}$ is

$$\delta\mathsf{G}_{ij}(\mathbf{x}, \xi) = \int_V \mathrm{d}^3\eta\, \mathbf{f}_i \cdot \delta\mathsf{G}_j(\eta, \xi)\,\delta(\mathbf{x}-\eta), \tag{11.1.18}$$

where we have suppressed the dependence on $\mathfrak{m}$ to avoid clutter. To eliminate $\delta\mathsf{G}_j$ from equation (11.1.18) we use the variation of equation (11.1.17),

$$\delta\mathcal{L}_\eta \mathsf{G}_j(\eta, \xi) + \mathcal{L}_\eta \delta\mathsf{G}_j(\eta, \xi) = \mathbf{0}. \tag{11.1.19}$$

We now introduce a test field $\mathbf{u}^\dagger(\eta')$. We multiply (11.1.19) by the test field and then integrate over the volume. We then add the result to equation (11.1.18) to give

$$\begin{aligned}\delta\mathsf{G}_{ij}(\mathbf{x}, \xi) &= \int_V \mathrm{d}^3\eta\, \mathbf{f}_i \cdot \delta\mathsf{G}_j(\eta, \xi)\,\delta(\mathbf{x}-\eta) \\ &+ \int_V \mathrm{d}^3\eta\, \mathbf{u}^\dagger(\eta)\cdot[\delta\mathcal{L}_\eta \mathsf{G}_j(\eta, \xi)] + \int_V \mathrm{d}^3\eta\, \mathbf{u}^\dagger(\eta)\cdot[\mathcal{L}_\eta \delta\mathsf{G}_j(\eta, \xi)].\end{aligned} \tag{11.1.20}$$

We introduce the adjoint operator $\mathcal{L}^\dagger_\eta$ to $\mathcal{L}_\eta$, which satisfies

$$\int_V \mathrm{d}^3\eta\, \mathbf{u}^\dagger(\eta)\cdot[\mathcal{L}_\eta \delta\mathsf{G}_j(\eta, \xi)] = \int_V \mathrm{d}^3\eta\, \delta\mathsf{G}_j(\eta, \xi)\cdot[\mathcal{L}^\dagger_\eta \mathbf{u}^\dagger(\eta)]. \tag{11.1.21}$$

We can then rearrange the third term in equation (11.1.21) to the form

$$\begin{aligned}\delta\mathsf{G}_{ij}(\mathbf{x}, \xi) &= \int_V \mathrm{d}^3\eta\, \delta\mathsf{G}_j(\eta, \xi)\cdot\left[\mathbf{f}_i\delta(\mathbf{x}-\eta) + \mathcal{L}^\dagger_\eta \mathbf{u}^\dagger(\eta)\right] \\ &+ \int_V \mathrm{d}^3\eta\, \mathbf{u}^\dagger(\eta)\cdot[\delta\mathcal{L}_\eta \mathsf{G}_j(\eta, \xi)]\,.\end{aligned} \tag{11.1.22}$$

We now specify the test function $\mathbf{u}^{\dagger}(\eta)$ to be the solution of the *adjoint equation*

$$\mathcal{L}_{\eta}^{\dagger}\mathbf{u}^{\dagger}(\eta) = -\mathbf{f}_{i}\delta(\mathbf{x}-\eta), \tag{11.1.23}$$

so that $\mathbf{u}^{\dagger}(\eta)$ is the negative adjoint Green's function for a unit force in the ith-direction at position $\mathbf{x}$, i.e.,

$$\mathbf{u}^{\dagger}(\eta) = -\mathsf{G}_{i}^{\dagger}(\eta, \mathbf{x}). \tag{11.1.24}$$

With the aid of equations (11.1.23) and (11.1.24), the variation of the Green's function (11.1.22) simplifies to

$$\delta\mathsf{G}_{ij}(\mathbf{x}, \xi) = -\int_{V} d^{3}\eta\, \mathsf{G}_{i}^{\dagger}(\eta, \mathbf{x}) \cdot [\delta\mathcal{L}_{\eta}\mathsf{G}_{j}(\eta, \xi)]\,. \tag{11.1.25}$$

Equation (11.1.25) gives the variation of the Green's function in terms of a scalar product of the Green's function itself and its adjoint. The integral kernel of the scalar product is the variation of the forward modelling operator $\delta\mathcal{L}_{\mathbf{x}}$.

In terms of the full Green's tensor G, the variation due to a modification of structure takes the form

$$\delta\mathsf{G}(\mathbf{x}, \xi) = -\int_{V} d^{3}\eta\, \left[\mathsf{G}^{\dagger}(\eta, \mathbf{x})\right]^{T} \delta\mathcal{L}_{\eta}\mathsf{G}(\eta, \xi). \tag{11.1.26}$$

With appropriate choices for $\delta\mathcal{L}$ corresponding to variations of the elastic moduli and density, the full suite of derivatives of the Green's tensor can be extracted through the interaction of the direct and adjoint fields (see Chapter 16).

11.1.4 Perturbations of Reflection Operators

The disadvantage of the representations of the influence of heterogeneity we have just discussed is that they require knowledge of the Green's function for the entire domain of interest. With the dominant vertical variation of seismic properties in the Earth's interior, the nature of heterogeneity will change with depth and it is helpful to be able to separate the influence of different zones. One way of achieving this goal is to work with the concept of reflection and transmission operators introduced in Section 3.3, and look at the way that these operators are modified as structure is varied.

We consider a region between the levels z_A and z_B with a background stratified structure on which is imposed distributed heterogeneity. For the stratified medium we can construct a reflection operator $\hat{\mathbf{R}}_{D}^{AB}$ and transmission operator $\hat{\mathbf{T}}_{D}^{AB}$ for downward incident waves from above A, in a similar way to the treatment of a single interface in Section 3.3.1 by Fourier transformation over horizontal wavenumber. The full operators for the zone including heterogeneity can then be written as

$$\mathbf{R}_{D}^{AB} = \hat{\mathbf{R}}_{D}^{AB} + \mathbf{r}_{D}^{AB}, \qquad \mathbf{T}_{D}^{AB} = \hat{\mathbf{T}}_{D}^{AB} + \mathbf{t}_{D}^{AB}. \tag{11.1.27}$$

The additional contributions $\mathbf{r}_{D}^{AB}$, $\mathbf{t}_{D}^{AB}$ are produced by the effects from the

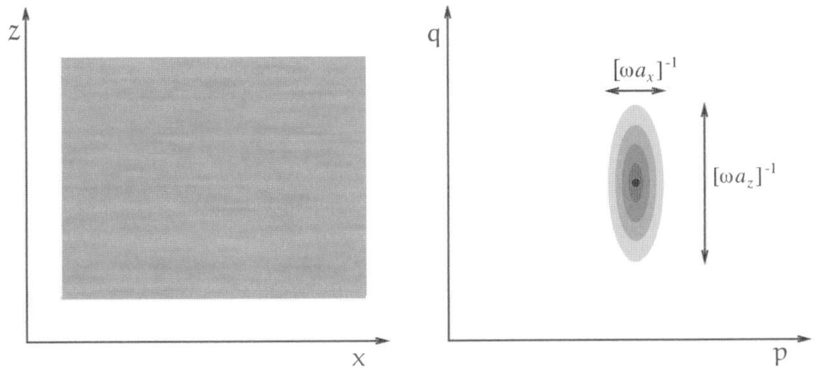

Figure 11.1 Mapping between the spatial representation of heterogeneity with much longer horizontal than vertical correlation length and its effects in the slowness domain. The slowness response is convolved with the Fourier transform of the heterogeneity and so there is tight coupling in horizontal slowness, but a broad distribution in vertical slowness.

heterogeneity. These heterogeneity contributions may also be constructed with wavenumber synthesis, but as in (3.4.3) there is a wavenumber convolution between the transform of the deviations from stratification and the representation of the wavefield in terms of upgoing and downgoing waves (Kennett, 1986). When the heterogeneity level is small and all the physical properties have the same functional dependence on position, the influence of the heterogeneity can be represented with a a single wavenumber function together with a set of scaling terms associated with the deviation from stratification. The influence of the heterogeneity contribution depends on its wavenumber transform as illustrated in Figure 11.1. The width of interaction in horizontal slowness varies inversely with horizontal scale, so that a quasi-laminate structure leads to compact interaction horizontally but stronger effects on vertical slowness.

At each level h within AB the wavenumber effects from heterogeneity can be projected onto upgoing and downgoing waves via a local scattering matrix **S**. For weak heterogeneity, or a thin zone, the situation can be simplified with a first-order Born treatment. To this approximation, the reference wavefield maintains its wavenumber identity through the full zone. The resulting expressions are quite complicated, but can be readily generated through tracking the sequence of physical processes, as in Figure 11.2 for reflection. Transmission to the level h with possible reverberation within AB with the initial wavenumber k_i is followed by conversion of the downgoing waves to the new wavenumber k via the action of the scattering elements $\mathbf{S}_{UD}$, $\mathbf{S}_{DD}$ and following reflection from below h there is a similar operation on upgoing waves from $\mathbf{S}_{UU}$, $\mathbf{S}_{DU}$. These interactions produce new secondary sources U^s_h, D^s_h with propagation at the transformed wavenumber k. Once again we have transmission to A from h with possible reverberation within AB. A comparable treatment can be made for transmission to B with just a modification or the second stage of propagation at wavenumber k. The effect of the

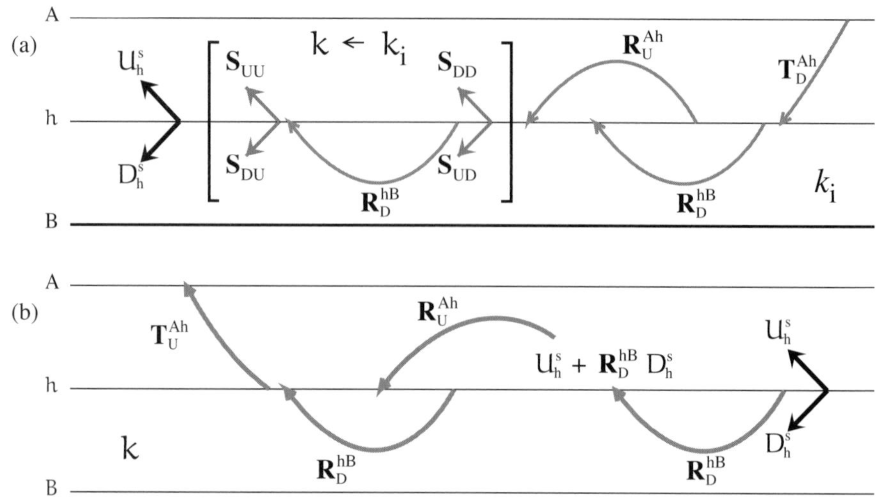

Figure 11.2 Representation of the physical processes associated with the presence of heterogeneity superimposed on a stratified structure: (a) from initial wavenumber k_i to wavenumber k by the effect of heterogeneity through the **S** matrix terms, (b) propagation at wavenumber k from the effective heterogeneity sources U_h^s, D_h^s.

entire layer of heterogeneity is then produced by integration over h from z_A to z_B. The complex process induced by the impact of heterogeneity in the wavenumber domain, mirrors the escalating complexity effects seen in numerical calculations as waves encounter a disturbed zone.

In the case of a uniform zone AB with no contrast with its surroundings, the internal reflection terms vanish and we are left with transmission to the level h, wavenumber change via scattering through the elements S_{UD} and S_{DD} followed by transmission to either A for reflection, or B for transmission.

The effect of multiple layers can be introduced by combining the reflection and transmission operators using the composition relations (3.3.30) and (3.3.32). Thus, for a region AC divided at level B, the overall reflection operator

$$
\begin{aligned}
\mathsf{R}_D^{AC} &= \mathsf{R}_D^{AB} + \mathsf{T}_U^{AB}\mathsf{R}_D^{BC}[\mathsf{I} - \mathsf{R}_U^{AB}\mathsf{R}_D^{BC}]^{-1}\mathsf{T}_D^{AB}, \qquad (11.1.28)\\
&= \left(\hat{\mathsf{R}}_D^{AB} + \mathsf{r}_D^{AB}\right) + \left(\hat{\mathsf{T}}_U^{AB} + \mathsf{t}_U^{AB}\right)\mathsf{X}^{ABC}\left(\hat{\mathsf{R}}_D^{BC} + \mathsf{r}_D^{BC}\right)\left(\hat{\mathsf{T}}_D^{AB} + \mathsf{t}_D^{AB}\right),
\end{aligned}
$$

where we have introduced the separation into stratified and heterogeneous parts, as in (11.1.27). The reverberation operator

$$
\begin{aligned}
\mathsf{X}^{ABC} &= \left[\mathsf{I} - \hat{\mathsf{R}}_D^{BC}\hat{\mathsf{R}}_U^{AB} + \left\{\hat{\mathsf{R}}_D^{BC}\hat{\mathsf{r}}_U^{AB} + \hat{\mathsf{r}}_D^{BC}\hat{\mathsf{R}}_U^{AB} + \hat{\mathsf{r}}_D^{BC}\hat{\mathsf{r}}_D^{BC}\right\}\right]^{-1}\\
&= \left[\mathsf{I} - \hat{\mathsf{R}}_D^{BC}\hat{\mathsf{R}}_U^{AB} + \mathcal{H}^{AC}\right]^{-1}. \qquad (11.1.29)
\end{aligned}
$$

Using the operator identity (2.3.18) we can extract the reverberations associated with the stratification which modulate the influence of heterogeneity so that

$$\mathbf{X}^{ABC} = [\mathbf{I} - \hat{\mathbf{R}}_D^{BC}\hat{\mathbf{R}}_U^{AB}]^{-1} + [\mathbf{I} - \hat{\mathbf{R}}_D^{BC}\hat{\mathbf{R}}_U^{AB}]^{-1}\mathcal{H}^{AC}[\mathbf{I} - \hat{\mathbf{R}}_D^{BC}\hat{\mathbf{R}}_U^{AB}]^{-1} + \ldots. \quad (11.1.30)$$

From (11.1.28) and (11.1.30) we can extract the contribution from the underlying stratified medium, with a residuum of contributions induced by the heterogeneity. These will include two possibilities of scattering in transmission through AB and two in reflection, one above and one below B. Even if we only use a first-order Born representation in any layer, the composite effect of successive layers will introduce multiple scattering contributions.

Thus, for example, the different nature of heterogeneity in the crust and mantle will interact in surface observations for regional phases such as *Pn*, *Sn* (cf. § SWII:19.2). Waves scattered by the small-scale features in the crust will enter the upper mantle to undergo further scattering from the larger-scale heterogeneity in that region. Following refraction back from the upper mantle, transmission through the crust adds extra complexity to the seismograms.

11.2 Coupling of Modal Fields by Heterogeneity

For surface waves, which have strong sampling of the near surface, the effects of heterogeneity can be introduced through coupling between the modes of reference structures.

The most effective way of representing the surface-wave portion of the seismic wavefield is via the superposition of modal contributions. For a laterally homogeneous medium such modes propagate independently, but when structural boundaries are encountered, as in Section 4.6.3, there has to be a redistribution of energy between modes. Similar concepts apply to the evolution of the field of surface waves as it passes through a changing medium that may also display smaller-scale heterogeneity. Because the energy in the lower-order modes of Rayleigh and Love waves is concentrated near the surface where the variations in seismic wavespeed are strongest, the potential exists for substantial disruption of the wavefield. Fortunately, the relatively long wavelengths of the surface waves help to smooth out the influence of heterogeneity, so the dominant effects come from larger-scale features as exploited in surface-wave tomography. We will first consider 2-D problems for which techniques exploiting coupling between modes are well developed. We then turn attention to the fully 3-D case where the evolution of the modal field is more complex, and scattering techniques are often used. For stochastic heterogeneity we can relate the attenuation of surface waves to the properties of the wavespeed distribution.

11.2.1 2-D Propagation

We build on the work of Kennett (1984c), Maupin (1988) to develop a description of the horizontal evolution of the modal field in terms of position-dependent coefficients that allow for transfer between modes. We consider a scenario with a smoothly varying reference structure on which is imposed finer-scale heterogeneity. We then need to describe the behaviour of the modes in the reference model and their interaction with the full structure.

In a 2-D situation we can represent the modal field as a sum of contributions that separate vertical dependence in terms of modal eigenvectors and a phase term varying with horizontal position x. We express the displacement as

$$\mathbf{u}(x,z,\omega) = \sum_K c_K(x,\omega)\mathbf{u}_K^e(x,z,\omega)\exp\left\{i\omega\int^x d\xi\, p_K(\xi,\omega)\right\}. \tag{11.2.1}$$

Here $\mathbf{u}_K^e(x,z,\omega)$ is the displacement eigenfunction for the Kth local mode in the reference structure at position x, i.e., the Kth mode for a horizontally stratified structure with the same vertical wavespeed profile as in the reference model at x. $p_K(x,\omega)$ is the corresponding phase slowness. We have a comparable development for the horizontal traction $\hat{\mathbf{t}}_x = [\tau_{xx}, \tau_{xy}, \tau_{xz}]^T$,

$$\hat{\mathbf{t}}_x(x,z,\omega) = \sum_K c_K(x,\omega)\hat{\mathbf{t}}_K^e(x,z,\omega)\exp\left\{i\omega\int^x d\xi\, p_K(\xi,\omega)\right\}, \tag{11.2.2}$$

where $\hat{\mathbf{t}}_K^e$ is the traction eigenfunction for mode K. The displacement and horizontal traction are required to be continuous across each vertical plane. We allow horizontally varying interfaces of the form $z = h_i(x)$, and require both the displacement and normal traction to be continuous across such interfaces. This condition introduces equivalent forces at each interface proportional to the interface slope with a jump in horizontal traction.

We can express the horizontal evolution of the displacement and traction as a set of first-order differential equations in the horizontal coordinate x:

$$\frac{\partial}{\partial x}\begin{pmatrix}\mathbf{u}\\ \mathbf{t}_x\end{pmatrix} = \begin{pmatrix}\mathcal{A}_{uu} & \mathcal{A}_{ut}\\ \mathcal{A}_{tu} & \mathcal{A}_{tt}\end{pmatrix}\begin{pmatrix}\mathbf{u}\\ \mathbf{t}_x\end{pmatrix} - \begin{pmatrix}\mathbf{0}\\ \mathbf{f}\end{pmatrix}, \tag{11.2.3}$$

where $\mathcal{A}_{uu}$, $\mathcal{A}_{ut}$, $\mathcal{A}_{tu}$ and $\mathcal{A}_{tt}$ are partial differential operators that depend on $\partial/\partial y$, $\partial/\partial z$, frequency ω and the local properties of the medium. The term $\mathbf{f} = \sum_i \partial_x h_i [\mathbf{t}_x]_i \delta(z - h_i(x))$, with a sum over all interfaces, allows for the equivalent forces due to interface slope. For a 2-D medium, the evolution equation (11.2.3) separates into two distinct sets with separate propagation of Rayleigh and Love waves, even for oblique travel.

In terms of the displacement–traction vector $\mathbf{b} = [\mathbf{w}, \mathbf{t}]^T$, we can write (11.2.3) in the compressed form

$$\frac{\partial}{\partial x}\mathbf{b} = \mathcal{A}\mathbf{b} - \mathbf{v}, \tag{11.2.4}$$

where $\mathcal{A}$ includes contributions from both the reference structure $\mathcal{A}^{\text{ref}}$ and the heterogeneity $\Delta\mathcal{A}$. We now introduce the modal expansions (11.2.1), (11.2.2) for the displacement and traction components of $\mathbf{b}$ and exploit the orthonormality properties of the modal eigenvectors. For modes J, K we have the property (3.5.5)

$$i\int_0^{\infty} dz\, \left[\mathbf{u}_J^{eT}(-p_J, z, \omega)\hat{\mathbf{t}}_K^{e}(p_K, z, \omega) - \hat{\mathbf{t}}_J^{eT}(-p_J, z, \omega)\mathbf{u}_K^{e}(p_K, z, \omega)\right] = \delta_{JK}. \tag{11.2.5}$$

Since the relation (11.2.5) involves both forward and backward propagation, the modal sums in (11.2.1), (11.2.2) need to include both senses of propagation. We write (11.2.5) as an inner product between modal terms

$$\left\{\mathbf{b}_J, \mathbf{b}_K\right\} = i\int_0^{\infty} dz\, \left([\mathbf{u}_J^e]^{T*}\hat{\mathbf{t}}_K^e - [\hat{\mathbf{t}}_J^e]^{T*}\mathbf{u}_K^e\right) = \delta_{JK}, \tag{11.2.6}$$

with a normalisation to unit energy flux in the x direction. The complex conjugates of $\mathbf{u}_J^e$, $\hat{\mathbf{t}}_J^e$ are equivalent to the switch in the sense of propagation.

We write the modal field expansion in the form

$$\mathbf{b} = \sum_K c_K \mathbf{b}_K^e \exp\left[i\omega \int^x d\xi\, p_K(\xi, \omega)\right] = \sum_K c_K \mathbf{b}_K^e e_K. \tag{11.2.7}$$

The horizontal derivative of **b** is then given by

$$\begin{aligned} \frac{\partial}{\partial x}\left[\sum_K c_K \mathbf{b}_K^e e_K\right] &= \mathcal{A}\left[\sum_K c_K \mathbf{b}_K^e e_K\right] - \left[\sum_K c_K \mathbf{v}_K^e e_K\right], \\ &= \left[\sum_K \frac{\partial c_K}{\partial x}\mathbf{b}_K^e e_K\right] + \left[\sum_K c_K \frac{\partial \mathbf{b}_K^e}{\partial x} e_K\right] + \left[\sum_K c_K \mathbf{b}_K^e \frac{\partial e_K}{\partial x}\right]. \end{aligned} \tag{11.2.8}$$

With the aid of the orthonormality relation (11.2.6) we can extract a differential equation for the horizontal evolution of the modal coefficients

$$\begin{aligned} \frac{\partial c_L}{\partial x} &= \sum_K c_K \left[\left\{\mathbf{b}_L^e, \frac{\partial \mathbf{b}_K^e}{\partial x}\right\} - \left\{\mathbf{b}_L^e, \mathbf{v}_K^e\right\}\right] e^{i\omega\int^x d\xi\,[p_K(\xi,\omega) - p_L(\xi,\omega)]} \\ &\quad + \sum_K c_K \left\{\mathbf{b}_L^e, \Delta\mathcal{A}\mathbf{b}_K^e\right\} e^{i\omega\int^x d\xi\,[p_K(\xi,\omega) - p_L(\xi,\omega)]}. \end{aligned} \tag{11.2.9}$$

The first contribution on the right-hand side of (11.2.9) represents the evolution of the reference modes and the second the effects of deviations from the reference structure through $\Delta\mathcal{A} = \mathcal{A} - \mathcal{A}^{ref}$.

Maupin (1988) has demonstrated that the contribution to (11.2.9) from the reference modes for $K \neq L$ can be written in terms of horizontal derivatives of the structure rather than requiring derivatives of the eigenfunctions. When gradients in seismic structure are low and there are no deviations from the reference structure so $\Delta\mathcal{A} = 0$, there will be almost no transfer between modes. The modes then remain essentially independent and the wavefield is described directly by (11.2.7), which is equivalent to the results of Woodhouse (1974).

For stronger variation in structure, the local mode approach requires fewer modes to be considered than the use of a constant reference structure, but at the cost of more effort in calculating eigenfunctions. Fortunately, the shape of the eigenfunctions varies relatively slowly in a smooth structure, and so a limited number of regional models can be used with interpolation for local structure.

In all cases the modal sum needs to include sufficient modes to provide an adequate description of the wavefield behaviour under consideration. The exploitation of higher modes means that many classes of multiply reflected *S* waves can be well represented, but the full range of body-wave phenomena would require an excessive number of modes.

We have noted that the coupling equations involve both forward (+) and backward (−) propagation. We can emphasise this aspect by introducing vectors $\mathbf{c}^+, \mathbf{c}^-$ and writing the coupling equations (11.2.9) in the form

$$\frac{\partial}{\partial x}\begin{pmatrix}\mathbf{c}^+\\ \mathbf{c}^-\end{pmatrix} = \begin{pmatrix}\mathbf{B}^{++} & \mathbf{B}^{+-}\\ \mathbf{B}^{-+} & \mathbf{B}^{--}\end{pmatrix}\begin{pmatrix}\mathbf{c}^+\\ \mathbf{c}^-\end{pmatrix}, \tag{11.2.10}$$

where the elements of the matrices $\mathbf{B}^{++}$ are constructed from the coupling terms, both for the reference structure and the effect of heterogeneity. For a coupling region between x_L and x_R we can introduce reflection and transmission matrices connecting the vectors $\mathbf{c}^+$, $\mathbf{c}^-$ at the two sides, e.g.,

$$\mathbf{c}^+(x_R) = T^{++}(x_R, x_L)\mathbf{c}^+(x_L), \qquad \mathbf{c}^-(x_L) = R^{-+}(x_R, x_L)\mathbf{c}^+(x_L). \tag{11.2.11}$$

Kennett (1984c) has shown that nonlinear Ricatti differential equations can be constructed for the reflection and transmission properties in terms of the coupling matrices $\mathbf{B}^{++}$ etc. The coupled reflection and transmission equations are

$$\frac{\partial}{\partial x_L}T^{++} + T^{++}\mathbf{B}^{++}T^{++}\mathbf{B}^{+-}R^{-+} = 0, \tag{11.2.12}$$

$$\frac{\partial}{\partial x_L}R^{-+} + R^{-+}\mathbf{B}^{++} - \mathbf{B}^{--}R^{-+} + R^{-+}\mathbf{B}^{-+}R^{-+} - \mathbf{B}^{-+} = 0. \tag{11.2.13}$$

These equations are integrated starting at x_R, with progressive exposure of structure to the left. The initial conditions are that there is complete transmission and no reflection at the starting point

$$T^{++}(x_R, x_R) = I, \qquad R^{-+}(x_R, x_R) = 0. \tag{11.2.14}$$

With the choice of eigenfunction normalisation the transmission matrices satisfy

$$[T^{++}(x_R, x_L)]^{\mathrm{T}} = T^{--}(x_R, x_L), \tag{11.2.15}$$

so that knowledge of the transmission matrix for waves on one side determines the transmission properties for the other direction of incidence. If reflection effects are weak then a simplified equation can be employed for the transmission case, with neglect of the term involving the reflection matrix. However, when a fixed reference mode set is used, the reflection terms can be significant during the course of the calculation even through the net reflection is low. This arises because deviations from the reference structure force localised changes in the coupling system to match the nature of the heterogeneous structure.

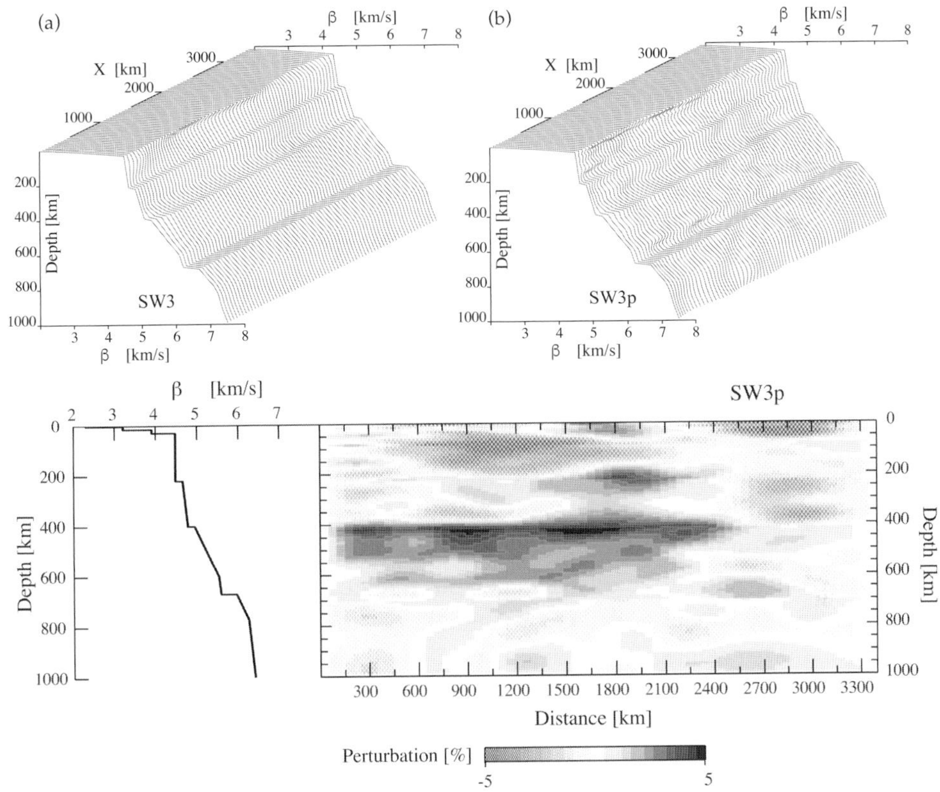

Figure 11.3 Heterogeneity models: (a) model SW3 uses the shear wavespeed model WEPL3 from Nolet (1990) over the zone from 200–3200 km with tapering back to the PREMC background at the ends; (b) model SW3p has the addition of 1% variability with a scale length around 400 km to give both medium- and broad-scale variations. The lower panel shows the deviations of SW3p from the PREMC reference model.

Illustration of heterogeneity effects

We here illustrate the way in which heterogeneity modifies the properties of surface waves using heterogeneity models spanning over 3000 km. The results build on the study of structure under western Europe by Nolet (1990) who studied the structure along the great circle through a linear array of stations (NARS) using earthquakes in Japan and Africa. With the use of a partitioned waveform inversion technique Nolet (1990) generated a shear-wavespeed model WEPL3 over the span of the array. The model SW3 coincides with the shear-wave structure of WEPL3 where this is defined, but which is graded back to the PREMC reference model at either edge of the region over a distance of 200 km (Figure 11.3). This enables the application of the Ricatti scheme for modal reflection and transmission matrices to examine the way in which the modal field interacts with the heterogeneity (Kennett & Nolet, 1990). A further heterogeneous model SW3p was created with the imposition of up to 1% additional heterogeneity with a scale length around 400 km on the wavespeeds of SW3 (Figure 11.3). In this model

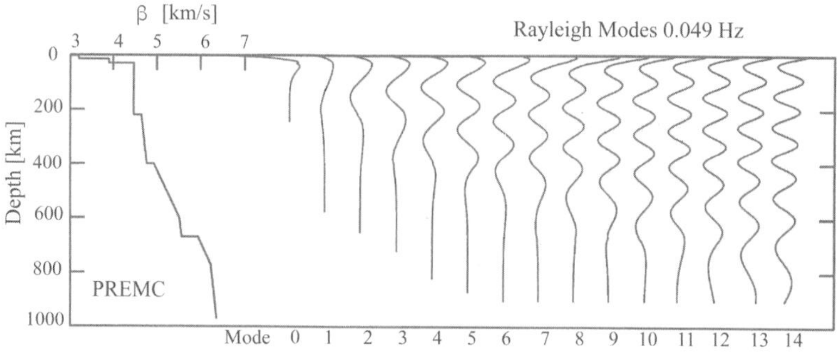

Figure 11.4 The PREMC reference model and vertical-component eigenfunctions for the first 15 Rayleigh modes at a frequency of 0.049 Hz.

the principal features of SW3 can still be distinguished, but are broken up and modified by the presence of the smaller-scale heterogeneity, notably around the 410 km discontinuity. Such a composite heterogeneity model is consistent with both body-wave and surface-wave behaviour (Nolet et al., 1994).

The reference model is the PREMC model of Nolet (1990) which is modified from PREM (Dziewonski & Anderson, 1981) by the inclusion of a thickened crust and a uniform S wavespeed of 4.47 km/s between 30 km and 220 km depth. These modifications were designed to give a good fit to fundamental mode Rayleigh wave propagation along the great circle from Japan to the linear NARS array in western Europe.

The vertical-component eigenfunctions for the first 15 Rayleigh modes on the PREMC model at 0.049 Hz are shown in Figure 11.4. The modes have appreciable amplitude from the surface down to depths a little deeper than the turning point for a ray with the same phase slowness. As a result, cross-coupling between modes can be induced by heterogeneity over a wide range of depths. However, unless the deviations from the reference model are large, energy will tend to be transferred to nearby modes with similar character in depth.

In Figure 11.5 we show results for the amplitude of the transmission matrix for the SW3p structure; there is sufficient deviation from the base model PREMC that a full calculation including reflection terms is required. The presence of the finer-scale structures in SW3p breaks up the large-scale features in SW3 and actually enhances transmission slightly. Surface-wave modes with group velocity below 4.2 km/s can be regarded as propagating independently up to 0.020 Hz. The body-wave group of modes with higher group velocity succumbs to moderate interaction above 0.040 Hz, but the approximation of largely independent propagation can be sustained to about 0.05 Hz. The errors introduced into the analysis methods by ignoring mode interactions above these frequency limits will depend on the distribution of energy across the modes imposed by the source, and the criterion used for waveform matching between observed and theoretical seismograms in tomographic inversion

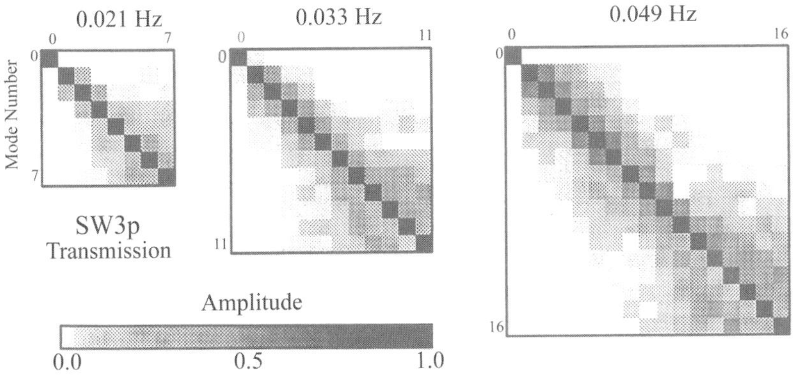

Figure 11.5 Representation of the amplitudes of the transmission matrix elements for model SW3p as a function of frequency, showing mild coupling between the higher modes.

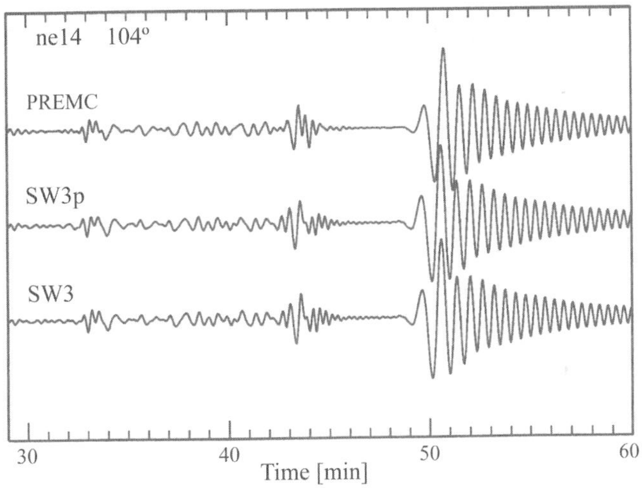

Figure 11.6 Comparison of theoretical seismograms for an event in Hokkaido recorded at station ne14 at 104° epicentral distance. The upper trace is for the PREMC reference model, the two lower traces include the effect of transmission through 3400 km of the heterogeneity models SW3 and SW3p.

Theoretical seismograms for the heterogeneity structures based on WEPL3 including full allowance for inter-mode coupling show a distinct phase shift for the fundamental mode when compared with the corresponding calculations for the reference model PREMC (Figure 11.6), as was observed at the NARS stations by Nolet (1990). The portions of the seismograms which show the largest influence from the presence of lateral heterogeneity are those which depend on interference phenomena, such as the *Sa* phase with group velocity around 4.5 km/s. The disruption of the phase patterns can have a profound influence on the appearance of the waveforms. The presence of small-scale heterogeneity has very little influence on the character of the seismograms below 0.035 Hz, but becomes more important as the frequency increases, especially for the body-wave phases.

Maupin (1992) has extended the concept of local mode coupling to a 2.5-D structure in the analysis of Rayleigh waves propagating along the swell associated with the Hawaiian island chain. She considers a structure with thinner lithosphere along the axis of the well connecting smoothly to normal oceanic lithosphere, 600 km away from the swell axis. Rayleigh waves propagating in a direction close to the axis of the swell get trapped in the low-velocity channel. The resulting interference between waves travelling on different paths leads to complex polarisation patterns after travelling for more than 2000 km along the swell.

11.2.2 3-D Propagation

In a 2-D situation conversions between modes stay inline, but with full 3-D variations we have to be able to describe processes in which the direction of propagation of modal contributions can change with propagation. Even in a smoothly varying medium the different modes follow distinct, frequency-dependent rays, that depend on the pattern of phase slowness for the mode (Woodhouse, 1974). Thus a suite of modes that leave the source in a similar direction can be spread out horizontally by the differences in phase slowness behaviour with focussing and defocussing, and variations in the vertical eigenfunctions influencing the amplitude of the displacement (Maupin, 2007). Often the fundamental Rayleigh wave will have a rather different path from the main group of higher modes.

Kennett (1998) has developed an extension of the concepts described in Section 11.2.1, with a representation of the seismic wavefield that depends only on first-order derivatives of displacement and traction with respect to the horizontal spatial coordinates. Once again an expansion can be made in terms of a sum of modal contributions with spatially varying weighting coefficients. In addition to a sum over modes, an angular spectrum of local plane waves has to be introduced to allow for the effects of gradients in material properties. The orthogonality relation between modes is now more complex, but it is possible to establish a coupled set of modal equations for the situation with a smoothly varying reference structure with superimposed heterogeneity. Horizontal gradients in seismic structure have the effect of changing the polarisation of the wavefield so that mutual coupling is induced between Rayleigh and Love modes, particularly for higher-frequency waves. The effect of the modal decomposition is to convert a fully 3-D problem into a linked set of 2-D propagation sheets for the individual modes.

For complex seismic structures, direct numerical calculation using fully numerical techniques such as spectral elements (e.g., Fichtner et al., 2009) are capable of producing the surface-wave component of the wavefield with less computational effort than 3-D mode coupling. Nevertheless, coupled mode tools are of value in understanding the nature of the complex wavefield and the way in which it develops. In Figure 11.7 we present a representation of mode coupling

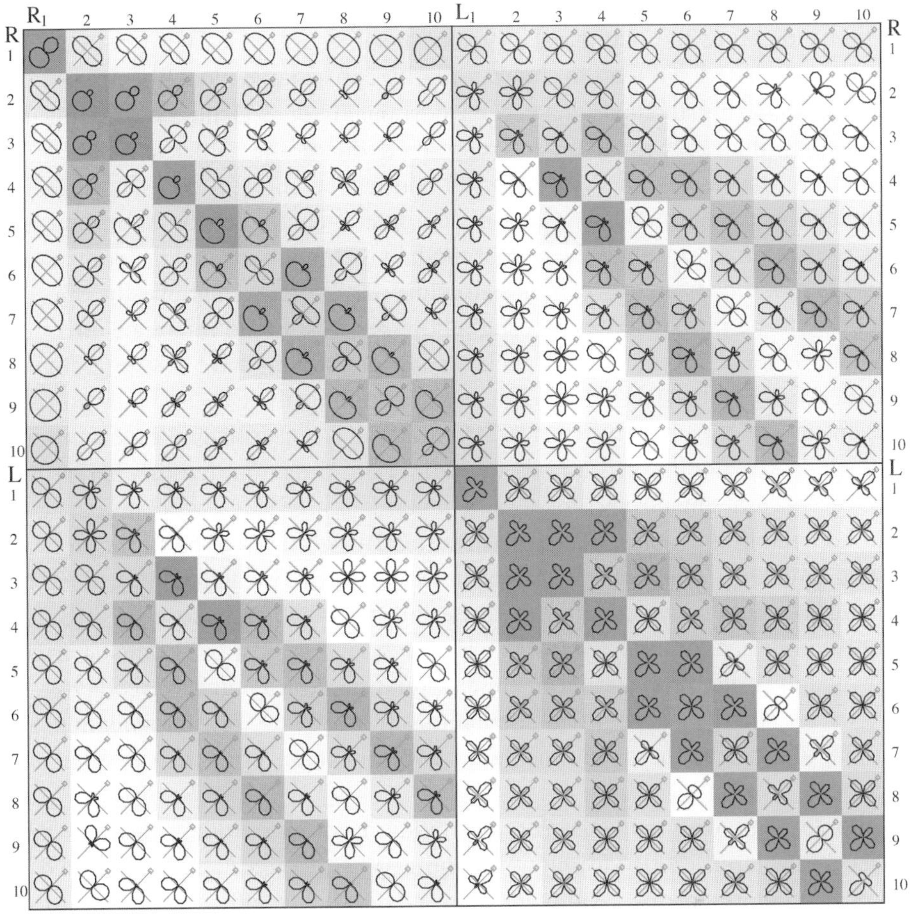

Figure 11.7 Modal coupling due to 3-D structure in a surface-wave tomography model for the Australian continent, using the modes of the reference structure PREMC at 0.078 Hz (12.8 s period). The matrix of coupling coefficients between the first 10 modes of Rayleigh and Love waves are shown for an incident wave at 45° azimuth. The shading indicates the amplitude of the mode, with darker stronger, and the rosette diagrams the associated angular distribution.

between Rayleigh and Love waves for 0.078 Hz waves using the modes of the PREMC reference structure (Figure 11.2.2) incident at an azimuth of 45° in a 3-D model for shear wavespeed beneath the Australian continent. The location at 20°S, 137°E in the north of central Australia lies on the edge of a zone of strong horizontal gradient in upper-mantle structure that induces complex behaviour. For this relatively high-frequency case, the fundamental modes show modest coupling to the higher modes. But, for both Rayleigh and Love waves, there is strong interaction between the first few higher modes whose eigenfunctions are concentrated in the top 200 km. Cross-coupling between Rayleigh and Love is more significant for mode 5 and above, and occurs in a band just beyond the dominant self coupling. For each of the coupling elements we show a rosette diagram indicating the azimuthal distribution of coupling. These patterns are complex and varied, but tend to show similarity across large blocks of modes.

Modal scattering methods have been developed for 3-D problems using the Born approximation with a modal representation of the Green's function for a reference structure (Snieder, 1986; Friederich et al., 1993). The coupling coefficients between modes for an isolated heterogeneity can be represented in terms of depth integrals over the pair of modal eigenfunctions and the contrasts in material properties from the reference structure. This approach allows the effect of anisotropy to be introduced as a perturbation on an isotropic reference structure. As we have seen in Figure 11.7, such modal interactions have a strong angular dependence. For Rayleigh–Rayleigh mode coupling there is a small isotropic component accompanied by θ and 2θ variation. For isotropic heterogeneity, Rayleigh–Love coupling has a node in the forward direction, but this is removed in the presence of anisotropy.

At moderate frequencies, scattering of surface waves tends to be strongest in the forward direction. This property has been exploited by Friederich et al. (1993) to construct a scheme in which multiple scattering is allowed in the forward direction, but only single scattering in the backward direction. Applications of this approach for inversion of the propagation of fundamental mode Rayleigh waves across an array of stations can account for deviations from great-circle propagation and focussing due the shape of the incident wavefront (e.g., Friederich & Wielandt, 1995).

Multiple scattering using successive iterations of the Born series has been applied by Maupin (2001) to allow for the Love-Rayleigh coupling induced by anisotropic heterogeneity. Although there is significant effort in setting up the calculations for the second iteration, all subsequent iterations have the same structure. Several orders of multiple scattering are required to stabilise the phase response, but quite strongly heterogeneous models can be handled in an efficient manner.

Single-scattering approaches have been used to look at the sensitivity of the wavefield of surface waves to perturbations of structure (e.g., Yoshizawa & Kennett, 2005). For multiple *S* waves, a multiple modal sum with inclusion of mode coupling allows the sensitivity kernels to track the appropriate ray path (e.g., Megnin & Romanowicz, 2000). Zhou et al. (2004) show that the full forms of sensitivity kernels for phase and polarisation for Rayleigh waves require the inclusion of coupling to Love waves even at relatively long periods (more than 100 s).

11.2.3 Attenuation

The attenuation of guided waves in stochastic media can be characterised through a coupled mode treatment of energy transport. For weak deviations in material properties from the reference structure ($< \pm 2\%$) we can work with a transmission approach so that evolution of the modal coefficients is governed by

$$\frac{\partial}{\partial x} c_r(x) = \sum_{s=0}^{N} B_{rs}(x) \exp\left[i\omega \int^x d\xi \{p_r(\xi) - p_s(\xi)\}\right], \tag{11.2.16}$$

in terms of the horizontal slownesses for the modes p_r, p_s. We require the scale of variation to not be rapid compared with the horizontal wavelengths for the frequency band being considered.

For an ensemble average over a collection of statistically similar heterogeneity models Kennett (1990) has shown that coupled modal equations can be developed for the ensemble averaged power in each mode,

$$\mathfrak{S}_m = \langle |c_m|^2 \rangle. \tag{11.2.17}$$

With the assumption that the different classes of physical properties have similar behaviour and that the heterogeneity values at widely separated points are uncorrelated, we can describe the statistical properties with a stationary random process associated with a horizontal correlation length a. We take the average heterogeneity level across the ensemble to be zero and then the ensemble average of the coupling terms will vanish,

$$\langle B_{mn}(x) \rangle = 0. \tag{11.2.18}$$

With neglect of third-order interaction terms, because the heterogeneity is small, we can extract coupled equations for $\mathfrak{S}_m$, where the intermode coupling depends on the horizontal correlation functions for the heterogeneity, and the integrated effect of the vertical variations in seismic properties and eigenfunctions. For frequency ω, the evolution equation for the power $\mathfrak{S}_m(x)$ in the mth mode, in propagation through the stochastic waveguide, takes the form

$$\begin{aligned}\frac{\partial}{\partial x}\mathfrak{S}_m(x,\omega) &= -\omega p_m [Q^{-1}]_m \mathfrak{S}_m(x,\omega) \\ &\quad + \sum_{n=0}^{N} H_{mn}(x,\omega)[\mathfrak{S}_m(x,\omega) - \mathfrak{S}_n(x,\omega)],\end{aligned} \tag{11.2.19}$$

where $[Q^{-1}]_m$ is the spatial loss factor for the mth mode from intrinsic attenuation, and p_m is the modal slowness. The power coupling term in (11.2.19)

$$H_{mn}(x,\omega) = \int_{-\infty}^{\infty} d\xi \, \langle B_{mn}(x,\omega) B^*_{nm}(x-\xi,\omega) \rangle e^{i\omega \int^{\xi} d\eta \, [p_n(\eta) - p_m(\eta)]}, \tag{11.2.20}$$

where $\langle \; \rangle$ indicates an ensemble average over the stochastic properties and $B_{mn}(x,\omega)$ represents a coupling integral over depth at position x, that involves the heterogeneity in seismic properties modulated by the modal eigenfunctions (Kennett, 1990). We note that (11.2.20) includes the ensemble average of a spatial convolution of coupling terms that can be related to the correlation properties of the heterogeneity.

We expect the heterogeneity in the crust to have a different character from that in the mantle, with at most very weak correlation between the two zones. In this case we can recast the power coupling term as

$$H_{mn} = |\langle B^{C}_{mn}(\omega)\rangle|^2 \langle |P^{C}(k_n - k_m)|^2\rangle + |\langle B^{M}_{mn}(\omega)\rangle|^2 \langle |P^{M}(k_n - k_m)|^2\rangle, \quad (11.2.21)$$

where $\langle |P^C|^2\rangle$, $\langle |P^M|^2\rangle$ are the power spectra of the horizontal correlation functions for the crust and mantle. In (11.2.21) $\langle B^C_{mn}\rangle$, $\langle B^M_{mn}\rangle$ represent ensemble averages of integrals over the crust and mantle zones. For further subdivisions into zones of uncorrelated heterogeneity, additional terms of a similar form would appear in (11.2.21). Although the heterogeneity is uncorrelated, the effects on the wavefield are not. For example, mode coupling induced in the mantle carries through into the crust to produce effects at the surface.

From the evolution equation (11.2.19), we can define a 'scattering' attenuation term for the mth mode associated with transfers of energy away from the original mode:

$$ {}_{s}Q^{-1}_{m}(\omega) = [\omega p_m]^{-1} \sum_{n}{}' H_{mn}(\omega), \qquad (11.2.22)$$

where $\sum'$ denotes a sum excluding $m = n$. The contribution to the scattering attenuation will depend strongly on the character of particular modes.

For example, for high-frequency regional phases the fundamental Rayleigh mode describes the *Rg* phase, the first few higher modes the complex *Lg* with phase velocity around 3.5 km/s, and the even higher modes describe *Sn* with upper-mantle propagation and a phase velocity around 4.5 km/s (cf. § SWII:20.1). Thus, the dominant term for *Lg* type modes will be $|\langle K^C_{mn}(\omega)\rangle|^2 \langle |P^C(k_n - k_m)|^2\rangle$, and since we would expect heterogeneity to be concentrated in the crust ${}_{s}Q^{-1}_{m}$ can readily exceed the intrinsic loss factor Q^{-1}_{m}. For *Sn* modes the main contribution will come from mantle heterogeneity. Estimates of scattering attenuation based on plausible crustal parameters tie well with observations.

The ensemble results can give a guide to the behaviour for propagation through a region, but have their closest correspondence to an average taken over a variety of paths across a single region (Kennett, 1990). In 3-D some additional loss can be expected due to intermode scattering with a significant change in the direction of propagation.

11.3 Interactions of Complex Seismic Wavefields

As we have seen, the larger scales of variation within the Earth can be imaged by tomographic techniques, because their primary influence on the seismic wavefield is to change the timing of seismic arrivals. The changes in amplitude due to focussing and defocussing of wavefronts complicate the assessment of seismic-wave attenuation within the Earth. With improvements in the inversion of seismic waveforms, amplitudes are playing a more important role in the extraction of seismic structure.

Once more detailed patterns of observation are established, the presence of smaller- scale features in Earth structure become apparent. Moderate levels of variability exist though much of the mantle, but the variation is stronger in zones about 800 km thick at the top and bottom of the mantle, with the strongest effects in the lithosphere and in the 300 km zone just above the core–mantle boundary (Figure 12.3). The multiple scales of variation within the Earth have a generally 'red' spectrum so that the largest-scale features dominate, and the overall behaviour is roughly a linear decrease with wavelength. Nevertheless, if interrogation is made with short wavelength probes considerable complexity results.

Structures associated with subduction zones, both present and past, form some of the strongest anomalies in seismic images. In the upper mantle these fast, low attenuation structures host most earthquake sources that occur beneath the crust, and so have a significant influence on the how the Earth is sampled. Since all seismic observations are made at or near the surface, information on the deeper Earth is contaminated by shallow structure. The most complex effects occur in long passage through heterogeneous structures, e.g., long-range propagation through the upper mantle. Seismic energy from the deepest sources can avoid interaction with the increased attenuation and heterogeneity on the source side, but still feels such effects on passage to receivers.

The style of heterogeneity changes with depth, e.g., between the crust and mantle, and as a result the seismic wavefield at surface observation points acquires characteristics that do not have a simple dependence on structure. With the aid of our representations of the effect of complex structure for body waves and surface waves we can see how different zones of heterogeneity interact to produce the full wavefield.

We can apply the reflection and transmission operator representations as in Section 3.3 with segmentation by zones of differing style of heterogeneity, and then recognise the importance of interaction between regions through the reverberation operators. The heterogeneous zones do not act in isolation, their effect spreads by modifying the wavefield impinging on their neighbours. This makes it particularly difficult to unravel the characteristics of lithospheric heterogeneity where multiple styles of variation occur. Similar interaction has been noted above for coupled modes.

We can illustrate the effects by using the operator representation (3.3.45) for a source at depth S,

$$\boldsymbol{w}_{\mathrm{o}} = \mathsf{W}_{\mathrm{U}}^{\mathrm{fS}}\mathsf{U}^{\mathrm{S}} + \mathsf{W}_{\mathrm{U}}^{\mathrm{fS}}[\mathsf{I} - \mathsf{R}_{\mathrm{D}}^{\mathrm{SL}}\mathsf{R}_{\mathrm{U}}^{\mathrm{fS}}]^{-1}\mathsf{R}_{\mathrm{D}}^{\mathrm{SL}}[\mathsf{D}^{\mathrm{S}} + \mathsf{R}_{\mathrm{U}}^{\mathrm{fS}}\mathsf{U}^{\mathrm{S}}], \tag{11.3.1}$$

with a decomposition of each operator of the form $\mathbf{R}_{\mathrm{D}}^{\mathrm{SL}} = \bar{\mathbf{R}}_{\mathrm{D}}^{\mathrm{SL}} + \mathbf{r}_{\mathrm{D}}^{\mathrm{SL}}$, where $\bar{\mathbf{R}}_{\mathrm{D}}^{\mathrm{SL}}$ describes the effects of broad-scale structure and $\mathbf{r}_{\mathrm{D}}^{\mathrm{SL}}$ the additional effects from smaller scales of heterogeneity. As in (11.1.30) we use an expansion of the reverberation operator:

$$\mathsf{X}^{fSL} = [\mathsf{I} - \mathsf{R}_D^{SL}\mathsf{R}_U^{fS}]^{-1} = \bar{\mathsf{X}}^{fSL} + \bar{\mathsf{X}}^{fSL}\mathcal{H}^{fL}\bar{\mathsf{X}}^{fSL} + \ldots. \tag{11.3.2}$$

with $\mathcal{H}^{fL} = \bar{\mathsf{R}}_D^{SL}\mathsf{r}_U^{fS} + \mathsf{r}_D^{SL}\bar{\mathsf{R}}_U^{fS} + \mathsf{r}_D^{SL}\mathsf{r}_U^{fS}$.

We now insert the split operators into (11.3.1), and then retaining only the dominant terms, we find

$$\begin{aligned}\boldsymbol{w}_o \approx\ & [\bar{\mathsf{W}}_U^{fS} + \mathsf{w}_U^{fS}]\mathsf{U}^S + [\bar{\mathsf{W}}_U^{fS} + \mathsf{w}_U^{fS}]\bar{\mathsf{X}}^{fSL}\bar{\mathsf{R}}_D^{SL}[\mathsf{D}^S + \bar{\mathsf{R}}_U^{fS}\mathsf{U}^S] \\ & + \bar{\mathsf{W}}_U^{fS}\{\bar{\mathsf{X}}^{fSL}\bar{\mathsf{R}}_D^{SL}\mathsf{r}_U^{fS}\mathsf{U}^S + \bar{\mathsf{X}}^{fSL}\mathsf{r}_D^{SL}[\mathsf{D}^S + \bar{\mathsf{R}}_U^{fS}\mathsf{U}^S]\} \\ & + \bar{\mathsf{W}}_U^{fS}\{\bar{\mathsf{X}}^{fSL}\mathcal{H}^{fL}\bar{\mathsf{X}}^{fSL}\bar{\mathsf{R}}_D^{SL}[\mathsf{D}^S + \bar{\mathsf{R}}_U^{fS}\mathsf{U}^S]\}.\end{aligned} \tag{11.3.3}$$

The first group of terms correspond to the influence of the broad-scale structure modulated by the heterogeneity effects just at the receiver. The remainder introduce additional small-scale contributions in reflection above or below the source that are carried through the rest of the structure by regular propagation processes. Different styles of heterogeneity may occur above and below the source, and so carry their distinct imprint to the wavefield.

From the nature of the surface-wave field, the coupled mode treatment puts most emphasis on the modification of the reverberation operator X^{fSL}. The representation (11.3.3) is roughly equivalent to using just transmission for mode coupling.

Explicit expressions for the effect of different zones of heterogeneity can be introduced by making a similar split operator substitution into the operator representation analogous to (2.3.9) with a division of the structure at J. The resulting expressions become rather complex because of interactions in the near-source and near-receiver zones, as well as with structure at depth. Nevertheless the overall character can be understood from the operator representations (4.1.1), (4.1.2) by thinking in terms of *fuzzy* operators that carry the imprint of smaller-scale structure. The impact of heterogeneity will depend on the statistical character of the parameter variations. Organised and well-developed patches of heterogeneity will have much greater significance than the same sort of heterogeneity distributed uniformly in a statistical sense.

In transmission to a receiver, the local variations with depth play an important role. Influences come from all scales of structure; crustal multiples between the surface and the Moho can arrive at the same time as more subtle arrivals whilst being modified by crustal heterogeneity. Comparable expressions to (11.3.3) can be developed to account for the interactions of multiple zones of heterogeneity with different character.

12

Scattering and Stochastic Waveguides

The presence of small-scale features in Earth structure becomes evident in the characteristics of the high-frequency wavefield. Interaction with such small-scale heterogeneity produces scattered waves that accompany the recognised seismic phases. In most cases the scattered energy follows the main phase as a coda of relatively incoherent energy. However, where the main phase is not a simple minimum in travel time, precursory scattered energy can emerge in advance of the regular phase arrival. Such energy is relatively easy to recognise, e.g. for precursors to *PKP*, and can provide strong constraints on possible scattering zones. In principle, the analysis of scattered energy can provide information on heterogeneity on much smaller scales than can be imaged by other means.

For much of the Earth scattering is relatively weak so the main phases remain distinct, accompanied by modest scattered trains that can often be described by single scattering. However, the complicated heterogeneity of the lithosphere produces multiple scattering and complex regional phases. In the presence of quasi-stratified media with much longer horizontal than vertical correlation length, guided waves can be trapped and carried to great distances by near-horizontal interaction with heterogeneity. Such *stochastic waveguides* play an important role in the lithosphere, and in subduction zones where the former oceanic lithosphere descends to depth.

12.1 Characterising Scattering

Scattered waves are evident in seismograms from their relatively chaotic variations at higher frequencies (Figure 12.1). The maximum effect of scattering occurs when the wavelength of seismic waves is comparable to the scale-length of the heterogeneous features. For scales of heterogeneity large compared with the seismic wavelength, forward scattering is the dominant process. When the seismic wavelengths are moderately large compared to the heterogeneous patches, scattering occurs in all directions and is often approximated as isotropic. Propagation through such complex scattering zones can be reasonably well described by the diffusion of seismic energy rather than a wave-based viewpoint.

In the Earth, scattering modulates the major seismic phases associated with the

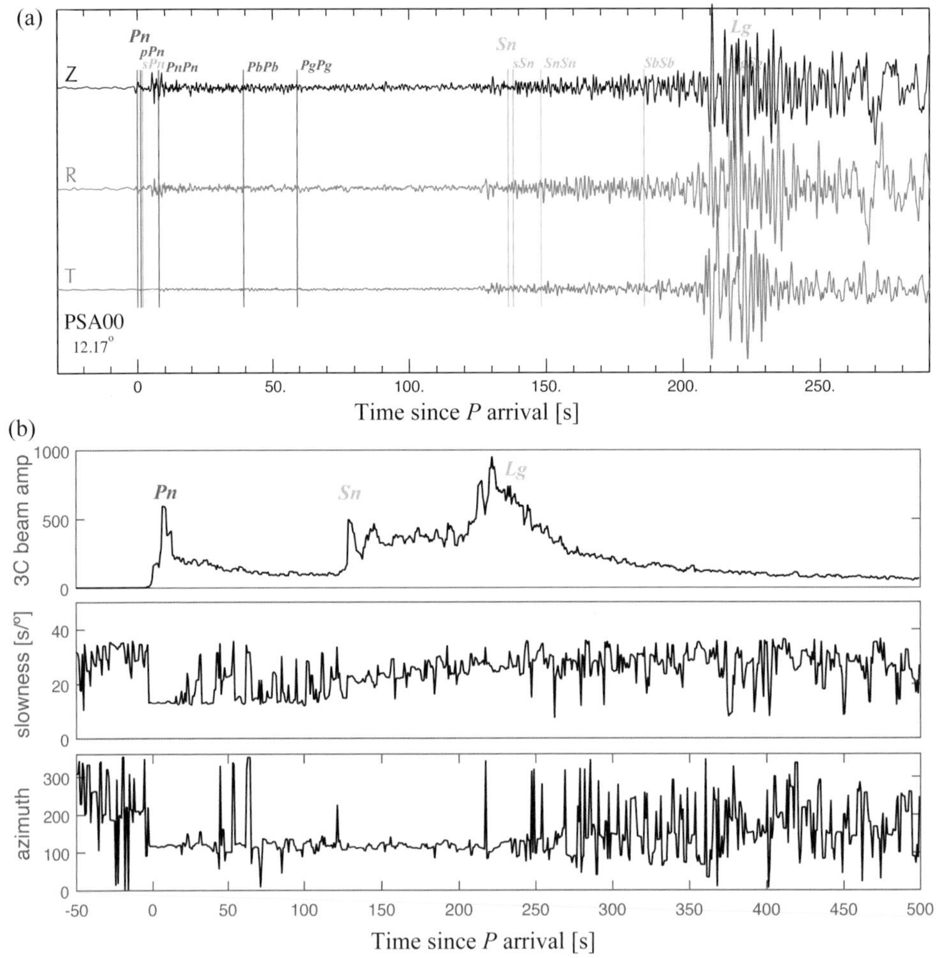

Figure 12.1 The Ernabella earthquake in central Australia recorded at the PSAR array in northwestern Australia. (a) Three-component seismograms at the central site PSA00, showing complex codas for both *Pn* and *Sn*. The phase timing is presented using the AK135 model, the arrivals for this cratonic path are earlier due to high lithospheric wavespeeds. (b) Array analysis of the character of the wavefield as a function of time: total energy in the three-component wavefield, horizontal slowness, and azimuth of incoming wavefront. [Courtesy of M. de Kool.]

main stratification with radius and the internal discontinuities, particularly as an extended coda following each phase. The situation is much different on the Moon, where the presence of a thick pulverised regolith with low seismic wavespeeds overlying intact rock leads to long complex seismograms with very slow decay (e.g., Dainty et al., 1974), and little indication of specific seismic phases. The situation on the planet Mars appears to be intermediate between the Earth and the Moon. The records of the largest Martian events on the InSight seismometer

(Giardini et al., 2020; Lognonné et al, 2020) show distinct phase arrivals that emerge from a background of extended scattering. This suggests a situation with a well-developed complex regolith near the surface, but not sufficient scattering to overwhelm the influence of stratification with depth.

12.1.1 Nature of Seismograms

We illustrate the nature of extended *P* and *S* wave coda with an Mb 5.2 earthquake in central Australia recorded at the Pilbara seismic array (PSAR) at an epicentral distance of 12.17° (Figure 12.1). The propagation path lies across the Precambrian cratons of western Australia that have very low intrinsic attenuation ($Q_P^{-1}, Q_S^{-1} < 0.001$). As a result we have minutes of scattered trains following the main arrivals extending into the dominant *Lg* crustal arrivals. At this distance the *Pn* and *Sn* arrivals have penetrated deep into the thick cratonic lithosphere (> 200 km thick), whereas the multiple reflections *PnPn*, *SnSn* have much shallower paths, though still within the lithospheric mantle. A feature of the seismograms, commonly associated with scattering, is that the very onsets of the *P* and *S* phases are lower frequency and then, after a delay of a couple of seconds, higher frequencies appear on all three components (cf. Figure 12.4). The presence of tangential T-component energy in the *P* wavetrain requires modification of the polarisation by scattering.

PSAR is a 13-element spiral-arm array with a diameter of 25 km and full three-component recording. This array provides good resolution for slowness and azimuth estimates for the arrivals from the Ernabella earthquake. The results from array analysis are summarised in Figure 12.1(b); the top trace indicates the total energy so that the main phases can be recognised and linked to the slowness and azimuth below.

The azimuth of the arrivals from the event is quite different from the preceding noise field, and is sustained until the main *Lg* group, with only a few localised deviations that are not associated with any prominent features in the seismograms. This means that the scattered energy is largely tracking the great circle between the source and the array. There is much more variability in the surface-wave portion of the wavefield, with both a wide range of azimuths and considerable local variability indicating scattered surface waves arriving from many different directions.

The slowness behaviour is somewhat more complicated. Within the *P* train the smallest slowness is set by the *Pn* arrival. Deviations link to azimuthal variations and suggest the presence of a secondary wavetrain with dominantly crustal characteristics interfering with the lower-amplitude segments of the wavefield from the Ernabella event. The situation for *S* shows less interaction, with a slow trend from mantle to crustal slownesses linking to *Lg*.

The PSAR array also provides additional information on the coherence of the wavefield from the Ernabella event In the 1–3 Hz frequency band, the cross-correlation of records between pairs of stations shows high coherence for *Pn* and the associated phases in the next 15 s across the full span of the array on both vertical (Z) and radial (R) components (Kennett & Furumura, 2016). Coherence is limited for the long *P* coda, and intriguingly only lifts slightly when *Sn* arrives. The pairwise cross-correlations for *S* and *SnSn* are only large out to 7–10 km station separation for the radial component, and are even less for the T-component. This difference in coherence arises from the shorter wavelength of *S* than *P*, and hence stronger interaction with heterogeneity.

The array results for scattering at PSAR are similar to those reported by Dainty & Toksöz (1990) for regional seismograms recorded at the Scandinavian arrays (NORES, FINES, ARCES). *P*-coda energy was concentrated in the on-azimuth direction, but appeared at a number of different slownesses, suggesting a variety of contribution mechanisms.

Array analysis can play an important role in the analysis of scattering, by being able to separate the slowness and azimuth of scattered energy from those for the major phases. Sophisticated stacking procedures (Rost et al., 2015) can achieve high precision for even small-aperture arrays, and so allow identification of the generation mechanism of different classes of recognised scattering. The power of array methods in array studies was recognised early, and played a major role in the identification of precursors to PKP_{DF} as due to scattering near the base of the mantle (Cleary & Haddon, 1972).

12.1.2 Modelling Scattering

It is hard to capture the full complexity of scattering processes in the Earth where the nature of heterogeneity changes with depth, and may be associated with local anisotropy. Wavefield simulations are possible using finite-difference or finite-element methods for complex media, but the computational demands for full 3-D models are large and restrict the range of practicable frequencies. Fortunately, as discussed in Section 12.3, the main features of the wavefield behaviour can often be extracted from 2-D simulations. When stochastic models are employed, in such simulations, interpretation is commonly based on a few representative structures rather than any significant ensemble. In consequence, as noted above, it is best to regard stochastic models as a parametric representation of structure.

Much effort has been expended in developing techniques for handling scattering problems for heterogeneous media using a variety of simplifying approximations. Sato, Fehler & Maeda (2012) provide a comprehensive treatment of studies relating to the lithosphere, particularly the crust, while Shearer (2017) puts more emphasis on the deep Earth. The background medium employed is commonly uniform or just radially stratified, and the heterogeneity described by a single correlation function. Where stochastic representations are used, isotropic correlations are usually assumed, since this reduces algebraic complexity. Often only perturbations in wavespeed are considered, with a scaling factor between *P*- and *S*-wavespeed variations. Yet, density can make a significant effect on the scattered field (Hong et al., 2004).

The first-order Born approximation for the scattered field does not conserve energy, and so even for long paths through weakly scattering it is desirable to include the effects of multiple scattering to rectify the energy imbalance. Many studies concentrate on the transport of seismic energy and employ some class of radiative transport theory (see, e.g., Margerin, 2004; Sato et al., 2012).

The use of Monte Carlo simulations of scattering based on radiative transport concepts has proved a powerful approach that avoids many of the restrictions imposed by analytical developments. Such methods can be applied to a fully 3-D situation without inordinate computational effort. By tracking the trajectories of 'particles' carrying seismic energy that interact with probabilities derived from random media theory, the envelopes of seismograms can be constructed. In this way it is possible to examine the the radiation of both P and S waves, intrinsic attenuation, and wavetype conversion. Multiple zones of heterogeneity with different characteristics can be included, with scattering properties derived from suitable auto-correlation functions. The major internal discontinuities and the free surface are handled with probabilistic particle properties designed to simulate reflection and transmission coefficients. Ray tracing establishes the basic particle trajectories between interactions, and it is possible to track S-wave polarisations. Shearer & Earle (2008) provide a comprehensive discussion of the computational implementation of such a Monte Carlo particle approach.

12.2 Scattering in the Earth

The most obvious evidence for scattering is the incoherent energy following phases such as P and S at regional and teleseismic distances. In a few instance, notably for core phases, scattered energy can arrive as a precursor to the main phase and so be quite prominent. The early arrival is possible because the main arrival does not represent a simple minimum in travel time.

In Figure 12.2 we summarise a number of the portions of the seismic wavefield where scattered high-frequency waves are visible (after Rost et al., 2015; Shearer 2017). We indicate the coda segments for S and P, as well as the coda of the diffraction P_{diff} and the precursors to the surface reflection PP. Most observations of high-frequency scattering are associated with P waves, since the higher attenuation of S waves removes the higher frequencies. For the core phases we indicate the symmetric scattering that comes in on-azimuth, and also off-azimuth scattering $P' \circ P'$ associated with the $P'P'$ phase reflected on the far side of the Earth, and the comparable $PK \circ KP$ zone. The various scattered portions are indicated by shading in travel time and epicentral distance to show their relation to the travel-time curves for the AK135 reference model; the most significant phases are identified in Figure 12.2.

The different scattering windows in Figure 12.2 are marked by circled numbers keyed to the portions of the Earth where they originate in Figure 12.3. Fine-scale heterogeneous structure is present throughout the Earth with the exception of the fluid outer core. The level of such small-scale heterogeneity as a function of depth

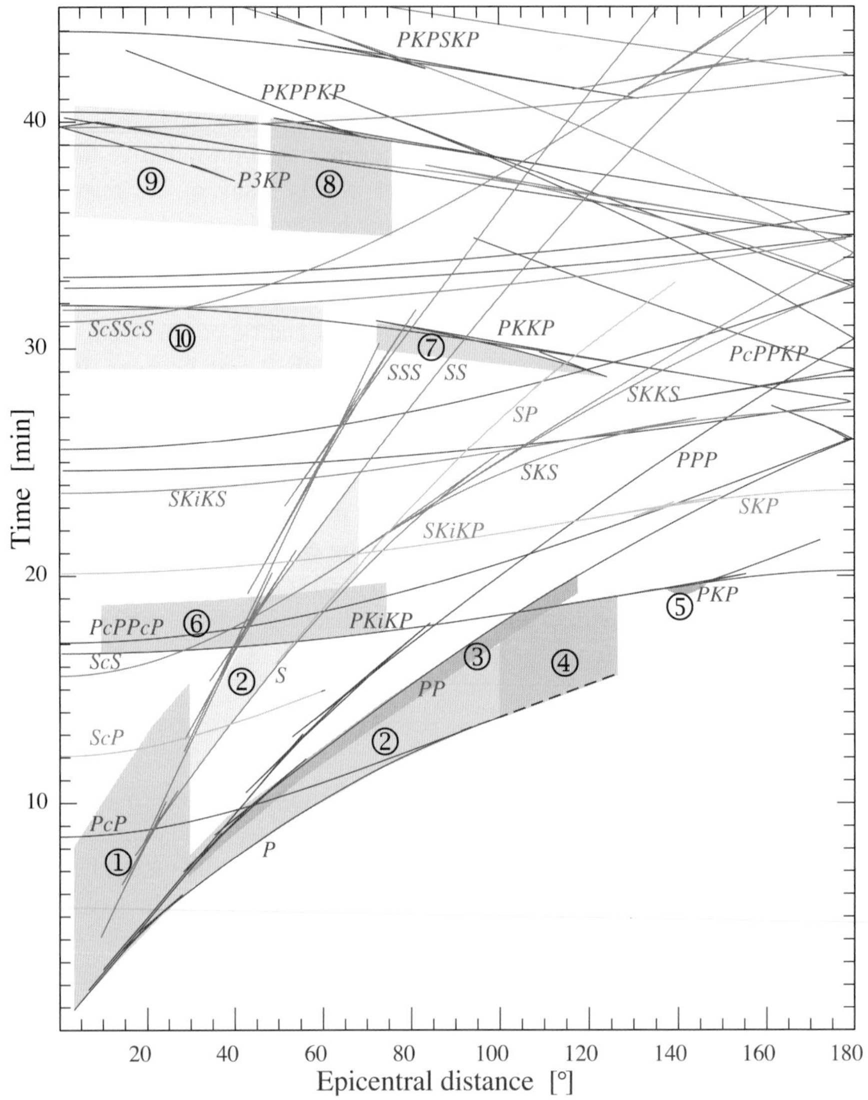

Figure 12.2 Travel-time curves for major seismic phases, with indications of some of the portions of the seismic wavefield that show significant scattering of high-frequency arrivals: (1) *P* and *S* coda at short distances, (2) teleseismic *P* and *S* coda, (3) precursors to *PP*, (4) P_{diff} coda, (5) precursors to *PKP*, (6) *PKiKP* coda, (7) precursors to *PKKP*, (8) precursors to *PKPPKP* ($P'P'$), (9) $P' \circ P'$ scattering, (10) $PK \circ KP$ scattering.

is represented schematically in Figure 12.3 by density of shading. The strongest heterogeneities are found in the lithosphere, in both the crust and the uppermost mantle (zones ①, ②). The effects of such structures, particularly near receiver stations, tend to mask the weaker scattering from deeper within the Earth. The portions of the wavefield indicated in Figure 12.2 (zones ③ and above) provide

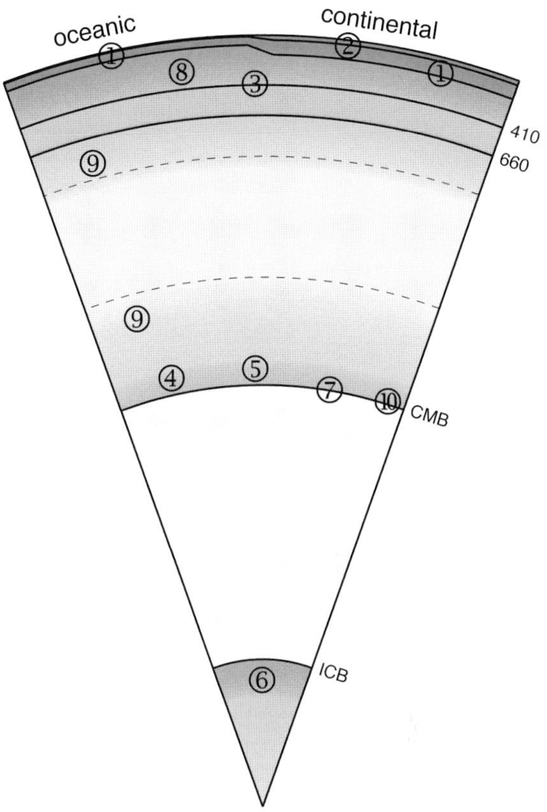

Figure 12.3 Schematic representation of small-scale heterogeneity within the Earth, with darker tones indicating stronger variation. The numbers are keyed to the wavefield features in Figure 12.2, and give an indication of the zones contributing most to the scattered field. These indicators are also used in the text to refer to the scattering features.

windows where deep scattering can largely be separated from near-surface effects, which produce the coda to the main phases.

12.2.1 Scattering and Coda

Local seismic events show a strong, complex coda with relatively rapid decay of seismic energy. Such observations led to the recognition of fine-scale crustal heterogeneity, and the development of concepts for estimating its strength (e.g., Aki & Chouet, 1975).

Strong scattering throughout the lithosphere becomes particularly manifest for long propagation paths. Striking observations of high-frequency *Pn* and associated coda have been observed from records of Soviet nuclear explosions to distances of more than 3000 km (e.g., Morozov et al., 1998). These paths mostly cover cratonic domains. The nature of the processes that give rise to such long-range

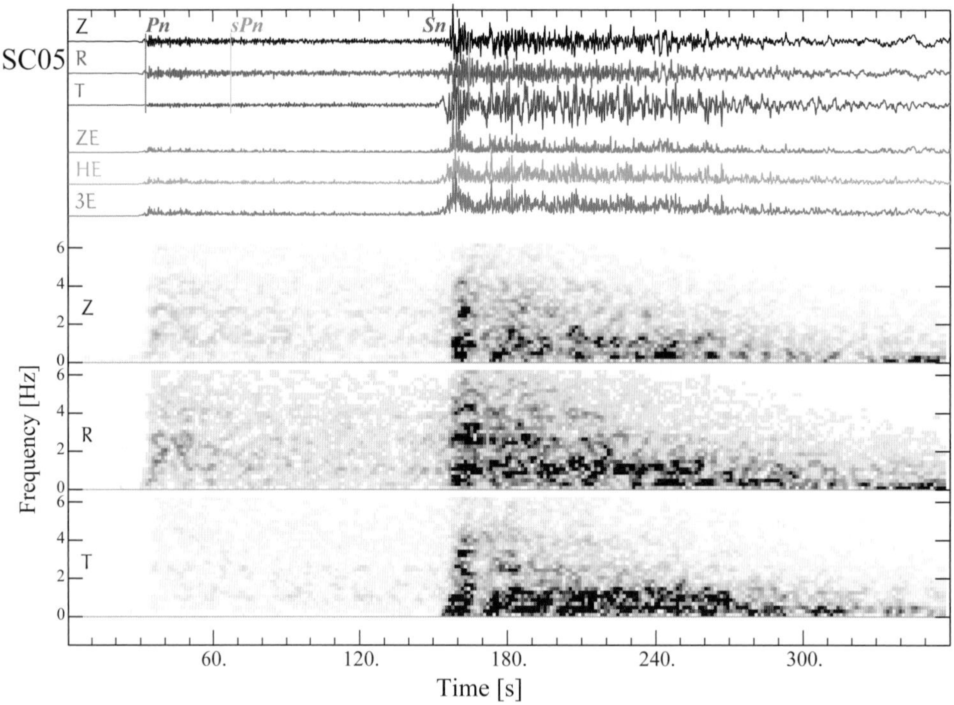

Figure 12.4 Flores Sea event recorded at a portable station SC05 in northern Australia at a distance of 12.0°, showing the three-component records rotated to the great-circle between source and station, and a suite of energy traces ZE – vertical component energy; HE – horizontal component energy; 3E – total energy. In the lower panel, a frequency–time analysis is performed for each component to show the slow decay of the high-frequency energy for both *P* and *S*. The vertical lines indicate the arrival times predicted for the AK135 model (Kennett et al., 1995).

Pn have been disputed, with emphasis placed either on 'whispering-gallery' effects with multiple underside reflections from the Moho, or on ducted energy due to multiple scattering from anisotropic heterogeneity (see, e.g., Nielsen et al., 2003). Crustal scattering is undoubtedly important, and some component of heterogeneity is needed in the lithospheric mantle particularly in the lower part of the cratonic lithosphere. Variations in the wavespeed gradients in the mantle will be influential: whispering-gallery phenomena are most effective with a moderate gradient, whereas stochastic waveguide effects are less dependent on the background gradient (see Section 12.4).

Seismograms for events in the Indonesian subduction zone recorded on paths to stations in northern Australia show an extended coda of high-frequency waves for both *Pn* and *Sn* waves – scattering zone ①. Such high-frequency arrivals are seen for largely cratonic paths, and where a significant stretch of oceanic lithosphere lies between the source and the continental margin (Kennett & Furumura, 2008). As we have seen in Figure 12.1, similar behaviour is found for the limited number

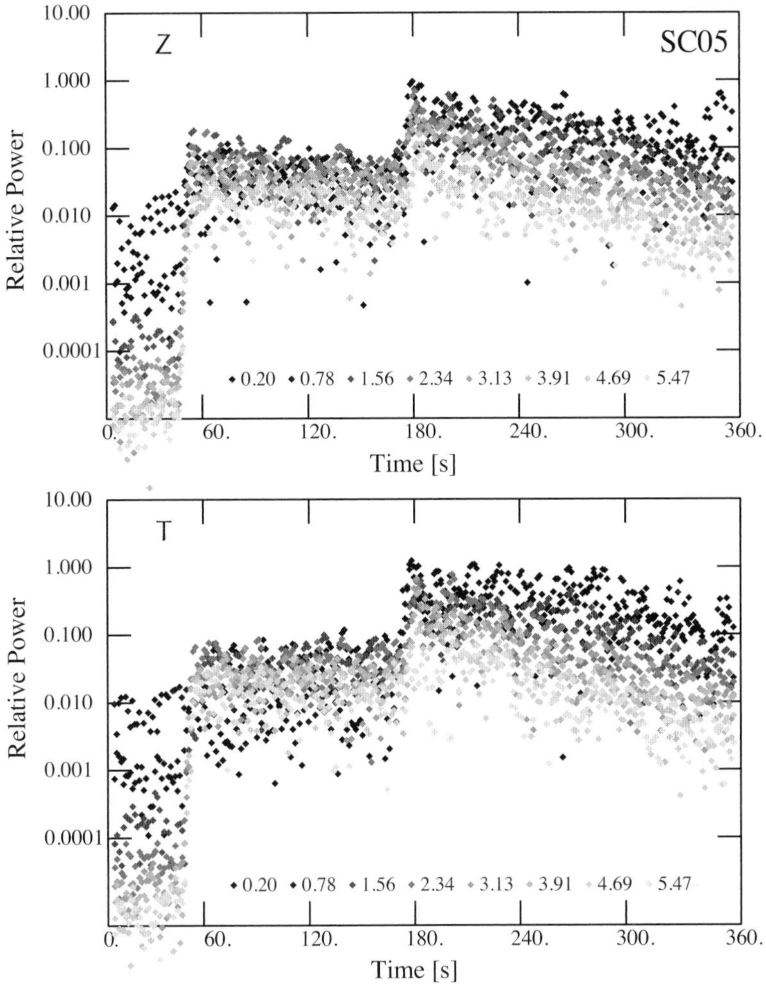

Figure 12.5 The variations of the power-spectral density in the records of Figure 12.4 as a function of frequency and time. The vertical (Z) and transverse (T) components are shown, and soon after the *P* arrival the energy on the transverse component approaches that on the vertical for the higher frequencies. The symbols grade to lighter shades as the frequency increases, as indicated in the key at the base of each panel.

of earthquakes in central Australia recorded on cratonic paths.

Commonly for such observations, the onset of *Pn* and *Sn* is marked by a lower-frequency arrival, but within a second or two the high frequencies dominate and are sustained for minutes (Figure 12.4). The seismograms have significant energy above 4 Hz and the rate of decay of this high-frequency energy is slow for all components of motion, as can be seen from the frequency–time analysis in Figure 12.4.

The initial onset of *P* shows negligible amplitude on the transverse component to the great-circle path, but the following high-frequency arrivals are of similar amplitude on all three components. Such behaviour is what would be expected of an environment with intense multiple scattering and a very high intrinsic Q. The high-frequency arrivals can persist to epicentral distances of 18° or more for paths within cratonic material where intrinsic attenuation is very low. Once paths penetrate into the lower Q environment of the Paleozoic fold belts of eastern Australia, the higher frequencies are rapidly suppressed.

From the Indonesian subduction zone, the characteristic extended coda is seen for events down to 160 km depth. Events deeper than 200 km have a shorter coda, most likely generated close to the receiver (Korn, 1988). Propagation paths for such deeper events will largely lie outside the heterogeneous lithosphere.

The rate of decay of the high-frequency components can be described by a simple coda-Q_c model for which the amplitude at frequency f as a function of travel time T is represented as (Xie & Nuttli, 1988)

$$A \propto T^{-\eta} \exp(-\pi f T/Q_c). \tag{12.2.1}$$

Here η describes the geometrical spreading of the waves, and T is the travel time for the relevant wave type with an appropriate coda Q_c. For crustal studies, particularly the analysis of *Lg* attenuation, the exponent η has typically been taken as 5/6, based on the model of an Airy phase in the crust. For propagation distances beyond 1000 km, there will be a substantial component of propagation within the lithospheric mantle so that the apparent geometric spreading is modified. For the seismograms shown in Figure 12.4, the frequency decay of the power-spectral density is illustrated in Figure 12.5. The patterns of relative decay as a function of frequency and time can be well represented with $\eta = 2/3$ and $Q_c > 1500$. With such a high coda Q it is difficult to get any control on possible frequency dependence of the coda Q_c. Comparable results are found for many different paths, although the envelope of the amplitude pattern varies with the position of the source along the subduction zone. Some time needs to elapse before a systematic coda decay is established.

The character of the teleseismic coda to *P* and *S* is more complex (zone ②). As we have seen in Section 9.2, exploitation of the early portion of the coda following the *P*-wave arrival can provide imaging of lithospheric structure exploiting both *P* reflections and conversion to *S* in transmission. The results suggest the presence of small-scale structures superimposed on perceptible lateral variations. A single style of multiple-scale heterogeneity provides a good description of both horizontally ducted propagation and reflection/transmission character (see Section 13.2).

Dainty (1990) reviewed array studies of teleseismic *P* coda, using arrays in North America and Scandinavia, where slowness and azimuth measurements allow the characterisation of the coda. Two classes of behaviour can be distinguished: 'coherent' coda that has nearly the same slowness and back azimuth as the main phase, e.g., P, and 'diffuse' coda where energy arrives from many different directions. Dainty (1990) suggests that the coherent coda is generated by shallow, near-source scattering in the crust, rather than deeper in the mantle, because it is absent or weak for deep-focus earthquakes. In contrast, much of the diffuse coda is produced by near-receiver scattering and has power concentrated at slownesses typical of surface shear waves (Lg). Scattering from deeper in the mantle can also contribute to the coda of teleseismic P, being generated as the wavefront passes through weak scattering zones, producing forward scattered waves that then track the main propagation path.

Beyond 98° epicentral distance, the direct P-wave grades into the diffraction P_{diff} guided along the base of the mantle. High-frequency P_{diff} can be observed to at least 115°, though the amplitude diminishes with increasing distance. The coda of P_{diff} – scattering zone ④ – becomes increasingly extended at greater distances and can last several hundred seconds (Astiz et al., 1996). The coda arrivals at large ranges are emergent with the largest amplitudes occurring tens of seconds after P_{diff}. The character of the P_{diff} coda is consistent with a dominant role from multiple scattering along the diffracted segment of the path (Bataille et al., 1990), potentially enhanced by lowered seismic wavespeeds near the core–mantle boundary.

12.2.2 Mantle Scattering

In addition to direct studies of teleseismic coda, there is additional evidence for complexity in the lithospheric zone down to around 220 km depth from the properties of steeply arriving phases linked to the core. The $P'P'$ ($PKPPKP$) phase reflects from the far side of the Earth and arrives in a relatively uncluttered portion of the wavefield – scattering zone ⑧. Distinct underside reflections from the major upper-mantle discontinuities at 660 and 410 km depth are often found at 150–90 seconds ahead of $P'P'$, but precursors close to the main phase come from shallower regions near the reflection point. Tkalčić et al. (2006) have examined array data for precursors to $P'P'$ at short epicentral distances, where the path at the receiver is near vertical. The timing of these precursors suggests an origin in the depth range 150–220 km.

The visibility of the phases PcP and $PKiKP$ at short distances shows strong anti-correlation irrespective of the nature of the source (earthquake or explosion). There is a significant separation in the zones of interaction for these two phases near the core–mantle boundary, so that a shallower cause needs to be sought. Tkalčić et al. (2010) have demonstrated that focussing and defocussing by upper-mantle heterogeneity can explain the anti-correlation. Both reflected phases are weak and so a minor change can drop a particular phase beneath the noise level. Simulations suggest that an anisotropic correlation with much longer horizontal than vertical correlation has the noticeably different effects on the two phases because of their differences in incidence angle. The dominant horizontal scale is of the order of 10–30 km, which fits with results from the nature of the coda of regional *P* and *S* (e.g., Kennett & Furumura, 2016).

Ahead of the *PP* phase there is distinct precursory energy that builds from the tail of the high-frequency coda of *P* and P_{diff} – scattering zone ③. Underside reflections from the upper-mantle discontinuities are observable at long periods, but have little role in the short-period band. Much of the scattered energy comes from the lithosphere around the reflection point, but is augmented by contributions from asymmetric scattering at about 20° from the source or the receiver where internal caustics enhance the amplitude of reflected waves and hence those of the scattered phase. The travel time for the *PP* phase is a maximum in the great-circle plane, but a mininum in the perpendicular direction. This property allows off-azimuth contributions to be significant in the development of the scattered field. Scattering close to the receiver gives rise to a broad distribution of arrival azimuths that narrows as the scattering zone moves towards the source area (§ SWII:29.3.4).

A consequence of the sharp drop in *P* wavespeed between the mantle and core is that *PKP* waves scattered in the lowermost mantle can arrive as apparent precursors to *PKIKP* (PKP_{DF}) in the distance range from 120° to 143° (Cleary & Haddon, 1972). Scattering at the core–mantle boundary or in the lowermost mantle diverts energy from the AB and BC branches of *PKP* so that it arrives at closer epicentral distances than the B caustic near 145°, and earlier than the PKP_{DF} (*PKIKP*) phase – scattering zone⑤. The geometry is illustrated in Figure 12.6(a), where locations of scattering at the entry and exit points of *PKP* from the core are indicated by stars. The earliest possible arrivals occur when scattering takes place just at the core–mantle boundary. Scattering from PKP_{DF} appears as a coda to this phase, and does not contribute to the precursors.

The precursory energy is somewhat variable in character, but can be quite distinct on individual seismograms, so that early interpretations invoked specific radial structure. Array analysis of the precursors to *PKIKP* indicate that not just the timing of arrivals but their incidence angles are consistent with scattering (e.g., Doornbos & Vlaar, 1973; King et al., 1974; Doornbos, 1976).

The zones that can produce precursory scattering have limited extent (Hedlin & Shearer, 2000). For scatterers at the core–mantle boundary these zones at entry and exit from the core extend at most $\pm 12°$ away from the great circle, with a distance span of less than 9° along the great circle, which tapers to zero at the azimuthal limits. The scattered waves will therefore display slowness and azimuthal deviations at the receiver associated with the location of their scattering points, which can be detected at seismic arrays. Scattering on the source side, on entry to the core, produces arrivals with larger slowness than the PKP_{DF} reference phase, with a rather narrow range of azimuth variation. In contrast, scattering after exit from the core produces precursory arrivals with a steeper path through the mantle than PKP_{DF} and thus smaller slowness; there will also be a broader span of azimuth deviations.

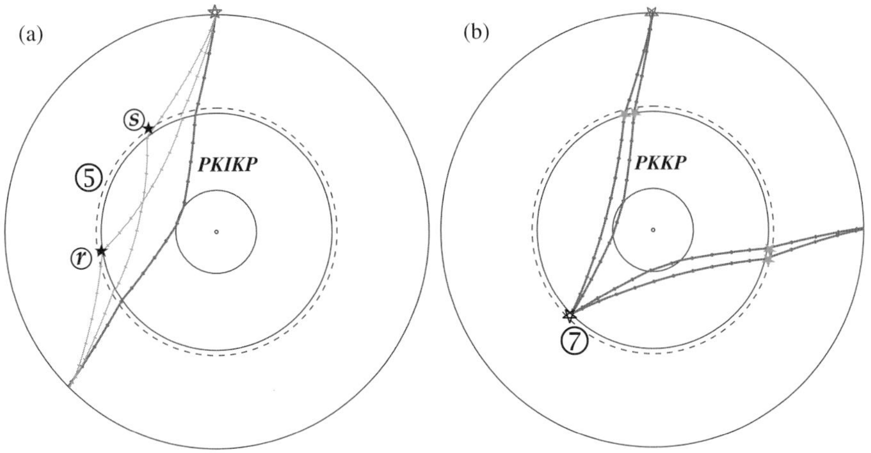

Figure 12.6 Scattering processes, with scattering locations marked by stars, and linked to the scattering zones from Figures 12.2, 12.3: (a) *PKIKP* (PKP_{DF}) precursors, typical ray paths for scattering are indicated for both source side ***s*** and receiver side ***r*** scattering; (b) symmetric *PKKP* precursors.

If scattering is generated above the core–mantle boundary, the domain of possible precursors to PKP_{DF} is reduced. The precursor window shrinks rapidly as the location of scattering moves above the core–mantle boundary and is almost extinguished for scatterers more than 1000 km above the boundary. Stacked precursors as a function of epicentral distance show an increase in amplitude towards the PKP_{DF} phase, which is consistent with scattering extending well into the mantle (see, e.g., Shearer, 2017).

Interpretations of the properties of the *PKP* precursors have used a variety of stochastic models that suggest rather modest scattering strength, with less than 0.2% wavespeed variations (e.g., Mancinelli & Shearer, 2013), and a possible contribution from patches of rough core–mantle boundary. Array studies of the *PKP* precursors suggest the presence of some strong isolated scatterers in the lowermost mantle that may be associated with fragments of past subduction or melt zones (e.g., Miller & Niu, 2008; Thomas et al., 2009).

The core phase *PKKP* with an underside reflection at the core–mantle boundary also has precursory arrivals for some branches – scattering zone ⑦. At epicentral distances beyond the B caustic (about 125°) $PKKP_{DF}$ precursors can arise from mantle scattering in an analogous fashion to the PKP_{DF} precursors. Scattering can occur from both entry and exit points to the outer core (Figure 12.6b). Doornbos (1974) used NORSAR recordings of Solomon Islands events and showed that the slownesses of the $PKKP_{DF}$ precursors were consistent with scattering from the deep mantle. For epicentral distances less than 125°, precursors to $PKKP_{BC}$ can arise from short-wavelength core–mantle boundary topography (e.g., Mancinelli & Shearer, 2013). The underside reflection point of *PKKP* at the core-mantle boundary is a maximum time point with respect to changes in bounce point, so scattered energy arrives earlier.

PKKP energy can also scatter from around the reflection point at the core–mantle boundary giving rise to off-azimuth scattering (*PK*∘*PK* – zone ⑩). This class of scattered arrival has been used to map lateral variations in scattering strength close to the core-mantle boundary (Rost & Earle, 2010).

A further probe for fine-scale structure comes from array analysis of off-azimuth $P'\circ P'$ – scattering zone ⑨. Rost et al. (2015) show how improvements in stacking procedures allow the separation of scattered energy from the main arrivals. $P'\circ P'$ provides definition of small-scale heterogeneities throughout the mantle.

Although there is a general consensus on the relative levels of heterogeneity in the mantle, as sketched in Figure 12.3, the estimates of wavespeed variation are difficult to compare because of a wide variety of techniques and assumptions. In many cases, a very simple background model has been assumed, and then all effects have to be induced by small-scale variations. When a full account is taken of the spectrum of heterogeneity, the level of small-scale features tends to diminish, to a level that is easier to reconcile with geochemical and geodynamic proceses.

12.2.3 Core Scattering

The earliest onset time for observed precursors to PKP_{DF} agrees closely with that predicted for scattering at the core–mantle boundary (e.g., Cleary and Haddon, 1972; Shearer et al., 1998). If small-scale structure existed in the outer core at some depth below the core–mantle boundary, scattered arrivals would be expected at earlier times than have been observed. Thus on the scales of seismic observation, the outer core appears to have no apparent fine-scale heterogeneity. The motions of the core fluid can be expected to produce homogenisation of structure. Though there could be differences inside a tangent cylinder to the inner core aligned with the axis of rotation, since the flow patterns differ from the outer zone.

The *PKiKP* phase reflected from the inner-core boundary has rather low amplitude and is often difficult to see on single stations, but can have a presence in the 1–5 Hz band (Poupinet & Kennett, 2004). With the aid of arrays, the phase characteristics emerge more clearly. Typically the *PKiKP* arrival is followed by a long and complex coda up to 200 s long (e.g., Vidale & Earle, 2000; Poupinet & Kennett, 2004; Koper et al., 2004). The slowness throughout the coda is close to that for the onset of *PKiKP*, as expected for an origin at depth – scattering zone ⑥.

The general properties of the coda can be explained with fine-scale heterogeneity confined within a few hundred kilometres of the inner-core boundary, where attenuation is low. Variation in wavespeeds are of the order of 1% with scale lengths of a few kilometres (e.g., Peng et al., 2008). There are some similarities in the appearance of the *PKiKP* coda and that for P_{diff}, so that energy channelling near the inner-core boundary may help to extend the coda wavetrain. As with many other aspects of the inner core, there appears to be a hemispheric dependence of scattering properties – with stronger effects from the Pacific and Asia than under the Atlantic Ocean.

12.3 2-D and 3-D Scattering Wavefields

Three-dimensional calculations for large models at high frequencies require considerable computational resources, and so it is difficult to explore wide ranges in parameters. Whereas, 2-D calculations can be carried out much more easily but do not capture all the effects of heterogeneity. With a 2-D model scattering effects are only included in the plane of computation and so out-of-plane scattering is missed, though it is evident in observations through, e.g., P coda on the transverse component. In addition, the source model employed in 2-D simulations assumes a line source extending infinitely in the out-of-plane direction, and thus the wave shape and geometrical spreading are different from that for a point source in 3-D. Thus, how far can 2-D simulations match the results in 3-D?

We here examine the influence of the 2-D approximations needed to secure long-range propagation at high frequencies, by comparisons with 3-D simulation of the scattering of seismic waves using the finite-difference method. We compare a full 3-D model with a 2.5-D case where the structure used in 2-D calculations is extended perpendicular to the reference plane.

12.3.1 Heterogeneous Model

The 3-D model covers a zone 204.8 km by 204.8 km horizontally and 45 km vertically. The meshing of the grid is uniform with horizontal grid spacing of 0.1 km and a vertical grid increment of 0.05 km. The result is a 4.3×10^9 grid point model (Figure 12.7). We employ a simple half-space model with an average P wavespeed 6.0 km/s, S wavespeed 3.5 km/s, and density 2.6 Mg/m^3. Free surface conditions are applied at the top and bottom of the 45 km thick model. Stochastic heterogeneities are introduced throughout the half-space model with a von Kármán model with Hurst number 0.5. The horizontal correlation lengths a_x, a_y are set at 10 km, and a smaller correlation length a_z of 1 km is applied in the vertical direction. The standard deviation of the fluctuations for P wavespeed, S wavespeed, and density are set at 6% of the values for the background model. Such strong heterogeneities produce significant scattering effects within the limited spatial dimensions of this model. Weak intrinsic attenuation (Q_p 2400, Q_s 1200) in the model is used to enhance the scattering of high-frequency signals caused by the strong heterogeneities in the model. To suppress any artificial reflections from the model boundaries, an absorbing boundary (Cerjan et al., 1985) is introduced in a zone 40 points thick around the entire 3-D model

The simulations were performed using a fully elastic parallel finite-difference code (Kennett & Furumura, 2008) using a 16th-order staggered grid in the horizontal direction and a fourth-order scheme in the vertical direction. Simulations can be carried to frequencies up to 5 Hz, with more than five grid points per wavelength for the highest frequencies in the horizontal direction. An

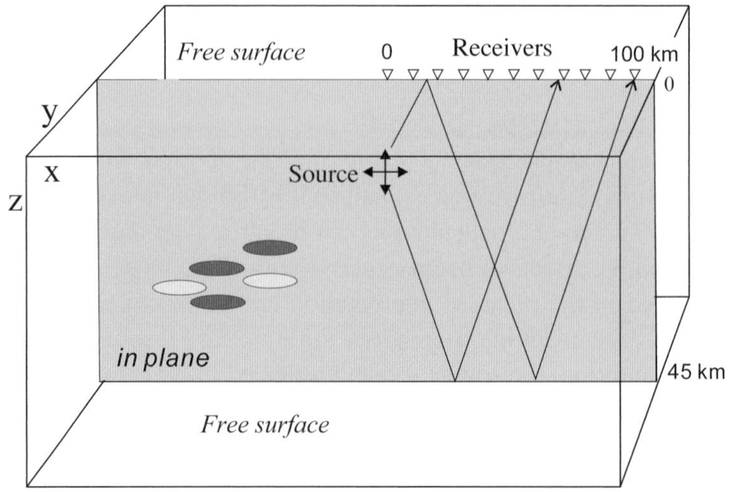

Figure 12.7 Model configuration for 3-D and 2-D finite-difference method simulation for a stochastic random heterogeneous model. Configuration of source and receivers are illustrated.

explosive source is placed on the middle line of the model at 11.25 km below the upper free surface. This source radiates high-frequency *P* waves isotropically into the 3-D model. The calculations were carried to 90 s after source initiation using over 30 000 time steps.

12.3.2 Simulation Results

The results of the 3-D simulations are shown in Figure 12.8(a) with displays of three-component ground motions for a linear set of stations along the x-direction. We also show the results for a point source embedded in a 2-D structure whose model properties do not change in the y-direction perpendicular to the station profile; these results are indicated as 2.5-D in Figure 12.8(b). The final case is a 2-D simulation of the radial and vertical ground motions in a model taken on the vertical plane through the source (Figure 12.8c). The structure used for the 2.5-D and 2-D calculations is the same, they differ in the use of a point source (2.5-D) and line source (2-D). A high-pass filter with a cut-off frequency of 5 Hz is applied to the synthetic seismograms to reduce the influence any numerical dispersion, which has somewhat different character in the 2-D and 3-D finite-difference calculations.

In Figure 12.8 we show record sections for the three-component ground motion (Z, R, and T components) for the three cases. Each trace is multiplied by the epicentral distance to make some compensation for geometrical spreading. More detailed comparisons of the synthetic seismograms at fixed distances of 30 km and 90 km from the source are shown in Figure 12.9. Multiple reflections of *P* waves between the free surfaces at the top and the bottom of the model, with a time separation of about 10 s, are very clear. Such reflection processes allow an

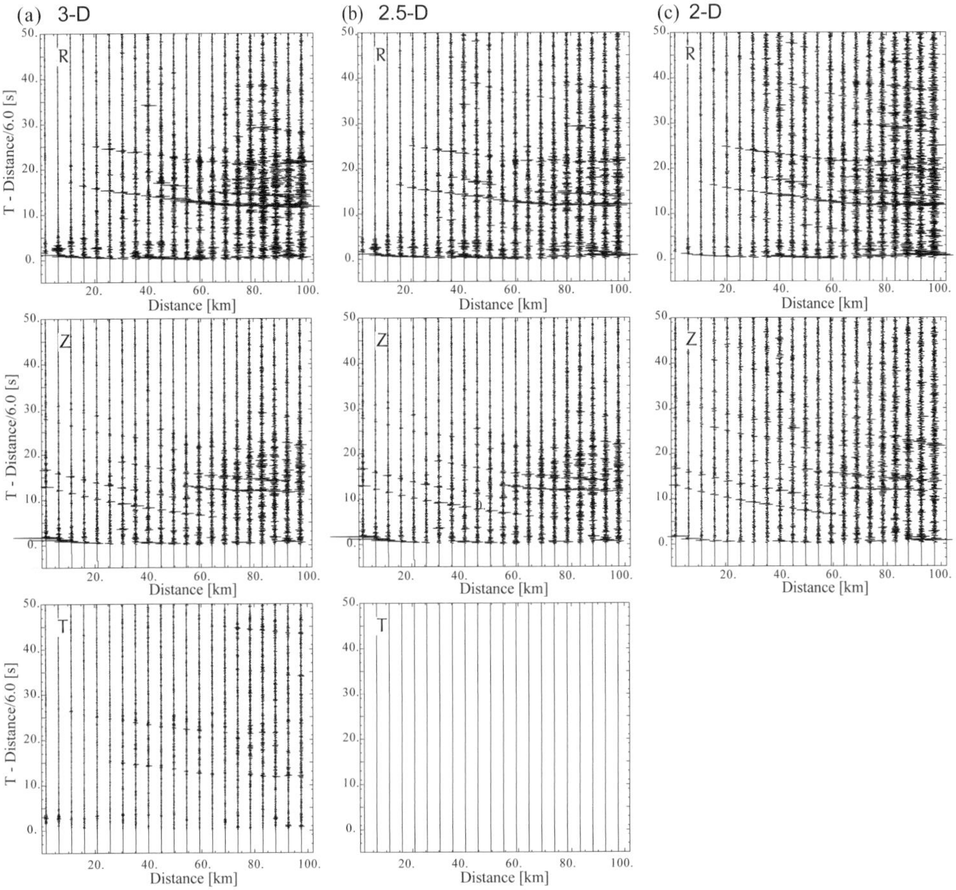

Figure 12.8 Synthetic seismograms of radial (R), vertical (Z) ,and transverse (T) components of velocity for propagation through the stochastic medium: (a) 3-D simulation model with stochastic random heterogeneities and a point source; (b) 2.5-D modelling with a point source assuming that model parameters are invariant in the out-of-plane direction; (c) 2-D simulation using the same structure as (b) but for a line source. In all cases we use 10 km horizontal correlation lengths, and a 1 km vertical correlation length.

effective continuation of the wavefield to a larger domain, and contribute to the higher-frequency scattering.

There is a good correlation between the scattering wavefields from the 2.5-D and 3-D finite-difference results, particularly for the characteristics of the *P* and *PS* converted signals and the long coda of high-frequency signals on both the radial and vertical components. The amplitude of coda at most stations is a little lower for the 2.5-D wavefield, and the decay rate of coda somewhat quicker. This occurs because the 2.5-D model does not account for the strong scattering of the seismic wavefield in the out-of-plane direction. Such 3-D scattering effects would

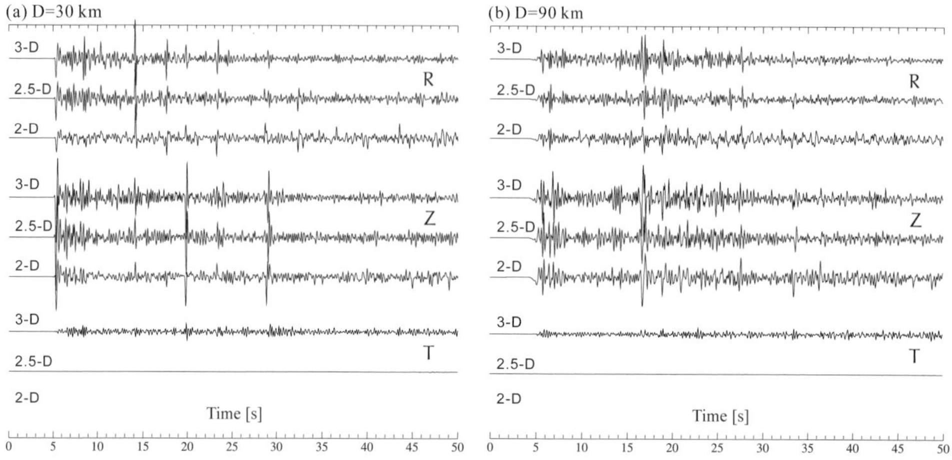

Figure 12.9 Expanded seismograms from Figure 12.3 comparing the waveforms and decay rate of synthetic seismograms from 3-D, 2.5-D, and 2-D simulations for fixed propagation distances, (a) 30 km and (b) 90 km.

be more significant with stronger horizontal heterogeneities, or shorter horizontal correlation lengths.

The relatively large amplitude of the tangential component (T) for the 3-D simulation indicates strong scattering of P waves, which remove some P wave energy from the main arrivals for in-plane motion. The energy of tangential motion increases gradually with increasing time as waves propagate further through the heterogeneous structure, so that the discrepancy between the 3-D and 2.5-D wavefields increases with epicentral distance. Tangential energy arising from out-of-plane scattering should not appear in the 2.5 D model. We see no T-component seismic signals in the 2.5-D simulation, confirming the accuracy of the calculations in 3-D.

The synthetic seismograms from the 2-D calculation show a similar character to the 2.5-D wavefield. Both have the same in-plane structure, and are invariant in the out-of-plane direction. However, the dominant period of the 2-D seismic waves produced by a line source is slightly longer than the point source used for the 2.5-D and 3-D calculations. Moreover, the geometrical spreading for a line source in 2-D as a function of distance R is given by $R^{-1/2}$, and so is weaker than the R^{-1} expected for a point source in 3-D. The result is the development of a longer coda in the 2-D simulation, since the later arrivals at larger propagation distances do not attenuate so quickly.

Nevertheless, the 2-D results provide a good representation of the main characteristics of the scattering wavefield in 3-D, even though out-of-plane scattering is excluded. The weaker geometrical spreading in 2-D tends to compensate for scattered energy returned from out of the plane of propagation.

Thus 2-D simulations can be used as a reliable guide to scattering behaviour in even complex models.

We note that the 2-D results, even with corrections, tend to have slightly lower amplitude than for the 3-D case with comparable parameters. In consequence, stronger heterogeneity would be needed compared with 3-D calculations to achieve the same amplitude level. Thus 2-D scattering modelling may tend to over-estimate levels of heterogeneity.

12.4 Stochastic Waveguides

A typical waveguide in the Earth comprises a region of reduced seismic wavespeed where multiple shallow angle reflections produce efficient transport of seismic energy. The entire crust can act as a waveguide for *S* waves with total reflection at the free surface and the crust–mantle boundary for slowness p greater than β_M^{-1}, where β_M is the shear wavespeed on the underside of the boundary. Such processes of efficient crustal reflections build up the *Lg* phase (see § SWII:20.1), which can also be viewed as a superposition of higher mode surface waves.

In contrast, a zone of elevated wavespeed compared to its surroundings does not, by itself, sustain long-distance propagation along the structure. This is because energy is shed from the region with high seismic wavespeed into the lower wavespeeds on the outside. Reflections at the boundaries will be sub-critical, and consequently energy is lost to transmission at each reflection point (an 'anti-waveguide' effect).

In the presence of organised heterogeneity, the fluctuations in seismic wavespeed around their background values produce localised zones of reduced wavespeed. It is then possible for multiply scattered energy to be ducted along the structure by means of the partially interconnected low-velocity zones. The effects are enhanced when the correlation properties of the heterogeneity are anisotropic, with a stochastic waveguide developing along the direction of extended correlation. Such guiding phenomena are evident in Figure 10.7 where distinct horizontally ducted energy (azimuth 0°) becomes more strongly developed as the ratio of horizontal to vertical correlation length increases. Waveguide effects are more marked for high-frequency waves, where multiple near-grazing reflections from gradients induced by the heterogeneity have maximal influence. Lower frequencies are less sensitive to the presence of the heterogeneity. Once the wavelength is noticeably larger than the longest correlation length, the composite wavefield effectively tunnels through the zones with higher wavespeeds as it progresses with strong wavefront healing.

When a zone of correlated heterogeneity is present within a structure it can act to enhance waveguide effects if the direction of elongated scatterers coincides with the orientation of the structure. Thus, heterogeneity inside a zone of generally reduced wavespeed acts to redistribute seismic energy and sustain waveguide effects to even greater distances. For example, modest crustal heterogeneity can increase the amplitude of *Lg*. Multiple scattering also has the effect of equalising the amplitudes on the different components, so that even for an explosive source the tangential component of motion will be comparable to the vertical component after a few hundred kilometres of propagation.

For a zone of fast seismic wavespeeds, internal heterogeneity can act to counteract the anti-waveguide effects and keep seismic energy within the zone. Internally, the oriented heterogeneity acts to channel energy along the structure. Though energy reaching the boundaries of the fast-wavespeed zone may well be lost to the outside.

The presence of such a stochastic waveguide effect within a zone of faster wavespeed provides a means of transporting seismic signals to great distances. Thus it is possible for high-frequency seismic energy from very deep earthquakes to produce significant ground motion at the surface above a shallowing subduction zone.

In the oceanic lithosphere, the influence of lithospheric structure is reinforced by multiple reverberations in the sea water to allow the transport of energy at 1 Hz and above for thousands of kilometres. Continental lithosphere is also capable of sustaining high-frequency energy for long distances, particularly as an elongated coda for passage across cratonic regions (Figures 12.1, 12.4).

12.4.1 Subduction Zones

The Pacific Plate subducting beneath northern Japan acts an efficient waveguide for high-frequency signals from deeper earthquakes that produce ground motion concentrated at the eastern seaboard (Figure 12.10). The waveform records in the region of high intensity show a low-frequency onset (< 0.25 Hz) for both P and S waves, followed by large high-frequency coda (> 2 Hz), e.g., Abers (2000), Furumura & Kennett (2005). Utsu (1966) identified the need for a high-wavespeed, low-attenuation zone to bring energy from deep sources to the east coast of Japan; this feature was soon recognised as the signature of the subducted Pacific plate. Similar transport of high-frequency energy from sources at depth has been seen for many other subduction zones, e.g., the Nazca plate (Martin et al., 2003) and the Cocos plate (Abers et al., 2003).

For shallow- and intermediate-depth events in the subducted plate, an important role can be played by the former oceanic crust that forms a thin zone of lowered wavespeed at the top of plate down to around 160 km depth. This oceanic zone can guide moderate-frequency energy with distinct dispersion (Abers, 2000; Abers et al., 2003; Martin et al., 2003), and excitation is efficient for earthquakes located inside the low-wavespeed layer. By 200 km depth, the lowered wavespeeds are expected to be suppressed by dehydration and conversion of basalts to eclogite. The former oceanic layer does not therefore provide an explanation for the transport of very high-frequency energy from depths greater than 200 km. Such transport can be achieved with a stochastic waveguide in which the correlation length along the subducted plate is much larger than the correlation length across the plate (Furumura & Kennett, 2005).

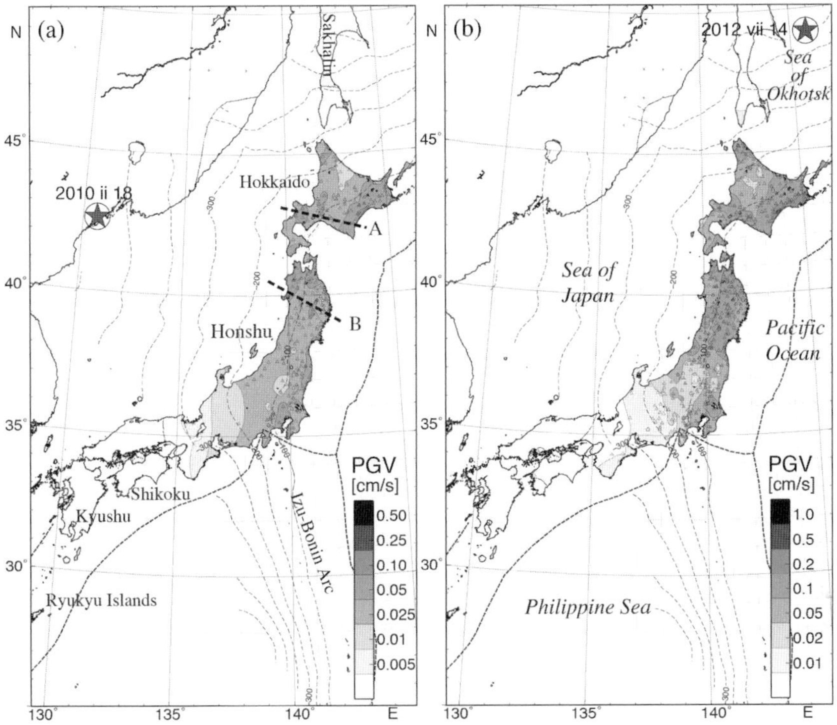

Figure 12.10 Peak horizontal ground velocity patterns for two very deep earthquakes: (a) 2010 February 18 under Russia, Mw 6.9, 590 km; (b) 2012 July 14 off Sakhalin, Mw 7.2, 590 km. The lines of stations displayed in Figure 12.11 are indicated as (A) and (B).

Figure 12.10 illustrates the patterns of peak horizontal ground velocity (PGV) for two very deep earthquakes recorded across the Japanese Islands. For the 2010 event (Mw 6.8, 598 km) beneath Russia, the PGV is concentrated on the east coast of Japan particularly in Hokkaido. This concentration is even more strongly developed for acceleration that emphasises the high frequencies. The 2012 event off Sakhalin (Mw 7.3, 590 km) produced significant ground motion along the eastern seaboard as far south as the Tokyo region where motion is enhanced by sediment amplification. This configuration of PGV arises from energy trapped in the Pacific Plate. Similar behaviour, with elongation of the PGV pattern to the north along the eastern seaboard, is seen for deep events in the Izu-Bonin segment of the subducting Pacific Plate.

The pattern of ground acceleration shows a distinct change for stations to the east of the volcanic front where the subduction zone approaches the surface (Figure 12.11), as demonstrated by record sections of vertical acceleration on profiles across Hokkaido and Tohoku for the Mw 6.9 event on 2002 June 28 at 566 km (43.7°, 130.7°E).

The action of the stochastic waveguide can be well illustrated with 2-D simulations of the wavefield produced by a relatively deep source. We consider the

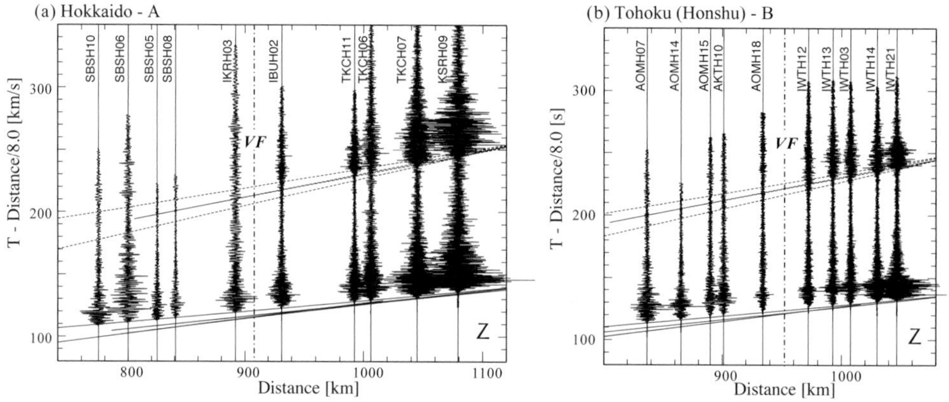

Figure 12.11 Record sections of the vertical component of ground acceleration across profiles in Hokkaido and Tohoku for the 2000 June 28 event at 566 km depth. To compensate for geometrical spreading, each trace is scaled by the epicentral distance. *VF* indicates the position of the volcanic front. The travel times are for the AK135 model.

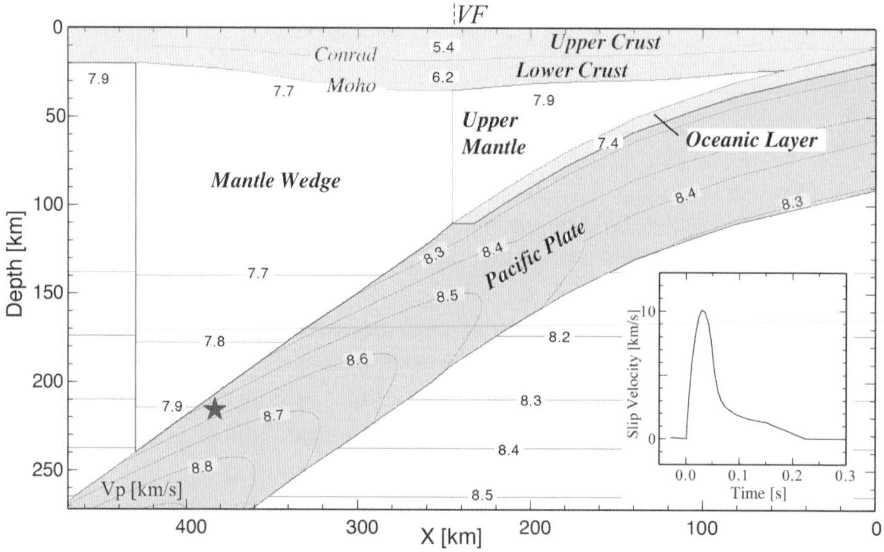

Figure 12.12 Variation of *P* wavespeed used for the 2-D simulation of propagation in the subduction-zone environment. The inset shows the source-time function employed for the double couple source.

plate scenario shown in Figure 12.12, where a simple thermal model is used for the subducted plate embedded in a lower-velocity and higher-attenuation upper-mantle model (Furumura & Kennett, 2005). A double-couple source is introduced 5 km below the plate surface, and the source-time behaviour is provided by the slip-velocity distribution for an in-slab event. Snapshots of the wavefield, and radial-component seismograms are shown in Figures 12.13 for the base model shown in Figure 12.12. Figure 12.14 shows the equivalent results for a model with superimposed heterogeneity in the subduction zone. This stochastic heterogeneity is described by a von Kármán distribution (10.3.3) with a standard deviation of the elastic wavespeeds from the background model ϵ set at 2%. The correlation length in the down-dip direction of the subducted plate is 10 km, and 0.5 km in thickness.

There is a striking difference in the character of the wavefield between the two cases shown in Figures 12.13 and 12.14. The base structure produces discrete pulses with a limited duration of coda. The wave trapped in the former oceanic layer O is very distinct and builds in relative amplitude for the stations further from the source. In contrast, the wavefield for the heterogeneous model is more complex, with a strong component of complex guided energy travelling up the slab and emerging towards the surface as the slab begins to bend. The result is complicated seismograms with a long coda comparable to the observations in Figure 12.11. There is considerable leakage of multiply scattered energy from the slab as can be readily seen in the snapshot at 33 s after source initiation. Nevertheless, sufficient scatted energy is retained to provide transport of seismic energy for long distances along the slab. For the heterogeneous model, the O phase in the former oceanic layer tends to get somewhat obscured by the coda of P, and scattered energy entering this low velocity channel.

The heterogeneity model employed in Figure 12.14 gives good compatibility with the character of recorded seismograms, but the specific stochastic model can only be regarded as a summary of the characteristics within the subduction zone, particularly since the simulation is only 2-D. In particular, long-distance propagation results provide little constraint on the distribution of heterogeneity within the slab, so simple models with a uniform distribution across the slab have been commonly assumed.

Furumura & Kennett (2005) tested a range of heterogeneous models, and suggested that the combination of correlation lengths 10:0.5 km gave the best match to the character of observed seismograms, particularly for P waves. Elongated heterogeneity providing a quasi-laminate structure is effective at retaining the high-frequency waves within the subduction zone. Even so, for very deep events some class of additional amplification is needed to match the amplitudes observed. Furumura et al. (2016) have investigated the role of lowered seismic wavespeeds induced by metastable olivine penetrating below the 410 km discontinuity within the subduction zone. They were able to demonstrate that such lowered wavespeeds can indeed enhance the high-frequency signal, but they were not able to constrain the geometry. Indeed it is possible that the main effect is an enhancement of the level of heterogeneity at depth.

The stochastic waveguide model can provide a good description of frequency selective wave propagation in many subduction-zone settings. The fast wavespeeds in the plate and the strong velocity gradients inside due to thermal effects allow low-frequency waves (0.3–0.5 Hz) to escape into the lower wavespeeds on the outside by refraction.

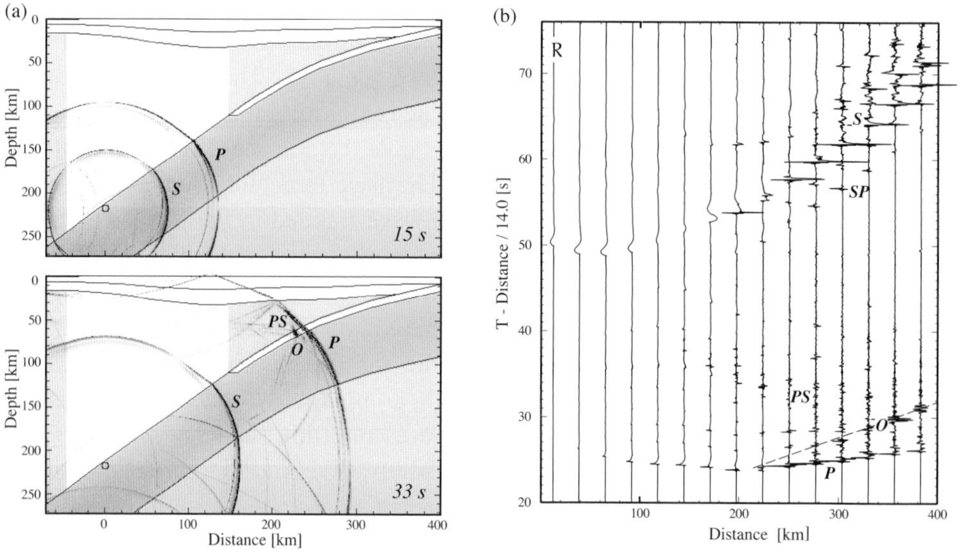

Figure 12.13 Snapshots of seismic-wave propagation at 15 s and 33 s after source initiation with the main phases marked, together with radial-component seismograms using the 2-D base model illustrated in Figure 12.12. The position of the volcanic front corresponds to the change of shading in the mantle wedge above the subduction zone.

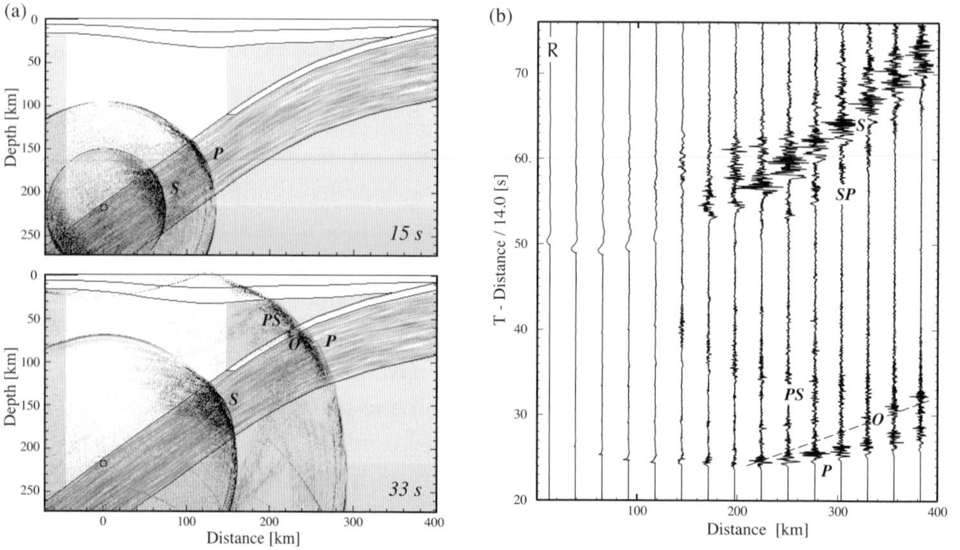

Figure 12.14 Snapshots of seismic-wave propagation at 15 s and 33 s after source initiation, together with radial-component seismograms for a 2-D model with stochastic heterogeneity superimposed on the base model of Figure 12.12. The correlation length along the plate is 10 km, and across the plate 0.5 km.

High-frequency waves (> 2 Hz) are ducted along the plate with the aid of multiple scattering. Low-frequency waves (< 0.15 Hz) with wavelengths longer than the plate thickness are not significantly affected by the presence of the plate or its internal properties.

Furumura & Kennett (2008) contrasted the guided wave behaviour in the thinner Philippine Plate with that for the Pacific Plate and showed that as the plate thins the efficiency of trapping is reduced. Sun et al. (2014) have used a model with elongated stochastic heterogeneity for the Calabrian subduction zone in Italy to model the character of observed seismograms, but need a stronger level of heterogeneity than suggested for the Pacific Plate. Shito et al. (2013) were able to use a single quasi-laminate model for both the subducted plate and oceanic lithosphere to simulate ocean-bottom seismometer records in the northwest Pacific. However, Garth & Rietbrock (2014) have examined relatively shallow events in the Hokkaido segment of the Pacific Plate and find they need a more complex structure to explain the details of seismograms. They suggest the presence of shallow inclined structures with hydration due to faulting in the bending of the plate as it begins to descend. The stochastic models are specified by just a few parameters, and so can capture the overall properties of propagation, but miss features that require more complex parametrisation.

12.4.2 Lithospheric Waveguides

The concept of stochastic models is also helpful for understanding long-distance propagation of higher-frequency waves in the lithosphere. The oceanic lithosphere links directly into subduction, but can also act as a conduit from a subduction zone source into the continental lithosphere (Kennett & Furumura, 2008).

Oceanic lithosphere

In many parts of the ocean high-frequency seismic energy is carried for thousands of kilometres from the source (e.g., Walker & Sutton, 1971) . The onsets of the *P* and *S* energy travel with speeds characteristic of the mantle lithosphere. These *Po* and *So* phases tend to be emergent with a gradual build-up of energy reaching a maximum at an apparent velocity close to that for the base of the crust. The coda is long and slowly decaying, so that the coda of *Po* overlaps with *So*. The nature of the observations implies a strong scattering environment.

Figure 12.15 illustrates the typical complex shape of *Po* and *So* arrivals for recordings from broadband ocean-bottom seismometers in the northwestern Pacific at the NWPAC site. We display a record section for a set of similar, shallow events along the Pacific Plate margin from Izu-Bonin to the northern end of the Kurile Arc, organised in terms of epicentral distance. The traces are aligned at the arrival of the initial *P* phase and the amplitude is normalized by the *S* wave. All the events at NWPAC1–3 display well developed *Po* and *So* wave trains with high-frequency content and very long coda, but there are notable differences in the relative proportions of *Po* and *So* between events. The complex character of *Po*

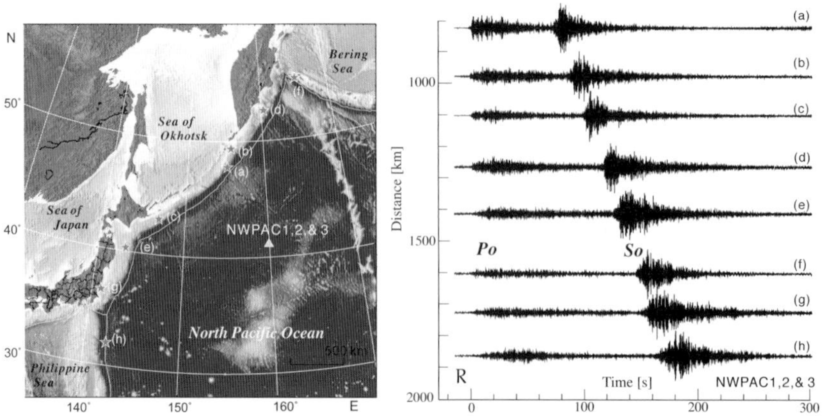

Figure 12.15 Record section of oceanic *Po*, *So* and their complex coda for shallow sources near the Pacific Plate trench. Left-hand panel: map of event locations. Right-hand panel: radial-component records from the ocean-bottom seismometer stations NWPAC 1, 2, & 3 that operated in the same location at different times.

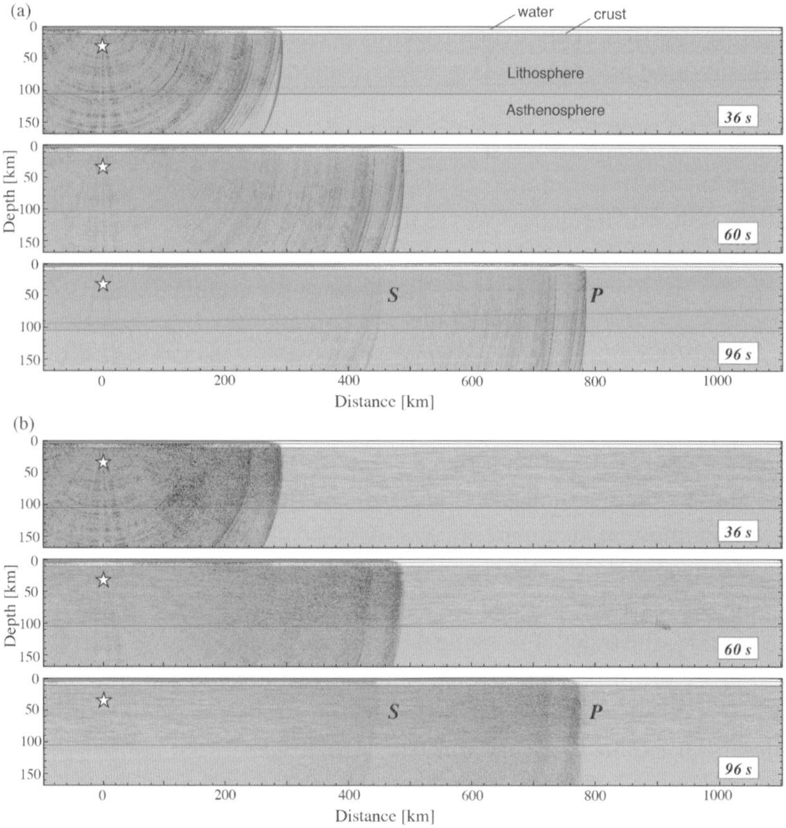

Figure 12.16 Comparative snapshots of the seismic wavefield for 2-D heterogeneity: (a) crustal heterogeneity only, with a dike-like configuration; (b) with the inclusion of mantle heterogeneity with much longer horizontal correlation length than vertical.

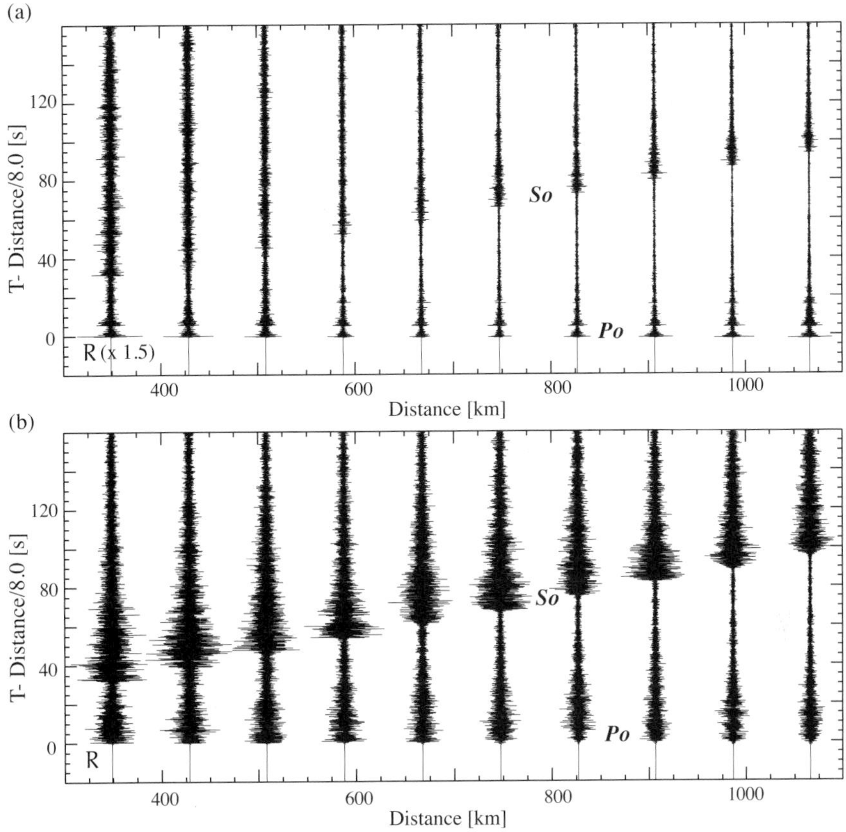

Figure 12.17 Record sections of radial-component seismograms for 2-D simulations: (a) crustal heterogeneity only, with a dike-like configuration (amplitude enhanced by factor of 1.5); (b) with the inclusion of mantle heterogeneity, with much longer horizontal correlation length than vertical.

and *So* with high-frequency content and long coda is independent of the nature of the focal mechanism.

Kennett & Furumura (2013) have shown that it is possible to explain the character of the *Po* and *So* phases with pervasive heterogeneity in the oceanic lithosphere with a horizontal correlation length (~10 km) much larger than the vertical correlation length (~0.5 km). Multiple scattered waves in the lithosphere interact strongly with reverberations in the sea water and sediment, as suggested by Sereno & Orcutt (1985). These reverberations play a major role in extending the coda of *Po* and *So*, but are modified by the interaction with heterogeneity to smear out any distinct arrivals.

The role of heterogeneity is illustrated in Figures 12.16 and 12.17, where we display snapshots of the wavefield and seismograms at the sea floor for two different models. In the first case, dike-like heterogeneity is introduced into the thin oceanic crust, and in the second, quasi-laminate structure is added in the

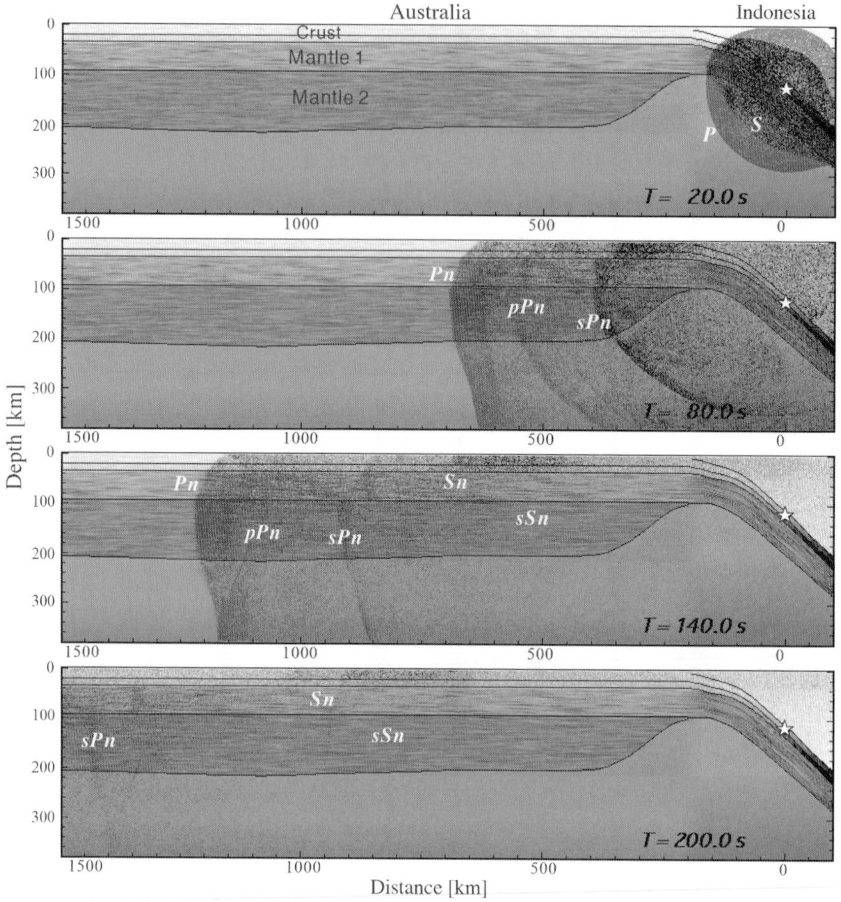

Figure 12.18 Snapshots of the seismic wavefield for a complex heterogeneity model linking Australian continental lithosphere to the Indonesian subduction zone. The source is placed at 120 km depth. Major seismic phases are marked.

lithospheric mantle. With just the crustal heterogeneity, we get sharp wavefronts followed by a suite of reverberations in the shallow structure, smeared slightly by interaction with the crust (Figure 12.16a). In contrast, with the inclusion of mantle heterogeneity the wavefronts are blurred and rapidly acquire a diffusive coda (Figure 12.16b). The differences are even more marked in the radial-component seismograms displayed in Figure 12.17. With crustal heterogeneity alone, the *Po* and *So* phases have a compact appearance, with discrete pulses of energy and relatively low amplitude. The addition of the quasi-laminate structure in the lithosphere has created a strong stochastic waveguide that enhances the amplitude of the *Po* and *So* phase and leads to an elongated coda. With the stochastic waveguide effect, enhanced by the water reverberations, the horizontal propagation of *Po* and *So* can be sustained to very large distances from the source.

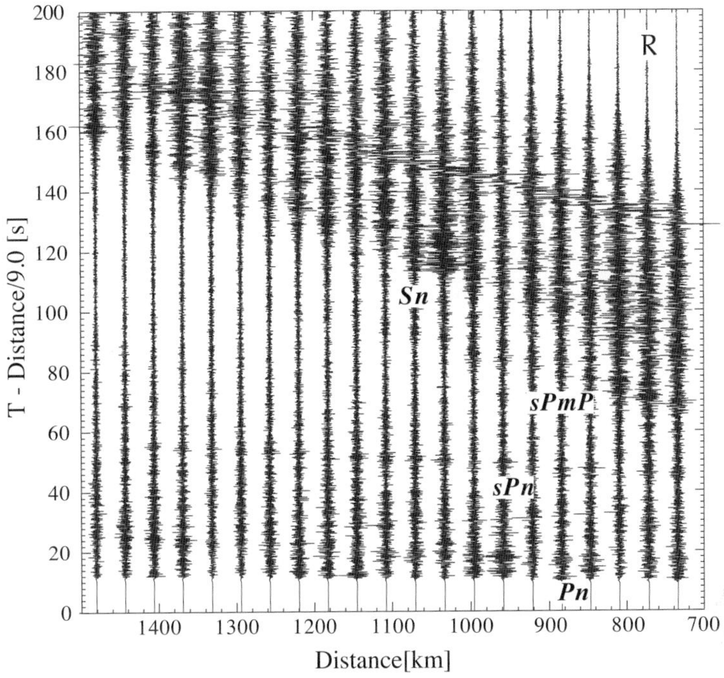

Figure 12.19 Radial component seismograms for stations in the epicentral distance range from 700 to 1500 km from a 120 km deep source in the Indonesian subduction zone for the structure illustrated in Figure 12.18.

Continental lithosphere

In Figures 12.1 and 12.4 we have illustrated the long high-frequency coda for *P*- and *S*-wave propagation across cratonic Australia. Similar behaviour is observed for both an earthquake in central Australia and an event in the Indonesian subduction zone. The character of the coda has resemblances to that seen for *Po* and *So*, and again suggests the presence of organised heterogeneity. Kennett & Furumura (2008) made a detailed study of the propagation of events from the subduction zone into northern Australia, and were able to show that the specific character of the arrivals depended on the position of the source along the Indonesian arc and the extent of oceanic lithosphere along the path. The most efficient propagation occurs when continental lithosphere abuts the subduction zone. In each case a good representation of the character of observed seismograms could be achieved with a quasi-laminate structure, with again a much longer horizontal correlation length than in the vertical direction.

Figure 12.18 displays snapshots of the seismic wavefield progressing through a composite stochastic waveguide from a source in the subduction zone. The background model is stratification with a modest low velocity zone at the base of the lithosphere. Based on the observations in northern Australia, low intrinsic attenuation is imposed in the lithospheric mantle (Q_P = 1600, Q_S = 800), with increased attenuation beneath the lithosphere. For the heterogeneity distribution in Figure 12.18, 2% variation is assumed for the Crust zone with correlation length a_x of 5.0 km and a_z of 0.25 km, 2% variation is imposed in Mantle1 with correlation lengths a_x of 20 km and a_z of 0.5 km, and a reduction in variation to 1% is made for Mantle2. No heterogeneity is included in the asthenosphere. These parameters produce seismograms (Figure 12.19) that have comparable properties to observations, but are not tightly constrained. Once again the processes of multiple scattering lead to ducting of seismic energy to considerable distances. The interactions between the different heterogeneous zones contribute to the development of a relatively diffuse wavefield with the entire lithosphere acting as a stochastic waveguide. Energy that has spent time in the crust acts to extend the coda, and can still return to the mantle.

13

Multi-Scale Heterogeneity

From the microscopic to the global scale, heterogeneity is pervasive in the Earth, though the balance of scales varies as we have seen in Section 12.2. The effects on seismic waves depend on the relation between the wavelength of the seismic waves and the scales of variation of the medium. When the wavelength is long compared with the variability of physical parameters seismic waves proceed unimpeded. A good description is then provided by an effective medium as in our treatment of *upscaling* in Section 10.1. Strong interactions occur when seismic wavelengths are comparable to the size of heterogeneity with complex scattering. If the scale of wavelength variation is very long then the seismic waves sense the local properties.

Because seismic waves span a broad spectrum of frequencies and hence wavelengths, any heterogeneous structures will be perceived in different ways by the various aspects of the wavefield, with significant difference in behaviour between body waves and longer-period surface waves. Such complications become most evident when a wide range of heterogeneity scales are present simultaneously, as in the lithosphere and at the base of the mantle.

Earth's lithosphere constitutes the upper thermal boundary layer of mantle convection, with strong 3-D variations that can be linked to the processes of creation and evolution across geological time. Oceanic lithosphere is recycled through the processes of sea-floor formation and subduction so that the oldest component is less than 200 Myr old. In contrast, the continental lithosphere remains mostly intact, buoyant enough to withstand recycling and therefore evolving slowly over geological time; fragments of Archean lithosphere are still found as the cores of the continents (see, e.g., Artemieva, 2011 for a general survey). Because of this dichotomy in the character of the oceanic and continental domains, the lithosphere is globally heterogeneous. This heterogeneity is manifested in various forms, including rheological, structural, and compositional variations (both lateral and vertical) inherited from both the time of formation and recent tectonic processes.

The superposition of multiple phases of deformation, growth, and destruction imposes a complex character on the lithosphere with multiple scales of heterogeneity. Such features are most evident in the crust (cf. Section 4.6), but persist into the upper mantle on somewhat longer scales (Sections 9.2, 9.3). Disentangling the various aspects of such heterogeneous structure requires care; so we first discuss the way in which seismic tomography techniques can be exploited to provide maximal definition of structure with an example of a portable seismograph experiment.

We then turn attention to the nature of the oceanic lithosphere, and what can be learned about its character from a variety of seismic probes. The thermal evolution of the oceanic lithosphere as it move away from mid-ocean ridges imposes large-scale structure that can be disrupted by, e.g., hotspot activity and smaller-scale features related to crustal generation. The properties of the high-frequency guided waves (*Po*, *So*) provide clues as to interaction of the lithosphere and the mantle beneath.

The continental lithosphere is more obviously heterogeneous with variations in surface geology and ages of continental rocks. Once again, multiple processes have helped to shape the current state and leave a complicated mixture of scales of variation. Although there is a natural tendency to assume that structures should be simple, the use of simplified models has the potential to be misleading when multi-scale heterogeneity is present. We look at the ways in which different scales of variation interact with the seismic wavefield and how this influences observations.

In the last section we consider wave propagation in and around subduction-zone structures with concentration on the Japanese region, where the high concentration of the seismic stations means that the seismic wavefield is exceptionally well sampled. As a result it is possible to tease out the origin of distinctive features of seismic ground motion due to earthquakes at depth.

13.1 Capturing Multi-Scale Structure

The presence of heterogeneity over a wide range of scales needs to be taken into consideration in the design of experimental configurations for structural studies. Even though there is a natural desire to focus attention on the primary target, the nature of the sampling provided by specific techniques needs to be taken into consideration.

In delay-time tomography and receiver function methods exploiting teleseismic arrivals, the propagation paths converge on the receiver. With good azimuthal coverage, each receiver contributes a cone-like sampling into the subsurface. the depth at which sampling merges into continuous coverage depends on the horizontal spacing of the stations. Even in the zones of overlap, the paths from the distant events will cross at relatively steep angles. There is therefore a tendency towards vertical smearing in imaging and tomography that is typically around 8–10% of the actual depth. A correction of this order brings body-wave and surface-wave results into better correspondence.

When horizontally travelling waves are employed, the spatial resolution is

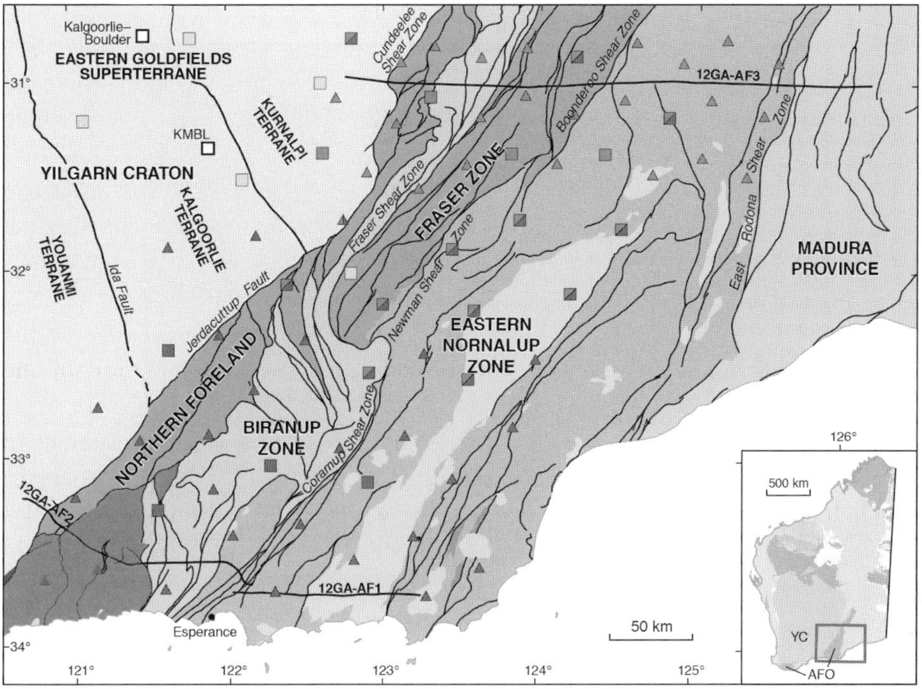

Figure 13.1 The ALFREX experiment in Western Australia designed to examine the relation of the Albany Fraser Orogen (AFO) to the Archean Yilgarn Craton (YC). The main geological units and faults are marked. Two deployments were made: (1) 2013 October – 2014 November, and (2) 2014 November – 2016 January, indicated by two different grey tones. Stations in place for the whole experiment are indicated by parti-toned symbols. Squares denote broad-band sensors and triangles shorter-period seismometers. The two permanent stations are shown in white, and prior temporary stations are indicated in light grey. Full-crustal reflection profiles carried out in 2012 are indicated by black lines (12GA-AF1, AF2, AF3). [Map courtesy of the Geological Survey of Western Australia.]

controlled by the pattern of crossing propagation paths within the region of interest. Thus, in ambient noise tomography it is desirable to have some stations surrounding the target zone so that maximal coverage is achieved between the different station pairs. Such a configuration with some external receivers is also useful with single-station methods exploiting regional sources, e.g., longer period surface wave tomography or tomography using regional phases such as *Pn*, *Sn*.

The availability of instrumentation and logistics can also play an important role in determining the actual station locations. Even where the planning concept is a regular grid, minor modulations of the configuration can be beneficial since they tend to soften the dominant inter-station directions.

As a specific example of a deployment of portable seismometers we consider the ALFREX experiment targetting the Albany Fraser Orogen in Western Australia (Figure 13.1). This experiment was carried out by the Australian National University in cooperation with the Geological Survey of Western Australia (Sippl

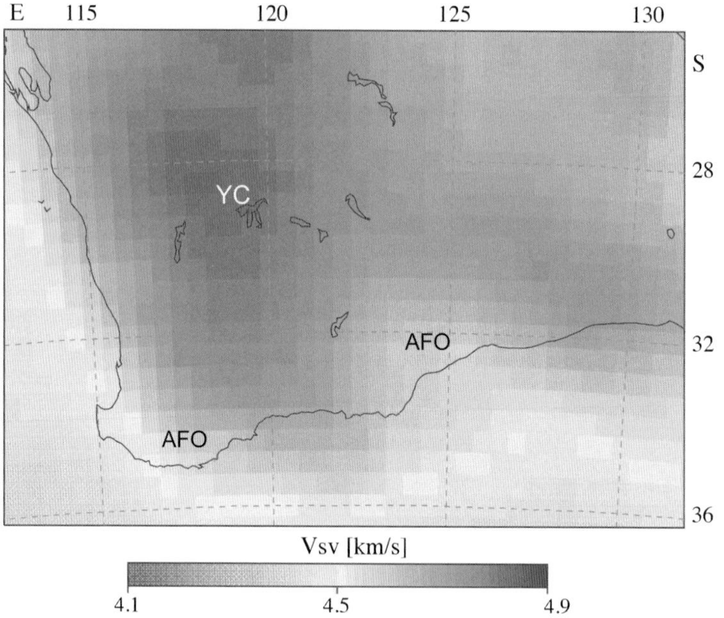

Figure 13.2 *SV* wavespeeds at 120 km depth from the AuSREM mantle model (Kennett et al., 2013). The Yilgarn Craton (YC) displays very high shear wavespeeds that reduce slightly in the Albany Fraser Orogen (AFO).

et al., 2017a,b). The stations were deployed across the eastern margin of the Archean Yilgarn Craton into the marginal orogen, which is marked by a large gravitational anomaly beneath the Fraser–Biranup zones (Figure 13.1). Full-crustal reflection profiles had been collected in 2012 in the southern and northern parts of the region (12GA-AF1, AF2, AF3). An objective of the ALFREX experiment deployed from 2013 October to 2016 January was to develop a model of 3-D structure, particularly in the crust, to link to the reflection profiles.

The deployment was carried out in two stages. The first stage from 2013 October to 2014 November was focussed on the northern reflection line, but extended into the centre of the region with a group of stations that remained in place throughout the experiment. The second deployment from 2014 November to 2016 January put more stations in the south, but also reoccupied a few northern stations to improve the inter-station path configuration. The eastern extent of the deployments was determined by the number of available instruments and the difficulties of access in this remote and uninhabited area.

Hints of structure linked to the eastern edge of the Yilgarn Craton can be seen in the shear-wavespeed distribution in the mantle (Figure 13.2). We show the *SV* wavespeed at 120 km from the AuSREM model derived using surface-wave tomography (Kennett et al., 2013). The Albany Fraser Orogen shows as a mild reduction of the rather high shear wavespeeds seen beneath the

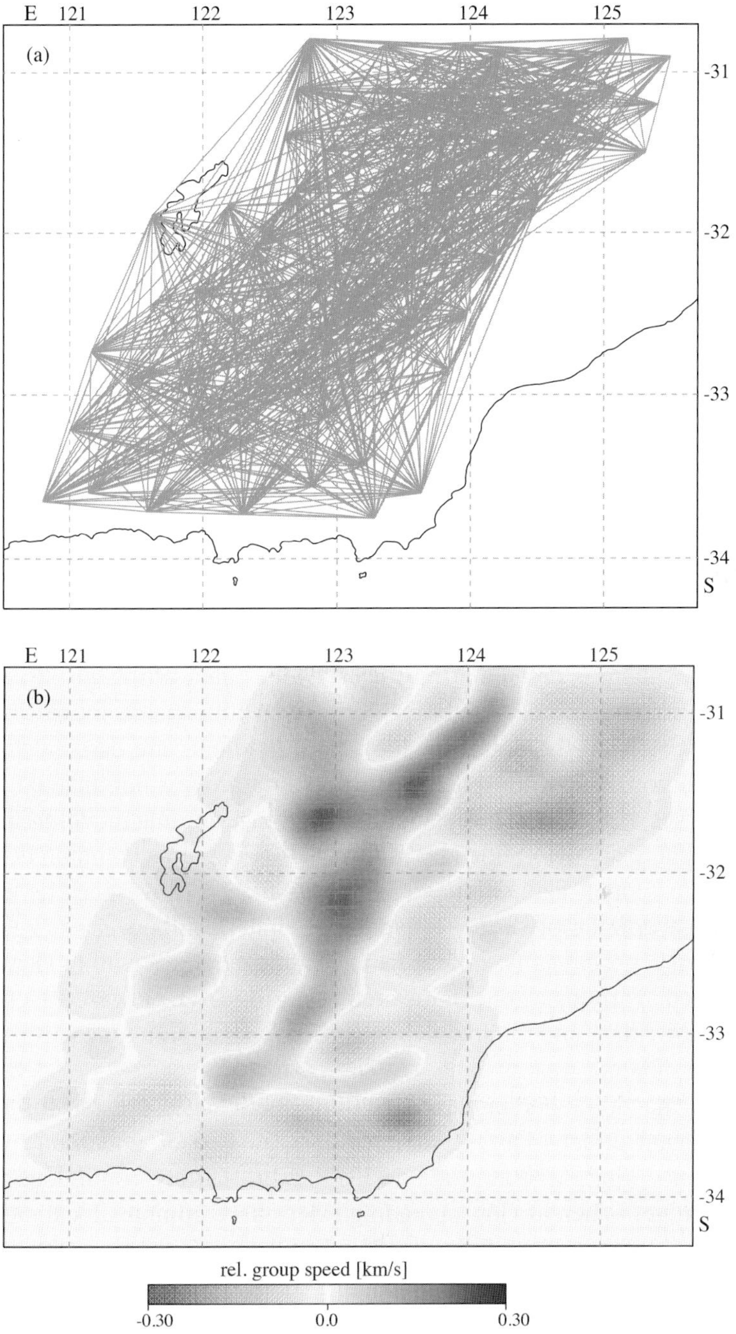

Figure 13.3 Ambient noise tomography using the ALFREX stations: (a) inter-station paths (b) group velocity variations at 5 s period sampling the upper crust. [Courtesy of Y. Chen.]

Yilgarn Craton. There are also indications of a change in the character of the lithosphere–asthenosphere transition with a distinct thinning just to the east of the study area (Yoshizawa, 2014). Prior crustal control in this region was very limited with just a few receiver function results from widely spaced stations.

The configuration of available teleseismic events is dominated by earthquakes in the subduction zones along the Pacific Rim, with events from the north through the Philippines, Japan, and the Kuriles as well as to the east from Tonga-Fiji. Other azimuths have limited coverage. The bias of events towards the north and east was taken into account in the experimental design, and helps to compensate for the limited extension of stations to the east, dictated by complex logistics.

The ALFREX experiment provided good definition of the complex behaviour of the Moho at the margin of the Yilgarn Craton with the aid of receiver function analysis (Sippl et al., 2017a). Good control on crustal structure was also achieved using ambient noise tomography (Sippl et al., 2017b). Figure 13.3 illustrates the dense network of inter-station paths achieved with just the temporary stations, and the resulting group velocity map at a period of 5 s sampling the upper crust. A distinct fast anomaly is associated with the Fraser zone (Figure 13.1).

13.2 Oceanic Lithosphere

Oceanic lithosphere is created at mid-ocean ridges and, ultimately, consumed in the process of subduction. As the newly-formed lithosphere moves away from the ridge it cools, so the dominant horizontal gradients are associated with the thermal evolution of the lithosphere in this cooling process dominated by thermal convection (see, e.g., § GC:12.2.1). With a half-space cooling model the properties of the lithosphere are determined by lithospheric age, with a common dependence on the square root of age. Figure 13.4 shows the *SV* wavespeed as a function of age, with averaging of surface-wave tomography results over age zones in the Pacific Ocean. These results show a distinct lithospheric lid with moderately high *SV* wavespeed overlying a zone of lowered wavespeed. The lowest wavespeeds are found near the ridge, and the low-velocity zone is least pronounced for the oldest oceanic lithosphere.

Although the dominant influence is lithospheric age, the rate of spreading has a significant effect. Fast spreading ridges tend to be underlain by shallow magma chambers (e.g., Kent et al., 1993) that are largely continuous along the mid-ocean ridge. Whereas slow spreading ridges have complex local morphology and considerable heterogeneity in magma production at any given time (Rubin et al., 2009).

The regular progression of physical properties away from the ridges are disrupted by the presence of hotspots and volcanic plateaus. Numerous island chains traverse the oceans where a relatively stationary plume interacts at intervals with the moving overlying plate. A well-known example is the Hawaiian–Emperor Sea Mount chain, which has a distinct change in orientation near Midway Island associated with a reorganisation of the tectonic plates in the Pacific.

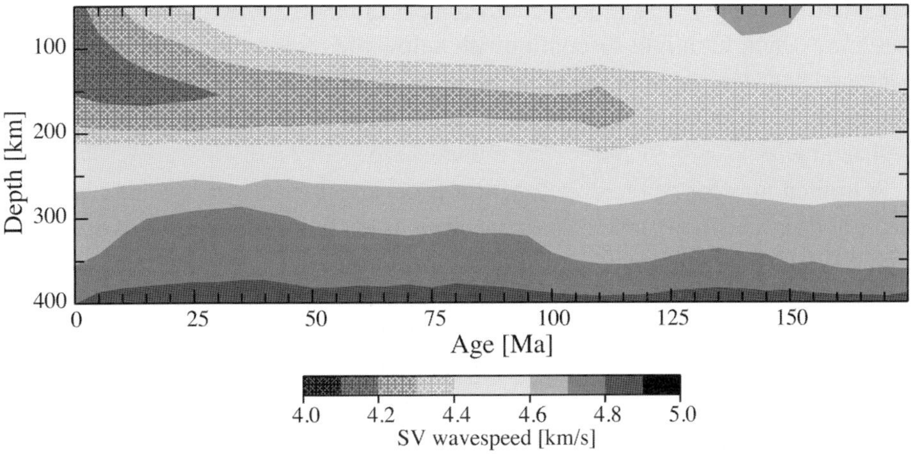

Figure 13.4 Averages of *SV* wavespeed in the Pacific Ocean as a function of plate age display a clear low-velocity zone below a faster lithospheric lid. The averages are derived from surface-wave tomography for the fundamental and first four higher modes of Rayleigh waves. [Courtesy of K. Priestley.]

Jogs in the the mid-ocean ridge configuration produce transform faults that extend across the ocean floor away from the ridge and bring lithosphere of different ages into contiguity, with consequent contrasts in physical properties (see, e.g., Van Avendonk et al., 2001). The motion between the two plates on either side of the ridge can be described by rotation about a pole, and the transforms lie along small circles about this pole of rotation. The patterns of magnetic stripes are laid down as the sea floor cools in the prevailing magnetic field at the ridge and so lie perpendicular to the current spreading direction. The complexity of the patterns of magnetic anomalies across the ocean floor attests to significant changes in plate movement in the last 120 Myr.

The oceanic lithosphere comprises a seismic 'lid' with high seismic wavespeeds, which extends across the whole tectonic plate. This lid is underlain by lower shear wavespeeds associated with the asthenosphere (see, e.g., Kawakatsu & Utada, 2017 for a broad review). The details of the seismic-wavespeed distribution in the lithosphere are not well known and are often assumed to be uniform through the mantle component. Thermal arguments suggest that there should be some decrease with depth, particularly for shear wavespeed. In the old oceanic lithosphere in the northwest Pacific, striking azimuthal anisotropy in *Pn* propagation is found in both active source experiments (e.g., Oikawa, 2010) and in studies using earthquake sources (Shintaku, 2014). *P*-wavespeed variations of almost 9% are reported with the fast direction approximately aligned with the fossil spreading direction.

Seismic body waves also intermittently detect a sharp velocity reduction near the base of the lithosphere, the Gutenberg (G) discontinuity, which cannot be explained by temperature alone. Schmer (2012) has exploited precursors to the phase *SS*, which highlight discontinuities below the region of the surface bounce point, and has identified considerable variations in the nature of the transition from lithosphere to asthenosphere, with an identification of the sharpest changes with localised melts.

A limited number of receiver function experiments has been carried out using ocean-bottom seismometers. To avoid noise issues only moderate frequencies, around 0.1 Hz, are employed and so there is limited vertical resolution. Receiver function methods are sensitive to significant contrasts in physical properties and have not registered any notable internal lithospheric discontinuities beside the Moho. From such receiver function studies, Kawakatsu et al. (2009) have proposed the presence of a 'mille-feuile' structure in the oceanic asthenosphere, with elongated pods of partial melt, as an explanation of the contrasts with the lithosphere. The quasi-laminate heterogeneity structures proposed for the oceanic lithosphere, discussed in Section 12.4.2 have a strong resemblance to these asthenospheric structures, but with reduced contrast as would be produced by freezing the melt.

The broad-scale structure of the oceanic regions has been investigated with a number of seismic tomography studies based on the analysis of surface-wave dispersion. Ekström (2011) presents global results for both Rayleigh and Love waves that display significant radial anisotropy in the ocean basins. In the asthenosphere, *SH* wavespeeds are higher than *SV*, which is characteristic of horizontal flow. The fast direction of *SV*-wave propagation inferred from the azimuthal anisotropy of Rayleigh waves is also largely aligned with absolute plate motion at depths around 120 km, particularly for fast moving plates.

The northwestern Pacific generally supports very efficient propagation of both *Po* and *So* phases crossing the old oceanic lithosphere of the Pacific plate ($>$ 100 Ma). However, in the northeast of the Pacific Basin, propagation is less efficient (Kennett et al., 2014). In this area there are considerable variations in the age of the oceanic lithosphere across major fracture zones. One of the few seismic stations in this area is the Hawaii-2 Observatory (H2O) that used the former ocean-bottom telephone cable HAW-2 from Hawaii to the US mainland for power supply and telemetry, and operated from 1998 to 2003 (Figure 13.5). The station sat on the sea bed between the Murray and Molokai fracture zones, on lithosphere which is somewhat younger than across the fracture zones to the north and south. A few volcanic earthquakes near Hawaii show well-developed *Po* and *So* phases at H2O, but from other directions *So* is weak or even suppressed below the noise level. The contrast in the character of the records is illustrated in Figure 13.6, where we compare seismograms and frequency–time analysis for an event from Hawaii recorded at H2O with an event from off the northern Californian coast that lies in young crust just to the north of the Mendocino fracture zone.

The propagation path from Hawaii to H2O crosses the multi-stranded Molokai

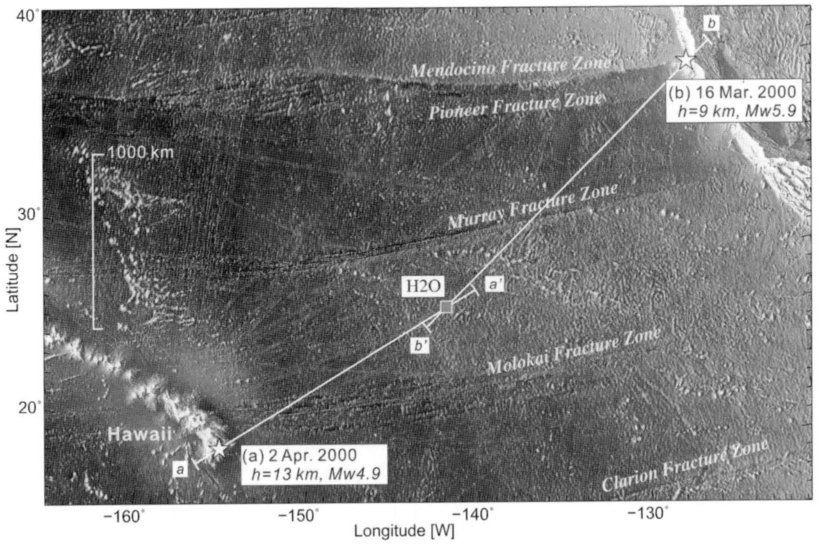

Figure 13.5 Configuration of events in Hawaii and off the Californian coast relative to the seismic station H2O in the northeast Pacific Ocean.

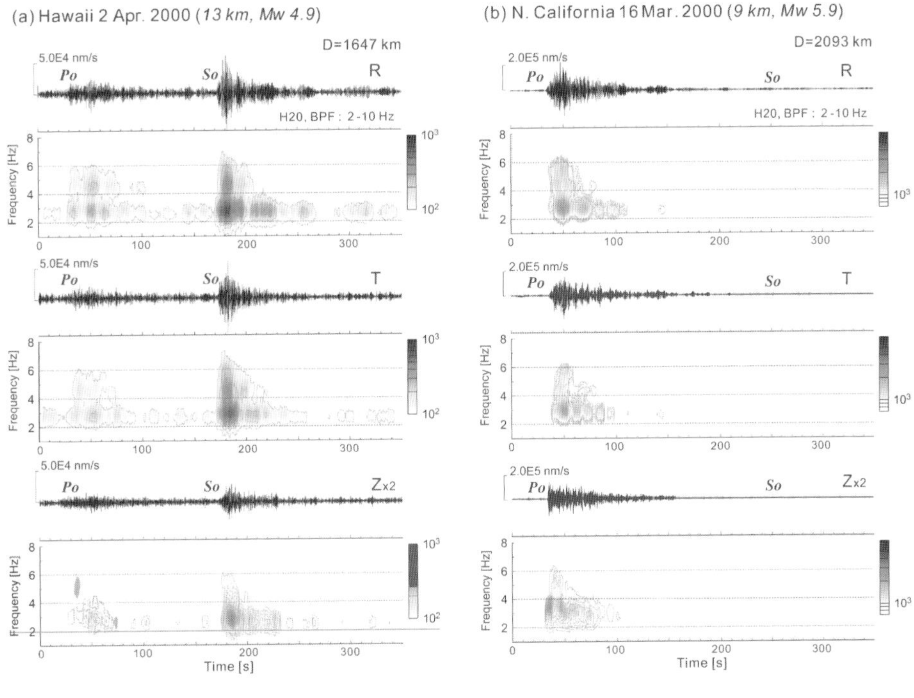

Figure 13.6 Character of events at H2O station, rotated three-component seismograms and frequency–time analysis: (a) Hawaiian event with significant *So*; (b) Dominant *Po* from the event in northern California. For clarity, the amplitudes of the vertical component (Z) are enhanced by a factor of two.

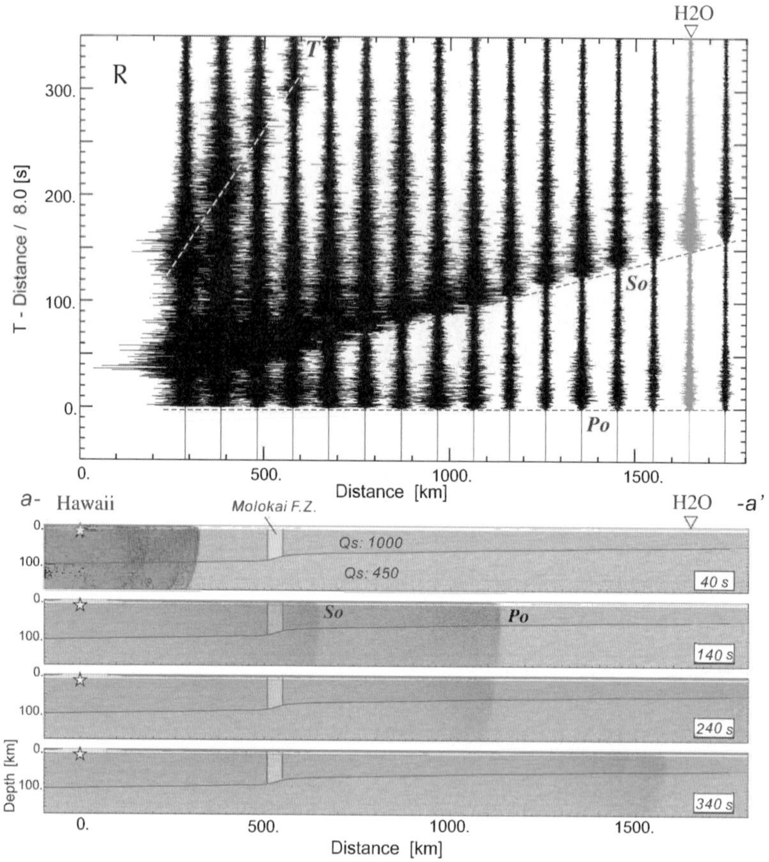

Figure 13.7 Finite-difference simulation of propagation to the H2O site in the NE Pacific along a 2-D profile from an event in Hawaii, showing synthetic seismograms and snapshots of the seismic wavefield. Despite crossing the Molokai fracture zone both *Po* and *So* are distinct for the path from Hawaii

fracture zone obliquely, but still produces a large-amplitude *So* phase in the frequency band above 3 Hz (Figure 13.6a). In contrast, the larger event from the Californian coast with an almost opposite azimuth shows a dominant *Po* phase and little discernible *So* (Figure 13.6b). The rate of decay of the *Po* coda is also larger than for the Hawaiian event.

The differences in the character of the propagation of *Po* and *So* on these two pages can be linked to the effects of large-scale structure in the lithosphere. We make a similar approach to that in Section 12.4.2, using finite-difference simulations building in the full range of complexity in the lithosphere, including stochastic quasi-laminar heterogeneity. We use 2-D simulations for profiles along the great-circle paths from the events in Hawaii and northern California to H2O (Figure 13.5). The models include sea-bed topography, and lithospheric thickness based on the estimate of the age of the plate. The complex properties of the

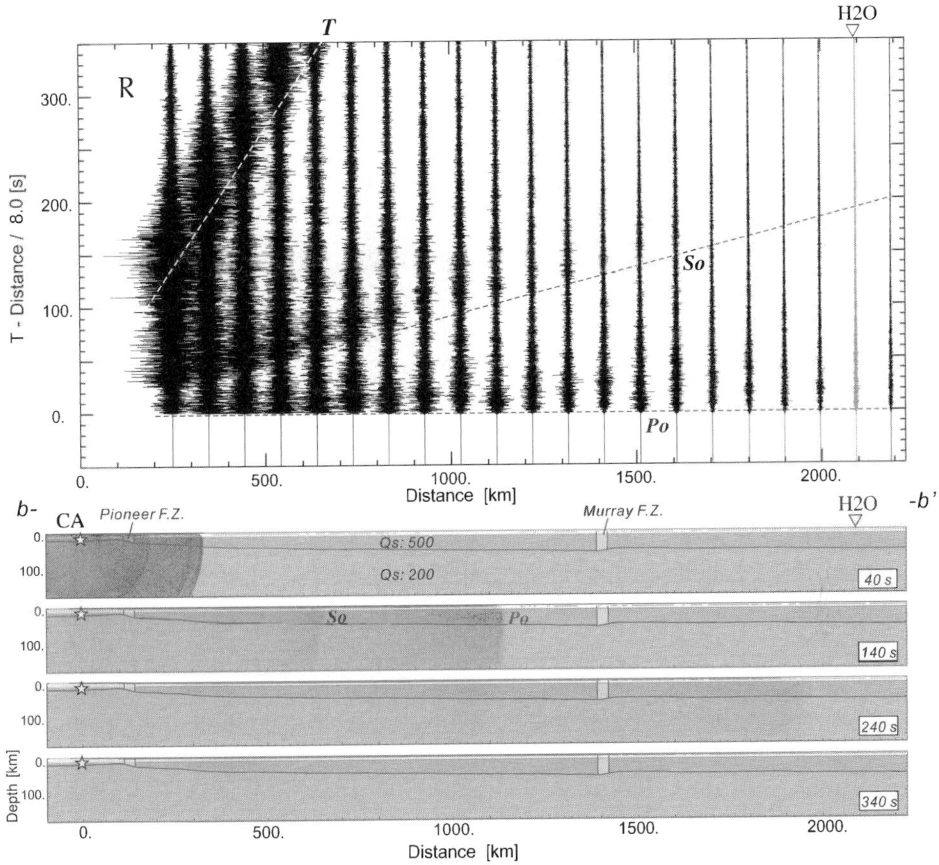

Figure 13.8 Finite-difference simulation of propagation to the H2O site in the northeast Pacific along a 2-D profile from an event off the coast of northern California, showing synthetic seismograms and snapshots of the seismic wavefield. The combination of young, thin and attenuative (Qs = 500) lithosphere and the crossing of the Pioneer and Murray fracture zones (Qs = 200; 10% reduction of shear wavespeed) leaves almost no energy in *So* and little in *Po*. The strong excitation of the *T* phase drains energy from the lithospheric wavefield.

transform faults are represented with reduced shear wavespeed and increased attenuation.

Along profile *a–a′* from Hawaii to H2O, the *Po* and *So* waves have to cross the Molokai fracture zone obliquely, and encounter progressive thinning of the lithosphere. The crossing of the fracture zone does not have a pronounced effect on the lithospheric waves, but there is some loss as the lithosphere continues to thin. Nevertheless, in Figure 13.7 we see clear *So* and *Po* wavetrains at the H2O location, comparable to those in Figure 13.6(a).

For propagation from the northern Californian coast to H2O along profile *b–b′*, the initial part of the path is on very young thin crust. The waves then encounter the Pioneer fracture zone, and after 1400 km the Murray fracture zone. We have

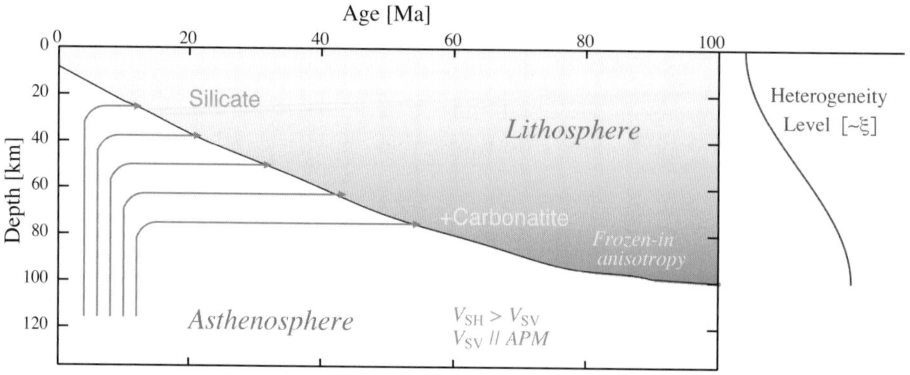

Figure 13.9 Schematic view of possible heterogeneity in the oceanic lithosphere associated with growth by cooling and influx of new material through the neighbourhood of the ridge crest. The heterogeneity level represents the likely amplitude of fluctuations in oceanic wavespeed with a quasi-laminate configuration, which would be linked to radial anisotropy ξ. Flow in the asthenosphere is dominantly horizontal, leading to strong radial anisotropy. The fast direction for *SV*-wave propagation is roughly parallel to absolute plate motion.

incorporated lowered Q_s of 500 in the lithosphere and 200 in the asthenosphere to represent the effect of the younger, and warmer, structure along this path. The combination of the steady change in structure along the path with the enhanced attenuation is able to suppress *So* whilst leaving perceptible *Po* (Figure 13.8), as observed in the records in Figure 13.6(b).

Both simulations include the *T* phase associated with energy that has spent most of its path as acoustic waves in the ocean. The excitation of the *T* phase in the simulations is much larger for the shallower northern Californian coast event than for that in Hawaii. Stronger radiation of seismic waves into the ocean will mean that less energy is available for ducting through the lithosphere, and contribute to the suppression of the *Po* and *So* phases.

Although the characteristics of *Po*, *So* propagation for long distances can be adequately represented with a uniform distribution of quasi-laminar heterogeneity in the lithospheric mantle (cf. Section 12.4.2), there is little direct control on the depth distribution of heterogeneity. Kennett & Furumura (2015) have shown that a modification of the distribution with stronger heterogeneity at the base of the lithosphere not only matches equally well to the observations but also fits better with petrological results. As the lithosphere ages it thickens and the presence of stronger laminar scatterers at the base can arise from underplating of extended melts or melt pockets. Upwelling near the ridge can bring in localised chemical heterogeneities that have low melting temperature (Hirschmann, 2010) and additional contributions can arise from plumes. Any melt pockets will be stretched by shear in the corner flow near ridges. Shear due to plate motion can produce shear-induced melt-rich bands in the asthenosphere (e.g., Kawakatsu et al., 2009) that can become part of the lithosphere as the plate cools away from the ridge.

The presence of enhanced heterogeneity at the base of the lithosphere will counteract any trends associated with the reduction of the background wavespeeds from thermal effects. The stochastic waveguide can still operate efficiently to trap high frequencies for long distances.

Figure 13.9 presents a schematic rendering of a plausible scenario for oceanic lithosphere heterogeneity combining both seismological and petrological information. Ascending flow near the ridge will divert sideways to impinge on the growing lithosphere, bringing melts and heterogeneities that are then incorporated into the lithospheric mantle. For younger lithosphere, underplating will be via silicate melts but, beyond 30 Ma, carbonatite melts can be added (Hirschmann, 2010) and can potentially be more fertile and hydrated. The change in chemical regime may well lead to a higher level of quasi-laminate heterogeneity at depth. The heterogeneity level can be viewed as a potential proxy for radial anisotropy (ξ), so that the change in gradient reflects the inflexion noted from surface-wave studies (Burgos et al., 2014).

13.3 Continental Lithosphere

Continental lithosphere displays a much greater range of structures than oceanic lithosphere in large part because of the wide range of crustal ages linked to the complex evolution of the continental masses. A limited amount of very ancient material is directly preserved, but much lithosphere resembles a palimpsest with successive events imprinting new structures on ancient foundations.

The full-crustal reflection results displayed in Section 4.6 indicate the complexity of crustal structure, with generally stronger variability in structure at the base of the crust. However, at the frequencies employed in reflection seismology, the mantle normally appears to be rather bland with no apparent reflectors. Nonetheless, there is significant evidence for the existence of heterogeneity within the mantle component of the lithosphere, but at larger scales.

As discussed in Section 10.2 the broad-scale features of the lithosphere–asthenosphere system are well mapped by exploiting the properties of surface waves and multiply reflected *S* body waves. Body-wave tomography exploiting relatively dense deployments of portable seismic recorders provides information on the lithospheric mantle at shorter scales. Potential horizontal resolution is limited by station spacing (Rawlinson et al., 2015), but vertical smearing due to the relatively narrow cone of incoming rays also limits resolution in the upper mantle.

Where dense deployments have been made there is evidence for considerable complexity in and beneath the crust (Figure 10.5), with medium-scale variations imaged in all classes of geological environments from cratons to fold belts. The presence of much finer-scale structure in the mantle is indicated by long trains of high-frequency energy following the phases *Pn* and *Sn* refracted back from the mantle, particularly for cratonic paths (Figure 12.1). The nature of this extended coda requires some form of distributed heterogeneity through the lithosphere (Section 12.4.2).

There is, in general, no sharp transition between the continental lithosphere and the asthenosphere beneath. The presence of the asthenosphere is, however, manifest in enhanced seismic attenuation. For example in northern Australia, fast lithospheric wavespeeds are accompanied by little loss of seismic energy, enabling high-frequency waves for both *P* and *S* to propagate readily from subduction zones into continental Australia (Kennett & Furumura, 2008) with extended scattered codas. The high-frequency *S* waves are suddenly lost when the seismic waves penetrate into the asthenosphere, because of its much higher attenuation of shear waves than the lithosphere (Gudmundsson et al., 1994). Unfortunately, such attenuation studies can only provide a broad-scale picture of the behaviour, but confirm the presence of enhanced attenuation below 210 km depth.

The transition from the lithosphere to the asthenosphere (LAT) means that it is difficult to assign a single depth to the base of the lithosphere. Often some level of perturbation from a reference wavespeed profile is used as an indicator. Such choices are necessarily subjective and depend strongly on the choices of parameters. Yoshizawa (2014) has made use of the vertical gradients of absolute shear wavespeed in the mantle to provide shallow and deep bounds on the extent of the LAT across the Australian continent. At each location a vertical profile of shear wavespeed is extracted from a 3-D model; the shallow bound on the LAT is taken at the point of maximum negative gradient in shear wavespeed and the deeper bound at the minimum in wavespeed with depth. This means that the entire span of the transition lies in a region of lower shear wavespeeds than the regions above and below.

All classes of results indicate that the effective thickness of the lithosphere is somewhat variable. These irregularities are likely to impose complex stress patterns in both the lithosphere and asthenosphere associated with the relative motions of the thick continental lithosphere and the more freely flowing asthenosphere. The presence of lithospheric ‘steps’ can produce complex flows around the trailing edges, particularly when these lie perpendicular to the direction of current plate motion. Such edge convection may produce localised volcanism (e.g., King & Anderson, 1998).

Evidence for discontinuities in physical properties within the mantle component of the lithosphere comes from studies of the structure in the vicinity of seismic stations. A commonly used approach exploits the wave conversions and reflections associated with the major seismic phases from distant earthquakes to construct receiver functions. Incident *S* waves have been favoured for discontinuity studies because the converted *P* waves arrive as precursors to *S*, though relatively low frequencies have to be used, which limits vertical resolution.

One of the striking features in *S*-wave receiver function analysis for the

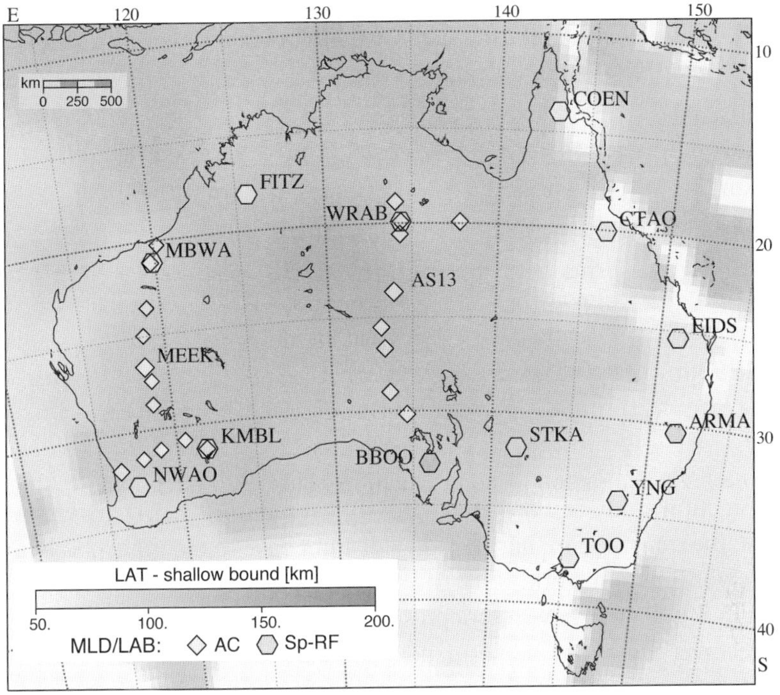

Figure 13.10 Estimated depths for the mid-lithosphere discontinuity (MLD) in Australia from receiver-based studies: AC – auto-correlation; Sp-RF – receiver functions for incident shear waves. The symbols at the various stations are superimposed on the shallower bound on the lithosphere–asthenosphere transition from Yoshizawa (2014).

Australian cratons is clear signals of discontinuities in the mid-lithosphere at around 70–90 km (Ford et al., 2010), which may indicate a rapid drop in seismic velocity or a change in the character of radial anisotropy in the middle part of the continental lithosphere where the wavespeed is generally at its fastest. The estimated depth of the enigmatic mid-lithosphere discontinuity (MLD) from receiver functions corresponds well with a rapid change in the strength of radial anisotropy derived from surface waves (Yoshizawa & Kennett, 2015).

An additional constraint on the nature of lithospheric heterogeneity comes from observations of high-frequency *P*-wave reflectivity profiles derived from the auto-correlograms of vertical-component records at seismic stations (Kennett, 2015; Sun et al., 2018). These *P*-reflectivity profiles suggest vertical changes in the character of the fine-scale structures in the Australian continent, indicating stronger reflectivity in the crust and modest reflectivity in the mantle component of the lithosphere underneath the cratons. Indications of mid-lithospheric discontinuities are provided by distinct changes in the frequency content of the reflectivity as a function of time. The gradational nature of the transition from lithosphere to asthenosphere would account for the absence of any clear signal from the base of the lithosphere beneath central and western Australia in the *Sp*

Table 13.1. *Multi-scale stochastic heterogeneity regimes: amplitudes and correlation lengths.*

Medium-scale			
Depth range	rms het	horiz correl.[km]	vert. correl. [km]
0–300 km	1%	100	24

Fine-scale			
Depth range	rms het	horiz correl. [km]	vert. correl. [km]
0–15 km	0.5%	2.6	0.4
15 km–Moho	1.5%	2.6	0.4
Moho–125 km	0.5%	10.0	0.5
125–200 km	1.0%	5.0	1.0
>200 km	1.0%	5.0	1.0

receiver function study of Ford et al. (2010), even though MLD arrivals are seen. In the east, the transition is sharper, and a distinct base to the lithosphere is imaged in the receiver function work.

In Figure 13.10 we show the depths of well-constrained discontinuities from S-wave receiver functions at permanent stations across the continent (Ford et al., 2010), auto-correlation studies using distant earthquakes (Sun et al., 2018) in Western Australia, and auto-correlation studies exploiting continuous seismic records in central Australia (Kennett et al., 2017; Kennett & Sippl, 2018). These depth estimates are compared in Figure 13.10 with the shallower bound on the LAT from the model of Yoshizawa (2014). In central and western Australia the discontinuities lie within the lithosphere (MLD); whereas beneath the Phanerozoic provinces in the east, and beneath station BBOO, the discontinuity most likely marks the base of the lithosphere (LAB). Although the receiver function and auto-correlation results are based on rather different frequency ranges, there is a good general correspondence. Sun et al. (2018) note that at a number of stations in the West Australian Craton there are indications of a second, deeper, discontinuity lying close to the depth of the shallower LAT bound shown in Figure 13.10. These results indicate the complexity of lithosphere structure revealed with higher-frequency probes (cf. Figure 13.11). Such observations of multiple discontinuities are consistent with the presence of fine-scale heterogeneity in the lithosphere superimposed on broader-scale wavespeed variations (Kennett et al., 2017). Where detailed studies have been made in southeastern Australia, there is clear evidence for variability on scales of 50–150 km, shorter than those seen by surface waves (Rawlinson et al., 2015). This intermediate-scale variation is seen beneath both cratonic and Phanerozoic regions and looks to be a pervasive component of lithospheric structure.

A variety of lines of evidence suggest that the finer-scale components of lithospheric structure have a quasi-laminate behaviour with much longer horizontal correlation than vertical. Such variations in physical properties have equivalent

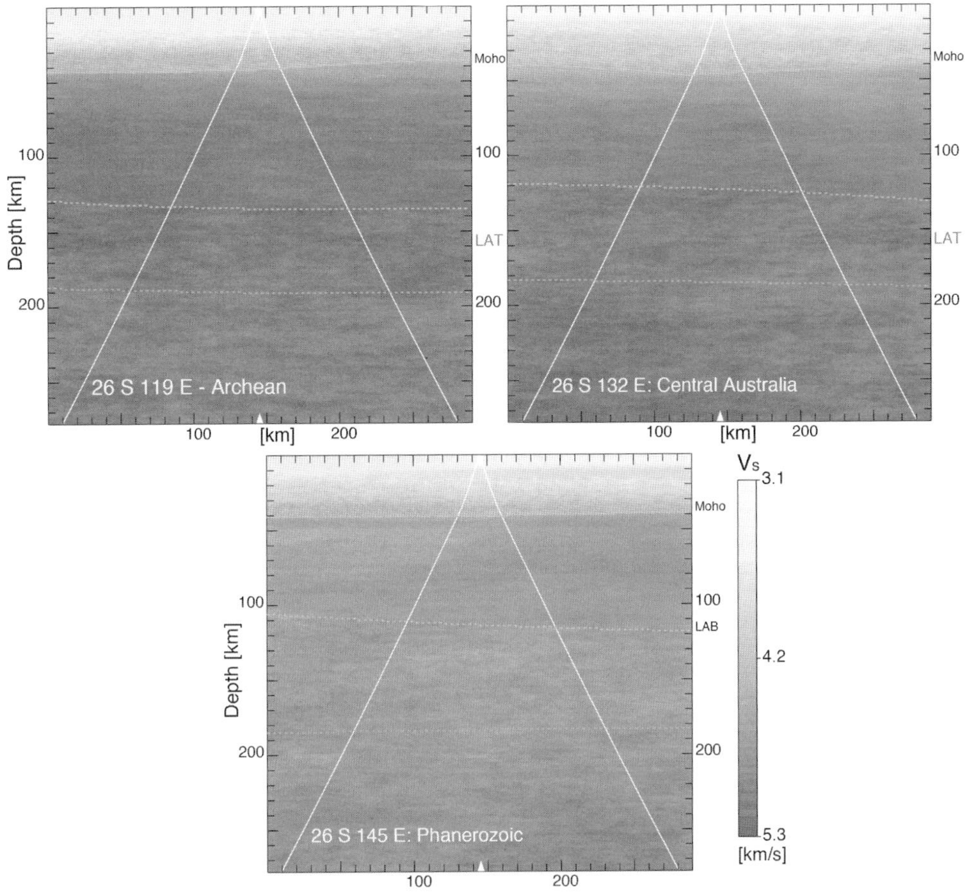

Figure 13.11 Simulations of lithospheric heterogeneity structure at three locations along latitude 26°S constructed using the multi-scale heterogeneity model described in Table 13.1. The 2-D heterogeneity segments are displayed at true scale, with indicative sampling limits for teleseismic arrivals at the central location. The interactions of the different scales of variation produce somewhat different patterns of structure between the Archean craton in the west and the Phanerozoic domain in the east, with intermediate character beneath the Proterozoic of central Australia.

effects to shape-preferred orientation of crystals, and so vertical variations in the character of the heterogeneity can produce changes in the effective radial anisotropy for long-wavelength surface waves as well as apparent discontinuities in the mid-lithosphere region for higher-frequency body waves (Kennett & Sippl, 2018).

Kennett et al. (2017) have shown that a multi-scale heterogeneity model with broad-scale structure from surface-wave tomography, and stochastic intermediate and fine-scale components (see Table 13.1), provides a good representation of many styles of seismological observations across Australia, including the character of regional seismic phases and the character of receiver functions.

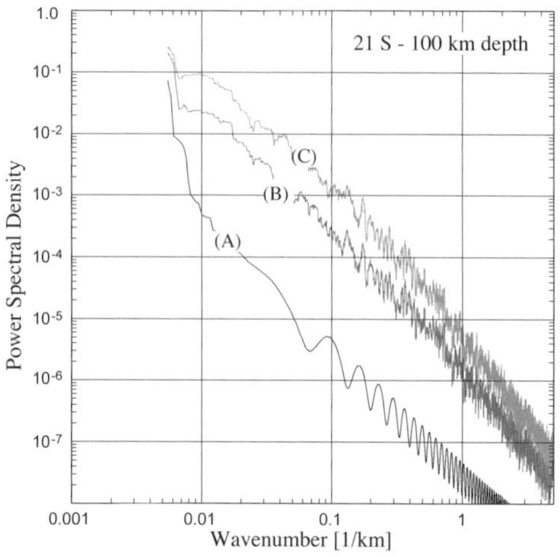

Figure 13.12 Wavelength spectra of *S*-wavespeed heterogeneity fluctuations for structural models taken at 100 km depth along 21°S. (A) AuSREM model from surface-wave tomography with broad-scale structure (Kennett et al., 2013); (B) with the addition of medium-scale heterogeneity; (C) full multi-scale heterogeneity.

This multi-scale model includes heterogeneity in both the crust and the mantle components of the lithosphere, with a change in style of heterogeneity in the LAT as suggested by Thybo (2008). The character of the heterogeneity in three distinct areas in Australia is indicated in Figure 13.11.

The addition of medium- and fine-scale heterogeneity to the broad-scale heterogeneity extracted from surface-wave models, such as AuSREM (Kennett et al., 2013) has the effect of producing a nearly self-similar wavenumber spectrum. In Figure 13.12 we show the wavenumber spectrum taken along 21°S at a depth of 100 km, for the surface-wave results alone (A), with the addition of medium-scale heterogeneity (B), and also fine-scale heterogeneity (C) using the distribution specified in Table 13.1.

The presence of the broad-scale structure means that the variations in the wavespeed gradients in the mantle lithosphere are taken into account. In zones with a strong vertical wavespeed gradient reflection from the upper mantle is efficient and can set up a 'whispering gallery' process for long-distance propagation as suggested by Nielsen et al. (2003). When vertical gradients are weak the presence of quasi-laminar features sustains the high-frequency field through stochastic waveguide effects. The inclusion of intermediate-scale features also acts to modulate the behaviour so that the multi-scale model approaches the complexity of the actual lithosphere. As discussed in Section 11.3, the seismic wavefield has a complex interaction with the various components of heterogeneity. The different scales of variation do not behave independently but modulate each other, with complex feedbacks by reverberations between regions with different properties. It

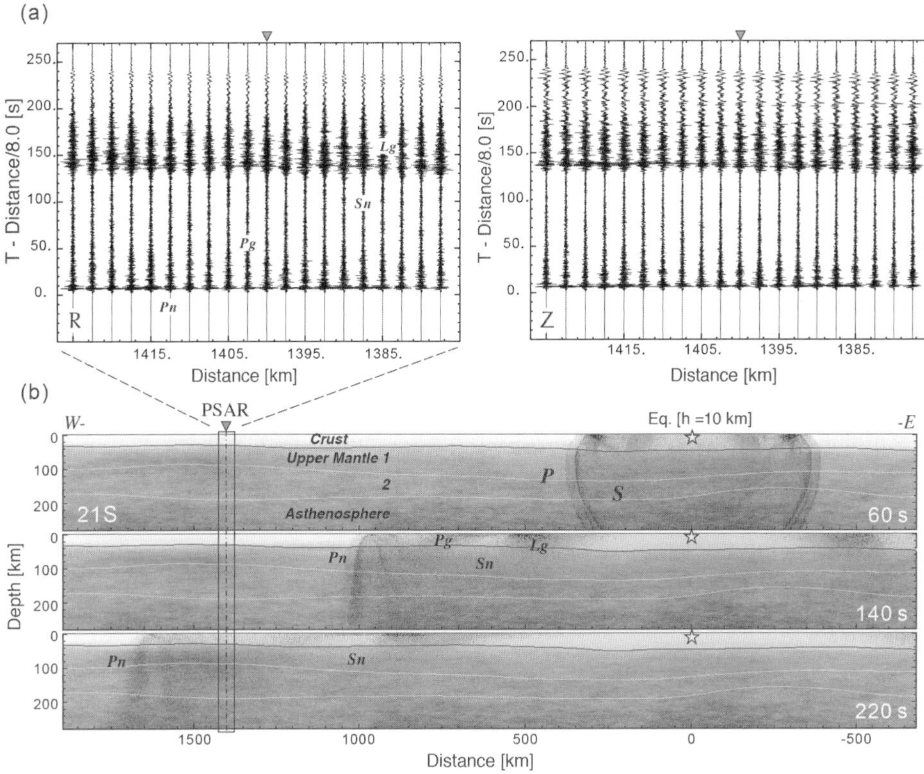

Figure 13.13 Simulation of the effect of propagation through a 2-D section of the multi-scale heterogeneous model for a path length comparable to that from the Ernabella event to PSAR, illustrated in Figure 12.1: (a) theoretical seismograms for closely spaced stations, (b) snapshots of the wavefield on passage to PSAR.

is therefore not possible to characterise the finer scales of variation in isolation, because their effect on the wavefield depends on context.

Nature of coda

In Section 12.1.1 we have discussed the complex nature of the Ernabella event in central Australia recorded at the PSAR array with strong coda and complex coherence properties. The multi-scale heterogeneity model is able to provide a good simulation of the observed behaviour. We make a 2-D FDM simulation along a model for 21°S, the latitude of PSAR, with a source set at the appropriate distance for the Ernabella event at a depth of 10 km. Since the simulation is in 2-D, we use a linear array of closely spaced stations at the PSAR location. The results are shown in Figure 13.13 for both radial (R) and vertical (Z) components in the distance range 1375–1425 km.

With only crustal heterogeneity there is only short-range coherence, but the addition of mantle heterogeneity extends the distance range of coherence. The full multi-scale model produces modulation of the patterns of arrivals with a scattered coda occupying a distance band of nearly 200 km for *S* waves. As in the observations, the spatial coherence of the *P* waves is much greater than for the *S* waves, particularly on the radial component. In this distance range the simulated waves have traversed many thousand wavelengths and so effects associated with the details of specific realizations of the stochastic media should be minimal.

Local sampling

We can gain insight into the way that observational probes sample the lithosphere from the multi-scale heterogeneity model. The contribution to the *P*-wave reflectivity estimates from distant earthquakes will sample a broad swath of structure beneath a station, as indicated by the pale lines in Figure 13.11 converging on the central point at the surface. When the apparent *P* reflectivity is estimated from stacked auto-correlograms segments there will be averaging of structure, which will encompass a wider domain at greater depth. In the crust and immediate uppermost mantle the effective structure will closely resemble the 1-D profile below the receiver. At greater depth, averaging will span similar, but not identical, structures and so the apparent variations in depth will be blurred. A natural consequence is that we expect the *P*-wave reflectivity to diminish in amplitude and to have slightly lower frequency at later times.

In Figure 13.14 we display the 1-D model for the central location in the heterogeneity spread around 26°S, 132°E, from Figure 13.11, along with the reflectivity estimated from autocorrelation of transmitted *P* waves. It is likely that these estimates overemphasise deeper structure relative to any observations for the same 2-D or 3-D heterogeneity structure. Nevertheless, the calculations will provide a good guide to the general behaviour.

We can use such 1-D models to look at the properties of receiver functions for such complex structures as well as the auto-correlation. We have therefore constructed *Sp* receiver functions using the same transmission response scheme as used for the *P*-wave reflectivity, with stacking over a bundle of slownesses. Since we are looking at the *P* conversions from an incident *S* wave we can tie the receiver function results to the reflection case through the depth of conversion. Accordingly we present reflectivity and *S* receiver function results for the 1-D models in Figure 13.14 on a common depth scale, using the AK135 model for conversion from time. When the *Sp* receiver functions are calculated for high frequencies they have very similar behaviour to the *P*-wave reflectivity, and indeed we can track the same major changes in time behaviour. Reflections and conversion pulses are commonly associated with locations where there is a change in the style of wavespeed variation rather than particular jumps.

Once progressively stronger low-pass filtering is imposed on the receiver functions we see a considerable change in the apparent behaviour. In the passage from the broadest band traces (0.05–2.0 Hz) to the lowest-frequency band (0.05–0.125 Hz) the receiver function records apparently simplify, and just a few apparent discontinuities emerge from an initially very complex pattern. However,

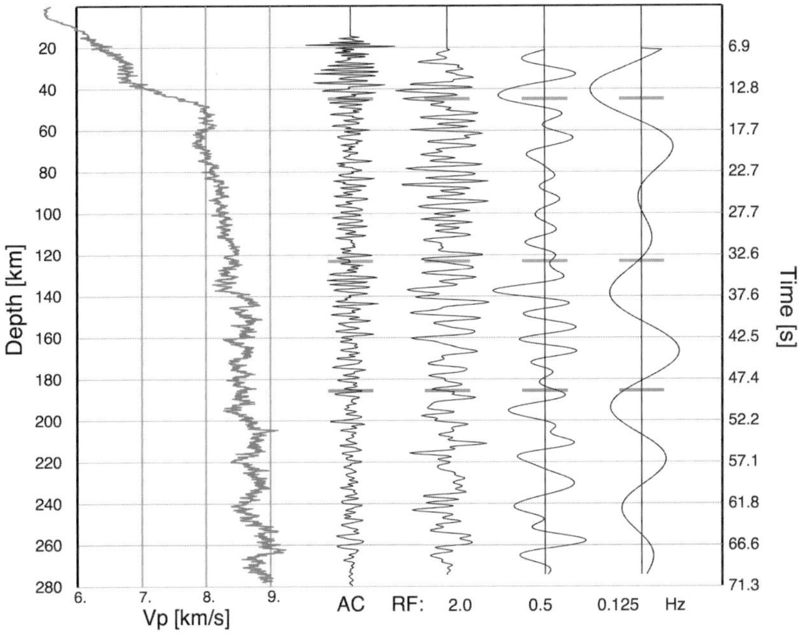

Figure 13.14 Simulations of *P*-wave reflectivity and *Sp* receiver functions for a 1-D model extracted at 26°S, 132°E from the multi-scale model displayed in Figure 13.11 are shown as a function of depth, with time conversion using the AK135 model. The *P*-wave reflectivity is extracted from the autocorrelation (AC) of the transmission response in the frequency band 0.6–3.0 Hz. The suite of *Sp* receiver functions (RF) are scaled by depth to match the reflectivity in time, and are shown for a range of band-pass filter settings: 0.05–2.0 Hz; 0.05–0.5 Hz; 0.05–0.125 Hz.

the seeming simplicity does not come from distinct changes in seismic wavespeed, but rather the interference of many subtly different arrivals from each of the tiny contrasts in the original wavespeed distribution.

The choice of a 0.05–0.125 Hz frequency band is typical of filtering used in *S*-wave receiver function studies directed at lithospheric studies (Ford et al., 2010). The time relationships of such filtered traces can indeed be well represented by simple wavespeed distributions with large jumps, but that is not their true origin. Hints of the actual complexity are likely to arise when higher frequencies are considered and then the single pulses break up into multiple sub-pulses. Such effects have been recognised in recent studies with suggestions of multiple mid-lithospheric discontinuities (e.g., Hopper et al., 2014), though it can be difficult to get enough high-frequency energy in incident *S*.

With the presence of broad-scale and medium-scale structure we can expect modulation of the fine-scale structure so that nearby stations may see broadly similar structures. This will mean that there can be continuity of apparent structure when seen with closely spaced stations, even though there may not be a simple ‘discontinuity’ being mapped.

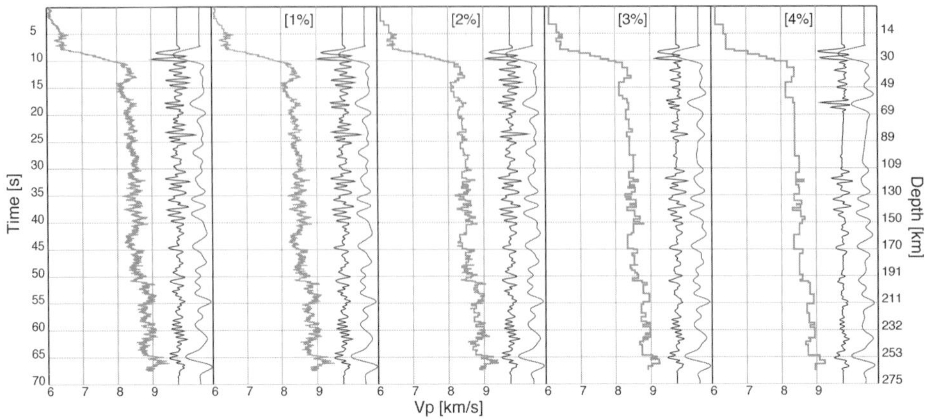

Figure 13.15 Progressive simplification of a complex P-wavespeed profile by filtering by impedance contrasts, indicated at the top of each panel. The left-hand trace in each panel is for the frequency band 0.5–3.0 Hz, and the right-hand trace for the frequency band 0.05–0.50 Hz.

Simplifying structure

In Section 10.4 we have discussed the process of wavespeed upscaling as a means of reducing the complexity of seismic structure by emphasising the component with longer wavelength. In the context of assessing potential discontinuities, an alternative approach can be employed where we seek to retain the major contrasts in the model. We set an impedance threshold and combine all layers whose contrasts to their surroundings is less than this threshold, ensuring that the travel time across the layer is preserved. By this means a complex model can be reduced to one with just a few discontinuities. The process is illustrated in Figure 13.15 for a 1-D profile extracted from the multi-scale heterogeneity model. We calculate *Sp* receiver functions for two pass bands 0.5–3.0 Hz and 0.05–0.50 Hz. For impedance thresholds less than 4% the waveform for the lower-frequency band is hardly changed, whilst the higher-frequency band shows strong segmentation for larger thresholds. Once the threshold is taken above 4% the character of the waveforms is markedly changed since controlling interfaces are lost. These results indicate that the primary control on the response comes from changes in the style of heterogeneity that translate into distinct small impedance jumps.

Only high-frequency probes can reveal the presence of the finer-scale structures in the lithosphere directly, but they can have a profound role in shaping seismic observations at lower frequencies. Quite subtle changes in the style of velocity variation can induce reflections and conversions. With lower-frequency probes interference occurs between waves generated from nearby features, and an apparent discontinuity can appear as the aggregate of interactions between different scales of heterogeneity, and modifications of correlation properties. An apparent MLD can arise without there ever having been a major change in seismic wavespeeds (Kennett & Sippl, 2018).

13.4 The Subduction-Zone Environment

The immediate neighbourhood of subduction zones is one of the most heterogeneous regions of the globe. In addition to the presence of the complex high-wavespeed subducting plate there are reduced seismic wavespeeds and enhanced attenuation in the mantle wedge at the plate corner and variations in structure in the arc structure above the subduction zone. The subduction system also generates a wide variety of earthquakes whose radiated seismic waves have distinctive propagation characteristics.

A 2-D cross-section through a subduction zone is shown in Figure 12.12, with a representation of typical *P* wavespeed. But, an important aspect of the subduction-zone setting is its 3-D character including changes in the dip of the subducting slab. Figure 13.16 shows a perspective view of the subducting plate in the transition between the Honshu and Kurile arcs in northern Japan, together with contours of slab depth and the depth to the Moho. The change in morphology of the subducting plate creates crustal convergence in the Hidaka mountains of central Hokkaido where lower crustal rocks are brought to the surface. The convergence also produces large earthquakes in the immediate off-shore region.

Slip on the plate interface between the subducting and overlying plates is responsible for many events. Where strain has accumulated over a long period of time, it can be released in great earthquakes such as the 2004 Sumatra–Andaman event and the 2011 Off-Tohoku event in northern Japan. Each of these huge earthquakes had segmented slip as different parts of the plate interface failed, but their configuration is very different. The 2004 event extended over 1350 km along the arc with initiation off northern Sumatra and progressively later failure to the north. The 2011 event was more concentrated with only a 500 km interval of the plate boundary involved, but multiple patches of slip. In each case deformation extended to the offshore trench and the large slip (60 m for the Off-Tohoku event) displaced large volumes of water generating devastating local and distant tsunamis. The large surface waves produced by the events caused substantial ground motion, especially where amplified by sedimentary basins.

The strain associated with the subduction process extends into the volcanic arcs above, producing shallow events in the continental material. Such events share the characteristics of intraplate earthquakes and can produce the full range of regional phases (§ SWII:18.3) that can carry energy to significant distances, with crustally guided phases such as *Lg*. Where young oceanic lithosphere is subducted, as in central America, the dip of the subducted plate is low and events at the plate interface can couple strongly into the overlying crust, with reinforcement of regional phases due to reflections from the dipping slab (§,SWII:20.2.2). Such wedge effects above the slab help to carry significant energy to Mexico City from

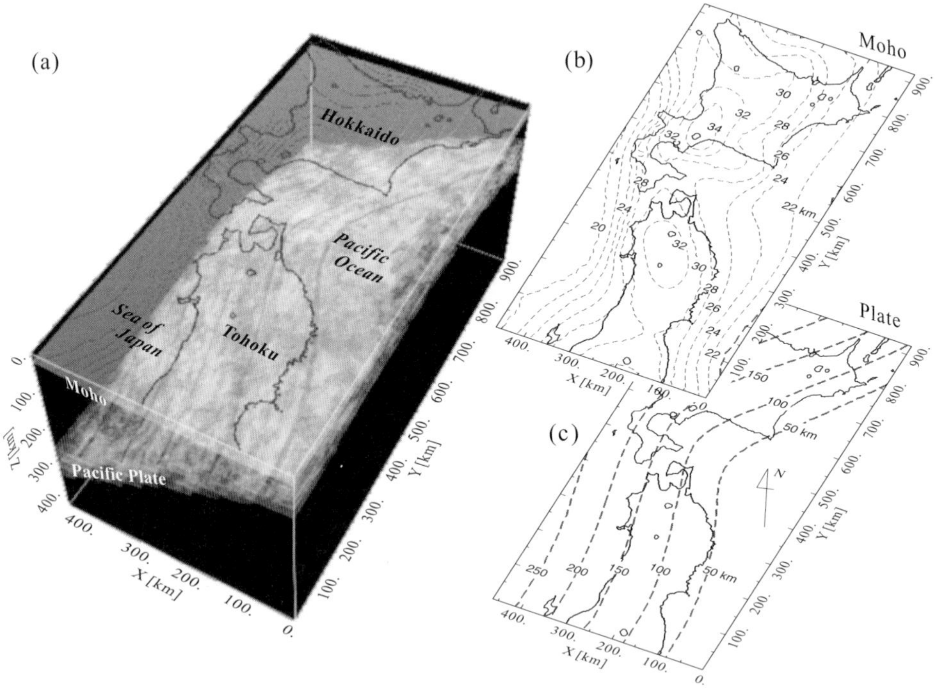

Figure 13.16 (a) 3-D configuration of the subducting Pacific plate at the junction between the Honshu and Kurile arcs in Japan; (b) Moho depth; (c) slab contours. [Courtesy of T. Furumura.]

events at the coast more than 350 km away, which are then amplified by very low seismic wavespeeds in the former lake bed underlain by nearly rigid bedrock.

The influence of events in the subducting slab on surface ground motion depend strongly on depth. Down to about 150 km the events occur below the overlying arc, but upward propagation occurs through the mantle wedge with strong attenuation so effects at the surface are muted. Indeed because propagation through the slab and the oceanic lithosphere are very efficient, small events in the Banda arc of Indonesia can be well recorded in northern Australia yet pass unnoticed in their immediate vicinity. The stochastic waveguide properties of the slab starts to make a significant influence on the character of propagation for events below 150 km depth. For shallower depths the former oceanic crust can act as an additional low-wavespeed guide to trap energy. Energy escapes from the slab as its curvature changes towards the arc, from both the former oceanic crust and the main body of the subduction zone. The loss of high frequencies from the slab allows such arrivals to reach the seismic stations on the fore-arc side of the volcanic front, whereas propagation to stations behind the volcanic front passes through the mantle wedge, and high frequencies are suppressed.

The character of the transition from back-arc to fore-arc stations is illustrated in Figure 13.17 for a profile of stations in-line with an earthquake at 230 km depth in

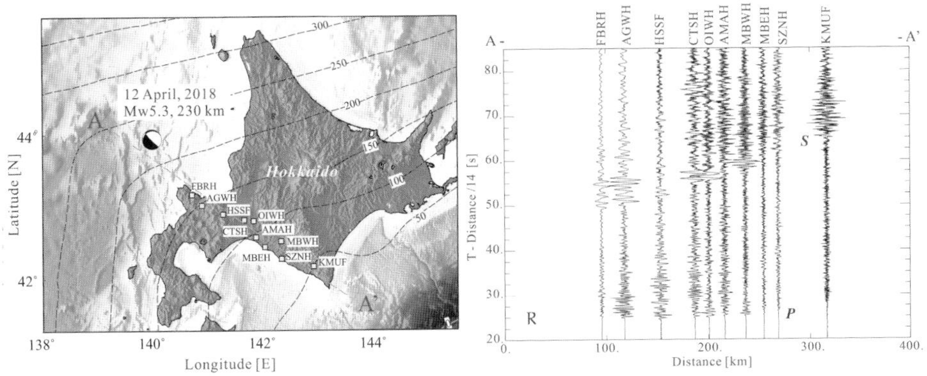

Figure 13.17 Seismic profile of broad-band radial ground velocity across Hokkaido, Japan, along the profile A–A′ from an earthquake at 230 km in the subducting Pacific Plate. [Courtesy of T. Furumura.]

the subducting Pacific Plate. This profile A–A′ crosses from behind the volcanic front to the fore-arc through the major sedimentary basin in the centre of Hokkaido. The westernmost stations show only low-requency direct arrivals. Sedimentary resonances enhance the central stations, but the high-frequency components build to the east. The easternmost station KMUF shows rather strong high-frequency content, particularly for *S* from slab-guided waves.

Seismic energy travelling outside the slab from deep earthquakes can make significant contributions to the overall ground motion (Chen M. et al. 2007), especially for lower frequencies ($<$ 1 Hz). The onset of lower-frequency energy travels by Fermat paths, i.e., the fastest possible path between source and receiver, and so constitutes the first arrival for each phase. For deep events the slab acts as a stochastic waveguide (Section 12.4.1) and considerable high-frequency energy can be carried to the surface, both directly up-slab and also traversing the subducting plate obliquely. Guided waves normally follow the lower-frequency phase onsets because they take a longer path even though most of their travel is in a higher-wavespeed medium.

The nature of the guided waves depends on the slab geometry. In the relatively simple configuration of the subducting Pacific plate in the Japanese region there is little change in the dip of the slab and the guided waves emerge near the surface as the plate bends towards the horizontal. In South America there are much stronger changes in slab morphology and also in the dip of individual sections of the subducting plate. As a result the role of guided waves in the former oceanic crust becomes much more important (Martin & Rietbrock, 2006).

As the hypocentral depth of the earthquakes increases a variety of phases at moderate epicentral distances can be produced from seismic waves that are radiated upwards from the source (Figure 13.18). For very deep events, there is a range of distances at which upgoing *S* can convert into *P* waves that travel in the

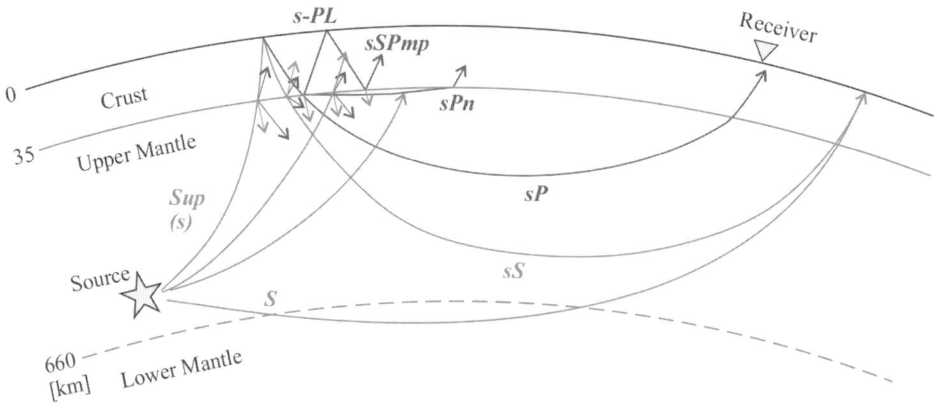

Figure 13.18 Schematic illustration of the raypaths for major seismic phases from a very deep source: upgoing *S* waves (*Sup*, *s*), downgoing *S* wave, surface reflected *sS* and *sP*, *sPn* waves, and multiple *sSPmp* reflections in the crust are indicated. Horizontally travelling *sSPmp* multiples in continental crust can couple with incoming *S* waves to develop the *s-PL* phase

crust as *sPg* or in the upper mantle as the *sPn* phase. For an event at 600 km depth, *sPn* becomes a precursor to *S* from about 8°, and can have significant amplitude if the source radiation pattern is favourable. These conversions to shallow travelling *P* have similar travel times to *S*, and can have notable interference with the *S* waves producing complex wavetrains on both vertical and radial components. Where the locus of conversion falls on thicker continental crust, *S* waves can be coupled into partially trapped *P* waves in the crust to produce long-period shear-coupled *PL* (*s-PL*). The influence of *sP* conversion on observed seismograms and the development of the other phases depend strongly on the properties of the crust in the conversion zone, which expands with increasing source depth (Kennett & Furumura, 2019).

Longer-period phases generated by large, very deep earthquakes can make a major contribution to sustaining substantial ground motion to considerable distances from the source, e.g., for the 2015 Ogasawara event at 680 km that produced JMA Intensity 1 or greater across the whole span of the Japanese islands.

As an illustration of the variety of behaviour seen from a single event we consider the very deep 2012 July 14 Off-Sakhalin event for which the peak horizontal ground velocity is displayed in Figure 12.10. Significant ground motion extends as far south as Tokyo, concentrated on the eastern sea board of Hokkaido and Honshu. In Figure 13.19 we display a record section from a chain of broadband F-net stations along the eastern seaboard of the Japanese islands extending down to Kyushu in the south. For this deep earthquake at 598 km depth, the path through the Pacific plate subduction zone to Japanese stations is oblique to the dip of the zone. High-frequency guided waves through the subducted slab are prominent in the Kurile section of the slab in Hokkaido, but are disrupted by the

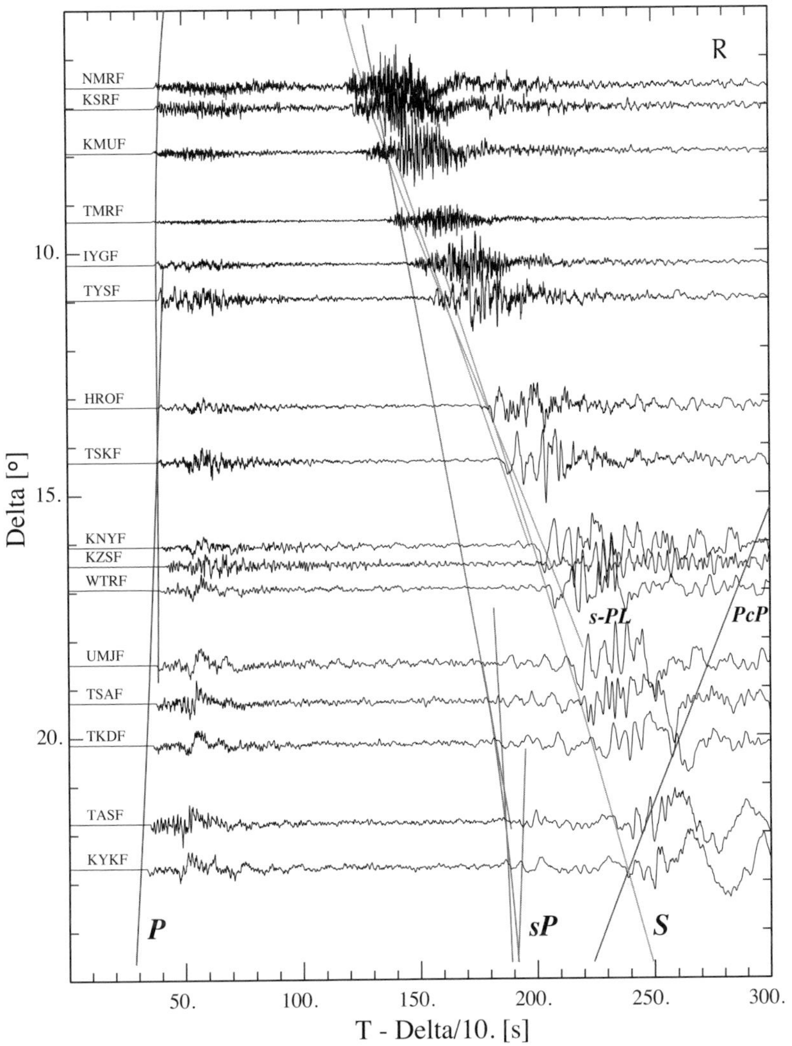

Figure 13.19 Radial-component seismograms for the 2012 Off-Sakhalin event (Mw 7.7, 598 km depth) at a chain of F-net seismic stations along the eastern coast of the Japanese islands. Traces are normalised by the maximum three-component energy in the time window. The travel times from the AK135 model are indicated.

complex geometry of the transition into the Honshu section (Miller et al., 2006). Guided *S* is the main contributor to ground motion in the east of Hokkaido, but in the west the *P*-wave amplitude exceeds *S* and produces the largest ground motion.

In Figure 13.19 we see high-frequency slab-guided waves at these eastern stations to around 12° from the source (at around 37°N), but at greater distances a variety of other seismic phases become prominent. Noticeable changes in seismic waveforms of *P* and *S* occur at around 14° from the source, as a switch occurs from the earlier arrival of energy that left the source upwards to that which has been refracted back from below the source level. The downward radiated waves interact with the '660 km' discontinuity to produce triplications in the arrivals.

The conversion zone for *sP* phases lies in the back-arc zone of the Pacific Plate descending below the Kurile Arc in the Sea of Okhotsk. The nature of the *sP* conversion depends on the location. Where this occurs closer to the Kurile Arc there can be strong attenuation of the *S*-wave leg below the Moho and the shallow part of the converted *P* wave path. The *sP* arrivals are quite strong on the southern stations in Figure 13.19, beyond 18° epicentral distance. Sufficient upgoing *S*-wave energy with a similar slowness to crustal *P* impinges on the continental crust of northern Hokkaido to produce the *s-PL* phase, whose low-frequency character develops by interference of many styles of multiples. Across Honshu (stations TMRF–TASF), the *s-PL* phase builds with distance and shows large amplitude relative to both *S* and *P*.

Part IV

Inversion for Earth Structure

14

Inference for Structure

In Part IV we consider the extraction of information about Earth structure from seismological results. The emphasis will be on the direct use of seismic waveforms, through many of the concepts discussed can also be applied to attributes of waveforms, such as travel times or amplitudes.

The early developments of Earth models relied primarily on the use of the travel times for major seismic phases (e.g., Dziewonski et al., 1975), so that it was necessary to develop in parallel methods of characterising seismic events in terms of position and origin time. With the steady enhancement of the seismological networks across the globe it became possible to consider 3-D models rather than summary models of radial structure (e.g., Aki & Lee, 1976; Dziewonski et al., 1977). Such seismic tomography using travel-time information, with the assumption of propagation along seismic rays, has played a major role in the understanding of the interior structure of the Earth, yet exploits only a small fraction of the information contained in seismograms. More information becomes available when the effects of the finite frequency of seismic arrivals are considered, and considerable effort has been expended to understand the sensitivity of such data to Earth structure (e.g., Yomogida, 1992; Dahlen et al., 2000; Yoshizawa & Kennett 2005). Rather than being confined to a narrow zone around the apparent propagation path, the influence zone encompasses a broader region whose size is inversely proportional to frequency. In the presence of triplications in seismic arrivals the shape of the influence zone becomes rather complicated.

The first class of tomographic inversions that started to use seismic waveforms at the global scale worked with perturbations from a radial reference model and used modal summation techniques, coupled to simple path-average approximations. When coupling between mode branches due to the effects of 3-D variations are included, it is possible to achieve an improved rendering of influence zones, and hence better structural results (e.g., Li & Romanowicz, 1995).

Improvements in computational power have meant that the goal of working directly with 3-D models using 3-D computations has become feasible. For the global scale, spectral-element methods have proved very effective (e.g., Komatitsch & Tromp, 2002a,b; Afanasiev et al., 2019). Though much work on waveform inversion for reflection data exploits finite-difference techniques (e.g., Moczo et al., 2000).

The goal now is to develop Earth models for which the observed seismic wavefield is adequately matched by simulations. The correspondence is assessed by some measure of misfit in conjunction with regularisation assumptions about the character of suitable models. Most current approaches depend on an optimisation procedure in a high-dimensional parameter space exploiting gradients of the misfit. Efficient derivative calculations using *adjoint* methods allow the development of large-scale inversions.

For waveform inversion, an important role is played by the numerical representation of the model and the frequency content, since these factors control how much computation is required for a single simulation of observations. Often assumptions are made about the relationships between, e.g., P wavespeed and S wavespeed to reduce the size of the model space. Where possible, it is preferable to limit such assumptions since they have the potential for imposing structure on the Earth.

14.1 Bayesian Framework

14.1.1 Data and Models

Consider a source of seismic waves recorded at many seismic stations across the globe (Figure 14.1). The observed seismograms contain coded information about the structures encountered as seismic waves spread out through the Earth, and across its surface. The objective of waveform inversion is to decipher this code, and for this we require information from many sources and stations so that structure is adequately sampled.

We can consider the complete suite of digitally recorded, i.e., discretely sampled, seismograms as a vector $\mathbf{d}^{\text{obs}}$ in a data space of dimension $\mathrm{N_D}$. We now want to find a representation of the Earth that is able to adequately reproduce these observations. We can consider our description of the physical properties of the Earth to be described by a continuous distribution of model parameters $\mathfrak{m}$. Yet, whatever computational procedure we may use, we will ultimately describe an Earth model with a finite number of discrete parameters that constitute a vector $\mathbf{m}$ in an $\mathrm{N_M}$ dimensional model space. The vector $\mathbf{m}$ can include parameters corresponding to different physical quantities, e.g., P wavespeed, S wavespeed, and density. For each realisation of the model $\mathbf{m}$, we can carry out a forward calculation to produce simulated data $\mathbf{d}$.

The data vector $\mathbf{d}$ will, in general, be a complicated functional of $\mathbf{m}$,

$$\mathbf{d} = \mathbf{G}[\mathbf{m}], \tag{14.1.1}$$

and the mapping between model and data space implied by (14.1.1) will be nonlinear. In order to assess how far the simulated $\mathbf{d}$ matches the observations $\mathbf{d}^{\text{obs}}$, we need to introduce some measure $\Phi(\mathbf{d}, \mathbf{d}^{\text{obs}})$ of the discrepancy that

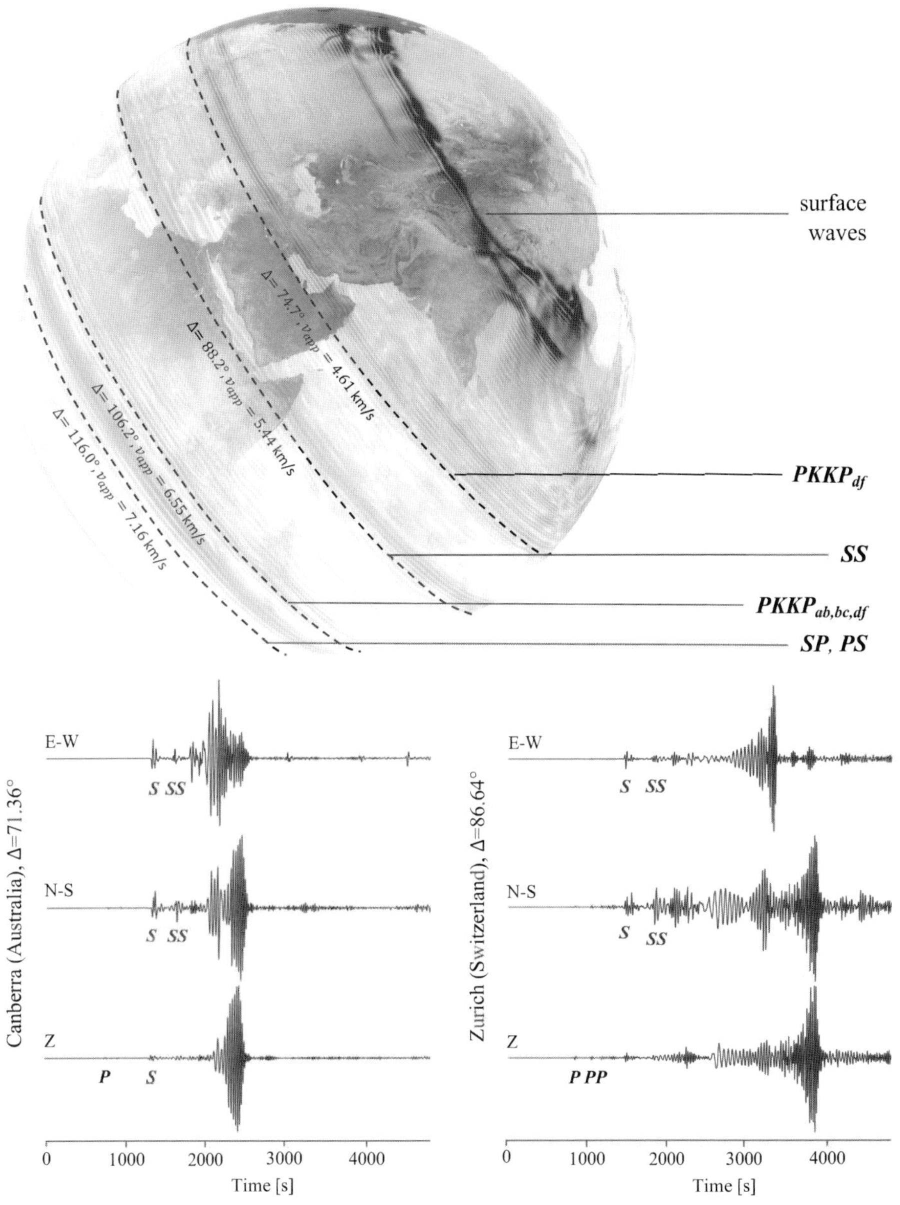

Figure 14.1 Wavefield snapshot and time series computed with a spectral-element solver of the seismic wave equation (Afanasiev et al., 2019) for the 2011 Tohoku (Japan) earthquake with a dominant period of 30 s. Topography and crustal depth (Laske et al., 2013), superimposed in the spherically symmetric AK135 model (Kennett et al., 1995), are taken into account in the simulation. The snapshot, taken 30 min. after the event origin time, shows the wavefield amplitude at the surface. It is dominated by high-amplitude surface waves, the *SS*, *SP*, and *PS* phases, as well as by several *PKKP* branches. Dashed curves are great circles at selected epicentral distances, and corresponding to various apparent velocities v_{app}. The three-component time series are for stations in Canberra (Australia) and Zurich (Switzerland).

builds on our understanding of the character of the observations and their inter-dependencies. A wide variety of forms can be chosen for Φ depending on the way in which we seek to compare observed and simulated seismograms (cf. § 5.3). We can illustrate the process by introducing the $N_D \times N_D$ data covariance matrix $\mathbf{C}_D$, whose diagonal entries are just the variances associated with individual observations and whose off-diagonal terms allow for cross-coupling. A simple measure of data match is then provided by the quadratic form

$$\Phi(\mathbf{d}, \mathbf{d}^{obs}) = \tfrac{1}{2}(\mathbf{d} - \mathbf{d}^{obs})^T \mathbf{C}_D^{-1} (\mathbf{d} - \mathbf{d}^{obs}). \tag{14.1.2}$$

Now when we undertake a simulation for a model $\mathbf{m}$, we can associate it with a data-mismatch value $\Phi(\mathbf{G}[\mathbf{m}], \mathbf{d}^{obs})$. For multiple realisations of the model, we can think of a cluster of points in model space whose colour is specified by Φ, e.g., red for a good model and blue for a poor match. Better data match will show up as a red zone, but so far no restrictions have been placed on the character of $\mathbf{m}$. Not all models whose simulations provide a close match to data will have acceptable properties. We need therefore to build on our *a priori* understanding of suitable models. We may well have a preferred prior model $\mathbf{m}^{prior}$, and can build around this a prior probability distribution. We introduce a function $\Psi(\mathbf{m}, \mathbf{m}^{prior})$ to judge the suitability of a proposed model. Once again, a simple measure is to assign the quadratic form

$$\Psi(\mathbf{m}, \mathbf{m}^{prior}) = \tfrac{1}{2}(\mathbf{m} - \mathbf{m}^{prior})^T \mathbf{C}_M^{-1} (\mathbf{m} - \mathbf{m}^{prior}), \tag{14.1.3}$$

where now we build in our requirements on the models via an $N_M \times N_M$ model covariance matrix $\mathbf{C}_M$. The use of (14.1.3) is equivalent to taking a Gaussian prior probability in model space.

Returning to our cloud of points in model space, we can add the attribute Ψ by modulating the intensity of the points so that large Ψ corresponds to weak illumination. We then seek bright points with a red hue, since they meet our combined specifications for good data match and suitable model character. Indeed, the entire model cloud can then be regarded as a visualisation of an *a posteriori* probability distribution, combining data constraints and our prejudices about the nature of the model based on the composite misfit function

$$\chi(\mathbf{d}, \mathbf{m}) = \Phi(\mathbf{d}, \mathbf{d}^{obs}) + \Psi(\mathbf{m}, \mathbf{m}^{prior}). \tag{14.1.4}$$

14.1.2 Application of Bayes' Theorem

We can impose our requirements on the model $\mathbf{m}$ by a prior probability distribution

$$P(\mathbf{m}) = k_m e^{-\Psi(\mathbf{m}, \mathbf{m}^{prior})}, \tag{14.1.5}$$

which expresses the distribution associated with multiple realisations of the model $\mathbf{m}$. The constant k_m is chosen to normalise the integral of $P(\mathbf{m})$ over all models $\mathbf{m}$. In a similar way we can envisage selecting multiple sets of observations $\mathbf{d}^{obs}$ from the available set of seismic sources so that, in principle, we can specify a probability distribution $P(\mathbf{d}^{obs})$.

With the mapping from model to data space through $\mathbf{G}[\mathbf{m}]$, we can assign a conditional probability of producing the observed data given $\mathbf{m}$,

$$\mathsf{P}(\mathbf{d}^{\mathrm{obs}} \,|\, \mathbf{m}) = \mathsf{k}_{\mathsf{d}} \mathsf{e}^{-\Phi(\mathbf{G}[\mathbf{m}],\mathbf{d}^{\mathrm{obs}})}, \tag{14.1.6}$$

where k_{d} is again a normalisation constant. This probability distribution depends on the measure of data match $\Phi(\mathbf{G}[\mathbf{m}], \mathbf{d}^{\mathrm{obs}})$ introduced in (14.1.2). We can think of (14.1.5) as specifying a prior probability distribution in data space.

From a complete suite of realisations of the model $\mathbf{m}$ we map out the influence of the observations on model space to produce a conditional probability $\mathsf{P}(\mathbf{m} \,|\, \mathbf{d}^{\mathrm{obs}})$ on the model $\mathbf{m}$ given the observations $\mathbf{d}^{\mathrm{obs}}$. Now invoking Bayes' theorem we can express this conditional probability in terms of the probability in data space and the prior constraints on the model,

$$\mathsf{P}(\mathbf{m} \,|\, \mathbf{d}^{\mathrm{obs}}) = \frac{\mathsf{P}(\mathbf{d}^{\mathrm{obs}} \,|\, \mathbf{m})\mathsf{P}(\mathbf{m})}{\mathsf{P}(\mathbf{d}^{\mathrm{obs}})}. \tag{14.1.7}$$

We identify the conditional probability $\mathsf{P}(\mathbf{m} \,|\, \mathbf{d}^{\mathrm{obs}})$ with the *a posteriori* probability distribution for $\mathbf{m}$,

$$\mathsf{P}(\mathbf{m} \,|\, \mathbf{d}^{\mathrm{obs}}) = \mathsf{C}\mathsf{e}^{\Phi(\mathbf{G}[\mathbf{m}],\mathbf{d}^{\mathrm{obs}})+\Psi(\mathbf{m},\mathbf{m}^{\mathrm{prior}})} = \mathsf{C}\mathsf{e}^{-\chi(\mathbf{m},\mathbf{d}^{\mathrm{obs}})}, \tag{14.1.8}$$

where the constant C is associated with $\mathsf{P}(\mathbf{d}^{\mathrm{obs}})$. The composite misfit function

$$\chi(\mathbf{m}, \mathbf{d}^{\mathrm{obs}}) = \Phi(\mathbf{G}[\mathbf{m}], \mathbf{d}^{\mathrm{obs}}) + \Psi(\mathbf{m}, \mathbf{m}^{\mathrm{prior}}) \tag{14.1.9}$$

depends on the match between observations $\mathbf{d}^{\mathrm{obs}}$ and simulations $\mathbf{d}$ computed from $\mathbf{m}$ via the solution of the forward modelling equations, and also the influence of our assumptions about the nature of acceptable models.

Though (14.1.8) and (14.1.9) have been derived for the special case of Gaussian probabilities, other functional forms may be used to describe observational errors or prior information on the model more adequately.

14.1.3 Bayesian Inference via Monte Carlo Techniques

The posterior $\mathsf{P}(\mathbf{m} \,|\, \mathbf{d}^{\mathrm{obs}})$ is an N_{M}-dimensional probability density function (PDF) that cannot normally be represented explicitly. Hence, quantities of interest, such as means, variances, covariances, and marginal PDFs, are typically approximated by point-wise evaluations of $\mathsf{P}(\mathbf{m} \,|\, \mathbf{d}^{\mathrm{obs}})$ for specific test models $\mathbf{m}$.

In principle, the systematic evaluation of $\mathsf{P}(\mathbf{m} \,|\, \mathbf{d}^{\mathrm{obs}})$ using a regular grid of test models, provides an arbitrarily close approximation of the posterior distribution, or any quantity derived from it. However, with increasing model-,space dimension, the number of test models grows exponentially, and with it the computational requirement for such a grid search. Furthermore, the likelihood of finding plausible models that explain the observations acceptably well, decreases super-exponentially with increasing dimension; a phenomenon colloquially referred to as the *curse of dimensionality*.

The infeasibility of grid search for problems whose size is not even close to the ones relevant for seismic inversion has been a major motivation for the development of techniques where test models are drawn randomly. Alluding to the randomness of success when gambling in a casino, such approaches are commonly referred to as Monte Carlo techniques. Their advantage, compared to grid search, lies in the ability to draw random models preferentially in regions of model space where the more probable models are located. This *importance sampling* therefore helps to avoid the squandering of computational resources in the less-relevant regions of model space.

In the context of geophysical inverse problems, numerous Monte Carlo methods have been developed in the past few decades, including adaptations of the Metropolis-Hastings algorithm (Mosegaard & Tarantola, 1995), the Neighbourhood algorithm (Sambridge, 1999), or the Reversible-Jump algorithm for the sampling of model spaces with variable dimension (Sambridge, 2006).

Despite progress in Monte Carlo algorithms and computational power, the dimension of model spaces that we can explore by random sampling is often limited to around 100, as an order of magnitude. Since this is far below the dimension relevant for most seismic inverse problems, the following sections will focus on deterministic (non-random) methods that aim to find a model that is compatible with the observations. While being computationally less demanding than Monte Carlo sampling, such deterministic methods generally provide less complete (or no) information on the quality and uniqueness of that model. A possible future direction, where random and deterministic approaches are combined, will be presented in Section 19.4.

14.2 Linear and Linearizable Problems

14.2.1 The Least-Squares Solution

If we can adopt the simple quadratic forms (14.1.2) and (14.1.3) for the data match and model suitability, we have Gaussian probability distributions in data and model space. With the further assumption that the relation between data space and model space is linear so that $\mathbf{d} = \mathbf{Gm}$, all probability distributions in model space are Gaussian and we achieve relatively simple results, corresponding to a least-squares treatment.

The posterior distribution from Bayes' theorem (14.1.8) for this linear situation has the form

$$\mathsf{P}(\mathbf{m}\,|\,\mathbf{d}^{\mathrm{obs}}) = \mathsf{C}e^{-\frac{1}{2}(\mathbf{m}-\mathbf{m}^{\mathrm{prior}})^{\mathrm{T}}\mathbf{C}_{\mathsf{M}}^{-1}(\mathbf{m}-\mathbf{m}^{\mathrm{prior}})-\frac{1}{2}(\mathbf{Gm}-\mathbf{d}^{\mathrm{obs}})^{\mathrm{T}}\mathbf{C}_{\mathsf{D}}^{-1}(\mathbf{Gm}-\mathbf{d}^{\mathrm{obs}})}, \tag{14.2.1}$$

where C is a constant. For multiple reasons, (14.2.1) is a remarkable equation. Firstly, we note that the maximum-likelihood model $\tilde{\mathbf{m}}$ may be found by minimising the *misfit function*

$$\chi(\mathbf{m}) = \tfrac{1}{2}(\mathbf{m}-\mathbf{m}^{\mathrm{prior}})^{\mathrm{T}}\mathbf{C}_{\mathsf{M}}^{-1}(\mathbf{m}-\mathbf{m}^{\mathrm{prior}}) + \tfrac{1}{2}(\mathbf{Gm}-\mathbf{d}^{\mathrm{obs}})^{\mathrm{T}}\mathbf{C}_{\mathsf{D}}^{-1}(\mathbf{Gm}-\mathbf{d}^{\mathrm{obs}}). \tag{14.2.2}$$

This approach corresponds to finding the minimum of the weighted square of the differences between observed data $\mathbf{d}^{\text{obs}}$ and synthetically computed data,

$$(\mathbf{Gm} - \mathbf{d}^{\text{obs}})^{\text{T}}\mathbf{C}_{\text{D}}^{-1}(\mathbf{Gm} - \mathbf{d}^{\text{obs}}),$$

plus an additional quadratic term that captures our preference for the prior model,

$$(\mathbf{m} - \mathbf{m}^{\text{prior}})\mathbf{C}_{\text{M}}^{-1}(\mathbf{m} - \mathbf{m}^{\text{prior}}).$$

In other words, finding the maximum-likelihood model corresponds to the solution of a *least-squares problem*.

Secondly, we note that (14.2.2) is quadratic in $\mathbf{m}$ and may therefore be rewritten in the form

$$\chi(\mathbf{m}) = \tfrac{1}{2}\,(\mathbf{m} - \tilde{\mathbf{m}})^{\text{T}}\,\tilde{\mathbf{C}}_{\text{M}}^{-1}\,(\mathbf{m} - \tilde{\mathbf{m}}) + \text{c}, \tag{14.2.3}$$

where the constant c absorbs all terms that do not depend on the model parameter vector $\mathbf{m}$. Now the posterior distribution $\text{P}(\mathbf{m}\,|\,\mathbf{d}^{\text{obs}}) = \text{C}e^{-\chi(\mathbf{m})}$ is once again a pure Gaussian in the model parameters $\mathbf{m}$, with *posterior model mean* $\tilde{\mathbf{m}}$ and *posterior model covariance* $\tilde{\mathbf{C}}_{\text{M}}$.

Being able to compute $\tilde{\mathbf{m}}$ and $\tilde{\mathbf{C}}_{\text{M}}$ exactly is particularly convenient, because these two quantities fully describe the posterior distribution. Hence the posterior distribution can be expressed analytically in closed form. This property eliminates the need to approximate the posterior distribution, for instance, by sampling with the help of Monte Carlo methods, see Section 14.1.3. Explicitly, the posterior mean $\tilde{\mathbf{m}}$ and the posterior covariance $\tilde{\mathbf{C}}_{\text{M}}$ are given by

$$\tilde{\mathbf{m}} = \left(\mathbf{G}^{\text{T}}\mathbf{C}_{\text{D}}^{-1}\mathbf{G} + \mathbf{C}_{\text{M}}^{-1}\right)^{-1}\left(\mathbf{G}^{\text{T}}\mathbf{C}_{\text{D}}^{-1}\mathbf{d}^{\text{obs}} + \mathbf{C}_{\text{M}}^{-1}\mathbf{m}^{\text{prior}}\right), \tag{14.2.4}$$

and

$$\tilde{\mathbf{C}}_{\text{M}} = \left(\mathbf{G}^{\text{T}}\mathbf{C}_{\text{D}}^{-1}\mathbf{G} + \mathbf{C}_{\text{M}}^{-1}\right)^{-1}. \tag{14.2.5}$$

The proof of (14.2.4) and (14.2.5) directly from (14.2.2) and (14.2.3) is a rather tedious exercise. Instead, we work in the opposite direction, and insert (14.2.4) and (14.2.5) into (14.2.3) to extract recognisable forms. With an expansion of (14.2.3), we obtain

$$\chi(\mathbf{m}) = \tfrac{1}{2}\mathbf{m}^{\text{T}}\tilde{\mathbf{C}}_{\text{M}}^{-1}\mathbf{m} - \mathbf{m}^{\text{T}}\tilde{\mathbf{C}}_{\text{M}}^{-1}\tilde{\mathbf{m}} + \text{c}. \tag{14.2.6}$$

The substitution of (14.2.4) and (14.2.5) into (14.2.6), yields

$$\chi(\mathbf{m}) = \tfrac{1}{2}\mathbf{m}^{\text{T}}\left(\mathbf{G}^{\text{T}}\mathbf{C}_{\text{D}}^{-1}\mathbf{G} + \mathbf{C}_{\text{M}}^{-1}\right)\mathbf{m} - \mathbf{m}^{\text{T}}\left(\mathbf{G}^{\text{T}}\mathbf{C}_{\text{D}}^{-1}\mathbf{d}^{\text{obs}} - \mathbf{C}_{\text{M}}^{-1}\mathbf{m}^{\text{prior}}\right) + \text{c}', \tag{14.2.7}$$

where c' is a new constant containing all terms that are independent of $\mathbf{m}$. Now we note that an expansion of (14.2.2) gives exactly the same expression, which demonstrates the correctness of (14.2.4) and (14.2.5). Although (14.2.4) and

(14.2.5) are formally correct, there are various more useful and interpretable forms. For instance, we may rewrite $\tilde{\mathbf{m}}$ as follows:

$$\begin{aligned}
\tilde{\mathbf{m}} &= \tilde{\mathbf{C}}_\mathrm{M}\left(\mathbf{G}^\mathrm{T}\mathbf{C}_\mathrm{D}^{-1}\mathbf{d}^\mathrm{obs} + \mathbf{C}_\mathrm{M}^{-1}\mathbf{m}^\mathrm{prior}\right) = \tilde{\mathbf{C}}_\mathrm{M}\mathbf{G}^\mathrm{T}\mathbf{C}_\mathrm{D}^{-1}\mathbf{d}^\mathrm{obs} + \tilde{\mathbf{C}}_\mathrm{M}\mathbf{C}_\mathrm{M}^{-1}\mathbf{m}^\mathrm{prior}, \\
&= \tilde{\mathbf{C}}_\mathrm{M}\mathbf{G}^\mathrm{T}\mathbf{C}_\mathrm{D}^{-1}\mathbf{d}^\mathrm{obs} + \tilde{\mathbf{C}}_\mathrm{M}\left(\mathbf{G}^\mathrm{T}\mathbf{C}_\mathrm{D}^{-1}\mathbf{G} + \mathbf{C}_\mathrm{M}^{-1} - \mathbf{G}^\mathrm{T}\mathbf{C}_\mathrm{D}^{-1}\mathbf{G}\right)\mathbf{m}^\mathrm{prior}, \\
&= \tilde{\mathbf{C}}_\mathrm{M}\mathbf{G}^\mathrm{T}\mathbf{C}_\mathrm{D}^{-1}\mathbf{d}^\mathrm{obs} + \tilde{\mathbf{C}}_\mathrm{M}\left(\tilde{\mathbf{C}}_\mathrm{M}^{-1} - \mathbf{G}^\mathrm{T}\mathbf{C}_\mathrm{D}^{-1}\mathbf{G}\right)\mathbf{m}^\mathrm{prior}, \\
&= \mathbf{m}^\mathrm{prior} + \tilde{\mathbf{C}}_\mathrm{M}\mathbf{G}^\mathrm{T}\mathbf{C}_\mathrm{D}^{-1}\left(\mathbf{d}^\mathrm{obs} - \mathbf{G}\mathbf{m}^\mathrm{prior}\right). \qquad (14.2.8)
\end{aligned}$$

We see that the final form in (14.2.7) expresses the posterior mean $\tilde{\mathbf{m}}$ in terms of an update to the prior mean $\mathbf{m}^\mathrm{prior}$. The update is controlled by the differences between observed and calculated data, $\mathbf{d}^\mathrm{obs} - \mathbf{G}\mathbf{m}^\mathrm{prior}$. When the prior measurement errors are large, so that the entries of $\mathbf{C}_\mathrm{D}^{-1}$ are small, the update will be small, and vice versa. Using the matrix identities (14.A.2) and (14.A.3) from Appendix 14.A, we arrive at a set of equivalent expressions for the posterior mean

$$\begin{aligned}
\tilde{\mathbf{m}} &= \left(\mathbf{G}^\mathrm{T}\mathbf{C}_\mathrm{D}^{-1}\mathbf{G} + \mathbf{C}_\mathrm{M}^{-1}\right)^{-1}\left(\mathbf{G}^\mathrm{T}\mathbf{C}_\mathrm{D}^{-1}\mathbf{d}^\mathrm{obs} + \mathbf{C}_\mathrm{M}^{-1}\mathbf{m}^\mathrm{prior}\right), \\
&= \mathbf{m}^\mathrm{prior} + \tilde{\mathbf{C}}_\mathrm{M}\mathbf{G}^\mathrm{T}\mathbf{C}_\mathrm{D}^{-1}\left(\mathbf{d}^\mathrm{obs} - \mathbf{G}\mathbf{m}^\mathrm{prior}\right), \\
&= \mathbf{m}^\mathrm{prior} + \mathbf{C}_\mathrm{M}\mathbf{G}^\mathrm{T}\left(\mathbf{C}_\mathrm{D} + \mathbf{G}\mathbf{C}_\mathrm{M}\mathbf{G}^\mathrm{T}\right)^{-1}\left(\mathbf{d}^\mathrm{obs} - \mathbf{G}\mathbf{m}^\mathrm{prior}\right), \qquad (14.2.9)
\end{aligned}$$

and the posterior covariance,

$$\begin{aligned}
\tilde{\mathbf{C}}_\mathrm{M} &= \left(\mathbf{G}^\mathrm{T}\mathbf{C}_\mathrm{D}^{-1}\mathbf{G} + \mathbf{C}_\mathrm{M}^{-1}\right)^{-1}, \\
&= \mathbf{C}_\mathrm{M} - \mathbf{C}_\mathrm{M}\mathbf{G}^\mathrm{T}\left(\mathbf{C}_\mathrm{D} + \mathbf{G}\mathbf{C}_\mathrm{M}\mathbf{G}^\mathrm{T}\right)^{-1}\mathbf{G}\mathbf{C}_\mathrm{M}. \qquad (14.2.10)
\end{aligned}$$

As in (14.2.7), equation (14.2.10) represents the posterior quantity (the covariance) as a modification of the prior quantity. We note that for any unit vector $\mathbf{e}$,

$$\mathbf{e}^\mathrm{T}\tilde{\mathbf{C}}_\mathrm{M}\mathbf{e} = \mathbf{e}^\mathrm{T}\mathbf{C}_\mathrm{M}\mathbf{e} - (\mathbf{G}\mathbf{C}_\mathrm{M}\mathbf{e})^\mathrm{T}\left(\mathbf{C}_\mathrm{D} + \mathbf{G}\mathbf{C}_\mathrm{M}\mathbf{G}^\mathrm{T}\right)^{-1}(\mathbf{G}\mathbf{C}_\mathrm{M}\mathbf{e}) < \mathbf{e}^\mathrm{T}\mathbf{C}_\mathrm{M}\mathbf{e}, \quad (14.2.11)$$

because $\left(\mathbf{C}_\mathrm{D} + \mathbf{G}\mathbf{C}_\mathrm{M}\mathbf{G}^\mathrm{T}\right)$ is a positive definite matrix. This result ensures that the posterior variances $\mathbf{e}^\mathrm{T}\tilde{\mathbf{C}}_\mathrm{M}\mathbf{e}$ will always be smaller than the prior variances $\mathbf{e}^\mathrm{T}\mathbf{C}_\mathrm{M}\mathbf{e}$. The reduction in variance is what we expect from the addition of new information.

It may seem excessive to provide a range of different expressions for $\tilde{\mathbf{m}}$ and $\tilde{\mathbf{C}}_\mathrm{M}$ as in (14.2.9) and (14.2.10). However, there can be significant computational advantages in choosing particular forms to suit a specific situation, depending on the balance of data and model parameters. For a data vector of dimension $\mathrm{N_D}$ and model vector of dimension $\mathrm{N_M}$, we note that $\left(\mathbf{G}^\mathrm{T}\mathbf{C}_\mathrm{D}^{-1}\mathbf{G} + \mathbf{C}_\mathrm{M}^{-1}\right)$ in (14.2.10) is an $\mathrm{N_M} \times \mathrm{N_M}$ matrix that includes the inverse of the prior model covariance $\mathbf{C}_\mathrm{M}$ that is itself an $\mathrm{N_M} \times \mathrm{N_M}$ matrix. In contrast, $\left(\mathbf{C}_\mathrm{D} + \mathbf{G}\mathbf{C}_\mathrm{M}\mathbf{G}^\mathrm{T}\right)$ is a matrix of dimension $\mathrm{N_D} \times \mathrm{N_D}$ that can be computed without any matrix inversion. Thus, depending on the relative sizes of the data and model spaces, different expressions will minimise computational demands.

All the equations we have given for the maximum-likelihood model and the posterior covariance involve matrix inverses. We have so far implicitly assumed that such inverses exist. But, if we make choices of $\mathbf{C}_\mathrm{D}$ and $\mathbf{C}_\mathrm{M}$ that honestly reflect our prior knowledge, we cannot guarantee that the matrices can be inverted. To ensure that the matrix forms are invertible and well-conditioned, prior distributions are often designed pragmatically in order to secure a definite result. For example Tarantola & Valette (1982) proposed a prior model covariance matrix of the form

$$(\mathbf{C}_\mathrm{M})_{ij} = \sigma_\mathrm{M}^2 e^{-|\mathbf{x}_i-\mathbf{x}_j|^2/2\lambda^2}. \tag{14.2.12}$$

Here, the *damping parameter* σ_M controls the extent to which we force $\tilde{\mathbf{m}}$ to resemble $\mathbf{m}^\mathrm{prior}$, and the *smoothing parameter* λ can be used to make $\tilde{\mathbf{m}}$ smoother or rougher. Replacing our full understanding of the character of models by an expression such as (14.2.12) in order to ensure the numerical invertibility of a matrix is referred to as a *regularisation* procedure.

The introduction of regularisation is generally an act of pragmatism, but is frequently unavoidable when the more holistic Bayesian approach is computationally too expensive, even when combined with Monte Carlo methods. Regularisation is almost by definition *ad hoc*. For instance, the choice of the functional form of the prior model covariance and the regularisation parameters is subjective. This subjectivity leaves an imprint on the results and should be taken into account when the results are interpreted. One of the undesirable effects of regularisation is that the posterior uncertainties may not be entirely meaningful. Injecting artificial prior knowledge to make a matrix invertible tends to decrease the apparent uncertainties. Specifically in the case of travel-time tomography, slowness in a cell that is not crossed by any ray is entirely undetermined. Still, through regularisation, it is possible to find a finite posterior variance. In summary, regularisation helps us to produce some model even though we have introduced a subjective component in the inversion. To quote words of the British statistician George Box (1919–2013), 'All models are wrong, but some are useful.'

14.2.2 Model and Data Resolution

Model resolution

To gain further understanding of the least-squares solution, we make the assumption that the observations $\mathbf{d}^\mathrm{obs}$ can be found from the solution of a forward problem with a target model $\mathbf{m}^\mathrm{target}$ as input,

$$\mathbf{d}^\mathrm{obs} = \mathbf{G}\mathbf{m}^\mathrm{target} + \varepsilon. \tag{14.2.13}$$

The additional term ε captures both observational and modelling errors. The errors in modelling arise from the simplifications introduced into the mathematical model $\mathbf{G}$ compared with the true physical processes that generate $\mathbf{d}^\mathrm{obs}$. We introduce $\mathbf{d}^\mathrm{prior}$ as the synthetic data computed with the prior mean model,

$$\mathbf{d}^\mathrm{prior} = \mathbf{G}\mathbf{m}^\mathrm{prior}. \tag{14.2.14}$$

Then,

$$\Delta \mathbf{d} = \mathbf{G}\,\Delta \mathbf{m}^{\text{target}} + \varepsilon, \tag{14.2.15}$$

with the data residuals $\Delta \mathbf{d} = \mathbf{d}^{\text{obs}} - \mathbf{d}^{\text{prior}}$, and the difference from the target model $\Delta \mathbf{m}^{\text{target}} = \mathbf{m}^{\text{target}} - \mathbf{m}^{\text{prior}}$. Furthermore, from the third option in (14.2.9), we find that the difference between the model estimate we can recover and the prior $\Delta \tilde{\mathbf{m}} = \tilde{\mathbf{m}} - \mathbf{m}^{\text{prior}}$ and the data difference $\Delta \mathbf{d}$ are related by

$$\Delta \tilde{\mathbf{m}} = \mathbf{G}^{-g}\,\Delta \mathbf{d}, \tag{14.2.16}$$

in terms of the *generalised inverse* $\mathbf{G}^{-g}$ defined as

$$\mathbf{G}^{-g} = \mathbf{C}_M \mathbf{G}^T \left(\mathbf{C}_D + \mathbf{G}\mathbf{C}_M \mathbf{G}^T\right)^{-1}. \tag{14.2.17}$$

Substituting $\Delta \mathbf{d}$ from (14.2.15) into (14.2.17), we find an explicit relation between the differences between the model estimate and the target model in terms of the prior model,

$$\Delta \tilde{\mathbf{m}} = \mathbf{G}^{-g}\mathbf{G}\,\Delta \mathbf{m}^{\text{target}} + \mathbf{G}^{-g}\varepsilon = \mathbf{R}_M\,\Delta \mathbf{m}^{\text{target}} + \mathbf{G}^{-g}\varepsilon. \tag{14.2.18}$$

We have here introduced the *model resolution matrix*

$$\mathbf{R}_M = \mathbf{G}^{-g}\mathbf{G} = \mathbf{C}_M \mathbf{G}^T \left(\mathbf{C}_D + \mathbf{G}\mathbf{C}_M \mathbf{G}^T\right)^{-1} \mathbf{G}, \tag{14.2.19}$$

sometimes termed simply the *resolution matrix*. This matrix establishes a linear relation between the target model $\Delta \mathbf{m}^{\text{target}}$ that we wish to recover, and the usually imperfect estimate $\Delta \tilde{\mathbf{m}}$ that is obtained from inversion as a result of limited data coverage. The second term on the right-hand side in (14.2.18) represents *error propagation*, that is, the direct imprint of observational errors on the estimated model.

In the hypothetical case of perfect recovery, we have $\Delta \tilde{\mathbf{m}} = \Delta \mathbf{m}^{\text{target}}$, so that the resolution matrix is equal to the identity matrix, $\mathbf{R}_M = \mathbf{I}$, and then the trace of the resolution matrix is equal to the number of model parameters, $\operatorname{tr} \mathbf{R}_M = N_M$. In realistic scenarios, the resolution matrix $\mathbf{R}_M$ differs from $\mathbf{I}$, and $\operatorname{tr} \mathbf{R}_M$ is interpreted as the *effective number of parameters that can be resolved*.

To provide a more detailed interpretation of the resolution matrix, we temporarily omit the error propagation term $\mathbf{G}^{-g}\varepsilon$. We now write (14.2.18) in index notation,

$$\Delta \tilde{m}_i = \sum_{j=1}^{N_M} (\mathbf{R}_M)_{ij}\,\Delta m_j^{\text{target}}. \tag{14.2.20}$$

Thus the i-component of the estimated model difference, $\Delta \tilde{m}_i$, is a sum over the target model differences $\Delta m_j^{\text{target}}$, with the entries in the ith row of $\mathbf{R}_M$ acting as weights. In this sense, $\Delta \tilde{m}_i$ is a weighted average of the $\Delta m_j^{\text{target}}$, with the rows of the resolution matrix acting as *averaging kernels*. In addition we can assign a specific meaning to the columns of $\mathbf{R}_M$. Consider the scenario where we wish to recover a model that differs from the prior model only in one parameter, that is, $\Delta m_j^{\text{target}} = m_j^{\text{target}} - m_j^{\text{prior}} = \delta_{jk}$, for a fixed index k. Inserting $\Delta m_j^{\text{target}}$ into (14.2.20), we find

$$\Delta \tilde{m}_i = (\mathbf{R}_M)_{ik} \,. \qquad (14.2.21)$$

Hence the recovered model, $\Delta \tilde{m}_i$, has non-zero components for indices $i \neq k$, unless we are in the situation of perfect resolution with $\mathbf{R}_M = \mathbf{I}$. The target $\Delta m_j^{\text{target}}$ would be non-zero only for index $j = k$, but is *blurred* or *smeared* into a broader distribution by the action of the kth column of the resolution matrix $\mathbf{R}_M$. The columns of $\mathbf{R}_M$ are therefore known as *point-spread functions*.

The concept of model resolution deserves a little more discussion. Firstly, the relation (14.2.16) has broad application. In our specific context, we defined the generalised inverse $\mathbf{G}^{-g}$ as in (14.2.17), which follows from the least-squares solution. However, other methods to estimate a model may lead to different expressions for $\mathbf{G}^{-g}$. In these cases, the model resolution matrix is still given generically by $\mathbf{R}_M = \mathbf{G}^{-g}\mathbf{G}$.

Secondly, (14.2.19) indicates that the computation of the resolution matrix requires the computation of a matrix inverse, $\left(\mathbf{C}_D + \mathbf{G}\mathbf{C}_M\mathbf{G}^T\right)^{-1}$. Once again, when we make an honest assessment of prior knowledge – which may not be particularly strong – the required inverse may not exist. Hence, as in the case of the posterior mean model, we may be required to regularise the problem by choosing artificial values for $\mathbf{C}_D$ and $\mathbf{C}_M$. Such choices mean that the regularised resolution may be overly optimistic and not represent well the true state of resolution.

Finally, we remark that the concept of model resolution, as introduced above, only applies directly to synthetic inversions, meaning that we build in an *inverse crime*. Equation (14.2.13) does not strictly apply to actual observed data, since we rely on necessarily simplified forward modelling theory $\mathbf{G}$. We have to assume that we can take the results from the simulated case to the real observations, so resolution estimated with the help of $\mathbf{R}_M$ may be rather optimistic.

Data resolution

With the solution of a least-squares problem, we may compute estimated data residuals $\Delta\mathbf{d}^{\text{est}}$ by applying the forward modelling matrix $\mathbf{G}$ to the estimated model difference $\Delta\tilde{\mathbf{m}}$, in a similar way to (14.2.15),

$$\Delta\mathbf{d}^{\text{est}} = \mathbf{G}\,\Delta\tilde{\mathbf{m}}. \qquad (14.2.22)$$

Substituting $\tilde{\mathbf{m}}$ from (14.2.16), we obtain

$$\Delta\mathbf{d}^{\text{est}} = \mathbf{G}\mathbf{G}^{-g}\,\Delta\mathbf{d}^{\text{obs}} = \mathbf{R}_D\Delta\mathbf{d}^{\text{obs}}, \qquad (14.2.23)$$

where the *data resolution matrix*

$$\mathbf{R}_D = \mathbf{G}\mathbf{G}^{-g} = \mathbf{G}\,\mathbf{C}_M\mathbf{G}^T\left(\mathbf{C}_D + \mathbf{G}\mathbf{C}_M\mathbf{G}^T\right)^{-1}. \qquad (14.2.24)$$

Formally, the data resolution matrix can be interpreted in a similar way to the model resolution matrix. The diagonal elements of $(\mathbf{R}_D)_{ii}$ of $\mathbf{R}_D$

are of particular interest. These elements are termed *data importance*, since they indicate the way in which an observed datum $\Delta\mathbf{d}_i^{obs}$ influences an estimated or predicted datum $\Delta\mathbf{d}_i^{est}$. Ideally, the set of data importance values $(\mathbf{R}_D)_{ii}$ have approximately equal values for all $i = 1, \ldots, N_D$, meaning that no single data point is much more or much less important than the others. Experimental configurations that achieve this goal are referred to as *equileverage designs*.

Random probing techniques to estimate resolution properties

For linear inverse problems, the resolution and posterior covariance matrices contain most of the information needed to assess the quality of the maximum-likelihood model $\tilde{\mathbf{m}}$. However, when the model-space dimension N_M is large, these matrices may be impossible to compute or store explicitly. Though it has been argued that growing computational resources may come to the rescue (e.g., Soldati et al., 2006), resolution and posterior covariance matrices are still rarely being computed. To some extent, this may be the result of an interesting psychological effect: we seem to be driven towards the construction of models that are as big (and detailed) as allowed by our currently available resources. As a consequence, the even larger resources needed to assess if these details are actually resolved, are frequently not available. A corollary is that our supercomputers will never be large enough to compute the resolution matrices of real interest.

The problem of having to compute properties of a matrix that is too big to compute or store, is the central theme of *matrix probing* methods. While the family of matrix probing methods has grown rapidly in recent years, they all share the common concept of estimating specific matrix properties, for instance, the trace or large eigenvalues, through the multiplication of random vectors with the matrix. In many cases, the matrix-vector product can be computed efficiently, without having to compute the matrix explicitly. An extensive summary of matrix probing methods can be found in Halko et al. (2011). Examples of applications to resolution analysis in inverse problems have been presented by An (2012), Trampert et al. (2013), and Fichtner & van Leeuwen (2015).

To illustrate the concept, our focus will be on a specific random probing method, known as *Hutchinson's method* (Hutchinson, 1990), which can be used to estimate the trace of a matrix. Since the trace of the model resolution matrix $\mathbf{R}_M$ has the concrete interpretation as the number of resolved model parameters, we take $\mathbf{R}_M$ as a working example.

As a first step, we generate a set of N *Rademacher vectors* $\mathbf{a}^n$ ($n = 1, \ldots, N$), that is, vectors where the entries are randomly either 1 or -1, and uncorrelated in the sense

$$\lim_{N\to\infty} \frac{1}{N} \sum_{n=1}^{N} a_i^n a_j^n = 0 \quad \text{for } i \neq j. \tag{14.2.25}$$

The multiplication of $\mathbf{a}^n$ with $\mathbf{R}_M$ from the left and the right, gives

$$(\mathbf{a}^n)^T \mathbf{R}_M \mathbf{a}^n = \sum_{i=1}^{N_M} R_{M,ii} (a_i^n)^2 + \sum_{i=1}^{N_M} \sum_{j=1, j\neq i}^{N} a_i R_{M,ij} a_j$$
$$= \sum_{i=1}^{N_M} R_{M,ii} + \sum_{i=1}^{N_M} \sum_{\substack{j=1 \\ j\neq i}}^{N_M} a_i R_{M,ij} a_j, \tag{14.2.26}$$

where we have used the property $(a_i^n)^2 = 1$ for any i and n. Taking the average over all realisations of random vectors, we obtain

$$\frac{1}{N} \sum_{n=1}^{N} (\mathbf{a}^n)^T \mathbf{R}_M \mathbf{a}^n = \sum_{i=1}^{N_M} R_{M,ii} + \sum_{i=1}^{N_M} \sum_{\substack{j=1 \\ j\neq i}}^{N_M} R_{M,ij} \left[\frac{1}{N} \sum_{n=1}^{N} a_i^n a_j^n \right]. \tag{14.2.27}$$

With the aid of (14.2.25), the right-hand side condenses to

$$\frac{1}{N} \sum_{n=1}^{N} (\mathbf{a}^n)^T \mathbf{R}_M \mathbf{a}^n \approx \sum_{i=1}^{N_M} R_{M,ii} = \operatorname{tr} \mathbf{R}_M, \tag{14.2.28}$$

where the $\approx$ sign results from considering only a finite number of random realisations N. Equation (14.2.28) provides a simple recipe for the estimation of $\operatorname{tr} \mathbf{R}_M$. We multiply $\mathbf{R}_M$ from both sides with a realisation of a random Rademacher vector and then average over several realisations. What makes (14.2.28) efficient and practical is that the computation of $\mathbf{R}_M \mathbf{a}^n$ actually does not require an explicit version of $\mathbf{R}_M$. In fact, we have

$$\mathbf{R}_M \mathbf{a}^n = \mathbf{G}^{-g} \mathbf{G} \mathbf{a}^n = \mathbf{G}^{-g} \mathbf{d}_a, \tag{14.2.29}$$

where $\mathbf{d}_a$ are the data corresponding to the random vector $\mathbf{a}^n$, which is here interpreted as a model with random entries of 1 or -1. According to (14.2.29), the computation of $\mathbf{R}_M \mathbf{a}^n$ is equal to the solution of the linear inverse problem with synthetic data $\mathbf{d}_a$, which does not require explicit matrices.

The method represented by (14.2.28) can be modified to estimate the trace of a submatrix of $\mathbf{R}_M$. For instance, multiplying $\mathbf{R}_M$ from the right with random realisations of

$$\mathbf{a} = (\underbrace{0, \ldots, 0}_{k}, a_1, \ldots, a_s, \underbrace{0, \ldots, 0}_{N_M - k - s})^T, \tag{14.2.30}$$

and from the left with

$$\mathbf{a} = (\underbrace{0, \ldots, 0}_{N_M - k - s}, a_1, \ldots, a_s, \underbrace{0, \ldots, 0}_{k})^T, \tag{14.2.31}$$

gives an approximation of the trace of the $s \times s$ submatrix located between columns k and $k + s$, and between rows $N_M - k - s$ and $N_M - k$.

The quality of the approxmation (14.2.28) depends on the number of random realisations N; and the number of realisations needed to obtain a useful approximation generally depends on properties of $\mathbf{R}_M$. When $\mathbf{R}_M$ is diagonally dominant, the cross terms on the right-hand side of (14.2.27) tend to vanish more quickly than in cases where $\mathbf{R}_M$ is a dense matrix with entries of nearly equal size.

In many tomographic inverse problems, the structure of $\mathbf{R}_M$ is indeed favourable. Often, the Earth model is parameterised in terms of N_v vertical and N_h horizontal basis functions, so that $N_M = N_v N_h$. The principal diagonal indicates how well the amplitude of a given parameter can be recovered. After N_h horizontal entries along a row, the next diagonal indicates how much a parameter at a given horizontal position is correlated with the one at the same horizontal position but at the following depth index, and so on.

An example for this parameterisation is the global seismic shear-velocity model S40RTS by Ritsema et al. (2011), built laterally from spherical harmonics up to degree 40 (1681 coefficients) and vertically from 21 cubic splines. As illustrated in Figure 14.2(b), the resolution matrix of S40RTS has dominant entries along the main diagonal, showing a progressive decay with increasing spherical harmonic degree. For any specific row of the resolution matrix, the off-diagonal terms are comparatively small, except for the same horizontal parameter at different depth indices. Figure 14.2(c) shows trace estimates of the main diagonal and all the minor diagonals (the diagonals shifted by a certain index from the main diagonal) for $N = 1$, that is, the extreme case where only a single realisation of random vectors was used. For index 1 (main diagonal), the random probing estimate of the trace is 8009. The exact value, which we can compute because $\mathbf{R}_M$ is actually small enough to be computed explicitly, is 8007. This illustrates that one random vector may be sufficient when the matrix is strongly diagonally dominant. The correlation at position 2 corresponds to the trace of the next diagonal, etc. The next significant diagonal is at position 1682, meaning that there is a correlation with the following depth layer. Eventually, the modulo-1681 diagonals disappear in the background noise, which may be taken as a rough estimate of the accuracy of the method.

14.3 Nonlinear Inversion and Optimisation Methods

We have so far considered two methodological extremes: In Section 14.1 we briefly introduced the very general and holistic framework of Bayesian inference and its practical implementation using Monte Carlo methods. Apart from computational difficulties related to the potentially large number of samples required to approximate the posterior, Bayesian inference has no inherent limitations. This was contrasted in Section 14.2 with the least-squares method for linear or nearly linear forward problems. Often, linearity is merely a convenient approximation, and the explicit solution of least-squares problems may require the injection of artificial prior knowledge, that is, regularisation.

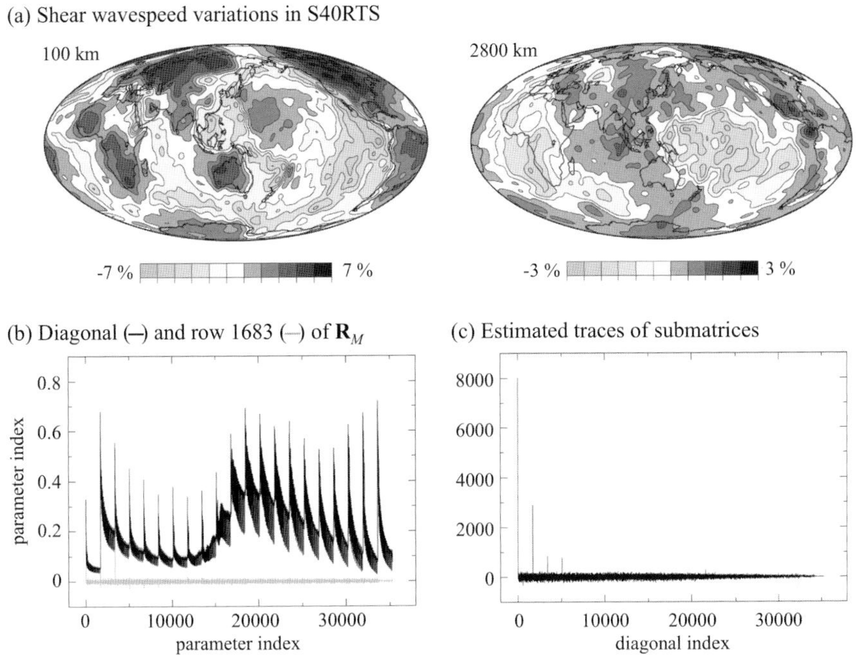

Figure 14.2 Random probing of the resolution matrix of the global shear velocity model S40RTS (Ritsema et al., 2011). (a) Relative lateral shear-velocity variations at 100 km and at 2800 km depth. (b) Diagonal (black) and row number 1683 of the resolution matrix $\mathbf{R}_\mathrm{M}$. (c) Random probing estimates of the traces of the main diagonal and all minor diagonals of $\mathbf{R}_\mathrm{M}$.

In the following paragraphs we will consider approaches that are located in between the extremes, and that do not require strict linearity of the forward problem. We consider the data vector $\mathbf{d}$ to be a function of the model vector $\mathbf{m}$:

$$\mathbf{d} = \mathbf{G}(\mathbf{m}), \tag{14.3.1}$$

and allow weak forms of nonlinearity. The price to pay for more flexibility is a reduction of scope. Instead of trying to characterise the complete posterior distribution, we will mostly limit ourselves to finding an approximation of the maximum-likelihood model which minimises some misfit functional, for instance, the least-squares misfit

$$\chi(\mathbf{m}) = \frac{1}{2}\left[\mathbf{G}(\mathbf{m}) - \mathbf{d}^{\mathrm{obs}}\right]^{\mathrm{T}} \mathbf{C}_\mathrm{D}^{-1} \left[\mathbf{G}(\mathbf{m}) - \mathbf{d}^{\mathrm{obs}}\right] + \frac{1}{2}(\mathbf{m} - \mathbf{m}^{\mathrm{prior}})^{\mathrm{T}}\mathbf{C}_\mathrm{M}^{-1}(\mathbf{m} - \mathbf{m}^{\mathrm{prior}}). \tag{14.3.2}$$

Additionally, within a neighbourhood of the maximum-likelihood model, we may find an approximation of the posterior covariance.

During the past few decades, the field of nonlinear optimisation has received tremendous attention, not only in the context of geophysical inverse problems, but also in mathematical finance, machine learning, optimal design, and many other domains. Exhaustively describing the diversity of methods that have been developed, often for specific applications, is far beyond the scope of any single textbook, and so we will limit ourselves to the subset of methods that are relevant for seismic waveform inversion covered in Chapters 17 and 18. For a more comprehensive and general account of nonlinear optimisation methods, the reader is referred to the classic book by Nocedal & Wright (1999).

Iterative linearisation

Probably the most straightforward approach to the least-squares solution of an inverse problem with nonlinear foward modelling equations is brute-force *linearisation*. A first-order Taylor expansion of $\mathbf{G}(\mathbf{m})$ from (14.3.1) around an initial model $\mathbf{m}_0$, yields

$$\mathbf{G}(\mathbf{m}) \doteq \mathbf{G}\left(\mathbf{m}_0\right) + \mathbf{J}_0\left(\mathbf{m} - \mathbf{m}_0\right), \tag{14.3.3}$$

where $\doteq$ denotes a first-order approximation, and $\mathbf{J}_0$ is the Jacobian of $\mathbf{G}$ evaluated at $\mathbf{m}_0$. The components of $\mathbf{J}_0$ are given by

$$J_{0,ij} = \left.\frac{\partial G_i}{\partial m_j}\right|_{\mathbf{m}=\mathbf{m}_0}. \tag{14.3.4}$$

Substituting the approximation (14.3.4) into the least-squares misfit (14.3.2), yields

$$\begin{aligned}\chi(\mathbf{m}) \;\doteq\; & \tfrac{1}{2}(\mathbf{m} - \mathbf{m}^{\text{prior}})^T\mathbf{C}_M^{-1}(\mathbf{m} - \mathbf{m}^{\text{prior}}) \\ & + \tfrac{1}{2}(\mathbf{J}_0\mathbf{m} - \mathbf{d}_0^{\text{obs}})^T\mathbf{C}_D^{-1}(\mathbf{J}_0\mathbf{m} - \mathbf{d}_0^{\text{obs}}),\end{aligned} \tag{14.3.5}$$

where we have defined

$$\mathbf{d}_0^{\text{obs}} = \mathbf{d}^{\text{obs}} + \mathbf{J}_0\mathbf{m}_0 - \mathbf{G}\left(\mathbf{m}_0\right). \tag{14.3.6}$$

Structurally, (14.3.5) is identical to the least-squares misfit for linear forward problems, given in (14.2.2). The only differences are that the forward modelling matrix $\mathbf{G}$ is replaced by the Jacobian matrix $\mathbf{J}_0$, and that $\mathbf{d}^{\text{obs}}$ is replaced by the modified observations $\mathbf{d}_0^{\text{obs}}$. Thus, we may employ the machinery derived for linear forward problems in Section 14.2 in order to compute a least-squares solution. In particular, using the third of the three equivalent expressions in (14.2.9), we find

$$\mathbf{m}_1 = \mathbf{m}^{\text{prior}} + \mathbf{C}_M\mathbf{J}_0^T\left(\mathbf{C}_D + \mathbf{J}_0\mathbf{C}_M\mathbf{J}_0^T\right)^{-1}\left(\mathbf{d}_0^{\text{obs}} - \mathbf{J}_0\mathbf{m}_0\right). \tag{14.3.7}$$

Depending on how close $\mathbf{G}$ is to being linear, $\mathbf{m}_1$ may or may not be a good approximation of the maximum-likelihood model $\tilde{\mathbf{m}}$. Improvements may be obtained by iteratively updating the reference model used in the linear approximation (14.3.3). Replacing $\mathbf{m}_0$ in the approximation by $\mathbf{m}_1$ computed via

(14.3.7), we obtain a simple iterative scheme:

$$\begin{aligned} \mathbf{m}_2 &= \mathbf{m}^{\text{prior}} + \mathbf{C}_{\text{M}}\mathbf{J}_1^{\text{T}}\left(\mathbf{C}_{\text{D}} + \mathbf{J}_1\mathbf{C}_{\text{M}}\mathbf{J}_1^{\text{T}}\right)^{-1}\left(\mathbf{d}_1^{\text{obs}} - \mathbf{J}_1\mathbf{m}_1\right), \\ \ldots &= \ldots \\ \mathbf{m}_{n+1} &= \mathbf{m}^{\text{prior}} + \mathbf{C}_{\text{M}}\mathbf{J}_n^{\text{T}}\left(\mathbf{C}_{\text{D}} + \mathbf{J}_n\mathbf{C}_{\text{M}}\mathbf{J}_n^{\text{T}}\right)^{-1}\left(\mathbf{d}_n^{\text{obs}} - \mathbf{J}_n\mathbf{m}_n\right). \end{aligned} \tag{14.3.8}$$

The convergence of this scheme depends critically on the properties of the forward problem, but there is also a strong dependence on the choice of the initial model $\mathbf{m}_0$.

The linearisation approach has important properties and limitations. Most importantly, the iteration is *local*, meaning that all iterative solutions are within the same misfit valley as the initial model. As a consequence, the iteration may stall in a *local minimum*, without any possibility to escape from the local valley into the neighbouring valley where the global minimum may be located. This property is in contrast to Monte Carlo methods, which perform *global optimisation*.

The strong dependence of the iterative procedure on the choice of the initial model also gives a more precise meaning to the term *weakly nonlinear* relative to the choice of initial model. A strongly nonlinear and oscillatory misfit function may still be tractable with iterative linearisation, provided that the initial model is sufficiently close to the global minimum. Finding a suitable initial model is therefore of critical importance for nonlinear inverse problems.

Many methods exist for the approximation of the maximum-likelihood model or various other aspects of the solution of an inverse problem. Frequently, one method is claimed to be superior to others. While this may seem intuitively obvious, the opposite can be shown rigorously in the form of the *No-Free-Lunch theorem*, sketched in Section 15.7. At least statistically, all sampling or optimisation methods are equally inefficient. Efficiency only arises in the context of specific problems for which we have prior knowledge about the properties of the situation. This prior knowledge guides the choice of a suitable algorithm that may then indeed be more efficient than another one.

The iterative linearisation of the least-squares solution, outlined in the previous paragraphs, is only one member of a large family of *descent methods*. We discuss the properties of such methods in the context of waveform inversion in the following chapter.

Appendix 14.A: Covariance Matrix Identities

When dealing with the posterior mean and posterior covariance in (14.2.4) and (14.2.5), two matrix identities turn out to be useful. First, we note that

$$\begin{aligned} \mathbf{G}^{\text{T}} + \mathbf{G}^{\text{T}}\mathbf{C}_{\text{D}}^{-1}\mathbf{G}\mathbf{C}_{\text{M}}\mathbf{G}^{\text{T}} &= \mathbf{G}^{\text{T}}\mathbf{C}_{\text{D}}^{-1}\left(\mathbf{C}_{\text{D}} + \mathbf{G}\mathbf{C}_{\text{M}}\mathbf{G}^{\text{T}}\right) \\ &= \left(\mathbf{G}^{\text{T}}\mathbf{C}_{\text{D}}^{-1}\mathbf{G} + \mathbf{C}_{\text{M}}^{-1}\right)\mathbf{C}_{\text{M}}\mathbf{G}^{\text{T}}. \end{aligned} \tag{14.A.1}$$

Multiplying by $\left(\mathbf{G}^{\mathrm{T}}\mathbf{C}_{\mathrm{D}}^{-1}\mathbf{G}+\mathbf{C}_{\mathrm{M}}^{-1}\right)^{-1}$ from the left, and by $\left(\mathbf{C}_{\mathrm{D}}+\mathbf{G}\mathbf{C}_{\mathrm{M}}\mathbf{G}^{\mathrm{T}}\right)^{-1}$ from the right, gives

$$\left(\mathbf{G}^{\mathrm{T}}\mathbf{C}_{\mathrm{D}}^{-1}\mathbf{G}+\mathbf{C}_{\mathrm{M}}^{-1}\right)^{-1}\mathbf{G}^{\mathrm{T}}\mathbf{C}_{\mathrm{D}}^{-1}=\mathbf{C}_{\mathrm{M}}\mathbf{G}^{\mathrm{T}}\left(\mathbf{C}_{\mathrm{D}}+\mathbf{G}\mathbf{C}_{\mathrm{M}}\mathbf{G}^{\mathrm{T}}\right)^{-1}. \tag{14.A.2}$$

Making use of (14.A.2), we furthermore find

$$\begin{aligned}
&\mathbf{C}_{\mathrm{M}}-\mathbf{C}_{\mathrm{M}}\mathbf{G}^{\mathrm{T}}\left(\mathbf{C}_{\mathrm{D}}+\mathbf{G}\mathbf{C}_{\mathrm{M}}\mathbf{G}^{\mathrm{T}}\right)^{-1}\mathbf{G}\mathbf{C}_{\mathrm{M}}\\
&\quad=\mathbf{C}_{\mathrm{M}}-\left(\mathbf{G}^{\mathrm{T}}\mathbf{C}_{\mathrm{D}}^{-1}\mathbf{G}+\mathbf{C}_{\mathrm{M}}^{-1}\right)^{-1}\mathbf{G}^{\mathrm{T}}\mathbf{C}_{\mathrm{D}}^{-1}\mathbf{G}\mathbf{C}_{\mathrm{M}}\\
&\quad=\left(\mathbf{G}^{\mathrm{T}}\mathbf{C}_{\mathrm{D}}^{-1}\mathbf{G}+\mathbf{C}_{\mathrm{M}}^{-1}\right)^{-1}\left[\left(\mathbf{G}^{\mathrm{T}}\mathbf{C}_{\mathrm{D}}^{-1}\mathbf{G}+\mathbf{C}_{\mathrm{M}}^{-1}\right)\mathbf{C}_{\mathrm{M}}-\mathbf{G}^{\mathrm{T}}\mathbf{C}_{\mathrm{D}}^{-1}\mathbf{G}\mathbf{C}_{\mathrm{M}}\right].
\end{aligned} \tag{14.A.3}$$

Noticing that the term in square brackets is equal to the unit matrix **I**, we finally obtain

$$\left(\mathbf{G}^{\mathrm{T}}\mathbf{C}_{\mathrm{D}}^{-1}\mathbf{G}+\mathbf{C}_{\mathrm{M}}^{-1}\right)^{-1}=\mathbf{C}_{\mathrm{M}}-\mathbf{C}_{\mathrm{M}}\mathbf{G}^{\mathrm{T}}\left(\mathbf{C}_{\mathrm{D}}+\mathbf{G}\mathbf{C}_{\mathrm{M}}\mathbf{G}^{\mathrm{T}}\right)^{-1}\mathbf{G}\mathbf{C}_{\mathrm{M}}. \tag{14.A.4}$$

15

Gradient Methods for Nonlinear Inversion

In the previous chapter we have introduced a variety of general concepts regarding inversion, and now we turn to specific methods for extracting models with suitable properties in situations where we have very large numbers of structural parameters, as in waveform inversion. We consider the situation where we have a composite misfit functional

$$\chi(\mathbf{d}, \mathbf{m}) = \Phi(\mathbf{d}, \mathbf{d}^{\text{obs}}) + \Psi(\mathbf{m}, \mathbf{m}^{\text{prior}}),$$

which combines a measure $\Phi(\mathbf{d}, \mathbf{d}^{\text{obs}})$ of the data misfit between observed seismograms $\mathbf{d}^{\text{obs}}$ and simulations $\mathbf{d}$ for a specific model $\mathbf{m}$, and a measure of model suitability $\Psi(\mathbf{m}, \mathbf{m}^{\text{prior}})$. In general $\chi(\mathbf{d}, \mathbf{m})$ is a complex nonlinear function of the model parameters $\mathbf{m}$, so that the process of seeking the minimum of $\chi(m)$ is not straightforward. Even if only a single wavespeed is specified, regional models require many thousands of parameters. Ideally we would like to resolve the full set of isotropic parameters (*P* wavespeed, *S* wavespeed, and density) and even include the possibility of anisotropy. Each additional parameter magnifies the size of the problem of searching in a high-dimensional model space for a global minimum of $\chi(\mathbf{m})$, since the properties of the model dependence can beguile the attempt at optimisation and lead it to be trapped in purely local minima.

Rather than attempt any class of parameter search, effective algorithms concentrate on descending the gradients of the misfit function $\chi(\mathbf{m})$ with the aim of reaching the global minimum. In the context of waveform inversion it is important to aid the computational process as much as possible, by making good choices of starting model, using a suitable form of the misfit function, and appropriate handling of the observations. Thus, it is desirable to start with lower frequencies and aim to extract the broad-scale features of the Earth model before expanding the filter band with the aim of introducing additional detail (see Section 17.2). The misfit measure needs to be tolerant to possible misalignment of seismograms, particularly in the early stages of an iterative process,

There are a wide range of *descent methods* that vary in their computational demands. We start by taking a brief look at some basic properties of such algorithms and then examine a number of specific approaches.

15.1 General Descent Methods

We will work with a general misfit function $\chi(\mathbf{m})$ that allows for non-Gaussian observational errors and non-Gaussian prior knowledge in model space. We look at the way that the gradient properties of χ can be exploited to find new models for which the misfit value is reduced.

Descent directions

Our general goal is to improve an initial model $\mathbf{m}_0$ by adding to it an update $\gamma_0\,\mathbf{h}_0$, where $\mathbf{h}_0$ is the *descent direction* and $\gamma_0 > 0$ is the *step length*:

$$\mathbf{m}_1 = \mathbf{m}_0 + \gamma_0\,\mathbf{h}_0. \tag{15.1.1}$$

For the construction of a suitable descent direction, we require that the misfit is actually reduced during the step. We therefore require

$$\chi(\mathbf{m}_1) = \chi(\mathbf{m}_0 + \gamma_0\,\mathbf{h}_0) < \chi(\mathbf{m}_0), \tag{15.1.2}$$

for all positive step lengths γ_0 that are smaller than a maximum step length $\gamma_{\max}$. Rearranging (15.1.2) and taking the limit $\gamma_0 \to 0$, we obtain

$$\mathbf{h}_0^T \nabla\chi(\mathbf{m}_0) = \lim_{\gamma_0\to 0} \frac{1}{\gamma_0}\left[\chi(\mathbf{m}_0 + \gamma_0\,\mathbf{h}_0) - \chi(\mathbf{m}_0)\right] < 0, \tag{15.1.3}$$

where $\mathbf{h}_0^T \nabla\chi(\mathbf{m}_0)$ is the directional derivative of the misfit χ in the direction of $\mathbf{h}_0$.

The condition $\mathbf{h}_0^T \nabla\chi(\mathbf{m}_0) < 0$ can be considered as the definition of a local descent direction, that is, a direction that indicates the way we need to move away from $\mathbf{m}_0$ in order to reduce the misfit. It follows immediately that a descent direction always exists as long as $\nabla\chi(\mathbf{m}_0) \neq 0$. To see this, we choose

$$\mathbf{h}_0 = -\nabla\chi(\mathbf{m}_0), \tag{15.1.4}$$

and then

$$\mathbf{h}_0^T \nabla\chi(\mathbf{m}_0) = -[\nabla\chi(\mathbf{m}_0)]^2 < 0. \tag{15.1.5}$$

The result (15.1.5) means that choosing $\mathbf{h}_0$ from (15.1.4) to be the negative gradient of the misfit satisfies condition (15.1.3). Many other descent directions are possible. Indeedt, when $\mathbf{A}$ is a positive definite matrix, the vector

$$\mathbf{h}_0 = -\mathbf{A}\,\nabla\chi(\mathbf{m}_0), \tag{15.1.6}$$

is also a descent direction because

$$\mathbf{h}_0^T \nabla\chi(\mathbf{m}_0) = -\nabla\chi(\mathbf{m}_0)^T \mathbf{A}^T \nabla\chi(\mathbf{m}_0) < 0. \tag{15.1.7}$$

In an iterative minimisation we successively update models by going from the current model $\mathbf{m}_i$ along a descent direction

$$\mathbf{h}_i = -\mathbf{A}_i \nabla\chi(\mathbf{m}_i) \tag{15.1.8}$$

towards the improved model $\mathbf{m}_{i+1}$. This general descent algorithm can be summarised as follows:

(1) Choose an initial model, $\mathbf{m}_0$, and set $i = 0$.

(2) Compute the descent direction $\mathbf{h}_i = -\mathbf{A}_i \nabla\chi(\mathbf{m}_i)$.

(3) Update $\mathbf{m}_i$ according to $\mathbf{m}_{i+1} = \mathbf{m}_i + \gamma_i \mathbf{h}_i$, with a suitable step length γ_i that ensures $\chi(\mathbf{m}_{i+1}) < \chi(\mathbf{m}_i)$.

(4) Set $i \to i+1$ and go back to (2).

The process is repeated until the misfit is as small as permitted by the errors in the data. A critical element of the above algorithm is the gradient $\nabla\chi$, which may be expensive to compute, but can often be aided by adjoint methods (Chapter 16). The choice of the positive definite matrices $\mathbf{A}_i$ characterises different descent methods, and controls the speed of convergence towards the optimum $\tilde{\mathbf{m}}$. Given the availability of sufficient computational resources, the minimum of χ can be approximated arbitrarily closely with the help of descent methods. However, $\tilde{\mathbf{m}}$, may explain the observed data so well that the remaining residuals are smaller than the uncertainty in the data.

Optimal step length

To complete the description of the general descent method, it remains to construct an efficient step length for each iteration. Ideally, we would like to find an optimal choice of γ_i such that $\chi(\mathbf{m}_{i+1}) = \chi(\mathbf{m}_i+\gamma_i \mathbf{h}_i)$ is minimal. A necessary condition for the optimality of γ_i is then

$$\frac{d}{d\gamma_i}\chi(\mathbf{m}_i + \gamma_i \mathbf{h}_i) = 0. \tag{15.1.9}$$

Evaluating the derivative in (15.1.9), gives

$$\mathbf{h}_i^T \nabla\chi(\mathbf{m}_i + \gamma_i \mathbf{h}_i) = 0. \tag{15.1.10}$$

Equation (15.1.10) defines the optimal value of γ_i implicitly. To obtain an explicit approximation, we employ a first-order approximation of $\nabla\chi(\mathbf{m}_i + \gamma_i \mathbf{h}_i)$,

$$\nabla\chi(\mathbf{m}_i + \gamma_i \mathbf{h}_i) \doteq \nabla\chi(\mathbf{m}_i) + \gamma_i \mathbf{H}(\mathbf{m}_i)\mathbf{h}_i, \tag{15.1.11}$$

where the *Hessian* matrix contains the second derivatives of the misfit function χ,

$$H_{jk}(\mathbf{m}_i) = \left.\frac{\partial^2\chi}{\partial m_j \partial m_k}\right|_{\mathbf{m}=\mathbf{m}_i}. \tag{15.1.12}$$

Substituting (15.1.11) into (15.1.10), yields

$$0 = \mathbf{h}_i^T \nabla\chi(\mathbf{m}_i) + \gamma_i \mathbf{h}_i^T \mathbf{H}(\mathbf{m}_i)\mathbf{h}_i. \tag{15.1.13}$$

We can then solve 15.1.13 for γ_i to give

$$\gamma_i = -\frac{\mathbf{h}_i^T \nabla\chi(\mathbf{m}_i)}{\mathbf{h}_i^T \mathbf{H}(\mathbf{m}_i)\mathbf{h}_i}. \tag{15.1.14}$$

A commonly used alternative to (15.1.14), which allows us to circumvent the computation of the Hessian $\mathbf{H}(\mathbf{m}_i)$ acting on the vector $\mathbf{h}_i$, is a *line search*. We choose a small number of trial step lengths, $\gamma_i^{(k)}$ $(k = 1, ..., n)$, and evaluate the corresponding misfit functionals $\chi(\mathbf{m}_i + \gamma_i^{(k)}\,\mathbf{h}_i)$. These values are then used to generate an interpolating polynomial of degree $n-1$ to approximate $\chi(\mathbf{m}_i + \gamma_i\,\mathbf{h}_i)$. The minimum of the polynomial is then an estimate of the optimal step length. In favourable circumstances the misfit function, $\chi(\mathbf{m}_i + \gamma_i\,\mathbf{h}_i)$ may be sufficiently quadratic with respect to γ_i that a second-order polynomial yields a good approximation.

The choice of the trial step lengths is crucial for the success of a line search. The only obvious trial step length is $\gamma_i^{(0)} = 0$ because the corresponding misfit, $\chi(\mathbf{m}_i)$, is already known. Other step lengths $\gamma_i^{(k)}$ can be found most effectively with the help of intuition to provide a range of step lengths that lead to physically plausible models. In the specific context of tomographic inverse problems, trial step lengths may be chosen such that the resulting wavespeed variations in the Earth are physically acceptable.

15.2 The Method of Steepest Descent

We would like to reduce the value of the misfit $\chi(\mathbf{m})$ as much as possible in just the first iteration. So starting from $\mathbf{m}_0$ we try to find the descent direction, $\mathbf{h}_0$, that leads to the maximum decrease of χ for a small fixed step length, γ_0. We seek $\mathbf{h}_0$, with $|\mathbf{h}_0| = 1$, such that

$$\chi(\mathbf{m}_1) - \chi(\mathbf{m}_0) = \chi(\mathbf{m}_0 + \gamma_0\,\mathbf{h}_0) - \chi(\mathbf{m}_0) \approx \gamma_0\,\mathbf{h}_0^T \nabla\chi(\mathbf{m}_0) \tag{15.2.1}$$

is minimal. The assumption that $\mathbf{h}_0$ is a descent direction in the sense of (15.1.3), yields the inequality

$$\gamma_0\,\mathbf{h}_0^T \nabla\chi(\mathbf{m}_0) \geq -\gamma_0\,|\nabla\chi(\mathbf{m}_0)|\,|\mathbf{h}_0| = -\gamma_0\,|\nabla\chi(\mathbf{m}_0)|, \tag{15.2.2}$$

meaning that the quantity that we seek to minimise, $\gamma_0\,\mathbf{h}_0^T \nabla\chi(\mathbf{m}_0)$, is always larger than or equal to $-\gamma_0\,|\nabla\chi(\mathbf{m}_0)|$. The minimum of $\gamma_0\,\mathbf{h}^T \nabla\chi(\mathbf{m}_0)$ corresponds to the direction $\mathbf{h}_0$ for which the equal sign holds in (15.2.2). This is the case for

$$\mathbf{h}_0 = -\frac{\nabla\chi(\mathbf{m}_0)}{|\nabla\chi(\mathbf{m}_0)|}, \tag{15.2.3}$$

which is the *direction of steepest descent*.

Since $\mathbf{h}_0$ as defined in (15.2.3) leads to the most rapid decrease in χ for a given small step length γ_0, we can repeat this procedure for multiple model estimates. We iteratively move from a current model $\mathbf{m}_i$ along the local descent direction

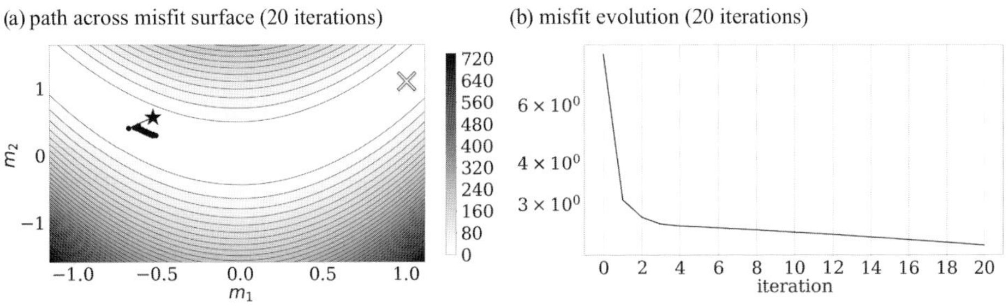

Figure 15.1 Illustration of the steepest-descent optimisation of the Rosenbrock function (15.2.4) using 20 iterations. (a) Path across the misfit surface from the initial model (star) towards the global minimum (cross). Each of the black dots represents a new update. Since the optimal step length in the direction of steepest descent is so small, individual dots can hardly be distinguished. (b) As a consequence of slow movement across the misfit surface, misfit decreases very slowly, after a short initial phase of strong misfit reduction (only the first two iterations).

$-\nabla\chi(\mathbf{m}_i)$ towards an updated model $\mathbf{m}_{i+1}$. This is the concept used in the steepest-descent algorithm:

(1) Choose an initial model, $\mathbf{m}_0$, and set $i = 0$.

(2) Compute the gradient for the current model, $\nabla\chi(\mathbf{m}_i)$.

(3) Update $\mathbf{m}_i$ according to $\mathbf{m}_{i+1} = \mathbf{m}_i - \gamma_i \nabla\chi(\mathbf{m}_i)$, with a suitable step length γ_i that ensures $\chi(\mathbf{m}_{i+1}) < \chi(\mathbf{m}_i)$.

(4) Set $i \to i + 1$, go back to (2) and repeat until the data are explained sufficiently well.

While being conceptually simple and attractive, the steepest-descent method is rarely used in practice because it tends to converge rather slowly towards an acceptable model. The slow convergence arises because a succession of locally-optimal descent directions tend not be be optimal from a global perspective. A specific symptom of this problem is illustrated in Figure 15.1, where the steepest-descent method is used to minimise the two-dimensional Rosenbrock function (Rosenbrock, 1960),

$$\chi(\mathbf{m}) = (1 - m_1^2) + 100\,(m_2 - m_1^2)^2, \tag{15.2.4}$$

which is widely used to test the performance of optimisation algorithms. The Rosenbrock function is characterised by a long curved valley that descends very slowly towards the global minimum at $(m_1, m_2) = (1, 1)$, where $\chi(\mathbf{m}) = 0$. The 20 iterations of the steepest-descent procedure shown in Figure 15.1 are obviously not sufficient to approach the minimum, despite always moving in the direction where the misfit decreases most rapidly. This is in sharp contrast to other, more elaborate, optimisation methods that we will cover in the following paragraphs.

15.3 Newton's Method and its Variants

In the steepest-descent algorithm, the direction $\mathbf{h}_i$ is determined only from first-derivative information at the current iterate $\mathbf{m}_i$. *Newton's method* modifies the steepest-descent direction with the help of second-derivative information, which may lead to faster convergence. The algorithm exploits the equivalence of finding an extremum of the misfit, χ, and finding a zero of its gradient, $\nabla\chi$. Using the necessary condition that the gradient vanishes at the optimum, $\nabla\chi(\tilde{\mathbf{m}}) = \mathbf{0}$, the linear approximation of $\nabla\chi(\tilde{\mathbf{m}})$ around a nearby model $\mathbf{m}$ is given by

$$\mathbf{0} = \nabla\chi(\tilde{\mathbf{m}}) \approx \nabla\chi(\mathbf{m}) + \mathbf{H}(\mathbf{m})\,(\tilde{\mathbf{m}} - \mathbf{m}). \tag{15.3.1}$$

Formally solving for $\tilde{\mathbf{m}}$, we obtain

$$\tilde{\mathbf{m}} \approx \mathbf{m} - \mathbf{H}^{-1}(\mathbf{m})\,\nabla\chi(\mathbf{m}). \tag{15.3.2}$$

Equation (15.3.2) suggests the use of the solution of

$$\mathbf{H}(\mathbf{m}_i)\,\mathbf{h}_i = -\nabla\chi(\mathbf{m}_i), \tag{15.3.3}$$

as the descent direction, which leads to the following algorithm:

(1) Choose an initial model, $\mathbf{m}_0$, and set $i = 0$.

(2) Compute the gradient for the current model, $\nabla\chi(\mathbf{m}_i)$.

(3) Determine the descent direction, $\mathbf{h}_i$, as the solution of $\mathbf{H}(\mathbf{m}_i)\,\mathbf{h}_i = -\nabla\chi(\mathbf{m}_i)$.

(4) Update $\mathbf{m}_i$ according to $\mathbf{m}_{i+1} = \mathbf{m}_i + \mathbf{h}_i$.

(5) Set $i \to i+1$, go back to (2) and repeat as often as needed.

In each iteration, Newton's method requires the computation of the Hessian acting on vectors and the solution of the linear system (15.3.3), which is known as the *Newton equation*. The solution for the descent direction $\mathbf{h}_i$ can be found, for instance, with the help of iterative matrix solvers (e.g., Quarteroni et al., 2000). We note that Newton's method corresponds to choosing the matrix $\mathbf{A}_i$ from (15.1.8) as

$$\mathbf{A}_i = \mathbf{H}^{-1}(\mathbf{m}_i). \tag{15.3.4}$$

Provided that $\mathbf{H}(\mathbf{m}_i)$ is positive definite, $-\mathbf{H}^{-1}(\mathbf{m}_i)\,\nabla\chi(\mathbf{m}_i)$ is a descent direction. The beauty of Newton's method becomes apparent when it is applied to a misfit function that is actually quadratic, so that the approximation in (15.3.1) is exact. In fact, using the least-squares misfit for the linear forward problem from (14.2.2), we find

$$\nabla\chi(\mathbf{m}_0) = \mathbf{C}_M^{-1}\left(\mathbf{m}_0 - \mathbf{m}^{\text{prior}}\right) + \mathbf{G}^T\mathbf{C}_D^{-1}\left(\mathbf{G}\mathbf{m}_0 - \mathbf{d}^{\text{obs}}\right), \tag{15.3.5}$$

and

$$\mathbf{H}(\mathbf{m}_0) = \mathbf{C}_M^{-1} + \mathbf{G}^T\mathbf{C}_D^{-1}\mathbf{G}. \tag{15.3.6}$$

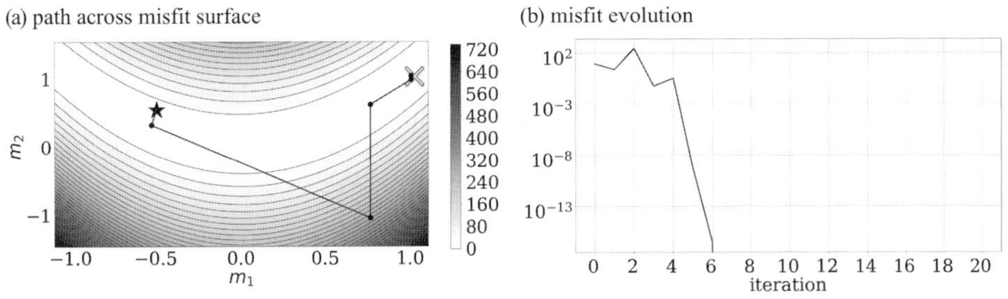

Figure 15.2 Illustration of the Newton optimisation of the Rosenbrock function (15.2.4). (a) Path across the misfit surface from the initial model (star) towards the global minimum (cross). Each of the black dots represents a new update. (b) Within six iterations, the minimum is reached almost exactly. Compare to the much slower convergence of the steepest-descent method, shown in Figure 15.1.

Using (15.3.3), the descent direction $\mathbf{h}_0$ is then given by

$$\begin{aligned}\mathbf{h}_0 &= -\left(\mathbf{C}_M^{-1} + \mathbf{G}^T\mathbf{C}_D^{-1}\mathbf{G}\right)^{-1}\left[\mathbf{C}_M^{-1}\left(\mathbf{m}_0 - \mathbf{m}^{\text{prior}}\right) + \mathbf{G}^T\mathbf{C}_D^{-1}\left(\mathbf{G}\mathbf{m}_0 - \mathbf{d}^{\text{obs}}\right)\right]\\ &= -\left(\mathbf{C}_M^{-1}+\mathbf{G}^T\mathbf{C}_D^{-1}\mathbf{G}\right)^{-1}\left[\left(\mathbf{C}_M^{-1} + \mathbf{G}^T\mathbf{C}_D^{-1}\mathbf{G}\right)\mathbf{m}_0 - \mathbf{C}_M^{-1}\mathbf{m}^{\text{prior}} - \mathbf{G}^T\mathbf{C}_D^{-1}\mathbf{d}^{\text{obs}}\right]\\ &= -\mathbf{m}_0 + \left(\mathbf{C}_M^{-1} + \mathbf{G}^T\mathbf{C}_D^{-1}\mathbf{G}\right)^{-1}\left(\mathbf{C}_M^{-1}\mathbf{m}^{\text{prior}} - \mathbf{G}^T\mathbf{C}_D^{-1}\mathbf{d}^{\text{obs}}\right). \end{aligned} \tag{15.3.7}$$

From the application of the Newton algorithm, the first update, $\mathbf{m}_1$, is given by

$$\mathbf{m}_1 = \mathbf{m}_0 + \mathbf{h}_0 = \left(\mathbf{C}_M^{-1} + \mathbf{G}^T\mathbf{C}_D^{-1}\mathbf{G}\right)^{-1}\left(\mathbf{C}_M^{-1}\mathbf{m}^{\text{prior}} - \mathbf{G}^T\mathbf{C}_D^{-1}\mathbf{d}^{\text{obs}}\right). \tag{15.3.8}$$

Comparing with (14.2.4), we see that already the first update is indeed equal to the minimum of the least-squares misfit function for a linear forward problem, that is

$$\mathbf{m}_1 = \tilde{\mathbf{m}}. \tag{15.3.9}$$

Though (15.3.9) is strictly valid only in the case of a linear forward problem, it still suggests fast convergence in cases where the misfit function is nearly quadratic. The comparatively fast convergence of Newton's method is illustrated in Figure 15.2, where we again consider the two-dimensional Rosenbrock function (15.2.4). In contrast to the steepest-descent method, where 20 iterations were far from sufficient to reach the minimum (see Figure 15.1), Newton's method reaches the minimum almost exactly within only six iterations.

Newton's method relies critically on the appropriateness of the local approximation from (15.3.1). When the current iterate, $\mathbf{m}_i$, is far from the optimum, the Hessian, $\mathbf{H}(\mathbf{m}_i)$, may have negative eigenvalues and be ill-conditioned or even singular. This can result in very slow convergence, movement in non-descent directions, and heavy oscillations that diverge from the solution. Such behaviour is indeed visible in Figure 15.2, where the second iteration leads to an over-shooting towards a larger misfit because $\mathbf{H}$ after the first iteration is nearly singular.

As a consequence, many variants of Newton's method have been developed. For instance, the Hessian may be pre-multiplied by a step length smaller than 1 to prevent over-shooting (damped Newton method), or a regularisation term may be added to the Hessian to ensure invertability (regularised Newton method). The Gauss–Newton and Levenberg–Marquardt variants replace the full Hessian by a first-order approximation, for which the computation may require fewer resources. A quantitative comparison of Newton methods in the context of seismic waveform inversion can be found in Pratt et al. (1998).

15.4 The Conjugate-Gradient Method

Newton's method and its variants often converge faster than the steepest-descent method, but at the expense of having to compute the inverse of the (approximate) Hessian applied to a vector. The steepest-descent method often converges slowly because the algorithm tends to walk repeatedly in directions where it has walked before. This deficiency motivates the *conjugate-gradient method.* Instead of reusing descent directions that have been used in previous iterations, the conjugate-gradient method constructs a sequence of descent directions that are mutually orthogonal. In the following, we will first derive the method for purely quadratic misfit functions and then propose several generalisations for arbitrary misfit functions.

Derivation of the method for quadratic misfit functions

The conjugate-gradient method for quadratic misfit functions can be understood as a method for the direct inversion of the vector-matrix equation $\mathbf{Gm} = \mathbf{d}$. Left-multiplication by $\mathbf{G}^T$, gives $\mathbf{G}^T\mathbf{Gm} = \mathbf{b} = \mathbf{G}^T\mathbf{d}$ which we write as

$$\mathbf{Hm} = \mathbf{b}, \tag{15.4.1}$$

in terms of the symmetric matrix $\mathbf{H} = \mathbf{G}^T\mathbf{G}$, and the constant vector $\mathbf{b} = \mathbf{G}^T\mathbf{d}$. We assume for the moment that $\mathbf{H}$ is positive definite, though this may not be the case in practice. First, we note that the solution of (15.4.1) is equivalent to the minimisation of the quadratic misfit function

$$\chi(\mathbf{m}) = \tfrac{1}{2}\mathbf{m}^T\mathbf{Hm} - \mathbf{m}^T\mathbf{b}, \tag{15.4.2}$$

where $\mathbf{H}$ is the Hessian of χ. Indeed, forcing the gradient of χ to $\mathbf{0}$ returns (15.4.1), which has a unique solution as a consequence of the assumed positive definiteness of the matrix $\mathbf{H}$. To minimise χ using a descent method, we update the current model $\mathbf{m}_i$ along a descent direction $\mathbf{h}_i$ with some step length γ. This modifies the misfit from $\chi(\mathbf{m}_i)$ to $\chi(\mathbf{m}_i + \gamma\mathbf{h}_i)$. Differentiating $\chi(\mathbf{m}_i + \gamma\mathbf{h}_i)$ with respect to γ and setting the derivative to 0, yields the optimal step length

$$\gamma_i = \frac{\mathbf{h}_i^T\mathbf{r}_i}{\mathbf{h}_i^T\mathbf{H}\mathbf{h}_i}, \tag{15.4.3}$$

where we define the *residual* $\mathbf{r}_i$ as

$$\mathbf{r}_i = \mathbf{b} - \mathbf{H}\mathbf{m}_i. \tag{15.4.4}$$

The expression (15.4.3) is a special case of the optimal step length (15.1.14) for non-quadratic misfit functions. We arrive at $\mathbf{m}_i$ through a sequence of updates in some directions $\mathbf{h}_j$, with $j = 0, ..., i-1$. To avoid convergence problems as in the steepest descent method, we want to ensure that $\mathbf{m}_i$ cannot be updated in any of the previous directions without actually increasing the misfit. Thus, we require

$$\chi(\mathbf{m}_i) \leq \chi(\mathbf{m}_i + \gamma\mathbf{h}_j), \quad \text{for any } \gamma \text{ and } j = 0, ..., i-1. \tag{15.4.5}$$

An iterate $\mathbf{m}_i$ that satisfies (15.4.5) is said to be *optimal* with respect to the directions $\mathbf{h}_j$. Clearly, (15.4.5) implies that $\chi(\mathbf{m}_i + \gamma\mathbf{h}_j)$ has a minimum at $\gamma = 0$. Thus, its derivative must vanish at $\gamma = 0$, which is equivalent to

$$\mathbf{h}_j^T\mathbf{r}_i = 0, \quad j = 0, ..., i-1. \tag{15.4.6}$$

Equation (15.4.6) means that $\mathbf{m}_i$ is optimal with respect to all the previous directions when they are orthogonal to the current residual $\mathbf{r}_i = \mathbf{b} - \mathbf{H}\mathbf{m}_i$. Interestingly, we have already computed an optimal model without wanting to during the first iteration of the steepest-descent method. The first steepest-descent update $\mathbf{m}_1$ is optimal with respect to the initial descent direction $-\nabla\chi(\mathbf{m}_0) = \mathbf{h}_0 = \mathbf{r}_0$. This property can be deduced from

$$\mathbf{h}_0^T\mathbf{r}_1 = \mathbf{r}_0^T(\mathbf{b} - \mathbf{H}\mathbf{m}_0 - \gamma_0\mathbf{H}\mathbf{h}_0) = \mathbf{r}_0^T\mathbf{r}_0 - \gamma_0\mathbf{r}_0^T\mathbf{H}\mathbf{h}_0 = 0, \tag{15.4.7}$$

where the last identity follows from the subsitution of γ_0 by (15.4.3) for $i = 0$. Thus, in a steepest-descent method, the first update is, by design, optimal to the previous descent direction.

How then can we automatically construct a sequence of descent directions such that the current iterate is always optimal with respect to its predecessors? We consider a descent direction $\mathbf{h}_i$ that updates $\mathbf{m}_i$ to $\mathbf{m}_{i+1}$, so that

$$\mathbf{m}_{i+1} = \mathbf{m}_i + \mathbf{h}_i. \tag{15.4.8}$$

From (15.4.6), we find a necessary condition for $\mathbf{h}_i$,

$$\mathbf{h}_j^T\mathbf{r}_{i+1} = \mathbf{h}_j^T(\mathbf{b} - \mathbf{H}\mathbf{m}_i - \mathbf{H}\mathbf{h}_i) = \mathbf{h}_j^T(\mathbf{r}_i - \mathbf{H}\mathbf{h}_i). \tag{15.4.9}$$

With the assumption that (15.4.6) is satisfied for all $j = 0, ..., i-1$, we then find that $\mathbf{r}_{i+1}$ is also optimal to the previous descent directions, when

$$\mathbf{h}_j^T\mathbf{H}\mathbf{h}_i = 0, \quad j = 0, ..., i-1. \tag{15.4.10}$$

Vectors $\mathbf{h}_j$ and $\mathbf{h}_i$ that satisfy (15.4.10) are said to be *H-orthogonal*. Equation

(15.4.10) thus implies that descent directions must be mutually H-orthogonal in order to ensure that successive iterates remain optimal with respect to the earlier descent directions.

Now we are equipped with a criterion for the necessary character of the descent directions, we proceed with their actual construction. For simplicity, we try an ansatz where the next descent direction $\mathbf{h}_{i+1}$ is equal to the direction of steepest descent, $-\nabla\chi(\mathbf{m}_{i+1}) = \mathbf{r}_{i+1}$ plus a scaled version of the previous descent direction $\mathbf{h}_i$,

$$\mathbf{h}_{i+1} = \mathbf{r}_{i+1} - \beta_i \mathbf{h}_i. \tag{15.4.11}$$

The requirement that condition (15.4.10) holds for $\mathbf{h}_i$ and $\mathbf{h}_{i+1}$, gives an expression for the scaling factor β_i:

$$\beta_i = \frac{\mathbf{h}_i^T \mathbf{H} \mathbf{r}_{i+1}}{\mathbf{h}_i^T \mathbf{H} \mathbf{h}_i}. \tag{15.4.12}$$

By induction we can show that property $\mathbf{h}_j^T \mathbf{H} \mathbf{h}_{i+1} = 0$ also holds for all $j = 0, ..., i-1$. For a detailed proof, the reader is referred to Nocedal & Wright (1999) or Quarteroni et al. (2000).

We can collect the previous equations into a *conjugate-gradient* algorithm:

(1) Choose an initial model, $\mathbf{m}_0$. Set $i = 0$ and $\mathbf{h}_0 = \mathbf{r}_0$.

(2) Compute the optimal step length γ_i as $\gamma_i = \mathbf{h}_i^T \mathbf{r}_i / (\mathbf{h}_i^T \mathbf{H}\, \mathbf{h}_i)$.

(3) Update $\mathbf{m}_i$ via $\mathbf{m}_{i+1} = \mathbf{m}_i + \gamma_i\, \mathbf{h}_i$.

(3) Compute the residual (gradient) for the next iterate, $\mathbf{r}_{i+1}$.

(4) Compute the descent direction for the next iteration, $\mathbf{h}_{i+1}$, using $\beta_i = \mathbf{h}_i^T \mathbf{H} \mathbf{r}_{i+1} / (\mathbf{h}_i^T \mathbf{H} \mathbf{h}_i)$ and $\mathbf{h}_{i+1} = \mathbf{r}_{i+1} - \beta_i\, \mathbf{h}_i$.

(5) Set $i \to i+1$, go back to (2) and repeat as often as needed.

Proving that the conjugate-gradient method has beneficial convergence properties is tedious, but can be understood intuitively. Each iteration adds a new orthogonal descent direction. Thus, after N_M iterations no more orthogonal descent directions are left. Being unable to descend further, the algorithm terminates at the global minimum.

The conjugate-gradient method for non-quadratic misfit functions

The conjugate-gradient method for general, non-quadratic misfit functions is essentially a leap of faith from the quadratic version we have just considered. The method rests on the simple realisation that the residual is equal to the negative gradient of the misfit, that is,

$$\mathbf{r}_i = -\nabla\chi(\mathbf{m}_i). \tag{15.4.13}$$

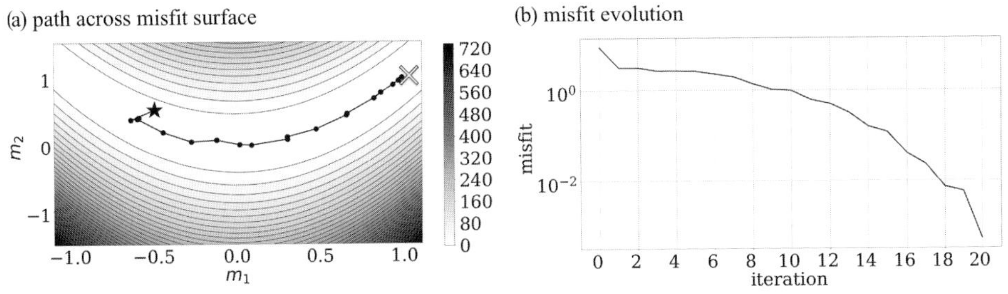

Figure 15.3 Illustration of the conjugate-gradient optimisation of the Rosenbrock function (15.2.4). (a) Path across the misfit surface from the initial model (star) towards the global minimum (cross). Each of the black dots represents a new update. (b) After 20 iterations, the initial misfit is reduced by nearly three orders of magnitude, and the minimum is approximated closely.

Thus, replacing residuals in the original algorithm by negative gradients, we obtain the conjugate-gradient algorithm for non-quadratic misfit functions:

(1) Choose an initial model, $\mathbf{m}_0$. Set $i = 0$ and $\mathbf{h}_0 = -\nabla\chi(\mathbf{m}_0)$.

(2) Compute the optimal step length γ_i, either according to (15.1.14) or using a line search.

(3) Update $\mathbf{m}_i$ via $\mathbf{m}_{i+1} = \mathbf{m}_i + \gamma_i\,\mathbf{h}_i$.

(4) Compute the gradient for the next iterate, $\nabla\chi(\mathbf{m}_{i+1})$.

(5) Compute the descent direction for the next iteration, $\mathbf{h}_{i+1}$, according to $\beta_i = -|\nabla\chi(\mathbf{m}_{i+1})|_2^2/|\nabla\chi(\mathbf{m}_i)|_2^2$ and $\mathbf{h}_{i+1} = -\nabla\chi(\mathbf{m}_{i+1}) - \beta_i\,\mathbf{h}_i$.

(6) Set $i \to i+1$, go back to (2) and repeat as often as needed.

Of course, for a non-quadratic misfit, the descent directions are not generally orthogonal, and the algorithm will not generally converge to the minimum within N_M iterations. Nevertheless, the conjugate-gradient method usually converges faster than the steepest-descent method. However, the use of just first derivatives, tends to result in slower convergence than for Newton's method. This trend is confirmed by the comparison of Figures 15.1, 15.2, and 15.3.

15.5 Quasi-Newton Methods

Quasi-Newton methods aim to circumvent the expensive calculation of the full Hessian matrix $\mathbf{H}$ with a sequence of approximations to the inverse Hessian. The general descent scheme (15.1.8) is used with matrices $\mathbf{A}_i$ that successively approximate the inverse Hessian $\mathbf{H}^{-1}$. Starting from some initial guess $\mathbf{A}_0$, the matrix is updated at each iteration to yield matrices $\mathbf{A}_1, \mathbf{A}_2, ...$, with each being hopefully a better approximation of $\mathbf{H}^{-1}$ than its predecessor. The descent directions are then chosen as $\mathbf{h}_i = -\mathbf{A}_i\,\nabla\chi(\mathbf{m}_i)$.

To illustrate the basic idea of how to obtain the sequence of matrices $\mathbf{A}_i$, we begin with the simplifying assumption that the misfit χ is perfectly quadratic in $\mathbf{m}$:

$$\chi(\mathbf{m}) = \tfrac{1}{2}\mathbf{m}^T\mathbf{H}\mathbf{m} + \mathbf{b}^T\mathbf{m}. \tag{15.5.1}$$

Using the first derivative of (15.5.1), we obtain

$$\mathbf{H}\,(\mathbf{m}_{i+1} - \mathbf{m}_i) = \nabla\chi(\mathbf{m}_{i+1}) - \nabla\chi(\mathbf{m}_i). \tag{15.5.2}$$

The result (15.5.1) suggests that, for $\mathbf{A}_{i+1}$ to be a reasonable approximation of $\mathbf{H}^{-1}$, we should to impose the condition

$$\mathbf{A}_{i+1}^{-1}\,(\mathbf{m}_{i+1} - \mathbf{m}_i) = \nabla\chi(\mathbf{m}_{i+1}) - \nabla\chi(\mathbf{m}_i). \tag{15.5.3}$$

Equation (15.5.3), known as the *quasi-Newton equation*, does not uniquely determine the matrix $\mathbf{A}_{i+1}$ or its inverse. Numerous options have been proposed, some of which will be explained in the following paragraphs.

The BFGS method

Simultaneously studied by C.G. Broyden, R. Fletcher, D. Goldfarb and D. Shanno (Broyden, 1970; Fletcher, 1970; Goldfarb 1970; Shanno, 1970), the method named after them applies specific modifications to the current matrix $\mathbf{A}_i$ to obtain an updated matrix $\mathbf{A}_{i+1}$ such that the quasi-Newton equation (15.5.3) remains satisfied.

To simplify the subsequent notation we set

$$\mathbf{s}_i = \mathbf{m}_{i+1} - \mathbf{m}_i, \qquad \mathbf{y}_i = \nabla\chi(\mathbf{m}_{i+1}) - \nabla\chi(\mathbf{m}_i). \tag{15.5.4}$$

With these definitions, the quasi-Newton equation takes the simplified form

$$\mathbf{A}_{i+1}^{-1}\mathbf{s}_i = \mathbf{y}_i. \tag{15.5.5}$$

As a first step, we compute an auxiliary matrix,

$$\tilde{\mathbf{A}}_i^{-1} = \mathbf{A}_i^{-1} - \frac{(\mathbf{A}_i^{-1}\mathbf{s}_i)(\mathbf{A}_i^{-1}\mathbf{s}_i)^T}{\mathbf{s}_i^T\mathbf{A}_i^{-1}\mathbf{s}_i}. \tag{15.5.6}$$

The auxiliary matrix in (15.5.6) constitutes a *rank-1 modification* of $\mathbf{A}_i^{-1}$ because the updated matrix $(\mathbf{A}_i^{-1}\mathbf{s}_i)(\mathbf{A}_i^{-1}\mathbf{s}_i)^T$ has rank 1. Multiplying $\tilde{\mathbf{A}}_i^{-1}$ with $\mathbf{s}_i$ from the right, we find $\tilde{\mathbf{A}}_i^{-1}\mathbf{s}_i = \mathbf{0}$. To satisfy the quasi-Newton equation, we add another rank-1 modification,

$$\mathbf{A}_{i+1}^{-1} = \tilde{\mathbf{A}}_i^{-1} + \frac{\mathbf{y}_i\mathbf{y}_i^T}{\mathbf{y}_i^T\mathbf{s}_i}. \tag{15.5.7}$$

By combining (15.5.6) and (15.5.7), we find the complete BFGS update formula

$$\mathbf{A}_{i+1}^{-1} = \mathbf{A}_i^{-1} - \frac{(\mathbf{A}_i^{-1}\mathbf{s}_i)(\mathbf{A}_i^{-1}\mathbf{s}_i)^T}{\mathbf{s}_i^T\mathbf{A}_i^{-1}\mathbf{s}_i} + \frac{\mathbf{y}_i\mathbf{y}_i^T}{\mathbf{y}_i^T\mathbf{s}_i}. \tag{15.5.8}$$

The matrices $\mathbf{A}_i^{-1}$ are approximations of the Hessian $\mathbf{H}$ in the sense of the quasi-Newton equation. They can be used to compute descent directions $\mathbf{h}_i =$

$-\mathbf{A}_i\,\nabla\chi(\mathbf{m}_i)$. To avoid the potentially expensive inversion of $\mathbf{A}_i^{-1}$, updates of $\mathbf{A}_i$ may also be computed directly using the iteration

$$\mathbf{A}_{i+1} = \left(\mathbf{I} - \rho_i \mathbf{s}_i \mathbf{y}_i^T\right) \mathbf{A}_i \left(\mathbf{I} - \rho_i \mathbf{y}_i \mathbf{s}_i^T\right) + \rho_i \mathbf{s}_i \mathbf{s}_i^T, \tag{15.5.9}$$

with

$$\rho_i = \frac{1}{\mathbf{y}_i^T \mathbf{s}_i}. \tag{15.5.10}$$

A demonstration of the vailidity of (15.5.9) can be shown via the rather tedious procedure of multiplying the equation by $\mathbf{A}_{i+1}^{-1}$ from (15.5.8).

We may now formulate the BFGS algorithm:

(1) Choose an initial model, $\mathbf{m}_0$, and an initial, positive definite matrix $\mathbf{A}_0$. Set $i = 0$.

(2) Compute the descent direction $\mathbf{h}_i = -\mathbf{A}_i\,\nabla\chi(\mathbf{m}_i)$.

(3) Compute the optimal step length $\gamma_i = -\mathbf{h}_i^T \nabla\chi(\mathbf{m}_i)/(\mathbf{h}_i^T \mathbf{H} \mathbf{h}_i)$.

(4) Update $\mathbf{m}_i$ via $\mathbf{m}_{i+1} = \mathbf{m}_i + \gamma_i\,\mathbf{h}_i$.

(5) Compute the new gradient, $\nabla\chi(\mathbf{m}_{i+1})$, and the differences $\mathbf{s}_i = \mathbf{m}_{i+1} - \mathbf{m}_i$ and $\mathbf{y}_i = \nabla\chi(\mathbf{m}_{i+1}) - \nabla\chi(\mathbf{m}_i)$.

(6) Update the matrix $\mathbf{A}_{i+1} = \left(\mathbf{I} - \rho_i \mathbf{s}_i \mathbf{y}_i^T\right) \mathbf{A}_i \left(\mathbf{I} - \rho_i \mathbf{y}_i \mathbf{s}_i^T\right) + \rho_i \mathbf{s}_i \mathbf{s}_i^T$ with $\rho_i = 1/\mathbf{y}_i^T \mathbf{s}_i$.

(7) Set $i \to i + 1$, go back to (2) and repeat as often as needed.

The update formulas (15.5.8) and (15.5.9) may seem somewhat complicated as a means of satisfying the quasi-Newton equation. However, for quadratic misfit functions it can be shown that these choices have additional benefits (e.g., Nocedal & Wright, 1999). Most importantly, the BFGS iteration produces descent directions that are mutually $\mathbf{H}$-orthogonal in the sense of (15.4.10). This orthogonality implies that the iteration terminates after at most N_M iterations with the exact optimal model, $\tilde{\mathbf{m}} = \mathbf{m}_{N_M}$; assuming, of course, that rounding errors can be neglected. Furthermore, the matrices $\mathbf{A}_i$ converge towards the inverse Hessian, and after N_M iterations we have $\mathbf{A}_{N_M} = \mathbf{H}^{-1}$.

As in the case of the conjugate-gradient method, the BFGS method for quadratic misfit functions is translated almost directly to general, non-quadratic misfit functionals. The only components of the algorithm that need adaptation are those where the Hessian $\mathbf{H}$ is required explicitly, that is, the computation of the optimal step length γ_i. The estimate of step length is most frequently replaced by a line search.

As in the examples from the previous sections, we again consider the minimisation of the Rosenbrock function (15.2.4), which is shown in Figure 15.4. While not converging as fast as the Newton method, the BFGS method still

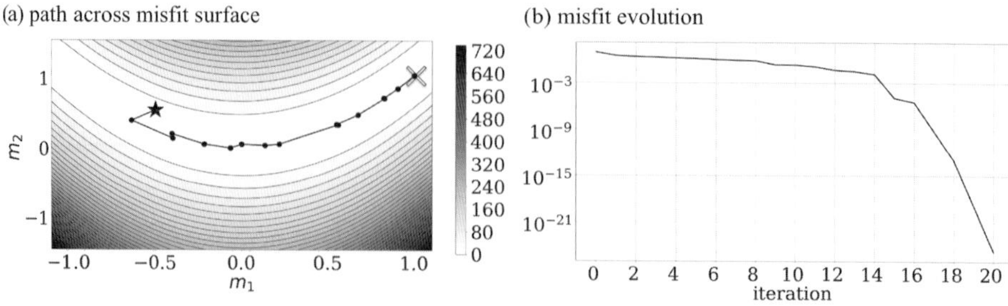

Figure 15.4 Illustration of BFGS optimisation of the Rosenbrock function (15.2.4). (a) Path across the misfit surface from the initial model (star) towards the global minimum (cross). Each of the black dots represents a new update. (b) After 20 iterations, the initial misfit is reduced by nearly 22 orders of magnitude, and the minimum is reached to within machine precision.

succeeds in reducing the initial misfit by nearly 22 orders of magnitude within only 20 iterations.

The L-BFGS method

A drawback of the BFGS method is the need to explicitly compute and store the inverse Hessian approximations $\mathbf{A}_i$. The dimension of $\mathbf{A}_i$ is $N_M \times N_M$. As a result the computational and storage requirements of BFGS quickly become prohibitive as the model space dimension increases. This deficiency motivates an approximate version of BFGS, known as *limited-memory BFGS* or the *L-BFGS method* (Nocedal, 1980).

Before explaining the details of L-BFGS, we demonstrate that the BFGS descent directions can in fact be computed through a sequence of vector–vector multiplications, without any need to explicitly compute or store a matrix. For this, we write the BFGS update of the matrix $\mathbf{A}_i$ in (15.5.9) in the more condensed form

$$\mathbf{A}_{i+1} = \mathbf{V}_i^T \mathbf{A}_i \mathbf{V}_i + \rho_i \mathbf{s}_i \mathbf{s}_i^T, \tag{15.5.11}$$

with the matrix $\mathbf{V}_i$ defined as

$$\mathbf{V}_i = \mathbf{I} - \rho_i \mathbf{y}_i \mathbf{s}_i^T. \tag{15.5.12}$$

Starting from the initial matrix $\mathbf{A}_0$, the repeated application of (15.5.11) yields

$$\begin{aligned}\mathbf{A}_i = &\left(\mathbf{V}_{i-1}^T \cdots \mathbf{V}_0^T\right) \mathbf{A}_0 \left(\mathbf{V}_0 \cdots \mathbf{V}_{i-1}\right)\\ &+ \rho_0 \left(\mathbf{V}_{i-1}^T \cdots \mathbf{V}_1^T\right) \mathbf{s}_0 \mathbf{s}_0^T \left(\mathbf{V}_1 \cdots \mathbf{V}_{i-1}\right)\\ &+ \rho_1 \left(\mathbf{V}_{i-1}^T \cdots \mathbf{V}_2^T\right) \mathbf{s}_1 \mathbf{s}_1^T \left(\mathbf{V}_2 \cdots \mathbf{V}_{i-1}\right)\\ &+ \ldots\\ &+ \rho_{i-1} \mathbf{s}_{i-1} \mathbf{s}_{i-1}^T.\end{aligned} \tag{15.5.13}$$

To see how we can obtain the descent direction $\mathbf{h}_i = -\mathbf{A}_i \nabla \chi(\mathbf{m}_i)$ via (15.5.13) without explicitly building a matrix, we start with the term

$\mathbf{A}_0\left(\mathbf{V}_{i-1}\cdots\mathbf{V}_0^T\right)\nabla\chi(\mathbf{m}_i)$, which can be computed with the following iterative algorithm (I):

(1) Set the vector $\mathbf{q}=\nabla\chi(\mathbf{m}_i)$, and the iteration index $n=i-1$.

(2) Compute the scalar $\alpha_n=\rho_n\mathbf{s}_n^T\mathbf{q}$, with ρ_n defined in (15.5.10).

(3) Set $\mathbf{q}\rightarrow\mathbf{q}-\alpha_n\mathbf{y}_n$.

(4) Decrease the iteration index by 1, that is, $n\rightarrow n-1$. If $n\geq 0$ go back to 2. Otherwise go to step 5.

(5) Compute the vector $\mathbf{r}=\mathbf{A}_0\mathbf{q}$.

As one can see by writing the algorithm explicitly for $n=i-1,...,0$, the following sequence of $\mathbf{q}$ vectors are generated:

$$\{\nabla\chi(\mathbf{m}_i),\ \mathbf{V}_{i-1}\nabla\chi(\mathbf{m}_i),\ \mathbf{V}_{i-2}\mathbf{V}_{i-1}\nabla\chi(\mathbf{m}_i),\ \ldots,\ \mathbf{V}_0\cdots\mathbf{V}_{i-1}\nabla\chi(\mathbf{m}_i)\}. \quad (15.5.14)$$

The final multiplication with $\mathbf{A}_0$ then yields $\mathbf{r}=\mathbf{A}_0\left(\mathbf{V}_0\cdots\mathbf{V}_{i-1}\right)\nabla\chi(\mathbf{m}_i)$, as desired. Since $\mathbf{q}$ from the individual iterations is subsequently not needed, its values are overwritten, which saves storage. Along the way, iteration I also produces a series of scalars α_n,

$$\begin{aligned}
\alpha_{i-1} &= \rho_{i-1}\mathbf{s}_{i-1}^T\nabla\chi(\mathbf{m}_i),\\
\alpha_{i-2} &= \rho_{i-2}\mathbf{s}_{i-2}^T\mathbf{V}_{i-1}\nabla\chi(\mathbf{m}_i),\\
\alpha_{i-3} &= \rho_{i-3}\mathbf{s}_{i-3}^T\mathbf{V}_{i-2}\mathbf{V}_{i-1}\nabla\chi(\mathbf{m}_i),\\
\ldots &= \ldots,\\
\alpha_0 &= \rho_0\mathbf{s}_0^T\mathbf{V}_1\cdots\mathbf{V}_{i-1}\nabla\chi(\mathbf{m}_i).
\end{aligned} \quad (15.5.15)$$

The comparison of this sequence with (15.5.13) suggests that the values of α_n can be reused to complete the computation of $\mathbf{A}_i\nabla\chi(\mathbf{m}_i)$. Indeed, this can be done using a second iterative algorithm (II):

(1) Set the iteration index $n=0$.

(2) Compute the scalar $\beta=\rho_n\mathbf{y}_n^T\mathbf{r}$.

(3) Set $\mathbf{r}\rightarrow\mathbf{r}+\mathbf{s}_n(\alpha_n-\beta)$.

(4) Increase the iteration index by 1, that is, $n\rightarrow n+1$. If $n\geq i-1$ go back to step 2. Otherwise stop.

In the first iteration, for $n=0$, the algorithm II transforms the initial $\mathbf{r}=\mathbf{A}_0\left(\mathbf{V}_0\cdots\mathbf{V}_{i-1}\right)\nabla\chi(\mathbf{m}_i)$ from iterative scheme I into the new value

$$\mathbf{r}=\mathbf{V}_0^T\mathbf{A}_0\left(\mathbf{V}_0\cdots\mathbf{V}_{i-1}\right)\nabla\chi(\mathbf{m}_i)+\rho_0\mathbf{s}_0\mathbf{s}_0^T\left(\mathbf{V}_1\cdots\mathbf{V}_{i-1}\right)\nabla\chi(\mathbf{m}_i). \quad (15.5.16)$$

In other words, the initial $\mathbf{r}$ is left-multiplied by $\mathbf{V}_0^T$ and the term $\rho_0\mathbf{s}_0\mathbf{s}_0^T\left(\mathbf{V}_1\cdots\mathbf{V}_{i-1}\right)\nabla\chi(\mathbf{m}_i)$ is added. In the next iteration, the algorithm then left-multiplies by $\mathbf{V}_1^T$ and adds $\rho_1 \mathbf{s}_1 \mathbf{s}_1^T (\mathbf{V}_2 \cdots \mathbf{V}_{i-1}) \nabla\chi(\mathbf{m}_i)$, and so on and so forth. The final $\mathbf{r}$ is then the negative descent direction in the ith BFGS iteration,

$$\mathbf{r} = \mathbf{A}_i \nabla\chi(\mathbf{m}_i) = -\mathbf{h}_i. \tag{15.5.17}$$

The only matrix–vector operation contained in iterative schemes I and II is the multiplication of $\mathbf{A}_0$ with $\mathbf{q}$ in $\mathbf{r} = \mathbf{A}_0\mathbf{q}$. However, choosing the initial matrix $\mathbf{A}_0$ to be diagonal, the computational and storage requirements of $\mathbf{A}_0\mathbf{q}$ are the same as for a vector–vector multiplication. Thus, we have already shown that the BFGS algorithm can be used without explicitly computing and storing a large, and usually dense, matrix.

The problem that remains to be solved is the increasing number of vectors $\mathbf{s}_i$ and $\mathbf{y}_i$ that need to be stored. In each BFGS iteration, a new pair of vectors is added. In the ideal case of a quadratic misfit function, where the BFGS iteration converges in N_M steps, we would need to store $2N_M$ vectors of dimension N_M, that is, $2N_M^2$ numbers. We are therefore back to our original problem of having to store a quantity of numbers that grows quadratically with model-space dimension.

At this point, the basic idea of L-BFGS is very simple: Instead of using all vector pairs $(\mathbf{s}_0, \mathbf{y}_0), \ldots, (\mathbf{s}_{i-1}, \mathbf{y}_{i-1})$ in iterative schemes I and II, we only use the last k vector pairs $(\mathbf{s}_{i-k}, \mathbf{y}_{i-k}), \ldots, (\mathbf{s}_{i-1}, \mathbf{y}_{i-1})$, where k is typically a small number between 3 and 20. After each completed model update, we delete the vector pair with the lowest index, and we store the next vector pair. For the first k iterations, all vector pairs are used, as in the original BFGS algorithm. Limiting the number of vector pairs to k, reduces the storage requirements to $2kN_N$. The price that has to be paid is an incorrect computation of the descent directions.

The L-BFGS algorithm can be summarised as follows:

(1) Choose an initial model, $\mathbf{m}_0$, a maximum number of vector pairs, k, and set the iteration index $i = 0$.

(2) If $i = 0$, compute the descent direction $\mathbf{h}_0 = -\nabla\chi(\mathbf{m}_0)$. Otherwise, compute the descent direction $\mathbf{h}_i$ by running iterations I and II with the k most recent vector pairs $(\mathbf{s}_{i-k}, \mathbf{y}_{i-k}), \ldots, (\mathbf{s}_{i-1}, \mathbf{y}_{i-1})$ and an initial, positive definite matrix $\mathbf{A}_{0,i}$.

(3) Compute the optimal step length, γ_i, using a line search.

(4) Update $\mathbf{m}_i$ via $\mathbf{m}_{i+1} = \mathbf{m}_i + \gamma_i \mathbf{h}_i$.

(5) Compute the new gradient, $\nabla\chi(\mathbf{m}_{i+1})$, and the differences $\mathbf{s}_i = \mathbf{m}_{i+1} - \mathbf{m}_i$ and $\mathbf{y}_i = \nabla\chi(\mathbf{m}_{i+1}) - \nabla\chi(\mathbf{m}_i)$.

(6) Update the set of vector pairs by deleting the one with the lowest index, and storing the new one, computed in the previous step.

(7) Set $i \to i + 1$, go back to (2) and repeat as often as needed.

An interesting aspect of the L-BFGS method is that the initial matrix $\mathbf{A}_{0,i}$ may be chosen differently for each iteration. The only requirements are that $\mathbf{A}_{0,i}$ be positive definite and nearly diagonal, so that $\mathbf{A}_{0,i}\mathbf{q}$ can be computed easily.

The results for the L-BFGS method applied to the Rosenbrock function are essentially indistinguishable from those shown in Figure 15.4 for the full BFGS method. Once again, strong minimisation is achieved in a moderate numbers of iterations.

15.6 Subspace Methods

In the gradient methods we exploit the local behaviour of the misfit function χ. All the forms we have considered exploit different aspects of the locally quadratic expansion about a current model $\mathbf{m}_c$ obtained by truncating the Taylor's series for the misfit function

$$\chi^Q(\mathbf{m}_c + \delta\mathbf{m}) = \chi(\mathbf{m}_c) + \mathbf{g} \cdot \delta\mathbf{m} + \tfrac{1}{2}\delta\mathbf{m} \cdot \mathbf{H} \cdot \delta\mathbf{m}, \tag{15.6.1}$$

in terms of the local gradient $\mathbf{g}$ and Hessian matrix $\mathbf{H}$:

$$\mathbf{g} = \nabla_m \chi(\mathbf{m}_c), \qquad \mathbf{H} = \nabla_m \nabla_m \chi(\mathbf{m}_c). \tag{15.6.2}$$

The gradient approaches exploit a single direction in each iteration, but there is potentially more information that can be used.

A different class of algorithms can be developed by minimising the quadratic approximation χ^Q in a small n-dimensional subspace of the full model parameter space (e.g., Kennett et al., 1988). A projection is made onto a set of n basis vectors $\{\mathbf{b}^j\}$ with the aid of a projection matrix $\mathbf{B}$ composed of the components of these vectors. We construct a perturbation to the current model in the subspace spanned by the $\{\mathbf{b}^j\}$:

$$\delta\mathbf{m} = \sum_{j=1}^{n} \mu_j \, \mathbf{b}^j. \tag{15.6.3}$$

The set of coefficients $\{\mu_j\}$ are then to be found by minimising χ^Q for this restricted class of perturbation. With the choice (15.6.3) we can express χ^Q as

$$\chi^Q(\mathbf{m}_c + \delta\mathbf{m}) = \chi(\mathbf{m}_c) + \sum_{j=1}^{n} \mu_j \, \mathbf{g} \cdot \mathbf{b}^j + \tfrac{1}{2} \sum_{j=1}^{n} \sum_{k=1}^{n} \mu_j \mu_k \, \mathbf{b}^k \cdot \mathbf{H} \cdot \mathbf{b}^j. \tag{15.6.4}$$

Now minimising χ^Q with respect to μ_j we require $\partial\chi^Q/\partial\mu_j = 0$, so that

$$\mathbf{g} \cdot \mathbf{b}^j + \sum_{k=1}^{n} \mu_k \, \mathbf{b}^k \cdot \mathbf{H} \cdot \mathbf{b}^j = 0. \tag{15.6.5}$$

In terms of the projection matrix $\mathbf{B}$, (15.6.5) can be rewritten as

$$\mathbf{B}^T\mathbf{g} + \mathbf{B}^T \, \mathbf{H} \, \mathbf{B}\mu = 0. \tag{15.6.6}$$

The coefficients of the model perturbation can therefore be determined from the projection of the gradient onto the subspace:

$$\mu = -[\mathbf{B}^{\mathrm{T}}\,\mathbf{H}\,\mathbf{B}]^{-1}\mathbf{B}^{\mathrm{T}}\mathbf{g}. \tag{15.6.7}$$

The projected Hessian is a small $n \times n$ matrix, which with suitable choices of the basis vectors $\{\mathbf{b}^j\}$ is generally well conditioned.

The model perturbation $\delta\mathbf{m}$ can be recovered by projecting back onto the full model space as

$$\delta\mathbf{m} = -\mathbf{B}[\mathbf{B}^{\mathrm{T}}\,\mathbf{H}\,\mathbf{B}]^{-1}\mathbf{B}^{\mathrm{T}}\mathbf{g}. \tag{15.6.8}$$

Suitable choices for the basis vectors $\{\mathbf{b}^j\}$ will normally be related to the gradient vector $\mathbf{g}$ and its rate of change, and then (15.6.8) can be regarded as combining gradient and matrix estimation procedures.

The subspace algorithm thus takes the form:

(1) Construct χ^Q for the current model $\mathbf{m}_c$.

(2) Extract a model update estimate $\delta\mathbf{m}$ from (15.6.8).

(3) Create a new model estimate $\mathbf{m} = \mathbf{m}_c + \delta\mathbf{m}$.

(4) If $\chi(\mathbf{m})$ satisfies a suitable termination criterion stop, otherwise return to (1).

The subspace method is very effective for problems where the computational cost of producing an approximation to the Hessian matrix is not too large, such as travel-time tomography (e.g., Rawlinson & Urvoy, 2006).

Subspace techniques for multiple parameter types

The subspace approach has particular merit in a situations where inversion is carried for multiple types of parameters since it is able to balance the contributions associated with the different parameter classes. With P different types of parameters, the full model space is a product space $\mathbf{M} = \mathbf{M}_A \times \mathbf{M}_B \times \mathbf{M}_C \times \ldots$, As the relative size of the units of the different parameter classes are changed, the direction of the gradient vector for the misfit function χ will be modified even though the physical scenario is the same. Similar effects arise with non-dimensional quantities that depend on the reference values employed. Such scaling issues can have a significant effect on the efficacy of direct gradient methods.

The subspace method provides a convenient way to exploit the dependence of the objective function χ on each parameter class by partitioning the gradient vector $\mathbf{g}$:

$$\mathbf{g} = [\mathbf{g}_A, \mathbf{g}_B, \mathbf{g}_C, \ldots], \tag{15.6.9}$$

where $\mathbf{g}_A = \partial\chi/\partial\mathbf{m}_A$. With the choice (15.6.3) we have assumed that the basis vectors $\{\mathbf{b}^j\}$ lie in model space, but the gradient $\mathbf{g}$ lies in dual space. With a model covariance matrix of the form

$$\mathbf{C}_{\mathrm{m}} = \begin{pmatrix} \mathbf{C}_{\mathrm{AA}} & \mathbf{C}_{\mathrm{AB}} & & 0 \\ \mathbf{C}_{\mathrm{BA}} & \mathbf{C}_{\mathrm{BB}} & & \\ & & \mathbf{C}_{\mathrm{CC}} & \\ 0 & & & \ddots \end{pmatrix} \tag{15.6.10}$$

we can extract gradient components for each parameter type for use as basis vectors. For parameter type K we define

$$\bar{\mathbf{g}}_{\mathrm{K}} = \mathbf{C}_{\mathrm{m}}[\ldots, 0, \mathbf{g}_{\mathrm{K}}, 0, \ldots]. \tag{15.6.11}$$

Using (15.6.10) the parameter contributions are

$$\begin{aligned} \bar{\mathbf{g}}_{\mathrm{A}} &= [\mathbf{C}_{\mathrm{AA}}\mathbf{g}_{\mathrm{A}}, \mathbf{C}_{\mathrm{AB}}\mathbf{g}_{\mathrm{B}}, 0, \ldots], \\ \bar{\mathbf{g}}_{\mathrm{B}} &= [\mathbf{C}_{\mathrm{BA}}\mathbf{g}_{\mathrm{A}}, \mathbf{C}_{\mathrm{BB}}\mathbf{g}_{\mathrm{B}}, 0, \ldots], \\ \bar{\mathbf{g}}_{\mathrm{A}} &= [0, 0, \mathbf{C}_{\mathrm{CC}}\mathbf{g}_{\mathrm{C}}, 0, \ldots], \\ & \ldots . \end{aligned} \tag{15.6.12}$$

The off-diagonal blocks in the model covariance allow for specific cross-coupling between parameter classes where this is a feature of the problem.

We now adopt the set of P gradient vectors $\bar{\mathbf{g}}_{\mathrm{K}}$ as the directions of the set of basis vectors

$$\mathbf{b}^{(1)} = \|\bar{\mathbf{g}}_{\mathrm{A}}\|^{-1}\,\bar{\mathbf{g}}_{\mathrm{A}}, \quad \mathbf{b}^{(2)} = \|\bar{\mathbf{g}}_{\mathrm{B}}\|^{-1}\,\bar{\mathbf{g}}_{\mathrm{B}}, \quad \ldots . \tag{15.6.13}$$

with a suitable normalisation. When cross-coupling exists between parameter sets as in (15.6.12) it is desirable to avoid linear dependence between the different basis vectors, which can be achieved by orthogonalisation – removing the component parallel to any other vectors.

When P is greater than 4, the gradient vectors for the different parameter classes are normally a sufficient basis. For a small number of parameters it is useful to add in additional vectors representing the rate of change of the $\bar{\mathbf{g}}_{\mathrm{K}}$ (Kennett et al., 1988). The perturbation to the current model is once again estimated once from (15.6.8):

$$\delta\mathbf{m} = -\mathbf{B}_{\mathrm{P}}[\mathbf{B}_{\mathrm{P}}^{\mathrm{T}}\,\mathbf{H}\,\mathbf{B}_{\mathrm{P}}]^{-1}\mathbf{B}_{\mathrm{P}}^{\mathrm{T}}\,\mathbf{g}, \tag{15.6.14}$$

using the multi-parameter projection matrix $\mathbf{B}_{\mathrm{P}}$.

The subspace method essentially performs a local least-squares inversion within the subspace, spanned by vectors that reflect the dependence on all the parameter classes. The model step is independent of the scaling of the particular parameter types. The weighting between parameter classes is determined solely by the behaviour of the misfit functional χ. This means that it is possible to avoid the bias that can be created when disparate parameter types are used to construct a single gradient vector. Sambridge et al. (1991) have shown that the subspace approach can provide a suitable way of estimating multiple types of physical parameters in the context of the inversion of seismic reflection waveforms, because it makes optimal use of the individual descent directions for the different parameters.

15.7 No Free Lunch

The multitude of optimisation methods, of which we could cover only a few, naturally raises the question of which one is the best. A distinct benefit of iterative linearisation, introduced in Section 14.3, is the possibility to compute the posterior covariance, the resolution matrix, or other related quality measures, in the vicinity of the optimum. When the misfit function is nearly quadratic, iterative minimisation often converges more quickly than steepest-descent and conjugate-gradient methods. This comes at the price of having to compute the Jacobian **J** explicitly, which may be impossible unless the forward operator **G** can be differentiated (semi-)analytically. In waveform inversion, the conjugate-gradient and L-BFGS methods discussed in this chapter are widely used. They avoid the explicit computation of the Jacobian or the inverse Hessian, and frequently they converge faster than steepest descent. However, this comes at the expense of not having information on the quality of the final model, e.g., a resolution or posterior covariance estimate. This brief discussion already suggests that the search for the best optimisation algorithm is complicated because it depends on the specifics of the problem (e.g., the possibility of easy differentiation of **G**) and on our personal priorities (e.g., if fast convergence is more important than a resolution estimate).

To formalise these considerations, we consider the model space $\mathbb{M}$ such that the components m_i of each N_M-dimensional model $\mathbf{m} \in \mathbb{M}$ can only take a finite number of values, for instance, the numbers that can be represented on a computer. In this sense, the model space consists of all possible combinations of N_M values. We denote the total number of these combinations by $|\mathbb{M}|$. For a simple illustration, we consider the case shown in Figure 15.5, where $N_M = 2$. Each of the two model parameters, m_1 and m_2, can only take the values 0 or 1. It follows that the total number of different models is $|\mathbb{M}| = 2^2 = 4$.

Then there is a set $\mathbb{S}$ of misfit values s. Since $\mathbb{M}$ is discrete, the attainable misfit values must be discrete as well. We denote the number of all possible distinct misfit values by $|\mathbb{S}|$. The discrete models and the misfit values are related by misfit functions χ, the ensemble of which forms the set $\mathbb{X}$. These misfit functions are in fact made of two components, namely the forward problem that relates a model to some synthetic data, and an actual measurement that compares observed and synthetic data to produce a misfit.

With both $\mathbb{M}$ and $\mathbb{S}$ being of finite dimension, it follows that the number of distinct misfit functions χ that map $\mathbb{M}$ to $\mathbb{S}$ has a finite dimension, too. In the example shown in Figure 15.5, the misfit values s have a binary representation with 3 bits, meaning that there are $|\mathbb{S}| = 8$ different ones. The first misfit function χ_1 maps any of the four possible models to the misfit value $s = (0, 0, 0)$. The

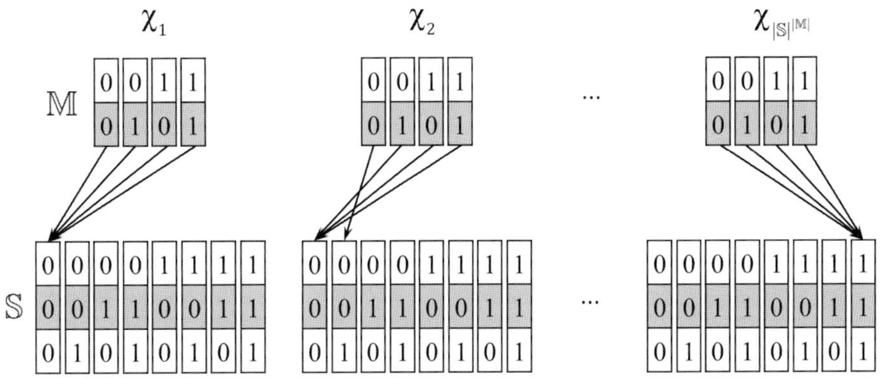

Figure 15.5 Illustration of the mapping between the finite-dimensional model space $\mathbb{M}$ to the finite-dimensional space of attainable misfit values $|\mathbb{S}|$. In the case shown here, models are represented by two bits, meaning that there are four different models. The attainable misfit values are represented by three bits, which allows for eight different values. In total, there are $|\mathbb{X} = 8^4 = 4096$ distinct misfit functions connecting models and misfit values.

second misfit function χ_2 maps the first model to the misfit value $s = (0, 0, 1)$ and all the other models to $s = (0, 0, 0)$, and so on. In total, there are $|\mathbb{X}| = 8^4 = 4096$ distinct misfit functions. More generally, we find

$$|\mathbb{X}| = |\mathbb{S}|^{|\mathbb{M}|}. \tag{15.7.1}$$

Now assume there is an algorithm $a(\chi)$ that works with a specific misfit function $\chi \in \mathbb{X}$ in order to generate a specific sequence of $m < |\mathbb{M}|$ models $M = \{\mathbf{m}_1, \ldots, \mathbf{m}_m\}$, corresponding to a specific sequence of misfit values $\{s_1, ..., s_m\}$. Depending on the algorithm, this sequence may be generated for different purposes, for instance, sampling a probability density or finding an optimal model that minimises the misfit. Since m is smaller than the total number of possible models, there must be other misfit functions χ' that produce the exact same sequence of misfit values $\{s_1, ..., s_m\}$ for the model sequence $M = \{\mathbf{m}_1, ..., \mathbf{m}_m\}$, while being distinct from χ. In fact, all misfit functions χ' that assign $\{s_1, ..., s_m\}$ to $\{\mathbf{m}_1, ..., \mathbf{m}_m\}$, and some other values $\{s_{m+1}, s_{m+2}, \ldots\}$ to models $\{\mathbf{m}_{m+1}, \mathbf{m}_{m+2}, \ldots\}$ that have not been drawn, are different from χ. In total, there are

$$|\mathbb{X}_M| = |\mathbb{S}|^{|\mathbb{M}|-m} \tag{15.7.2}$$

such misfit functions that also relate the specific sequence $M = \{\mathbf{m}_1, ..., \mathbf{m}_m\}$ to the same misfit values $\{s_1, ..., s_m\}$. So, in this sense, there are $|\mathbb{X}_M| = |\mathbb{S}|^{|\mathbb{M}|-m}$ misfit functions that are equivalent for this specific algorithm.

Now we blindly apply the algorithm a to randomly chosen misfit functions, assuming that each of the $|\mathbb{X}|$ misfit functions is drawn with equal probability. Then it follows that the probability of encountering the sequence $\{s_1, ..., s_m\}$, given the algorithm a and the number m, is given by

$$P(s_1, ..., s_m|a, m) = \frac{|\mathbb{X}_M|}{|\mathbb{X}|} = |\mathbb{S}|^{-m}. \tag{15.7.3}$$

An interesting aspect of equation (15.7.3) is that the right-hand side is independent of the algorithm used to produce $M = \{\mathbf{m}_1, ..., \mathbf{m}_m\}$. Thus, we could have used another algorithm a' to produce some $M' = \{\mathbf{m}_1', ..., \mathbf{m}_m'\}$, and the probability of finding the exact same sequence $\{s_1, ..., s_m\}$ would still be

$$P(s_1, ..., s_m|a', m) = \frac{|\mathbb{X}_M|}{|\mathbb{X}|} = |\mathbb{S}|^{-m}. \tag{15.7.4}$$

This result has profound implications. If we choose an algorithm and apply it randomly to whatever problem, characterised by its misfit function, then the probability of encountering a specific sequence of misfit values is independent of the actual algorithm.

We now turn our attention to measuring the performance of an algorithm. Typically, algorithm performance is assessed on the basis of the sequence of misfit values $\{s_1, ..., s_m\}$ that they produce. Denoting the performance measure by Υ, we symbolically have

$$\Upsilon = \Upsilon(s_1, ..., s_m). \tag{15.7.5}$$

For example, if the algorithm is intended to minimise misfit as fast as possible, then Υ would measure how quickly the numbers in the sequence decrease. Since we know already that the probability of actually finding the sequence $\{s_1, ..., s_m\}$ is independent of the algorithm, it follows that the probability of finding a certain performance is independent of the algorithm as well. In other words, assuming that we take some algorithm and randomly apply it to all kinds of problems (misfit functions) then the probability of achieving a certain performance is the same as if we had chosen another algorithm.

What we just expressed is the essential content of the No-Free-Lunch theorem, of which different variants can be found in the literature (Wolpert & Macready, 1997; Mosegaard, 2012). Statistically, all algorithms are equally good or bad. The beauty of the theorem is its counter-intuitive nature, because we seem to know that certain algorithms perform better than others.

The solution to the apparent paradox lies in the *blindness* in the choice of misfit functions. In fact, (15.7.3) critically relies on the assumption that each of all the misfit functions is drawn with equal probability. However, in practice this is not the case. What we do instead is to apply an algorithm only to well-chosen misfit functions (inverse problems) for which we empirically, analytically, or intuitively know that it works rather well. In this regard, we inject prior knowledge into the problem. It follows that no algorithm is particularly efficient by itself. What makes it efficient is its application to well-chosen problems!

In particular, we can make suitable choices of the data misfit function $\Phi(\mathbf{d}, \mathbf{d}^{obs})$

and the model suitability criterion $\Psi(\mathbf{m}, \mathbf{m}^{\text{prior}})$ to improve the performance of algorithms, by reducing the nonlinearities. In this context we need to avoid strong sensitivity to time alignment, which is a defect of a simple least-squares measure applied to the the comparison of seismograms.

16
Adjoint Methods and Sensitivity Analysis

In Chapters 14, 15 we have introduced a variety of ways of seeking a match between model predictions and observations by way of minimising a misfit function. In every case we need to be able to extract gradients of the misfit function with respect to variations in the model parameters. Since such gradients play a central role in the calculations needed for improving fit to data, it is critical that the gradients are extracted in a computationally efficient manner. Fortunately, we are able to exploit the properties of adjoint fields to reduce the computational effort and so for waveform inversion problems we can work with a forward calculation in time to evaluate residuals and a backward propagation with the adjoint system to establish the necessary gradients.

A further application of adjoint methods comes in the extraction of Fréchet kernels associated with the effect of variations of individual parameters on the gradients of the misfit function. Such sensitivity kernels provide considerable insight into the nature of the inversion process. We illustrate their character for misfit criteria based on the travel times and amplitudes of global seismic phases.

16.1 Properties of Adjoints

The concept of adjoint methods has wide application, we here exploit the properties of the adjoint field to extract representations of derivatives of wavefield properties. In the context of inversion we would like to evaluate the gradient $\nabla\chi$ of the misfit function in an efficient manner. We can construct the partial derivatives $\partial\chi/\partial m_i$ with respect to a discrete number of model parameters directly, but this requires multiple calculations of the forward problem. The introduction of the adjoint field provides an alternative, and better, formulation (e.g., Tarantola, 1988; Tromp et al., 2005; Fichtner, 2006).

Consider the suite of continuous model parameters $\mathbf{m}(\mathbf{x})$, where the vector $\mathbf{m}$ may contain, e.g., the continuous distributions of density $\rho(\mathbf{x})$ or the elastic moduli $c_{ijkl}(\mathbf{x}, t)$. The misfit functional χ depends on $\mathbf{m}$ via the wavefield $\mathbf{u}(\mathbf{m})$, meaning that $\chi = \chi[\mathbf{u}(\mathbf{m})]$. Thus, a small perturbation $\delta\mathbf{m}$ induces a corresponding change in the wavefield $\delta\mathbf{u}$, which can formally be written as a functional or Fréchet derivative

$$\delta \mathbf{u}(\mathbf{m}) = \lim_{\varepsilon \to 0} \frac{1}{\varepsilon} \left[\mathbf{u}(\mathbf{m} + \delta \mathbf{m}) - \mathbf{u}(\mathbf{m}) \right]. \tag{16.1.1}$$

The symbol δ denotes a functional derivative with respect to $\mathbf{m}$, where the differentiation direction $\delta\mathbf{m}$ is implicit. Our interest is in the functional derivative of the misfit $\delta\chi$ induced by $\delta\mathbf{m}$ via $\delta\mathbf{u}$. From the definition of a derivative, $\delta\chi$ is a linear functional in $\delta\mathbf{u}$, that by virtue of the Riesz representation theorem can generally be written in the form of a scalar product between $\delta\mathbf{u}$ and some other function $-\mathbf{f}^\dagger$,

$$\delta\chi = \Big(-\mathbf{f}^\dagger, \delta\mathbf{u} \Big). \tag{16.1.2}$$

The use of the minus sign in (16.1.2) is a matter of convention, whose advantage will soon become more obvious. The function $\mathbf{f}^\dagger$ is completely determined by the misfit functional chosen to measure the differences between observations $\mathbf{u}^{\text{obs}}(\mathbf{x}_r, t)$ and synthetic results $\mathbf{u}(\mathbf{x}_r, t)$, for receiver position $\mathbf{x}_r$.

As an illustration, we consider the L_2 waveform difference over some time interval $T = [t_0, t_1]$

$$\begin{aligned}
\chi &= \tfrac{1}{2} \int_{t_0}^{t_1} dt \, \left[\mathbf{u}(\mathbf{x}_r, t) - \mathbf{u}^{\text{obs}}(\mathbf{x}_r, t) \right]^2 \\
&= \tfrac{1}{2} \int_{t_0}^{t_1} dt \int_V d^3\mathbf{x} \, \left[\mathbf{u}(\mathbf{x}, t) - \mathbf{u}^{\text{obs}}(\mathbf{x}, t) \right]^2 \delta(\mathbf{x} - \mathbf{x}_r),
\end{aligned} \tag{16.1.3}$$

for which the Fréchet derivative is

$$\delta\chi = \int_{t_0}^{t_1} dt \int_V d^3\mathbf{x} \, \left[\mathbf{u}(\mathbf{x}_r, t) - \mathbf{u}^{\text{obs}}(\mathbf{x}_r, t) \right] \delta(\mathbf{x} - \mathbf{x}_r) \delta\mathbf{u}(\mathbf{x}, t). \tag{16.1.4}$$

On comparison with (16.1.2), we find that the appropriate scalar product is defined by the integral $\frac{1}{2} \int_{t_0}^{t_1} dt \int_V d^3\mathbf{x}$, and that

$$\mathbf{f}^\dagger(\mathbf{x}, t) = - \left[\mathbf{u}(\mathbf{x}_r, t) - \mathbf{u}^{\text{obs}}(\mathbf{x}_r, t) \right] \delta(\mathbf{x} - \mathbf{x}_r). \tag{16.1.5}$$

The function $\mathbf{f}^\dagger$ can be interpreted as the (negative) integral kernel of the Fréchet derivative, when written as a scalar product. $\mathbf{f}^\dagger$ is localised at the receiver position $\mathbf{x}_r$, and its time evolution is controlled by the linear waveform residuals, i.e., the difference between observations and synthetics, $\mathbf{u}(\mathbf{x}_r, t) - \mathbf{u}^{\text{obs}}(\mathbf{x}_r, t)$. We will see later in Section 16.3 that the function $\mathbf{f}^\dagger$ will contain some form of residuals, irrespective of the precise definition of the misfit functional.

While the expression (16.1.2) for the derivative is formally correct, it is of little practical use because it does not allow us to compute $\delta\chi$ directly for some arbitrary model perturbation $\delta\mathbf{m}$. Instead, it forces us to first compute $\delta\mathbf{u}$. This indirect computation of $\delta\chi$ can be prohibitively expensive when the number of model perturbations $\delta\mathbf{m}$ of interest is very large. Therefore, we wish to eliminate the explicit occurrence of $\delta\mathbf{u}$ from (16.1.2). We can achieve this goal by introducing the wave equation satisfied by $\mathbf{u}$, which we write in the symbolic form

$$\mathcal{L}\left[\mathbf{u}(\mathfrak{m}), \mathfrak{m}\right] = \mathbf{f}. \tag{16.1.6}$$

Depending on the particular choice of wave propagation physics, the operator $\mathcal{L}$ may describe waves in a lossless acoustic, a viscoelastic, or any other kind of medium. Irrespective of this choice, $\mathcal{L}$ is linear in the wavefield $\mathbf{u}$, and the forcing term $\mathbf{f}$ on the right-hand side does not depend on the model parameters $\mathfrak{m}$. However, $\mathcal{L}$ is not generally linear in $\mathfrak{m}$. The Fréchet derivative of (16.1.6) takes the form

$$\mathcal{L}\left[\delta\mathbf{u}(\mathfrak{m}), \mathfrak{m}\right] + \delta\mathcal{L}\left[\mathbf{u}(\mathfrak{m}), \mathfrak{m}\right] = \mathbf{0}. \tag{16.1.7}$$

The first term in (16.1.7) is the derivative of $\mathcal{L}$ with respect to the wavefield $\mathbf{u}$, and the second term is the explicit derivative of $\mathcal{L}$ with respect to the model parameters $\mathfrak{m}$. We take the scalar product of (16.1.7) with some arbitrary function $\mathbf{u}^\dagger(\mathbf{x}, t)$ and add in the relation (16.1.2), to produce

$$\delta\chi = \left(\mathbf{u}^\dagger, \mathcal{L}\left[\delta\mathbf{u}, \mathfrak{m}\right]\right) + \left(\mathbf{u}^\dagger, \delta\mathcal{L}\left[\mathbf{u}, \mathfrak{m}\right]\right) - \left(\mathbf{f}^\dagger, \delta\mathbf{u}\right), \tag{16.1.8}$$

where, for convenience, we have omitted the dependence of $\mathbf{u}$ on $\mathfrak{m}$. We extract $\delta\mathbf{u}$ from $\mathcal{L}$ in the first term on the right-hand side of (16.1.8), by the action of the adjoint operator $\mathcal{L}^\dagger$, defined through the equation

$$\left(\mathbf{u}^\dagger, \mathcal{L}\left[\delta\mathbf{u}, \mathfrak{m}\right]\right) = \left(\delta\mathbf{u}, \mathcal{L}^\dagger\left[\mathbf{u}^\dagger, \mathfrak{m}\right]\right). \tag{16.1.9}$$

Now, by combining (16.1.8) and (16.1.9) with reordering of terms, we find

$$\delta\chi = \left(\delta\mathbf{u}, \mathcal{L}^\dagger\left[\mathbf{u}^\dagger, \mathfrak{m}\right] - \mathbf{f}^\dagger\right) + \left(\mathbf{u}^\dagger, \delta\mathcal{L}\left[\mathbf{u}, \mathfrak{m}\right]\right). \tag{16.1.10}$$

Equation (16.1.10) is valid for any arbitrary function $\mathbf{u}^\dagger$. In order to achieve the goal of eliminating $\delta\mathbf{u}$, we force $\mathbf{u}^\dagger$ to be a solution of

$$\mathcal{L}^\dagger\left[\mathbf{u}^\dagger, \mathfrak{m}\right] = \mathbf{f}^\dagger. \tag{16.1.11}$$

The *adjoint equation* (16.1.11) has the solution $\mathbf{u}^\dagger$, which we term the *adjoint field*. The *adjoint source* $\mathbf{f}^\dagger$ on the right-hand side of (16.1.11) is determined by the misfit definition. Since $\mathbf{u}^\dagger$ solves (16.1.11) we can reduce the expression (16.1.10) for $\delta\chi$ to the form

$$\delta\chi = \left(\mathbf{u}^\dagger, \delta\mathcal{L}\left[\mathbf{u}, \mathfrak{m}\right]\right). \tag{16.1.12}$$

By this construction, $\delta\mathbf{u}$ is absent from (16.1.12). We can thus suggest a simple recipe for the calculation of the Fréchet derivative $\delta\chi$:

(i) Solve the forward problem (16.1.6) to obtain synthetic seismograms.
(ii) Compare observed and synthetic seismograms using a suitable misfit, which determines the adjoint source $\mathbf{f}^\dagger$.

(iii) Using $\mathbf{f}^\dagger$ on the right-hand side, solve the adjoint equation (16.1.11) for the adjoint field $\mathbf{u}^\dagger$.

(iv) Insert $\mathbf{u}^\dagger$ into (16.1.12) to compute the Fréchet derivative $\delta\chi$.

16.1.1 The Adjoint of the Viscoelastic Wave Equation

We now consider the special case of the viscoelastic wave equation, and so provide a specific example of the development in the previous section. The governing equation is

$$\rho(\mathbf{x})\ddot{\mathbf{u}}(\mathbf{x},t) - \nabla\cdot\boldsymbol{\sigma}(\mathbf{x},t) = \mathbf{f}(\mathbf{x},t), \tag{16.1.13}$$

which relates the elastic displacement field $\mathbf{u}$ to the density ρ, the stress tensor $\boldsymbol{\sigma}$, and an external force density $\mathbf{f}$. Assuming a linear viscoelastic rheology, the stress tensor $\boldsymbol{\sigma}$ is related to the displacement gradient $\nabla\mathbf{u}$ via the constitutive relation

$$\boldsymbol{\sigma}(\mathbf{x},t) = \int_{\tau=t_0}^{\infty} d\tau\,\dot{\mathbf{C}}(\mathbf{x},t-\tau) : \nabla\mathbf{u}(\mathbf{x},\tau), \tag{16.1.14}$$

in terms of a relaxation tensor $\mathbf{C}$; : denotes a tensor contraction. The wave operator $\mathcal{L}$ in (16.1.6) then has the form

$$\mathcal{L}(\mathbf{u},\mathfrak{m}) = \rho(\mathbf{x})\ddot{\mathbf{u}}(\mathbf{x},t) - \nabla\cdot\int_{\tau=t_0}^{t} d\tau\,\dot{\mathbf{C}}(\mathbf{x},t-\tau) : \nabla\mathbf{u}(\mathbf{x},\tau). \tag{16.1.15}$$

The continuous model parameter vector $\mathfrak{m}$ is

$$\mathfrak{m}(\mathbf{x}) = [\rho(\mathbf{x}), \mathbf{c}(\mathbf{x},t)]. \tag{16.1.16}$$

The operator $\mathcal{L}$ is complemented by the initial and boundary conditions

$$\mathbf{u}|_{t\le t_0} = \dot{\mathbf{u}}|_{t\le t_0} = \mathbf{0}, \quad \mathbf{n}\cdot\boldsymbol{\sigma}|_{\mathbf{x}\in\partial V} = \mathbf{0}, \tag{16.1.17}$$

with $\mathbf{n}$ defined as the outward-pointing normal vector on the boundary ∂V of the Earth model.

To compute the Fréchet derivative $\delta\chi$, we require the adjoint $\mathcal{L}^\dagger$ of $\mathcal{L}$, as defined in (16.1.9). To find $\mathcal{L}^\dagger$, we first write an explicit form for the left-hand side of (16.1.9):

$$\left(\mathbf{u}^\dagger, \mathcal{L}(\delta\mathbf{u},\mathfrak{m})\right) = \int_T dt \int_V d^3\mathbf{x}\,\mathbf{u}^\dagger\cdot\mathcal{L}(\delta\mathbf{u},\mathfrak{m}) \tag{16.1.18}$$
$$= \int_T dt \int_V d^3\mathbf{x}\,\rho\mathbf{u}^\dagger\cdot\delta\ddot{\mathbf{u}} - \int_T dt \int_V d^3\mathbf{x}\,\mathbf{u}^\dagger\cdot\left[\nabla\cdot\int_{\tau=t_0}^{t} d\tau\,\dot{\mathbf{C}}(t-\tau) : \nabla\delta\mathbf{u}(\tau)\right].$$

For simplicity, we have again chosen the scalar product defined by the integral $\int_T dt \int_V d^3\mathbf{x}$, and have omitted all explicit dependencies on $\mathbf{x}$. To isolate $\delta\mathbf{u}$ following the approach in (16.1.9)–(16.1.12), we start with the first term on the right-hand side of (16.1.18),

$$\left(\mathbf{u}^\dagger, \rho\delta\ddot{\mathbf{u}}\right) = \int_T dt \int_V d^3\mathbf{x}\,\rho\mathbf{u}^\dagger\cdot\delta\ddot{\mathbf{u}}. \tag{16.1.19}$$

Double integration by parts then yields

$$\int_T dt \int_V d^3\mathbf{x}\, \rho \mathbf{u}^\dagger \cdot \delta\ddot{\mathbf{u}} = \int_T dt \int_V d^3\mathbf{x}\, \rho \delta\mathbf{u} \cdot \ddot{\mathbf{u}}^\dagger + \int_V d^3\mathbf{x}\, \rho \delta\dot{\mathbf{u}} \cdot \mathbf{u}^\dagger \bigg|_{t=t_1} - \int_V d^3\mathbf{x}\, \rho \delta\mathbf{u} \cdot \dot{\mathbf{u}}^\dagger \bigg|_{t=t_1}, \tag{16.1.20}$$

where we use the initial conditions (16.1.17). Now we impose the *terminal conditions*

$$\mathbf{u}^\dagger|_{t\geq t_1} = \dot{\mathbf{u}}^\dagger|_{t\geq t_1} = \mathbf{0}, \tag{16.1.21}$$

and obtain the first contribution to the adjoint operator,

$$\left(\mathbf{u}^\dagger, \rho\delta\ddot{\mathbf{u}}\right) = \left(\delta\mathbf{u}, \rho\ddot{\mathbf{u}}^\dagger\right). \tag{16.1.22}$$

For the contribution that involves the spatial derivatives of $\delta\mathbf{u}$, we need to rearrange the expression

$$\Gamma = (\mathbf{u}^\dagger, \nabla \cdot \sigma) = \int_V d^3\mathbf{x} \int_T dt\, \mathbf{u}^\dagger \cdot \left[\nabla \cdot \int_{\tau=t_0}^{t} d\tau\, \dot{\mathbf{C}}(t-\tau) : \nabla\delta\mathbf{u}(\tau)\right]. \tag{16.1.23}$$

Recalling the symmetries of the relaxation tensor $\mathbf{C}$, we find the relation

$$\mathbf{u}^\dagger \cdot [\nabla \cdot (\dot{\mathbf{C}} : \nabla\delta\mathbf{u})] = \nabla \cdot (\mathbf{u}^\dagger \cdot \dot{\mathbf{C}} : \nabla\delta\mathbf{u}) - \nabla \cdot (\delta\mathbf{u} \cdot \dot{\mathbf{C}} : \nabla\mathbf{u}^\dagger) + \delta\mathbf{u} \cdot [\nabla \cdot (\dot{\mathbf{C}} : \nabla\mathbf{u}^\dagger)]. \tag{16.1.24}$$

Using (16.1.24), we can rewrite (16.1.23) as

$$\begin{aligned}\Gamma = &\int_V d^3\mathbf{x} \int_{t=t_0}^{t_1} dt \int_{\tau=t_0}^{t} d\tau\, \nabla \cdot [\mathbf{u}^\dagger(t) \cdot \dot{\mathbf{C}}(t-\tau) : \nabla\delta\mathbf{u}(\tau)] \\ &- \int_V d^3\mathbf{x} \int_{t=t_0}^{t_1} dt \int_{\tau=t_0}^{t} d\tau\, \nabla \cdot [\delta\mathbf{u}(\tau) \cdot \dot{\mathbf{C}}(t-\tau) : \nabla\mathbf{u}^\dagger(t)] \\ &+ \int_V d^3\mathbf{x} \int_{t=t_0}^{t_1} dt \int_{\tau=t_0}^{t} d\tau\, \delta\mathbf{u}(\tau) \cdot [\nabla \cdot (\dot{\mathbf{C}}(t-\tau) : \nabla\mathbf{u}^\dagger(\tau))].\end{aligned} \tag{16.1.25}$$

Invoking Gauss' theorem in combination with the integral identity

$$\int_{\tau=t_0}^{t} d\tau \int_{t=t_0}^{t_1} dt = \int_{t=\tau}^{t_1} dt \int_{\tau=t_0}^{t_1} d\tau, \tag{16.1.26}$$

we can write Γ such that two of the contributing integrands can be eliminated,

$$\begin{aligned}\Gamma = &\int_{\partial V} d^2\mathbf{x} \int_{t=t_0}^{t_1} dt\, \mathbf{u}^\dagger(t) \cdot \left[\left[\int_{\tau=t_0}^{t} d\tau\, \dot{\mathbf{C}}(t-\tau) : \nabla\delta\mathbf{u}(\tau)\right] \cdot \mathbf{n}\right. \\ &- \int_{\partial V} d^2\mathbf{x} \int_{\tau=t_0}^{t_1} d\tau\, \delta\mathbf{u}(\tau) \cdot \left[\left[\int_{t=\tau}^{t_1} dt\, \dot{\mathbf{C}}(t-\tau) : \nabla\mathbf{u}^\dagger(t)\right] \cdot \mathbf{n}\right. \\ &+ \int_V d^3\mathbf{x} \int_{\tau=t_0}^{t_1} d\tau\, \delta\mathbf{u}(\tau) \cdot \left[\nabla \cdot \int_{t=\tau}^{t_1} dt\, \dot{\mathbf{C}}(t-\tau) : \nabla\mathbf{u}^\dagger(t)\right].\end{aligned} \tag{16.1.27}$$

The first term in (16.1.27) vanishes because the expression in square brackets is the stress tensor $\boldsymbol{\sigma}$, and $\boldsymbol{\sigma} \cdot \mathbf{n} = \mathbf{0}$ on ∂V as a consequence of the boundary condition (16.1.17). Since $\mathbf{u}^\dagger$ is so far only constrained by the terminal conditions (16.1.21), we can impose a boundary condition that forces the second term to zero,

$$\mathbf{n} \cdot \boldsymbol{\sigma}^\dagger|_{\mathbf{x}\in\partial V} = \mathbf{0}, \tag{16.1.28}$$

where we define the *adjoint stress tensor* $\boldsymbol{\sigma}^\dagger$ as

$$\boldsymbol{\sigma}^\dagger(t) = \int_{\tau=t}^{t_1} d\tau\, \dot{\mathbf{C}}(\tau - t) : \nabla \mathbf{u}^\dagger(\tau). \tag{16.1.29}$$

The third integrand in (16.1.27) is already in the form that we require.

Thus, it only remains to assemble the complete adjoint operator $\mathcal{L}^\dagger$ by combining (16.1.18), (16.1.22) and (16.1.27) with (16.1.29). We thereby obtain

$$\left(\delta\mathbf{u}, \mathcal{L}^\dagger(\mathbf{u}^\dagger)\right) = \left(\delta\mathbf{u}, \rho\ddot{\mathbf{u}}^\dagger\right) - \Gamma = \left(\delta\mathbf{u}, \rho\ddot{\mathbf{u}}^\dagger - \nabla \cdot \boldsymbol{\sigma}^\dagger\right). \tag{16.1.30}$$

The result is that the adjoint operator $\mathcal{L}^\dagger$ is given by

$$\mathcal{L}^\dagger(\mathbf{u}^\dagger, \mathbf{m}) = \rho\ddot{\mathbf{u}}^\dagger - \nabla \cdot \boldsymbol{\sigma}^\dagger. \tag{16.1.31}$$

Hence, the complete adjoint equation is

$$\rho\ddot{\mathbf{u}}^\dagger - \nabla \cdot \boldsymbol{\sigma}^\dagger = \mathbf{f}^\dagger, \tag{16.1.32}$$

subject to the terminal and boundary conditions

$$\mathbf{u}^\dagger|_{t\geq t_1} = \dot{\mathbf{u}}^\dagger|_{t\geq t_1} = \mathbf{0}, \quad \mathbf{n} \cdot \boldsymbol{\sigma}^\dagger|_{\mathbf{x}\in\partial V} = \mathbf{0}. \tag{16.1.33}$$

In non-dissipative media, where the time dependence of the relaxation tensor is given by $\dot{\mathbf{C}}(t) = \mathbf{c}\delta(t)$, the operator $\mathcal{L}$ is self-adjoint, meaning that $\mathcal{L} = \mathcal{L}^\dagger$. This self-adjoint property is equivalent to the conservation of energy. In any case, the adjoint equation (16.1.32) has the form of a wave equation. A pleasant side effect is that the adjoint equation can be solved with the same numerical modelling code used already for the solution of the forward problem, that is, the regular wave equation.

The only difficulty is the presence of the terminal conditions (16.1.33). To implement these conditions, the adjoint equation is most commonly solved backward in time, meaning that the terminal condition at $t = t_1$ serves as an initial condition for a numerically time-reversed adjoint wavefield, propagating towards $t_0 < t_1$. Frequently, the time-reversed solution of the adjoint equation is termed *propagating the residuals backwards in time*, alluding to the fact that the adjoint source $\mathbf{f}^\dagger$, for any waveform misfit definition, generally contains some form of data residual that is injected into the medium.

16.1.2 Fréchet Kernels and Gradients

We have established in (16.1.12) that the Fréchet derivative of the misfit, $\delta\chi$, can be written as scalar product of the adjoint field $\mathbf{u}^\dagger$ and the Fréchet derivative of the wave operator, $\delta\mathcal{L}[\mathbf{u}, \mathfrak{m}]$.

We consider, for simplicity, a non-dissipative, elastic medium, and then

$$\delta\mathcal{L}(\mathbf{u}, \mathfrak{m}) = \delta\rho(\mathbf{x})\ddot{\mathbf{u}}(\mathbf{x}, t) - \nabla\cdot[\delta\mathbf{c}(\mathbf{x}) : \nabla\mathbf{u}(\mathbf{x}, t)]. \tag{16.1.34}$$

So that from (16.1.12), the Fréchet derivative of the misfit is given by

$$\begin{aligned}\delta\chi = &- \int_T dt \int_V d^3\mathbf{x}\, \delta\rho(\mathbf{x})\dot{\mathbf{u}}^\dagger(\mathbf{x}, t) \cdot \dot{\mathbf{u}}(\mathbf{x}, t)\\ &+ \int_T dt \int_V d^3\mathbf{x}\, \nabla\mathbf{u}^\dagger(\mathbf{x}, t) : \delta\mathbf{c}(\mathbf{x}) : \nabla\mathbf{u}(\mathbf{x}, t).\end{aligned} \tag{16.1.35}$$

Although (16.1.35) is specific to the non-dissipative, elastic wave equation, it has some general characteristics that reappear for any type of wave equation, and so for any possible rheology.

Firstly, as a consequence of defining the scalar product via the integral $\int_T dt \int_V d^3\mathbf{x}$, we note that (16.1.35) can be written in the more succinct form

$$\delta\chi = \int_V d^3\mathbf{x}\, K_\rho(\mathbf{x})\delta\rho(\mathbf{x}) + \int_V d^3\mathbf{x}\, \mathbf{K}_\mathbf{c}(\mathbf{x}) :: \delta\mathbf{c}(\mathbf{x}), \tag{16.1.36}$$

where :: denotes a tensor contraction. $K_\rho(\mathbf{x})$ and $\mathbf{K}_\mathbf{c}(\mathbf{x})$ are the integral kernels, or volumetric densities, of the Fréchet derivative $\delta\chi$. These quantities are commonly described as *Fréchet* or *sensitivity kernels*, and defined as

$$K_\rho(\mathbf{x}) = -\int_T dt\, \dot{\mathbf{u}}^\dagger(\mathbf{x}, t) \cdot \dot{\mathbf{u}}(\mathbf{x}, t),\ \mathbf{K}_\mathbf{c}(\mathbf{x}) = \int_T dt\, \nabla\mathbf{u}^\dagger(\mathbf{x}, t) \otimes \nabla\mathbf{u}(\mathbf{x}, t), \tag{16.1.37}$$

where $\otimes$ indicates a tensor product.

Fréchet kernels indicate where and how perturbations in density or elastic moduli affect measurements. As such, they are not only essential for numerical optimisation, but also serve as a major source of physical intuition that is indispensable for the meaningful solution of an ill-posed inverse problem. In Section 16.3 we present an analysis of sensitivity kernels for global seismic phases.

Secondly, we note that the Fréchet derivatives and kernels generally involve some product of the forward wavefield $\mathbf{u}(\mathbf{x}, t)$ and the adjoint wavefield $\mathbf{u}^\dagger(\mathbf{x}, t)$. Therefore, Fréchet kernels can only be non-zero in regions where $\mathbf{u}(\mathbf{x}, t)$ and $\mathbf{u}^\dagger(\mathbf{x}, t)$ actually overlap at some time t. As a consequence, all Fréchet kernels have a roughly similar shape. As illustrated in Figure 16.1, the configuration of the sensitivity kernels comprises a *primary influence zone* that connects source and receiver. Broadly speaking, this influence zone marks the effective path of a wave propagating at finite frequency, and can be considered a generalisation of the Fresnel zone concept from geometric optics.

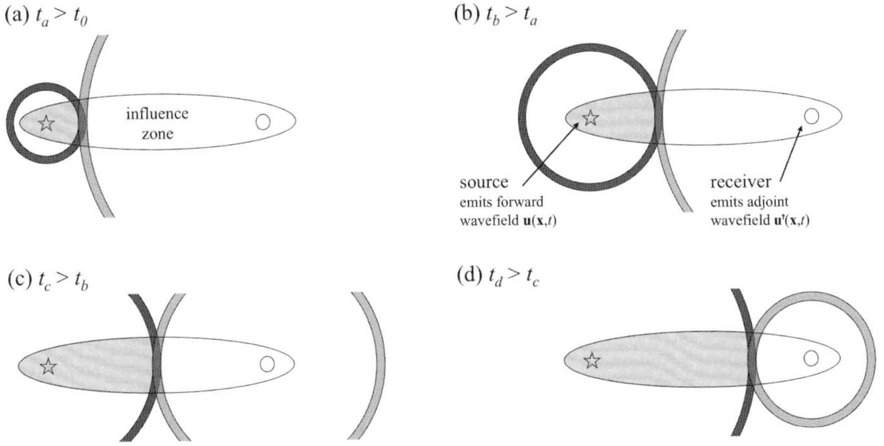

Figure 16.1 Schematic illustration of the propagating forward wavefield $\mathbf{u}(\mathbf{x}, t)$ (black circular wavefront) and adjoint wavefield $\mathbf{u}^\dagger(\mathbf{x}, t)$ (grey circular wavefront), respectively. Snaptshots (a) to (d) are for increasing times, meaning that $\mathbf{u}(\mathbf{x}, t)$ is emitted from the source (star), whereas $\mathbf{u}^\dagger(\mathbf{x}, t)$ collapses into the receiver (circle). In reverse time, when reading panels from (d) to (a), $\mathbf{u}^\dagger(\mathbf{x}, t)$ is emitted from the receiver, and $\mathbf{u}(\mathbf{x}, t)$ collapses into the source. The forward and the adjoint wavefields overlap within the influence zone, which is the volume where the Fréchet kernel is non-zero. For many seismic phases, the influence zone is roughly cigar shaped, connecting source and receiver. The shaded part of the influence zone marks the region that, at a certain point in time, has already been covered by the interaction of $\mathbf{u}(\mathbf{x}, t)$ and $\mathbf{u}^\dagger(\mathbf{x}, t)$.

We can specialise (16.1.36) and (16.1.37) to simpler rheologies, such as an isotropic medium where the elastic moduli components are given by

$$c_{ijkl} = \lambda \delta_{ij}\delta_{kl} + \mu\left(\delta_{ik}\delta_{jl} + \delta_{il}\delta_{jk}\right), \tag{16.1.38}$$

in terms of the Lamé parameters λ and μ. For isotropic elasticity the Fréchet derivative is composed of three terms,

$$\delta\chi = \int_V d^3\mathbf{x}\, K_\rho(\mathbf{x})\delta\rho(\mathbf{x}) + \int_V d^3\mathbf{x}\, K_\mu(\mathbf{x})\delta\mu(\mathbf{x}) + \int_V d^3\mathbf{x}\, K_\lambda(\mathbf{x})\delta\lambda(\mathbf{x}). \tag{16.1.39}$$

The Fréchet kernels take the form:

$$K_\rho = -\int_T dt\, \dot{\mathbf{u}}^\dagger \cdot \dot{\mathbf{u}}, \tag{16.1.40a}$$

$$K_\lambda = \int_T dt\, (\nabla \cdot \mathbf{u})(\nabla \cdot \mathbf{u}^\dagger), \tag{16.1.40b}$$

$$K_\mu = \int_T dt\, [(\nabla \mathbf{u}^\dagger) : (\nabla \mathbf{u}) + (\nabla \mathbf{u}^\dagger) : (\nabla \mathbf{u})^T]. \tag{16.1.40c}$$

Fréchet kernels for derived parameters, e.g., the *P* wavespeed $\alpha = [\lambda + 2\mu)/\rho]^{1/2}$ or the S-wave speed $\beta = [\mu/\rho]^{1/2}$, can be obtained by expressing $\delta\alpha$ and $\delta\beta$ in terms of $\delta\rho$, $\delta\lambda$ and $\delta\mu$, and then inserting these expressions into (16.1.39). After rearrangement of terms, the Fréchet derivative in terms of the new parameters takes the form

$$\delta\chi = \int_V d^3\mathbf{x}\, K'_\rho(\mathbf{x})\delta\rho(\mathbf{x}) + \int_V d^3\mathbf{x}\, K_\alpha(\mathbf{x})\delta\alpha(\mathbf{x}) + \int_V d^3\mathbf{x}\, K_\beta(\mathbf{x})\delta\beta(\mathbf{x}), \quad (16.1.41)$$

with Fréchet kernels

$$K'_\rho = K_\rho + (\alpha^2 - 2\beta^2)K_\lambda + \beta^2 K_\mu, \quad (16.1.42a)$$

$$K_\beta = 2\rho\beta K_\mu - 4\rho\beta K_\lambda, \quad (16.1.42b)$$

$$K_\alpha = 2\rho\alpha K_\lambda. \quad (16.1.42c)$$

We note that the density kernel K'_ρ differs from the density kernel K_ρ in (16.1.40a), meaning that the sensitivity to variations in density depends significantly on the choices made for the other model parameters. Thus, it is important that any sensitivity analysis is carried out in terms of the parameterisation that will be employed in an inversion, or otherwise misleading impressions can be gained.

Fréchet kernels represent derivatives with respect to the continuous distributions of model parameters, such as density, elastic moduli, or wave speeds. However, the numerical optimisation schemes described in Chapter 15 involve gradients. We therefore require partial derivatives with respect to a discrete number of model parameters. To make the transition from kernels to gradients, we recall the representation of a continuous model parameter $m(\mathbf{x})$ in terms of a finite number of basis functions $b_i(\mathbf{x})$,

$$m(\mathbf{x}) = \sum_{i=1}^{n} m_i b_i(\mathbf{x}), \quad (16.1.43)$$

where n is the model-space dimension. The expansion coefficients are collected into the discrete model vector $\mathbf{m} = (m_1, ..., m_n)^T$. Our interest is in the partial derivatives of the misfit, $\partial\chi/\partial m_i$, defined as

$$\frac{\partial\chi}{\partial m_i} = \lim_{\varepsilon\to 0} \frac{1}{\varepsilon} \left[\chi(...(m_i + \varepsilon)b_i...) - \chi(...m_i b_i...)\right]. \quad (16.1.44)$$

From (16.1.44) we see that the partial derivative $\partial\chi/\partial m_i$ equals the Fréchet derivative $\delta\chi$ for a model perturbation $\delta m(\mathbf{x}) = b_i(\mathbf{x})$. Inserting this form of the derivative into the representation of $\delta\chi$ in terms of Fréchet kernels, yields

$$\frac{\partial\chi}{\partial m_i} = \int_V d^3\mathbf{x}\, K(\mathbf{x}) b_i(\mathbf{x}). \quad (16.1.45)$$

Thus, the partial derivative of χ with respect to a coefficient m_i is the projection of the kernel $K(\mathbf{x})$ onto the corresponding basis function $b_i(\mathbf{x})$.

16.1.3 Born Approximation and Physics of Fréchet Kernels

In order to provide an understanding of the physics of adjoints beyond the formal analysis in the previous sections, we provide an alternative derivation of (16.1.12).

We introduce a model variation $\delta\mathbf{m}$, which produces a perturbation to the wavefield $\delta\mathbf{u}$. For the perturbed medium, the wave equation takes the form

$$\mathcal{L}[\mathbf{u}(\mathbf{m}+\delta\mathbf{m}),\mathbf{m}+\delta\mathbf{m}] = \mathbf{f}. \tag{16.1.46}$$

With a Taylor expansion of (16.1.46) in terms of $\delta\mathbf{m}$ using the first-order *Born approximation*

$$\mathcal{L}[\delta\mathbf{u}(\mathbf{m}),\mathbf{m}+\delta\mathbf{m}] \approx \mathcal{L}[\delta\mathbf{u}(\mathbf{m}),\mathbf{m}]. \tag{16.1.47}$$

We therefore have a wave equation for the perturbed wavefield $\delta\mathbf{u}$,

$$\mathcal{L}\left[\delta\mathbf{u}(\mathbf{m}),\mathbf{m}\right] \approx \delta\mathbf{f}, \tag{16.1.48}$$

with the wavefield source $\mathbf{f}$ defined as

$$\mathbf{f}(\mathbf{x},t) = -\delta\mathcal{L}\left[\mathbf{u}(\mathbf{m}),\mathbf{m}\right](\mathbf{x},t). \tag{16.1.49}$$

From (16.1.48) and (16.1.49), we can write $\delta\mathbf{u}$ explicitly using the representation theorem,

$$\delta\mathbf{u}(\mathbf{x},t) \approx -\int_{-\infty}^{\infty} d\tau \int_V d^3\xi\, \mathbf{G}(\mathbf{x},t-\tau;\xi)\cdot\delta\mathcal{L}\left[\mathbf{u}(\mathbf{m}),\mathbf{m}\right](\xi,\tau). \tag{16.1.50}$$

Invoking the general expression of the misfit derivative $\delta\chi = (-\mathbf{f}^\dagger,\delta\mathbf{u})$ from (16.1.2), this leads to

$$\delta\chi \approx \int_{-\infty}^{\infty} d\tau \int_{-\infty}^{\infty} dt \int_V d^3\xi \int_V d^3\mathbf{x}\, \mathbf{f}^\dagger(\mathbf{x},t)\cdot\mathbf{G}(\mathbf{x},t-\tau;\xi)\cdot\delta\mathcal{L}\left[\mathbf{u}(\mathbf{m}),\mathbf{m}\right](\xi,\tau). \tag{16.1.51}$$

We define the adjoint Green's tensor as the time-reversed Green's tensor,

$$\mathbf{G}^\dagger(\xi,t;\mathbf{x}) = \mathbf{G}(\xi,-t;\mathbf{x}), \tag{16.1.52}$$

and the adjoint wavefield as

$$\mathbf{u}^\dagger(\xi,\tau) = \int_{-\infty}^{\infty} dt \int_V d^3\mathbf{x}\, \mathbf{G}^\dagger(\xi,\tau-t;\mathbf{x})\cdot\mathbf{f}^\dagger(\mathbf{x},t). \tag{16.1.53}$$

With these forms (16.1.51) does indeed lead to an approximation to (16.1.12),

$$\delta\chi \approx (\mathbf{u}^\dagger,\delta\mathcal{L}\left[\mathbf{u},\mathbf{m}\right]). \tag{16.1.54}$$

The $\approx$ sign in (16.1.54) turns into an $=$ sign when the Born approximation (16.1.47) becomes exact. This will be the case when all interactions of the perturbed wavefield $\delta\mathbf{u}$ with the model perturbation $\delta\mathbf{m}$ are indeed zero, which requires second- and higher-order scattering to be absent. The profound implication of this condition is that the Fréchet derivative of any misfit functional strictly corresponds to first-order scattering. Therefore, Fréchet kernels correspond to those positions in space where first-order scattering from a perturbation $\delta\mathbf{m}$ has an effect on the measurements.

16.1.4 Fréchet Derivatives of Green's Functions

We present the nature of Fréchet kernels for concrete misfit functionals in Section 16.3. Here, we use the adjoint method to compute Fréchet derivatives of Green's functions, which will be particularly useful for Chapter 17.

We begin by analysing the response of a waveform component at a specific instant in time, $t = t_r$. For this purpose we define χ as

$$\chi = \mathbf{e}_i \cdot \mathbf{u}(\mathbf{x}_r, t_r) = \int_T dt \int_V d^3\mathbf{x}\, \mathbf{e}_i \cdot \mathbf{u}(\mathbf{x}, t)\delta(\mathbf{x} - \mathbf{x}_r)\delta(t - t_r), \tag{16.1.55}$$

where $\mathbf{e}_i$ is the unit vector in a predefined direction. Though this expression for χ does not strictly define a misfit function, the adjoint scheme developed above can still be applied to extract derivatives. The Fréchet derivative of χ from (16.1.55) is

$$\delta\chi = \int_T dt \int_V d^3\mathbf{x}\, \mathbf{e}_i \cdot \delta\mathbf{u}(\mathbf{x}, t)\delta(\mathbf{x} - \mathbf{x}_r)\delta(t - t_r). \tag{16.1.56}$$

Comparison with the generic representation (16.1.2) shows that the adjoint source corresponding to (16.1.55) is given by

$$\mathbf{f}^\dagger(\mathbf{x}, t) = -\mathbf{e}_i\delta(\mathbf{x} - \mathbf{x}_r)\delta(t - t_r). \tag{16.1.57}$$

This adjoint source is localised in time at the observation time $t = t_r$, and in space at the receiver position $\mathbf{x} = \mathbf{x}_r$, with an orientation in the observation direction $\mathbf{e}_i$. A consequence of (16.1.57) is that the adjoint wavefield $\mathbf{u}^\dagger$ is equal to the negative adjoint Green's function for a point source in the $\mathbf{e}_i$-direction,

$$\mathbf{u}^\dagger(\mathbf{x}, t) = -\mathbf{G}_i^\dagger(\mathbf{x}, \mathbf{t}; \mathbf{x}_r, t_r). \tag{16.1.58}$$

With the choice of a specific rheology, for instance, a perfectly elastic medium, we can combine (16.1.58) with (16.1.35) to obtain an explicit expression for the Fréchet derivative of the wavefield,

$$\begin{aligned} \mathbf{e}_i \cdot \delta\mathbf{u}(\mathbf{x}_r, t_r) = &- \int_T dt \int_V d^3\mathbf{x}\, \delta\rho(\mathbf{x})\, \dot{\mathbf{G}}_i^\dagger(\mathbf{x}, t; \mathbf{x}_r, t_r) \cdot \dot{\mathbf{u}}(\mathbf{x}, t) \\ &+ \int_T dt \int_V d^3\mathbf{x}\, \nabla\mathbf{G}_i^\dagger(\mathbf{x}, t; \mathbf{x}_r, t_r) : \delta\mathbf{c}(\mathbf{x}) : \nabla\mathbf{u}(\mathbf{x}, t). \end{aligned} \tag{16.1.59}$$

In the special case where the forward wavefield $\mathbf{u}(\mathbf{x}, t)$ is given by the Green's function $\mathbf{G}_j(\mathbf{x}, t; \mathbf{x}_s, t_s)$ for a source acting at position $\mathbf{x}_s$ and time t_s, (16.1.59) takes the form

$$\begin{aligned} \mathbf{e}_i \cdot \delta\mathbf{G}_j(\mathbf{x}_r, t_r; \mathbf{x}_s, t_s) = &- \int_T dt \int_V d^3\mathbf{x}\, \delta\rho(\mathbf{x})\, \dot{\mathbf{G}}_i^\dagger(\mathbf{x}, t; \mathbf{x}_r, t_r) \cdot \dot{\mathbf{G}}_j(\mathbf{x}, t; \mathbf{x}_s, t_s) \\ &+ \int_T dt \int_V d^3\mathbf{x}\, \nabla\mathbf{G}_i^\dagger(\mathbf{x}, t; \mathbf{x}_r, t_r) : \delta\mathbf{c}(\mathbf{x}) : \nabla\mathbf{G}_j(\mathbf{x}, t; \mathbf{x}_s, t_s). \end{aligned} \tag{16.1.60}$$

This is the time-domain version of *Green's theorem*, which expresses the Fréchet derivative of the Green's function in terms of the forward and adjoint Green's functions, and the variations in material parameters, $\delta\rho$ and $\delta\mathbf{c}$.

For a frequency-domain equivalent to (16.1.60), we consider the wavefield at position $\mathbf{x}_r$ and a specific frequency ω_r,

$$\chi = \mathbf{e}_i \cdot \mathbf{u}(\mathbf{x}_r, \omega_r). \tag{16.1.61}$$

The corresponding Fréchet derivative is

$$\delta\chi = \int_{-\infty}^{\infty} dt \int_V d^3\mathbf{x}\, \mathbf{e}_i \cdot \delta\mathbf{u}(\mathbf{x}, t) e^{-i\omega_r t}\, \delta(\mathbf{x} - \mathbf{x}_r). \tag{16.1.62}$$

Comparing (16.1.62) to the generic form (16.1.2), we infer that the adjoint source is now

$$\mathbf{f}^\dagger(\mathbf{x}, t) = -\mathbf{e}_i e^{-i\omega_r t} \delta(\mathbf{x} - \mathbf{x}_r). \tag{16.1.63}$$

Inserting (16.1.63) into the representation theorem, we recognise the adjoint wavefield as,

$$\mathbf{u}^\dagger(\mathbf{x}, t) = -\mathbf{G}_i^{\dagger *}(\mathbf{x}, \omega_r; \mathbf{x}_r) e^{-i\omega_r t}. \tag{16.1.64}$$

Once again using the example of a perfectly elastic medium as in Section 16.1.2, we substitute (16.1.64) into the Fréchet derivative expression (16.1.35). The time integral corresponds to a Fourier transform, and so we obtain

$$\delta\chi = \int_V d^3\mathbf{x}\, \delta\rho(\mathbf{x}) \dot{\mathbf{G}}_i^{\dagger *}(\mathbf{x}, \omega_r; \mathbf{x}_r) \cdot \dot{\mathbf{u}}(\mathbf{x}, \omega_r) - \int_V d^3\mathbf{x}\, \nabla \mathbf{G}_i^{\dagger *}(\mathbf{x}, \omega_r; \mathbf{x}_r) : \delta\mathbf{c}(\mathbf{x}) : \nabla \mathbf{u}(\mathbf{x}, \omega_r). \tag{16.1.65}$$

For the special case where the wavefield $\mathbf{u}$ is the Green's function $\mathbf{G}_j$, (16.1.65) transforms into the frequency-domain version of Green's theorem,

$$\mathbf{e}_i \cdot \delta\mathbf{G}_j(\mathbf{x}_r, \omega_r; \mathbf{x}_s) = \int_V d^3\mathbf{x}\, \delta\rho(\mathbf{x})\, \dot{\mathbf{G}}_i^{\dagger *}(\mathbf{x}, \omega_r; \mathbf{x}_r) \cdot \dot{\mathbf{G}}_j(\mathbf{x}, \omega_r; \mathbf{x}_s) - \int_V d^3\mathbf{x}\, \nabla \mathbf{G}_i^{\dagger *}(\mathbf{x}, \omega_r; \mathbf{x}_r) : \delta\mathbf{c}(\mathbf{x}) : \nabla \mathbf{G}_j(\mathbf{x}, \omega_r; \mathbf{x}_s). \tag{16.1.66}$$

16.2 Derivatives with Respect to Source Parameters

The properties of the seismic wavefield depend on the structure of the medium $\mathbf{m}$ but also on the source $\mathbf{f}$ that excites the wavefield in the first place. Neglecting the influence of the source may lead to unwanted artefacts in Earth models, especially in regions that are less well sampled (e.g, Valentine & Woodhouse, 2010). Thus, before covering seismic waveform inversion using event data in Chapter 17, we derive a simple modification of the general adjoint method recipe from Section 16.1 that allows us to compute partial derivatives with respect to source parameters.

In this context, we are interested in deriving an expression for the infinitesimal change of the misfit function, $\delta\chi$, induced by an infinitesimal change of the source, $\delta\mathbf{f}$, that does not explicitly contain the derivative of the wavefield $\mathbf{u}$ with respect to the source. This goal can be achieved most easily through the definition of a new operator

$$\mathcal{L}^{\mathrm{f}}\left[\mathbf{u}(\mathrm{m}), \mathrm{m}, \mathbf{f}\right] = \mathcal{L}\left[\mathbf{u}(\mathrm{m}), \mathrm{m}\right] - \mathbf{f}, \tag{16.2.1}$$

which is indeed just a simple rearrangement of (16.1.6). Applying the procedure outlined in Section 16.1 up to (16.2.1), we extract the adjoint equation already presented in (16.1.11). The Fréchet derivative with respect to **f** is

$$\delta\chi = -\left(\mathbf{u}^{\dagger}, \delta\mathbf{f}\right). \tag{16.2.2}$$

Equation (16.2.2) implies that derivatives with respect to source parameters can be computed with the same adjoint field $\mathbf{u}^{\dagger}$ as is used for derivatives with respect to the structural parameters m. However, the expression for the Fréchet derivative for source parameters is significantly simpler since it does not involve any interaction between the forward and the adjoint wavefields. The forward wavefield is absent from (16.2.2), and so it is sufficient to propagate the adjoint wavefield in reverse time and to evaluate its evolution at the position of the source perturbation $\delta\mathbf{f}$. This approach is reminiscent of time-reversal methods for the imaging of wavefield sources such as earthquakes (e.g., Larmat et al., 2006).

Many natural and human-made sources, including earthquakes and explosions, can be described by a moment-tensor point source,

$$\mathbf{f}(\mathbf{x}, \mathrm{t}) = -\nabla \cdot \left[\mathbf{M}\,\delta(\mathbf{x} - \mathbf{x}_0)\,\mathrm{s}(\mathrm{t} - \mathrm{t}_0)\right], \tag{16.2.3}$$

where s is a causal source-time function, $\mathbf{x}_0$ denotes the hypocentre location, and t_0 is the origin time. Applying (16.2.2) to the special source (16.2.3) yields the Fréchet derivative

$$\begin{aligned}\delta\chi = & -\int_{\mathrm{T}} \mathrm{dt}\,\delta\mathbf{M} : \nabla\mathbf{u}^{\dagger}(\mathbf{x}_0, \mathbf{t}_0) - \delta\mathbf{x}_0 \cdot \nabla \int_{\mathrm{T}} \mathrm{dt}\,\mathbf{M} : \nabla\mathbf{u}^{\dagger}\mathrm{u}(\mathbf{x}_0, \mathrm{t}) \\ & + \delta\mathrm{t}_0\,\mathbf{M} : \nabla\mathbf{u}^{\dagger}\mathrm{u}(\mathbf{x}_0, \mathrm{t}_0). \end{aligned} \tag{16.2.4}$$

In a similar way to the directional derivatives with respect to structure, the source derivatives of the misfit functionals can be used to drive the nonlinear optimisation schemes covered in Chapter 15 (e.g., Liu et al., 2004).

16.3 Misfit Functionals and Sensitivity Analysis

An interesting and practically very useful aspect of the Fréchet kernel expressions we have derived in this chapter is that they do not explicitly depend on the details of the measurement, that is, on the actual definition of the misfit. All kernels simply involve a product of the forward and the adjoint wavefield, occasionally embellished with temporal and spatial derivatives. The measurement process is entirely hidden in the definition of the adjoint source $\mathbf{f}^{\dagger}$, which can be derived by bringing $\delta\chi$ into the generic form (16.1.2).

The choice of misfit functional χ controls the essential properties of an inversion, including the type and amount of information that can be exploited, the structural and source parameters that can or cannot be resolved, and the extent to which the problem is nonlinear. A 'good' misfit functional adapts to the nature of the data, and achieves a delicate balance between extracting as much information as possible, whilst allowing a (computationally) tractable inversion algorithm.

It is therefore not surprising that the diversity of seismic data is reflected is a rich family of waveform misfit functionals, all of which fill a useful niche without generally being superior. Widely used examples include instantaneous phase and envelope misfits (Bozdağ et al., 2011), waveform correlation misfits (van Leeuwen & Mulder, 2010), optimal-transport distance measures (Métivier et al., 2016), robust L_1 norms (Brossier et al., 2009), or time–frequency phase and envelope misfits (Fichtner et al., 2008).

For the purpose of illustration, we will limit ourselves to simple examples. The windowed and filtered L_2 waveform misfit in Section 16.3.1 is to some extent of historical interest. This misfit was used in pioneering synthetic waveform inversion studies (e.g., Gauthier et al., 1986), but was soon found to be impractical for real-data applications with band-limited and noisy recordings. Yet, the L_2 waveform misfit allows us to conveniently demonstrate the effects of data processing, such as filtering and windowing. In Section 16.3.2 we consider the case of time-shift measurements, and in Section 16.3.3 amplitude measurements. Whilst these cases are simplistic, they share essential properties with more sophisticated time- and amplitude-like misfit measures.

16.3.1 L_2 *Waveform Misfit*

As a starting point, we return to the L_2 waveform misfit already introduced in (16.1.3). To be more realistic, we now consider the i-component of windowed and filtered seismograms,

$$\chi = \frac{1}{2}\int_T dt\, \left[w(t)\, F(t) * \left(u_i(\mathbf{x}_r, t) - u_i^{obs}(\mathbf{x}_r, t)\right)\right]^2 . \tag{16.3.1}$$

The convolutional filter $F(t)$, and the time windowing function $w(t)$ may serve, for instance, to isolate a specific seismic phase within a defined frequency band.

To obtain the adjoint source corresponding to this choice of misfit, the variation $\delta\chi$ with respect to an infinitesimal waveform change $\delta\mathbf{u}$ must be brought into the generic form (16.1.2). Using the identity

$$\int dt\, a(t)\, [F(t) * b(t)] = \int dt\, b(t)\, [F(-t) * a(t)] , \tag{16.3.2}$$

which holds for any functions $a(t)$ and $b(t)$, we find that

$$\delta\chi = \int_T dt \int_V d^3\mathbf{x}\, w^2(t) \left[F(-t) * F(t) * \left(u_i(\mathbf{x}_r, t) - u_i^{obs}(\mathbf{x}_r, t)\right)\right]\, \delta(\mathbf{x}_r - \mathbf{x})\mathbf{e}_i^T \delta\mathbf{u}(\mathbf{x}, t), \tag{16.3.3}$$

where $\mathbf{e}_i$ is the unit vector pointing in the i-direction. Comparing (16.3.3) with the generic form for adjoints (16.1.2), we infer that the adjoint source is

$$\mathbf{f}^\dagger(\mathbf{x}, t) = -w^2(t) \left[F(-t)*F(t)*\left(u_i(\mathbf{x}_r, t) - u_i^{obs}(\mathbf{x}_r, t)\right)\right] \delta(\mathbf{x}_r - \mathbf{x})\mathbf{e}_i^T. \quad (16.3.4)$$

This adjoint source has several interesting properties: it is localised at the receiver point $\mathbf{x}_r$, and is a vector force acting in the direction parallel to the wavefield component used to measure the misfit, that is, the i-component. The adjoint source-time function is a filtered version of the windowed residuals $u_i(\mathbf{x}_r, t) - u_i^{obs}(\mathbf{x}_r, t)$. The time window $w(t)$ is applied twice to the residual, and so appears squared. Further, the filter F is also applied twice, once forward in time and once in reverse time. Thus, the complete effective filter $F(-t) * F(t)$ has zero phase. This property indirectly ensures that the adjoint wavefield is properly aligned in time so that it interacts correctly with the forward wavefield in the calculation of sensitivity kernels.

While this L_2 waveform misfit serves well as an illustration of the derivation and key aspects of adjoint sources, its practical usefulness is rather limited. The L_2 waveform misfit function is susceptible to cycle skips, thereby introducing unwanted nonlinearity into the inverse problem. Furthermore, it mixes amplitude and phase information, which can cause difficulties in applications where measurements of wavefield amplitudes are less reliable. Separating amplitude and phase may also help to reduce the coupling between wave speeds and attenuation, thereby rendering the inverse problem more tractable.

16.3.2 Cross-Correlation Time Shifts

A milestone towards the explicit extraction of phase information in seismic waveform inversion was the method of Luo & Schuster (1991), which is based on the estimation of delay times by cross-correlation as in Section 5.1.1 and Section 5.3.1. For the i-component of an observed waveform $u_i^{obs}(\mathbf{x}_r, t)$ and a computed waveform $u_i(\mathbf{x}_r, t)$ at a receiver position $\mathbf{x}_r$, the cross-correlation time shift $\mathcal{T}$ is defined as the time where the cross-correlation function,

$$C_{\{u_i^{obs}, u_i\}}(\tau) = \frac{1}{2} \int_T dt\, u_i^{obs}(\mathbf{x}_r, t) u_i(\mathbf{x}_r, t + \tau), \quad (16.3.5)$$

attains its global maximum. This definition requires $\mathcal{T} > 0$ when the observed waveform arrives earlier than the simulation, and $\mathcal{T} < 0$ when the observations arrive later. In the interest of a more condensed notation, we implicitly assume that the waveforms have been properly windowed and filtered, meaning that we omit the time window function $w(t)$ and the filter $F(t)$ introduced in (16.3.1). The misfit functional that we wish to minimise, is now defined as

$$\chi = \frac{1}{2}\mathcal{T}^2. \quad (16.3.6)$$

Though (16.3.5) does not provide an explicit expression for $\mathcal{T}$ itself, we can derive an explicit expression for its variation. For this, we note that $C_{\{u_i^{obs}, u_i\}}(\tau)$ attains a maximum for $\tau = \mathcal{T}$, meaning that

$$0 = \frac{d}{d\tau} C_{\{u_i^{obs}, u_i\}}(\mathcal{T}) = -\int_T dt\, \partial_t u_i^{obs}(\mathbf{x}_r, t - \mathcal{T}) u_i(\mathbf{x}_r, t). \tag{16.3.7}$$

Using implicit function differentiation yields the variation of χ:

$$\delta\chi = \mathcal{T}\,\delta\mathcal{T} = \frac{\mathcal{T}\int_T dt\, \partial_t u_i^{obs}(\mathbf{x}_r, t - \mathcal{T})\, \delta u_i(\mathbf{x}_r, t)}{\int_T dt\, \partial_t^2 u_i^{obs}(\mathbf{x}_r, t - \mathcal{T})\, u_i(\mathbf{x}_r, t)}. \tag{16.3.8}$$

Making the simplifying assumption that the observed waveform u_i^{obs} and the computed waveform u_i are identical but shifted in time by $\mathcal{T}$, we find that $u_i^{obs}(\mathbf{x}_r, t - \mathcal{T}) = u_i(\mathbf{x}_r, t)$, and (16.3.8) condenses to

$$\delta\chi = -\frac{\mathcal{T}}{||\partial_t u_i||_2^2} \int_T dt\, \partial_t u_i(\mathbf{x}_r, t)\, \delta u_i(\mathbf{x}_r, t), \tag{16.3.9}$$

where the normalisation factor is defined as

$$||\partial_t u_i||_2^2 = \int_T dt\, [\partial_t u_i(\mathbf{x}_r, t)]^2. \tag{16.3.10}$$

A slight modification now allows us to bring (16.3.9) into the generic form (16.1.2),

$$\delta\chi = -\frac{\mathcal{T}}{||\partial_t u_i||_2^2} \int_T dt \int_V d^3\mathbf{x}\, \partial_t u_i(\mathbf{x}_r, t)\, \delta(\mathbf{x} - \mathbf{x}_r)\, \mathbf{e}_i^T \delta\mathbf{u}(\mathbf{x}, t). \tag{16.3.11}$$

Comparing (16.3.11) and (16.1.2), we infer that the adjoint source for the cross-correlation time shift misfit is given by

$$\mathbf{f}^\dagger(\mathbf{x}, t) = \frac{\mathcal{T}}{||\partial_t u_i||_2^2} \partial_t u_i(\mathbf{x}_r, t)\, \delta(\mathbf{x} - \mathbf{x}_r)\, \mathbf{e}_i. \tag{16.3.12}$$

As a result of the point-wise measurement, the adjoint source is again specifically localised at the receiver point. The adjoint source-time function is determined by the computed displacement velocity $\partial_t u_i(\mathbf{x}_r, t)$, and the normalisation factor $||\partial_t u_i||_2^2$ ensures that travel-time sensitivity does not depend on the amplitude of the forward wave field. The adjoint source, and therefore the sensitivity kernels, are quasi-independent of the data, except for the constant scaling factor $\mathcal{T}$. This is a result of our simplifying assumption that the observed and computed waveforms are sufficiently similar to allow for the replacement of $u_i^{obs}(\mathbf{x}_r, t - \mathcal{T})$ by $u_i(\mathbf{x}_r, t)$.

While this limited dependence on data may to some extent be counter-intuitive, there is still a difference if time shifts, or any other type of waveform misfit, are assessed from displacement, velocity, or acceleration seismograms. In fact, denoting the adjoint source in (16.3.12) by $\mathbf{f}^\dagger(\mathbf{u})$, in order to make the use of displacement seismograms explicit, an identical derivation with velocity seismograms $\mathbf{v} = \partial_t \mathbf{u}$ would have led to $\delta\chi = \left(-\mathbf{f}^\dagger(\mathbf{v}), \delta\mathbf{v}\right)$. This differs from the generic form (16.1.2) by the appearance of $\delta\mathbf{v}$ instead of $\delta\mathbf{u}$. Integrating by parts, we find

$$\delta\chi = \Big(-\mathbf{f}^\dagger(\mathbf{v}), \delta\mathbf{v} \Big) = \Big(\partial_t \mathbf{f}^\dagger(\mathbf{v}), \delta\mathbf{u} \Big), \tag{16.3.13}$$

from which we infer that the correct adjoint source for measurements on velocity seismograms is given by $-\partial_t \mathbf{f}^\dagger(\mathbf{v})$.

16.3.3 Amplitude Misfits

Again assuming that waveforms have been properly windowed and filtered, the observed amplitude of a seismic wave may be defined in terms of the L_2 norm, as in Section 5.3.1:

$$\mathcal{A}^{\text{obs}} = \left(\int_T dt\, [u_i^{\text{obs}}(\mathbf{x}_r, t)]^2 \right)^{\frac{1}{2}}. \tag{16.3.14}$$

Analogously, the amplitude of computed waveforms may be defined as

$$\mathcal{A} = \left(\int_T dt\, [u_i(\mathbf{x}_r, t)]^2 \right)^{\frac{1}{2}}. \tag{16.3.15}$$

Using (16.3.14) and (16.3.15) we define a quadratic amplitude misfit as

$$\chi = \frac{1}{2} \frac{(\mathcal{A} - \mathcal{A}^{\text{obs}})^2}{(\mathcal{A}^{\text{obs}})^2}. \tag{16.3.16}$$

For the variation of χ we find

$$\delta\chi = \left[\frac{\mathcal{A} - \mathcal{A}^{\text{obs}}}{(\mathcal{A}^{\text{obs}})^2} \right] \delta\mathcal{A} = \left[\frac{\mathcal{A} - \mathcal{A}^{\text{obs}}}{\mathcal{A}\,(\mathcal{A}^{\text{obs}})^2} \right] \int_T dt\, u_i(\mathbf{x}_r, t)\, \delta u_i(\mathbf{x}_r, t). \tag{16.3.17}$$

Writing (16.3.17) in vectorial form and including a spatial delta function, we obtain

$$\delta\chi = \left[\frac{\mathcal{A} - \mathcal{A}^{\text{obs}}}{\mathcal{A}\,(\mathcal{A}^{\text{obs}})^2} \right] \int_T dt \int_V d^3\mathbf{x}\, u_i(\mathbf{x}_r, t) \delta(\mathbf{x} - \mathbf{x}_r) \mathbf{e}_i^T\, \delta\mathbf{u}(\mathbf{x}_r, t), \tag{16.3.18}$$

which can be compared directly to the generic form (16.1.2) to obtain the adjoint source

$$\mathbf{f}^\dagger(\mathbf{x}, t) = -\left[\frac{\mathcal{A} - \mathcal{A}^{\text{obs}}}{\mathcal{A}\,(\mathcal{A}^{\text{obs}})^2} \right] u_i(\mathbf{x}_r, t) \delta(\mathbf{x} - \mathbf{x}_r) \mathbf{e}_i. \tag{16.3.19}$$

From (16.3.12) and (16.3.19) we see that the adjoint sources for travel-time and amplitude misfits are nearly identical. Apart from a constant scaling, they only differ by a time derivative of the adjoint source-time function.

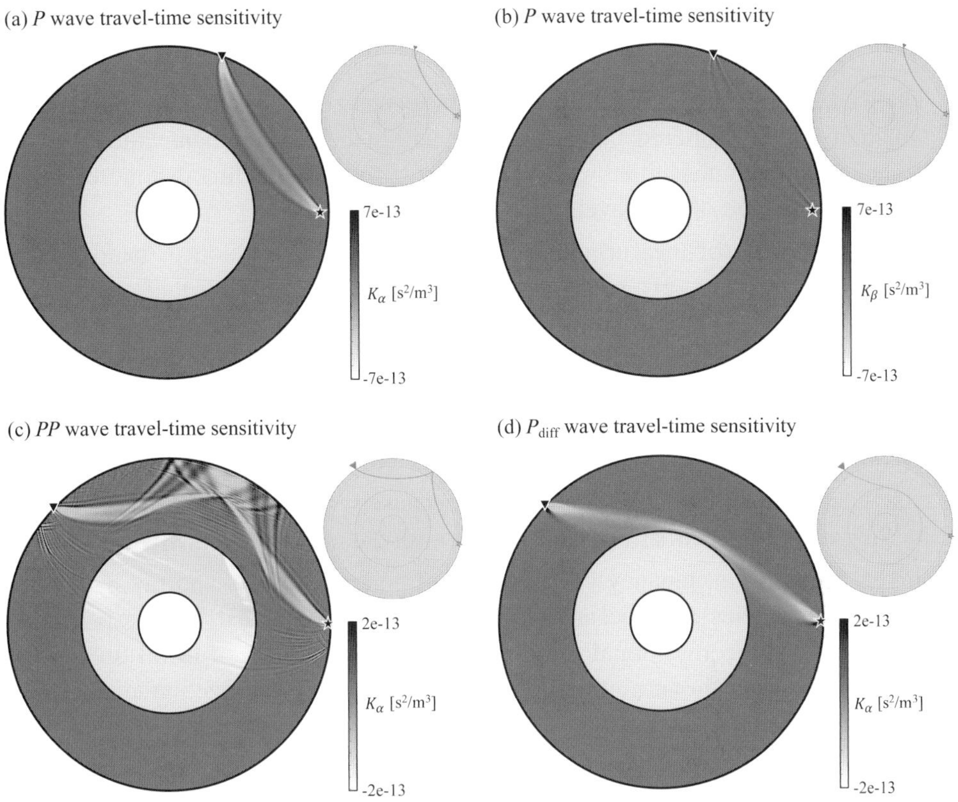

Figure 16.2 Sensitivity kernels for various *P* phases, recorded on the vertical component at a dominant period of 15 s. A double-couple source is placed at 200 km depth, with only the moment-tensor component $M_{r\phi}$ non-zero. The corresponding ray paths are shown in the smaller panels to the upper right. The source is marked by a star, and the receiver by a triangle. (a) The sensitivity K_α, with respect to *P* wavespeed, for a direct wave cross-correlation travel time at 70° epicentral distance. (b) Sensitivity for the same measurement as in (a) but with respect to *S* wavespeed. (c) The same as in (a) but for a *PP* wave at 135° epicentral distance. (d) The same as in (a) but for a P_{diff} wave at 135° epicentral distance.

16.3.4 Whole-Earth Kernel Gallery

The simplicity of the travel-time measure from Section 16.3.2 and the amplitude measure from Section 16.3.3, allow us to compute and analyse sensitivity kernels that are data-independent up to a scalar constant. For instance, if we divide the adjoint source for cross-correlation travel-time shifts in (16.3.12) by $\mathcal{T}$ we produce sensitivity kernels that are independent of an actual time-shift measurement.

Examples of such sensitivity kernels for travel-time shift are illustrated in Figure 16.2. The kernels are computed with a 2-D spectral-element wave propagation method (Afanasiev et al., 2019). The wavefield source is a double couple located at 200 km depth, with all moment-tensor components except $M_{r\phi}$ set to zero;

the dominant period for the calculation is 15 s. Cross-correlation time shifts are measured on the vertical component for various *P*-wave phases. Figure 16.2(a) shows the sensitivity kernel with respect to *P* wavespeed, K_α, for the direct *P* wave. The K_α kernel is non-zero within a broad influence zone, marking the region from which first-order scattered waves can arrive within the measurement window. Most of the influence zone is filled with negative sensitivity, indicating that increasing *P* wavespeed reduces the *P* travel time.

The procedure of measuring time shifts by cross-correlation of band-limited waveforms over an extended window rather than picking the onset of a high-frequency arrival, has a series of consequences often collectively referred to as finite-frequency effects. (1) Surrounding the broad region of negative sensitivity, there are narrow fringes of positive sensitivity where increasing *P* wavespeed leads to a slight increase in *P*-wave travel time. (2) There is a minimum of sensitivity along the geometric ray path. (3) Sensitivity with respect to *S* wavespeed is not exactly zero, as shown in Figure 16.2(b), because scattered *P* and *S* waves caused by *S*-wavespeed perturbations can arrive within the measurement window.

The apparent discrepancy between finite-frequency effects and ray theory for infinite frequencies arise because the sensitivity is calculated with respect to *P* wavespeed. The conventional result of maximum sensitivity of travel time in the neighbourhood of the turning point of a ray relates to wavespeed gradient rather than wavespeed.

Figure 16.2(c) illustrates an often unwanted finite-frequency effect. Although the measurement window is centred around the *PP* wave arriving at 135° epicentral distance on the vertical component, other phases interfere. In consequence, we obtain a sensitivity kernel that is dominated by the broad negative sensitivity of the *PP* wave, but also has contributions from *PPP* and various core phases that leak into the measurement window. The $P_{\rm diff}$ phase, in contrast, arrives more than one minute before all other phases at 135° epicentral distance, and can thus be isolated easily. The corresponding sensitivity kernel, shown in Figure 16.2(d), is not affected by interference with any other arrival.

In Section 16.3.3 we observed that adjoint sources for cross-correlation time shifts and amplitudes only differ by a time derivative and a constant scaling. This suggests that the corresponding sensitivity kernels should be rather similar. Indeed, the shape of the amplitude sensitivity kernel for the direct *P* wave, shown Figure 16.3(b), is nearly identical to the cross-correlation time shift kernel. The only major difference is a maximum of sensitivity along the geometric ray path. For Figure 16.3 we have used data-independent kernels, with the adjoint source in (16.3.19) divided by $(\mathcal{A} - \mathcal{A}^{\rm obs})/\mathcal{A}$.

Waveform sensitivity in general is frequency-dependent. Lowering the frequency leads to an elongation of waveforms, and hence to a broadening of the measurement window used to isolate a specific phase. Consequently, the influence zone from which scattered waves can arrive within the measurement

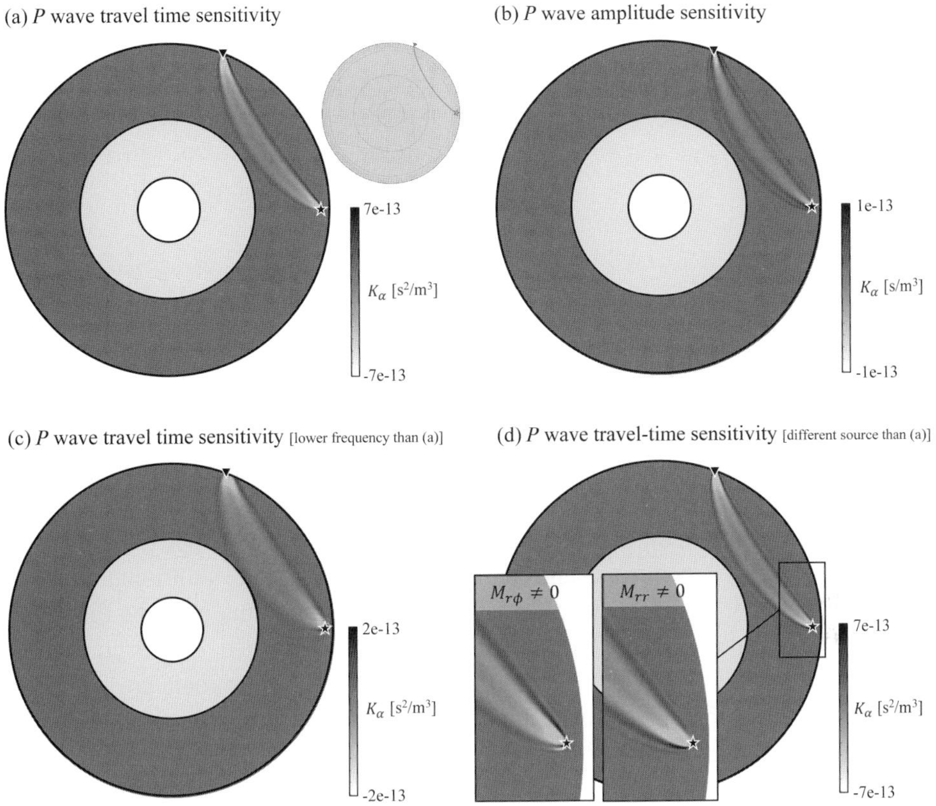

Figure 16.3 Effects of the measurement, frequency and source mechanism on *P*-wave sensitivity. The sensitivity kernel with respect to *P* wavespeed of the vertical-component direct *P* wave at 70° epicentral distance, shown in panel (a), serves as a reference. (This is identical to the kernel in Figure 16.2a.) (b) The measurement of a cross-correlation time shift is replaced by an amplitude measurement, as in Section 16.3.3. To obtain a data-independent kernel, the adjoint source in (16.3.19) is divided by $(\mathcal{A} - \mathcal{A}^{\text{obs}})/\mathcal{A}$. (c) The dominant period is increased from 15 s to 50 s. (d) The source mechanism is changed from $\mathsf{M}_{r\phi} \neq 0$ to $\mathsf{M}_{rr} \neq 0$.

window expands. This broadening effect is illustrated in Figure 16.3(c), where the dominant period is increased to 50 s rather than the 15 s used in Figure 16.3(b). The width of the influence zone is approximately proportional to $\sqrt{\text{period}}$.

The properties of the wavefield source tend to have more subtle effects that are most noticeable within a few wavelengths distance from the source. Changes of the source orientation modify the location of nodal planes on which sensitivity is small or even zero, which can have an effect on how well a specific phase can be isolated. An example of the influence of the source mechanism is shown in Figure 16.3(d), where the source mechanism is modified from $\mathsf{M}_{r\phi} \neq 0$ to $\mathsf{M}_{rr} \neq 0$. With the exception of small variations near the source, sensitivity within the primary influence zone is nearly unaffected.

17

Waveform Inversion of Event Data

From a purely practical perspective, the development of seismic waveform inversion comes from efforts to incorporate as many independent data as possible into tomographic inversions in order to maximise resolution. However, the fascination of this method is deeply rooted in the ideal concept of exploiting complete seismic waveforms, and so goes well beyond the traditional practice of extracting smaller pieces of information from seismograms, such as the travel times or amplitudes of a few seismic phases, or selected eigenfrequencies. Though the ideal is hardly achievable in practice, as waveform data are generally contaminated by ambient and instrumental noise, the concept motivates the widely used term *full-waveform inversion (FWI)*. Since it is by no means possible to use the full waveform, other names such as *adjoint tomography* or *full-wavefield inversion* have been suggested and can be found in the literature. Here, we prefer the hopefully less contradictory and less controversial term *waveform inversion*.

Although the basic mathematical ingredients had already been developed in the late 1970's and early 1980's (e.g., Bamberger et al., 1977; Tarantola, 1984), it took nearly 30 years before waveform inversion, in the sense described here, became a practical tool that provided new information about the 3-D internal structure of the Earth (e.g., Chen P. et al., 2007; Fichtner et al., 2009; Tape et al., 2010). This time delay largely arose from insufficient computational resources to permit realistic simulations of seismic-wave propagation in a 3-D heterogeneous Earth. However, there were also numerous technical details, not obvious at the beginning, which had to be investigated by laborious trial and error.

In this chapter, we will introduce waveform inversion of event data using a recent example from the Eastern Mediterranean (Blom et al., 2020). This will allow us to cover several important aspects such as the selection of data and measurements, the choice of an initial model, the treatment of the crust, the concept of nonlinear multi-scale optimisation, as well as resolution analysis and validation.

The Mediterranean has been a major focus of tomographic studies for several decades (e.g., Spakman et al., 1988; Piromallo & Morelli, 1997), due to its complex geology and tectonics with associated strong seismicity. The current tectonic setting is marked by the slow convergence between the African and Eurasian plates, at around 6 mm/yr. Much of this convergence is accommodated by the Alpine and Hellenic arcs. In contrast to the Western Mediterranean that mostly consists of oceanic lithosphere younger than 30 Ma, the Eastern Mediterranean is primarily composed of old African oceanic lithosphere. A notable exception is the Aegean Sea, where subduction beneath the Hellenic Arc has resulted in a young oceanic basin. The tectonics of the Eastern Mediterranean is further complicated by the presence of various interacting microplates, including the Arabian, Aegean and Anatolian plates.

17.1 Data and Measurements

17.1.1 Selection of Earthquake Data

The selection of data for waveform inversion differs in several aspects from data selection in ray-based travel-time tomography, mostly due to computational constraints. In particular, the high computational cost of numerical forward and adjoint simulations limits the number of events that can be employed much more strongly than in ray tomography. As a consequence, waveform inversion is typically only applied to events with high data quality where the presence of noise can almost be neglected. In cases where data quality is generally poor, the application of waveform inversion, which aims to exploit waveform details, may be hard to justify. Since the computational cost of adjoint simulations, introduced in Chapter 16, is largely independent of the number of stations, it is advisable to mainly select events that have been recorded by a large number of stations.

In the waveform inversion study presented here, we selected a total of 80 earthquakes that occurred in the Eastern Mediterranean region between 1998 and 2017. These provide a total of about 17 000 event–station pairs. The resulting data coverage is summarised in Figure 17.1.

17.1.2 Numerical Waveform Modelling

Strong lateral heterogeneities in the upper few hundred kilometres of the Earth, topography, and the presence of fluid layers are too complex for analytical solutions, and so numerical solutions of the seismic wave equation are required. As in most other physical sciences, finite-difference methods were the first and most straightforward family of numerical methods used for this purpose (Boore, 1970). Though higher-order finite-difference schemes (e.g., Moczo et al., 2014) continue to be widely used, especially in exploration seismology, there has been a gradual shift towards spectral-element methods (e.g., Faccioli et al., 1997; Komatitsch & Tromp, 2002a,b).

The choice of a suitable numerical method depends mostly on the complexity of the medium and the dominant wave types. The modelling of body waves in a medium with a limited number of interfaces, which is typical in exploration applications, may be achieved very efficiently using finite-difference methods. In

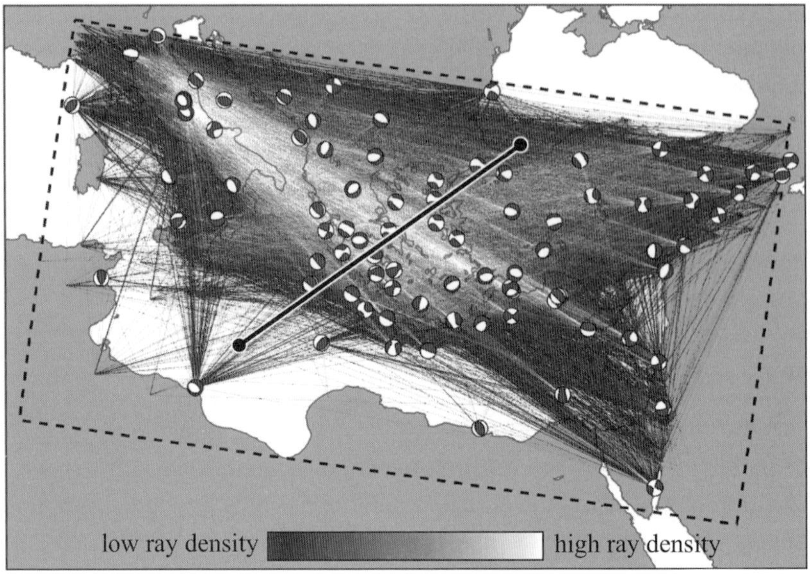

Figure 17.1 Summary of data coverage in the Eastern Mediterranean region in the form of great-circle rays connecting event and station locations. Earthquake source mechanisms are plotted as beachballs. The density of data coverage is proportional to the brightness of the grey tones. The black line in the centre marks the location of the vertical velocity profile in Figure 17.3(a).

contrast, the accurate modelling of surface waves or irregular topography tends to be more efficient with spectral-element methods, which have more geometric flexibility and naturally honour the free-surface boundary condition.

A complete treatment of this complex issue of the choice of numerical method is beyond the scope of this work. We therefore refer the reader to the excellent book of Igel (2016). For the application considered here, we employ a GPU-accelerated spectral-element method, specifically developed for regional-scale wave propagation in spherical subsections (Gokhberg & Fichtner, 2016).

17.1.3 Initial Model

Waveform inversion rests on the nonlinear minimisation of a misfit that quantifies the discrepancies between observed and calculated waveforms, typically using one of the methods described in Chapter 15. Since iterative nonlinear optimisation is prone to convergence towards meaningless local minima in the misfit, the choice of a suitable initial model is of major importance in waveform inversion.

In this context the term 'suitable' primarily means that the initial model is close enough to the optimum to avoid cycle-skipping, so that the alignment of observed and simulated seismograms is sufficiently close that the time segments directly correspond, rather than having a time offset. This requirement makes the quality of an initial model frequency-dependent. The experience from regional- to global-scale waveform inversion is that reasonable 1-D Earth models may already avoid cycle-skipping because earthquake data of sufficiently low frequency are often available. In seismic exploration, however, active data may not have enough low-frequency content, thus sparking the development of misfit functionals that circumvent cycle-skipping as far as possible (e.g., van Leeuwen & Mulder, 2010; Métivier et al., 2016).

A critical element of regional-scale to global-scale waveform inversion is the treatment of the crust, which is complicated for two reasons. (1) The crust, especially beneath the oceans, may be significantly thinner than the shortest seismic wavelength of interest. As a consequence, numerical simulations may become excessively expensive because small elements are needed to honour the Moho and other intra-crustal discontinuities. (2) For the frequencies that can be modelled numerically, waveform data may not contain sufficient information on the location of crustal discontinuities. An initial discontinuity implemented at an incorrect depth may therefore not be able to be moved during the iterative inversion, potentially leading to the contamination of upper-mantle structure.

A widely used solution to the crustal problem consists in the initial implementation of a smooth gradient instead of a discontinuity (Fichtner & Igel, 2008; French & Romanowicz, 2014). Gradients in material properties may cut through the interior of a finite element without degrading numerical accuracy, and the gradient can sharpen naturally during the inversion where this is required by the data. Care must of course be taken in the interpretation of the final results, as the absence of a sharp discontinuity in the model does not necessarily imply the absence of a such a sharp discontinuity in nature. For the Eastern Mediterranean waveform inversion we used an initial model that already contains smooth 3-D structure derived from previous inversions at larger scale (Fichtner et al., 2013b). A vertical profile through the initial model across the Helenic subduction zone is shown in the upper panel of Figure 17.3(a).

In contrast to ray-based modelling of travel times, waveform modelling requires moment tensors for the selected events as a component of the full suite of the initial model parameters. This naturally limits the set of usable sources to earthquakes that are large enough for a meaningful moment-tensor inversion. At the scales considered here, the minimum magnitude is around Mw 5. For this inversion for the Eastern Mediterranean region, all initial moment tensors were taken from the Global CMT Catalogue (www.globalcmt.org).

17.2 Nonlinear Optimisation of Waveform Misfit

During the past decade, numerous flavours of waveform inversion have been developed. They mostly differ in the choice of the misfit functional, used to quantify discrepancies between observed and calculated (synthetic) seismograms. A well-chosen misfit functional honours the nature of the data and wave propagation at different scales.

For regional-scale to global-scale inversion of earthquake data it is beneficial to design misfit functionals that emphasise phase information while giving less weight to amplitudes, as these depend strongly on factors that may be poorly known or difficult to constrain, such as earthquake magnitude and 3-D attenuation structure. Furthermore, the misfit functional should automatically balance information from low- and high-amplitude waveforms. For instance, a high-amplitude surface wave should not have more weight than a low-amplitude body wave, unless this is explicitly wanted. Misfit functionals that fulfil these requirements include cross-correlation time shifts (Luo & Schuster, 1991; Section 16.3.2), multi-taper phase measurements (Tape et al., 2010), and time–frequency phase misfits (Fichtner et al., 2008; Section 5.3.2). We use time-frequency phase misfits in the Eastern Mediterranean example presented here, since there is an additional benefit of not having to separate individual seismic phases, a difficult process for regional scales and lower frequencies.

The ideal situation for waveform inversion is the use of complete seismograms, from the first-arriving body wave to the last discernable oscillation of the surface-wave coda. In practice, this ideal can hardly be achieved. The amplitudes of some phases may not be significantly above the noise, and waveform differences for parts of the seismograms may be too large to confidently exclude the possibility of cycle skips. As a consequence, it has become common practice to window seismograms, so that misfit measurements are only made within selected time windows where data quality is sufficiently high and where cycle skips can be excluded. Because of the large number of seismograms included in a typical waveform inversion, several automatic windowing algorithms have been developed in recent years (e.g., Maggi et al., 2009; Krischer et al., 2015). Though their implementation details differ, these authors generally apply a range of waveform similarity measures combined with additional constraints, for instance, on the minimum and maximum window length, or the earliest and latest plausible arrival time of a seismic phase at a given epicentral distance.

Following the quantification of waveform misfits, gradients can be computed with the help of adjoint techniques (Chapter 16), and one of the iterative optimisation schemes from Chapter 15 can be used to minimise the misfit. In the specific context of waveform inversion, these optimisation schemes are typically modified to accelerate convergence towards the global minimum. The most important of these modifications is to use a multi-scale approach where the lowest available frequencies are included first, before progressing towards higher frequencies. This process is intended to avoid cycle-skipping problems that are likely to occur if the complete bandwidth of interest is included from the beginning (Bunks et al., 1995). Moving from a lower- to a higher-frequency band requires a reselection of the time windows employed, and effectively constitutes a restart of the optimisation scheme. To further improve convergence and to avoid artefacts, misfit gradients are typically preconditioned, often using *ad hoc* physical intuition.

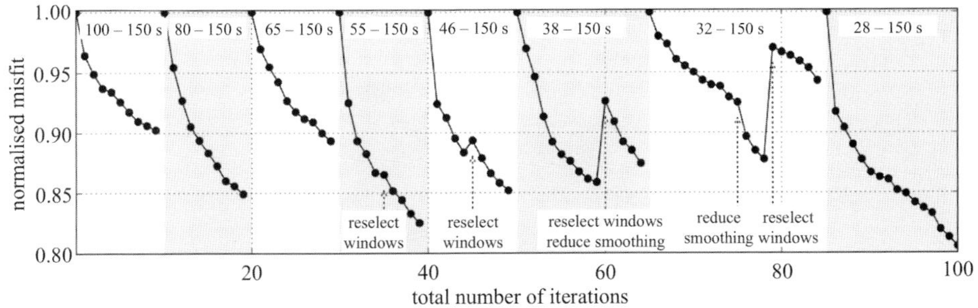

Figure 17.2 Evolution of waveform misfit per period band as a function of iteration. The initial misfit for each period band is normalised to 1. When the misfit improves significantly during the iteration, additional time windows may be selected even within a period band. Conversely, when the misfit stagnates, it may be justified to reduce the smoothing in order to allow for more structural heterogeneity.

By far the most common form of preconditioning is smoothing in order to exclude short-wavelength structure that is unlikely to be resolved.

In our Eastern Mediterranean example, we employ the conjugate-gradient method (Section 15.4), starting with the period band 100–150 s. The structural model parameters that are updated during the inversion are the *SH*- and *SV*-wavespeeds, β_h and β_v, the isotropic *P* wavespeed, α, and density, ρ. Other parameters could be included, but would be difficult to constrain. After 10 iterations, we move the lower limit of the period band to 80 s, reselect windows, and again perform 10 iterations. We then repeat this procedure, each time reducing the lowest period by around 20% until the broadest period band 28–150 s is reached. The number of iterations per period band is variable and controlled by two considerations. Firstly, there should be sufficient iterations to provide a misfit reduction that enables the transition to shorter periods without risking too many cycle skips. Secondly, the number of iterations should be small enough to avoid fitting noise. Clearly, these requirements impose a delicate balance that can only be achieved through a careful inspection of the data, or a representative data subset. We will revisit this aspect in the following section on model validation.

Figure 17.2 shows the evolution of the waveform misfit for the Eastern Mediterranean data as a function of iteration number. Overall, the misfit decreases within each period band. Sharp increases result from a reselection of windows, either when changing to the next period band or when the decreasing misfit makes a large number of new time windows available. When the misfit evolution stagnates, a reduction of the smoothing length may be justified. As can be seen around iteration 75 in Figure 17.2, the result can be a more rapid decrease of the misfit per iteration. Both the reselection of time windows within a period band and a change of regularisation are largely subjective choices. Although it is possible to devise algorithms to control these aspects of the waveform inversion more automatically and more objectively, the practical experience is that it is hard to

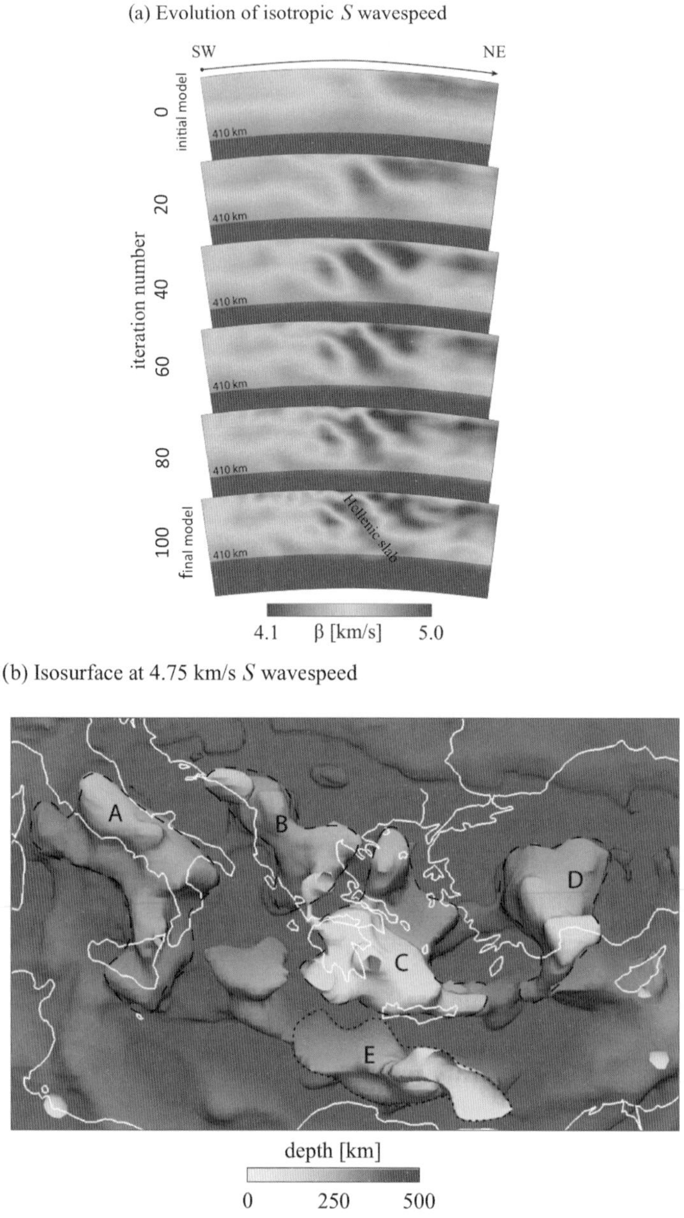

Figure 17.3 Model of isotropic S wavespeed β. (a) Evolution of β as a function of iteration number, shown along the vertical cross section marked in Figure 17.1.1. As the nonlinear optimisation progresses, and as shorter-period data are incorporated, more structural details appear. The central feature is the Hellenic slab, distinguished by anomalously high S wavespeeds, surrounded by comparatively low S wavespeeds. (b) Isosurface at $\beta = 4.75$ km/s, emphasising regions of high S-wave speed. Marked features are the Apennine–Calabrian slab (A), the Hellenic slab (B, C), the Cyprus slab (D), and a so-far unidentified feature beneath the southeastern Mediterranean (E).

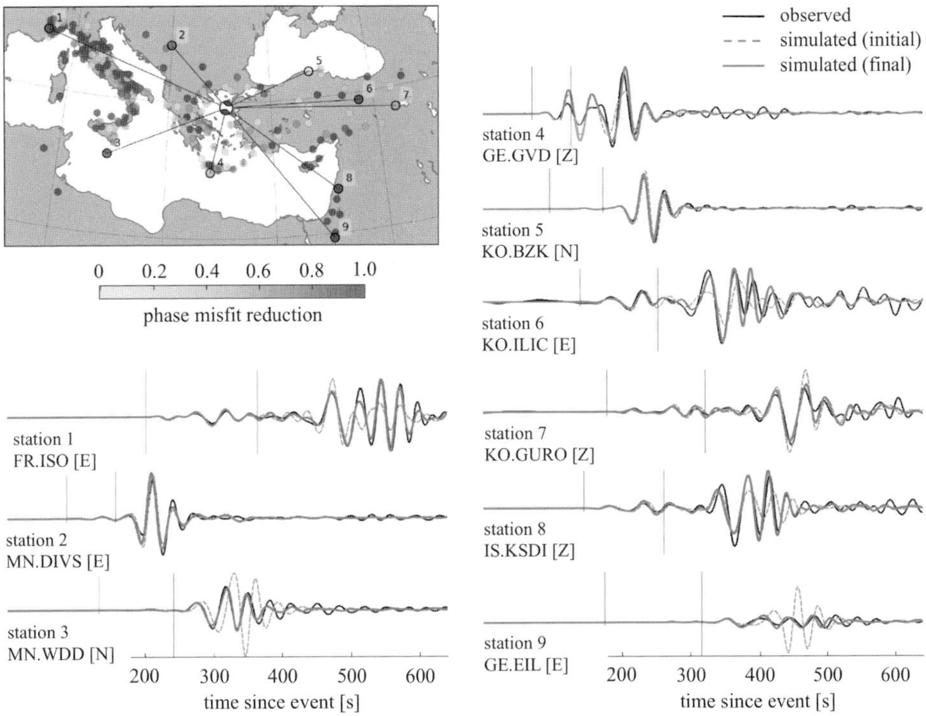

Figure 17.4 Examples of misfit reduction for an event in western Turkey. The phase misfit reduction from initial to final model is summarised for all recording stations in the upper-left panel. This is complemented by waveform plots for selected stations and components. Shown are the observed seismograms (black), the seismograms calculated for the initial model (grey dashed), and the seismograms calculated for the final model (grey solid). The period band is that of the last few iterations, 28–150 s.

outperform physical intuition.

The successive reduction in waveform misfit is accompanied by the appearance of smaller-scale heterogeneities in the Earth model, as illustrated in Figure 17.3(a). In this regard, the nonlinear optimisation is multi-scale in both model space and data space (increasingly shorter spatial scales as period decreases). The dominant structural features that emerge in the Eastern Mediterranean are well-defined subducting slabs, including the Hellenic and the Apennine–Calabrian slabs, marked in Figure 17.3(b).

The fit between observed and calculated seismograms can be improved substantially during the iterative optimisation, but this depends strongly on the choice of initial model. Selected examples of waveform fit are shown in Figure 17.4. In some cases, waveform differences for the initial model are too large to avoid cycle skips at the shortest period of 28 s, suggesting that an immediate waveform inversion with the full bandwidth would have stalled in a local minimum.

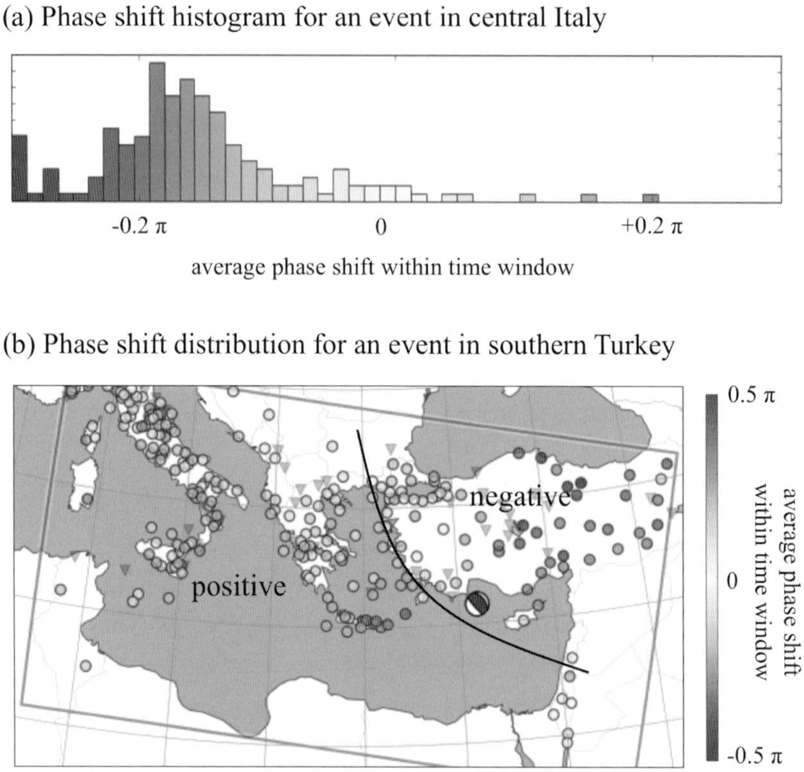

Figure 17.5 Examples of phase misfit patterns that indicate non-optimal earthquake source parameters. The average time–frequency phase difference within selected measurement time windows is displayed. (a) The phase misfit histogram for an event in central Italy reveals a consistent negative phase difference, indicating that the origin time is incorrect. Instead of systematically reinverting for all source parameters, this misfit can be more easily reduced by manually adjusting the event origin time. (b) The geographic phase shift distribution for an event in southern Turkey is consistently negative to the east and consistently positive to the west, suggesting a mislocation of the event.

Waveform inversion is a multi-parameter problem where both structural and source parameters should be constrained. While such an inversion can be formulated and implemented as a joint inverse problem, a careful analysis of the data can greatly improve efficiency and save computational resources. Non-optimal source parameters are often revealed through characteristic patterns, including consistent phase shifts in all recordings of an event that indicate an incorrect origin time (Figure 17.5a), or a bimodal geographic distribution of phase shifts that typically results from an incorrect location (Figure 17.5b). Thus, instead of systematically re-inverting for the source parameters of all events in each iteration, it is usually more efficient to do so only in cases where a clear misfit pattern already indicates that non-optimal source parameters are indeed the dominant contribution to the observed waveform misfit.

In a similar approach to the inversion for structural parameters, inversion for source parameters can be performed using one of the nonlinear optimisation methods from Chapter 15, combined with the adjoint techniques for source parameters introduced in Section 16.2.

17.3 Model Quality

Waveform inversion constitutes a nonlinear and computationally intensive inverse problem. As a consequence, the resolution matrix, introduced for linear or linearisable problems in Section 14.2.2, can usually not be computed; and if it could be computed, it may have little meaning. While Monte Carlo methods, briefly covered in Section 14.1.3, can provide meaningful measures of model quality in the presence of nonlinearity, their application to waveform inversion is still in its infancy, largely due to computational limitations. A promising research direction will be presented in Section 19.4.

With Monte Carlo sampling still being out of reach today for large 3-D waveform inversion problems, we must typically content ourselves with more easily computable proxies for model quality. These include the analysis of waveform fit for events not used in the inversion, and various types of recovery tests.

17.3.1 Validation and Analysis of Waveform Fit

The assessment of waveform fit for data that has not been included in the inversion has become common practice. Such assessments serve two related purposes. (1) Detection of over-fitting of waveforms in the inversion, i.e., the fitting of noise as a consequence of performing too many iterations. In the case of over-fitting, the fit between waveforms computed for the final model and observations not included in the inversion is significantly worse than the fit to data that were included. (2) The ability of the final model to explain waveforms from new events is critical for many applications, such as the modelling of earthquake-induced ground motion, and source inversion for earthquake and tsunami early warning.

Though an apparently simple procedure, the analysis of waveform fit for events not used in the inversion is complicated by the measurement process described in Section 17.2. Meaningful measurements of phase misfit, or most other phase- and time-like quantities, require a certain level of waveform similarity, for instance, the absence of cycle skips. The selection of measurement time windows serves the purpose of finding those parts of the observed and computed seismograms where comparisons can be made. As a consequence, the measurement of phase misfit for new events will be positively biased. Large waveform differences will simply be ignored because they would not be included in any time window.

Solutions to this problem can only be application-specific. For instance, in the context of ground motion modelling for seismic hazard assessment, one may wish to concentrate on the matching of surface-wave amplitudes at particular frequencies. For the Eastern Mediterranean example, we consider a normalised

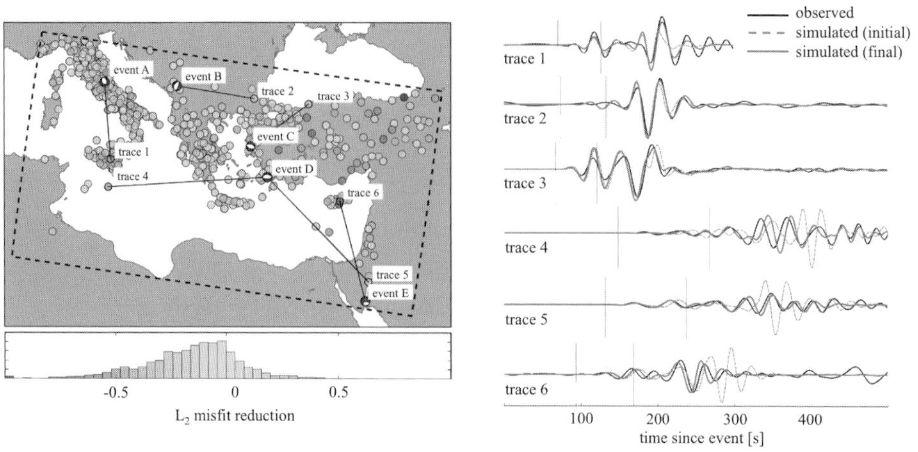

Figure 17.6 Summary of model validation using data from events not included in the waveform inversion. The distribution of events and L_2 misfit reductions per station are shown to the left. Example seismograms are displayed to the right.

L_2 waveform misfit that involves complete three-component seismograms, without any windowing,

$$\chi = \frac{\int dt\ (\mathbf{u} - \mathbf{u}_{\text{obs}})^2}{\left[\int dt\, \mathbf{u}^2\right]^{1/2} \left[\int dt\, \mathbf{u}_{\text{obs}}^2\right]^{1/2}}. \tag{17.3.1}$$

The data set that has not been used in the inversion comprises six events, distributed throughout the region of interest. Relative to the initial model, the misfit for the final model is reduced for 79% of these new recordings. Overall, the misfit reduction is 9%, compared to 13% for the data set that was included in the inversion. These L_2 misfit reductions are smaller than the phase misfit reductions shown in Figure 17.2, mostly because complete seismograms with portions dominated by noise also make a contribution. A summary of the model validation is presented in Figure 17.6.

17.3.2 Recovery Tests

The purpose of a recovery test is to provide proxy for model quality via the inversion of artificial data, computed for a known input structure. Despite being conceptually simple and apparently easy to interpret, most recovery tests suffer from being an *inverse crime*, i.e., a simplification of the inverse problem that may be considered inadmissible because nature, as far as we know, does not produce real data by numerically solving a partial differential equation. Nevertheless, recovery tests are useful when correctly interpreted as an analysis of the inversion scheme itself, under the idealised condition of perfect data. In this sense, recovery tests represent the best possible results that we may be able to achieve, and

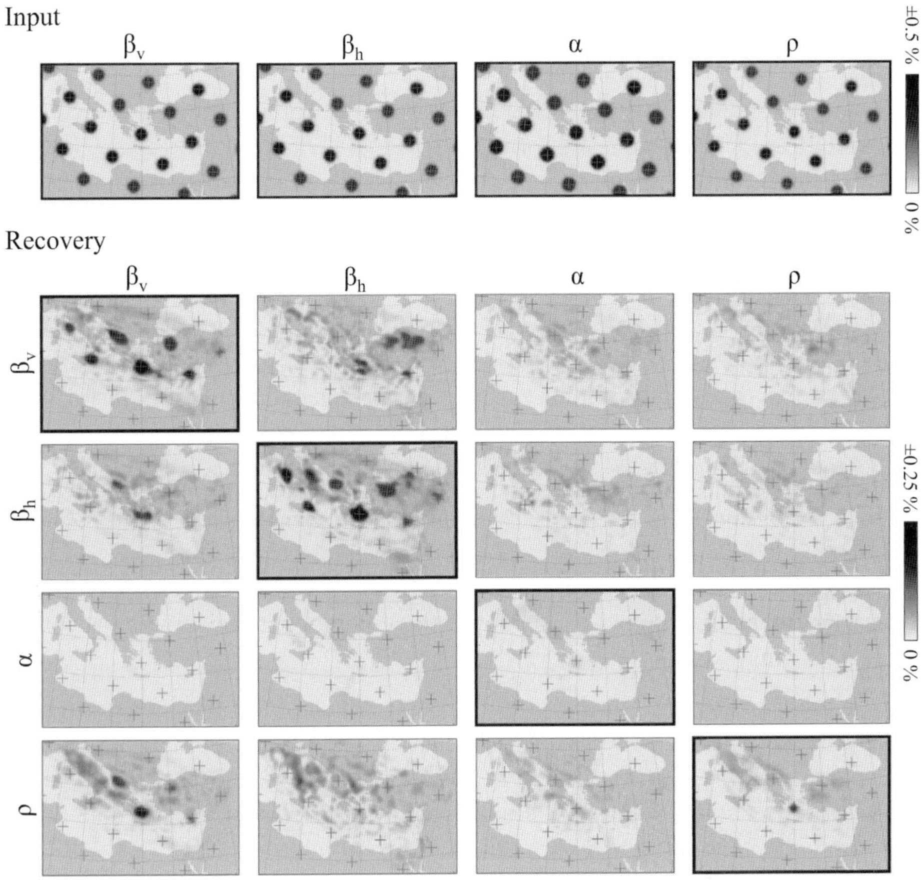

Figure 17.7 Recovery test using regularly spaced heterogeneities as the input model. Each column corresponds to the case where one model parameter is perturbed in the input model, while all others remain equal to the 1-D background model. Off-diagonals in the recovery plot thus correspond to inter-parameter trade-offs. All results are shown at 100 km depth.

often reveal interesting aspects of the physics that are at the root of unavoidably imperfect reconstructions.

Figure 17.7 summarises the results of four recovery tests to assess the waveform inversion example for the Eastern Mediterranean region. Each column corresponds to an input model where only one physical model parameter was perturbed relative to the spherically symmetric background. In the first column, for instance, only the *SV* wavespeed β_v is laterally heterogeneous. The results of the recovery test are shown in the four lower rows. They reveal, as expected, that β_v can be recovered only in the regions of dense coverage, shown in Figure 17.1 To some extent, variations in β_v leak into apparent variations of the *SH* wavespeed β_h and density ρ. While a similar recovery and leakage can be observed for β_h (second column),

α and ρ cannot be recovered. This is largely due to the dominance of long- to intermediate-period surface waves in our data set, and thus to an absence of strong sensitivity to density ρ and P wavespeed α.

In less computationally intensive travel-time ray tomography, recovery tests are widely used because a large variety of input structures, for instance, with variable characteristic length scales, can be considered (Rawlinson et al., 2014). Applications to waveform inversion are more limited due to the relatively high computational cost. As a consequence, alternative approaches to resolution analysis, e.g., through the analysis of second derivatives, have been developed on the basis of second-order adjoints. Instead of performing a larger number of recovery tests, the second-order adjoint method can provide 3-D volumes of resolution length and information on inter-parameter trade-off using just a small number of forward and adjoint simulations. For more details, the reader is referred to Fichtner & Trampert (2011b) and Fichtner & van Leeuwen (2015).

17.4 Practical and Computational Aspects

17.4.1 Automation

Much of the appeal of waveform inversion is associated with the potential for a fully automated process, in the sense of being a black box that transforms raw waveform data into high-resolution Earth models without any human intervention. The depression in waveform inversion research that separated the initial boom in the 1980's from actual 3-D applications in the mid 2000's is to some extent due to this automation concept being a misleading illusion.

Numerous technical details need to be decided prior to any waveform inversion. These include the misfit functional, the nonlinear optimisation scheme and regularisation, the inversion parameters, the type and setup of the numerical forward modelling technique, as well as the geometric Earth model parameterisation. These details critically depend on factors such as (1) the frequency content, spatial sampling, and noise level of the data, (2) the nature of the wavefield sources, (3) the strength and length scales of heterogeneities in the Earth, (4) the quality of our prior knowledge about all relevant model parameters, including both structural and source parameters, and (5) the hierarchy of more and less relevant parameters.

Although full automation that covers all conceivable scenarios may be possible in theory, it seems intuitively unlikely that such a general tool can be programmed or maintained. So far, high levels of automation have only been achieved for narrow classes of applications, for instance, at regional to global scales and at intermediate periods, where the nature of the wavefield is very well understood (e.g., Krischer et al., 2018; Modrak et al., 2018). Thus, in accord with the No-Free-Lunch theorem (Section 15.7), human interaction based on intuition generally enhances efficiency rather than introducing inefficiency.

17.4.2 Storage Requirements

As demonstrated in Section 16.1.2, the computation of sensitivity kernels involves the interaction of the forward and the adjoint wavefields. Since the adjoint wavefield must satisfy terminal conditions, it is typically computed backward in time. The direct consequence of this computational time reversal is that the forward and the adjoint wavefield are not naturally available at the same time t, unless the forward wavefield is stored during the forward wavefield simulation. The resulting storage requirements can quickly become prohibitive.

Storage requirements may be reduced by several orders of magnitude with the help of lossy compression methods, where losses in the accuracy of sensitivity kernels are balanced against the (iteration-dependent) accuracy requirements of gradient methods (Boehm et al., 2016). Several approaches may be employed together. In the most simple approach, the forward wavefield may not be stored at every numerical time step. Indeed, a rough analysis reveals that storing the forward wavefield only every 10 time steps hardly degrades the quality of the resulting sensitivity kernel (Fichtner et al., 2009). Substantial storage reductions can also be achieved by reducing the polynomial degree used to represent dynamic fields in spectral-element methods. For instance, storing the forward wavefield computed using degree-4 polynomials in the form of a degree-1 polynomial, lowers storage requirements by a factor of 16. Similarly, one may use fewer grid points in finite-difference methods. An approach that is more difficult to implement, but also efficient, is an amplitude-dependent reduction in the number of bits used to represent floating point numbers. Comparatively low-amplitude waves that contribute less to the overall sensitivity may be represented using less bits than a larger-amplitude wave that makes a more significant contribution.

An alternative to lossy compression of the forward wavefield is check-pointing (e.g., Griewank & Walther, 2000), where the forward wavefield is stored only at a very small number of time steps. Starting from these checkpoints, the forward wavefield is then numerically propagated backwards, alongside with the adjoint wavefield. The sensitivity kernel is then computed on the fly. In this manner, check-pointing trades the reduction of storage requirements for additional numerical simulations.

17.4.3 Extracting Additional Parameters

The development of waveform inversion based on fully numerical simulations and adjoint techniques sparked the hope that it might be possible to impose better constraints on 3-D variations of some physical parameters that are notoriously difficult to recover, including attenuation, density, and more complex forms of anisotropy that go beyond the commonly considered radial and azimuthal anisotropies. Though progress has undoubtedly been made (e.g., Zhu et al., 2013; Blom et al., 2017), the bandwidth that we can analyse with currently available computational resources is probably still too small to enable a big step forward.

In addition to enhanced computational power, a more profound understanding of waveform sensitivity to 3-D Earth structure seems essential. Principal component analysis (Sieminski et al., 2009) allows us to systematically construct linear combinations of physical model parameters to which seismic waveforms are most sensitive. Furthermore, the use of principal components establishes a hierarchy of model parameters that helps to steer and optimise the nonlinear optimisation scheme. For instance, the most important parameters may be included first, and additional parameters are only included when failure to reduce misfit to an acceptable level clearly indicates that this is necessary.

Principal component analysis may also be used constructively in order to design targeted misfit functionals. Starting from a large set of candidate misfits, analysis of the principal components can determine linear combinations of candidate misfits to produce a new misfit functional that is optimally sensitive to a pre-defined model parameter (e.g., density), while being optimally insensitive to all other parameters (Bernauer et al., 2014). Though such design concepts exist, their routine application in waveform inversion still lies in the future.

18

Waveform Inversion of Correlation Data

We have so far concentrated on the application of waveform inversion to the records of seismic events but, as demonstrated in Part II, the cross-correlations of seismic records depend on both the distribution of seismic sources and the structure in the vicinity of the path between the stations that are being correlated.

The differences between the segments of the correlograms corresponding to opposite senses of propagation between the stations provide information on source excitation, while the properties of the dominant arrivals are mainly sensitive to structure.

In this chapter we first discuss the general properties of the stacked station correlation waveforms and the way that these can be used to invert for sources and structure, without making any restrictions on source geometry. We then consider the specific example of seismic hum in the frequency band from 3–10 mHz, and show that it is possible to undertake a joint inversion for the distribution of seismic noise sources and long wavelength structure.

18.1 Modelling and Inverting the Correlation Wavefield

Most attempts to exploit ambient noise are based on the assumption that stacked noise correlations converge to an approximation of the Green's function between a pair of seismic stations. As we have seen in Section 6.1 even in ideal circumstances we do not directly recover the Green's function since the kernel involves the combination $\mathbf{G}(\omega, \mathrm{p})\mathbf{G}^*(\omega, \mathrm{p})$ rather than just $\mathbf{G}(\omega, \mathrm{p})$. This means that the amplitude patterns of higher modes and body waves relative to the fundamental mode are modified compared to direct propagation (e.g., Halliday & Curtis, 2008; Kimman & Trampert, 2010). A non-uniform source distribution also limits recovery of propagation characteristics since there may be inadequate cancellation of noise incident at high angle to the path between the stations and distortion of the resulting waveforms. Many authors have attempted to suppress source distribution effects by culling unsuitable azimuths of paths between receivers or by a specific choice of frequency band, but such efforts can lead to loss of up to 70% of an initial data set (Stehly et al., 2009). In contrast, we seek to work directly with the correlation wavefield and recover both the source distribution and structure.

18.1.1 Forward Modelling

We adopt a similar approach to that used in our discussion of generalised interferometry in Section 6.3 to establish a forward modelling procedure for correlations. For an ambient noise forcing field $\mathbf{N}(\mathbf{x})$, the resulting displacements in the frequency domain are given by the representation theorem (Section 3.2.1), in terms of an integral over the volume of the Earth as in (6.3.4):

$$\mathbf{u}(\mathbf{x}) = \int_V d^3\xi\, \mathbf{G}(\mathbf{m}; \mathbf{x}, \xi)\mathbf{N}(\xi). \tag{18.1.1}$$

Here, the Green's tensor depends on the structural model $\mathbf{m}$ and satisfies the equation,

$$\mathcal{L}[\mathbf{G}(\mathbf{m}; \mathbf{x}, \xi)] = \delta(\mathbf{x} - \xi). \tag{18.1.2}$$

For elastic waves the action of the operator $\mathcal{L}$ has the explicit form

$$\mathcal{L}_\mathbf{x}[\mathbf{u}(\mathbf{x})] = -\rho(\mathbf{x})\omega^2\mathbf{u}(\mathbf{x}) - \nabla_\mathbf{x}\cdot\boldsymbol{\sigma}, \tag{18.1.3}$$

where $\boldsymbol{\sigma}$ is the stress tensor.

From (18.1.1) the cross-correlation of components of the displacement field generated by noise at different locations $\mathbf{x}_1$, $\mathbf{x}_2$ is given by

$$\mathcal{C}_{ij}(\mathbf{x}_1, \mathbf{x}_2) = \int_V d^3\xi \int_V d^3\boldsymbol{\eta}\, \mathbf{G}(\mathbf{x}_1, \xi)\mathbf{N}(\xi)\mathbf{G}^*(\mathbf{x}_2, \boldsymbol{\eta})\mathbf{N}^*(\xi). \tag{18.1.4}$$

Since noise sources are transient, $\mathcal{C}_{ij}(\mathbf{x}_1, \mathbf{x}_2)$ varies with time interval. Taking a stack over many different time intervals, e.g. daily windows, we can approximate the expectation

$$\langle\mathcal{C}_{ij}(\mathbf{x}_1, \mathbf{x}_2)\rangle = \int_V d^3\xi \int_V d^3\boldsymbol{\eta}\, \mathbf{G}(\mathbf{x}_1, \xi)\mathbf{G}^*(\mathbf{x}_2, \boldsymbol{\eta})\langle\mathbf{N}(\xi)\mathbf{N}^*(\xi)\rangle. \tag{18.1.5}$$

Because the Earth is attenuative the response to any noise source will ultimately decay, so successive wavefield windows are almost independent.

If we are able to assume that the correlation between different noise sources decays relatively quickly with distance compared to the seismic wavelength, co-located sources will dominate. As in (6.3.11) we can then approximate the noise correlation in terms of a source power-spectral density

$$\langle N_n(\xi)N_m^*(\xi)\rangle = \mathcal{S}_{nm}(\xi, \omega)\delta(\xi - \boldsymbol{\eta}), \tag{18.1.6}$$

and then the integration in (18.1.5) reduces to

$$\langle\mathcal{C}_{ij}(\mathbf{x}_1, \mathbf{x}_2)\rangle = \int_V d^3\xi\, G_{in}(\mathbf{x}_1, \xi) \underbrace{\left[G_{jm}^*(\mathbf{x}_2, \xi)\mathcal{S}_{nm}(\xi, \omega)\right]}_{\text{distributed source}}, \tag{18.1.7}$$

where the term in square brackets can be thought of as an effective source for the correlation wavefield, cf. (6.3.12). In any event, the spatial correlation length for noise sources will be restricted and so the double integral could be evaluated in situations where there is sufficient information as to the nature of the sources.

We have here made the assumption of minimal processing in constructing the stacked correlation field. As discussed in Section 6.3, processing procedures can significantly modify the correlation response, and non-linear processing needs to be handled explicitly in modelling.

With the simplifying assumptions of uncorrelated noise sources and minimal processing, the computation of the forward correlation field then consists of the following steps:

(i) Using source–receiver reciprocity calculate the Green's tensor $\mathrm{G}_{jm}(\mathbf{x}_2, \xi, \omega)$, by placing a source at the receiver location $\mathbf{x}_2$ and evaluating the displacement fields in the regions where $\mathcal{S}_{nm}(\xi, \omega)$ is non-zero;
(ii) Combine the complex conjugate $\mathrm{G}^*_{jm}(\mathbf{x}_2, \xi, \omega)$ with the power-spectral density $\mathcal{S}_{nm}(\xi, \omega)$, which is equivalent to using the time-reversed Green's function; and
(iii) Model the correlation wavefield as a solution of the wave equation with $[\mathrm{G}^*_{jm}(\mathbf{x}_2, \xi, \omega)\mathcal{S}_{nm}(\xi, \omega)]$ as a distributed source, sampling the response at the positions of other receivers such as $\mathbf{x}_1$.

This approach to the simulation of the correlation wavefield depends on the spatial power-spectral distribution of the noise sources, it does not require knowledge of the phase behaviour of the noise forcing or long time series to extract statistical properties.

We can represent the modelling process for inter-station correlations as the solution of two separate propagation systems: from step (i),

$$\mathcal{L}[\mathrm{G}_{mj}(\mathbf{x}, \xi, \omega)] = \delta(\mathbf{x} - \xi), \tag{18.1.8}$$

and from step (ii),

$$\mathcal{C}(\mathbf{x}, \omega) = \mathcal{L}[\langle C_{ij}(\mathbf{x}, \mathbf{x}_r, \omega)\rangle] = \delta_{in}\mathrm{G}^*_{jm}(\mathbf{x}_r, \mathbf{x}, \omega)\mathcal{S}_{nm}(\mathbf{x}, \omega). \tag{18.1.9}$$

18.1.2 Inversion for Noise Sources and Earth Structure

From a network of stations we are able to make a large number of observations of the correlations between stations, and then the challenge is to extract information on both the source distribution and the structure of the Earth. The correlation wavefield (18.1.7) depends explicitly on the spatial distribution of the power-spectral density of noise excitation $\mathrm{S}_{nm}(\mathbf{x})$ and implicitly on the structural model $\mathbf{m}$ through the properties of the Green's tensor terms. Both factors have to be taken into account when inverting the correlation wavefield.

To measure the extent that any forward simulation matches observations we introduce a measure of data fit $\mathcal{X}(\mathbf{m}, \mathrm{S})$ that combines data fit information from the full suite of available station pairs. As in the treatment of event data we seek to minimise the discrepancy between observations and synthetics, using iterative minimisation algorithms to overcome the nonlinear dependence of $\mathcal{X}(\mathbf{m}, \mathrm{S})$ on $\mathbf{m}$ and S.

As we have seen in Chapter 15 such methods require directional derivatives with respect to the controlling parameters, and to miminise computation we employ the adjoint approach developed in Chapter 16. A full derivation of the adjoint results for both first- and second-order derivatives is provided by Sager et al. (2018), so here we summarise the results.

Source derivatives

As the source distribution is modified, the misfit function will change and we seek to head in a direction where $\mathcal{X}(\mathbf{m}, \mathrm{S})$ is reduced. Using the general properties established in Section 16.1, we can exploit adjoint equations to simplify the calculation of directional derivatives with respect to S.

When we consider a narrow frequency band we can expect to extract a common frequency dependence for the power-spectral density so that

$$\mathcal{S}_{nm}(\mathbf{x}, \omega) = \mathrm{s}(\omega)\mathrm{S}_{nm}(\mathbf{x}). \tag{18.1.10}$$

The directional derivative of the overall misfit function $\mathcal{X}(\mathbf{m}, \mathrm{S})$ with respect to the source distribution then takes the form

$$\nabla_{\mathrm{S}}\mathcal{X}\, \delta\mathrm{S}(\mathbf{x}, \omega) = -\int_V \mathrm{d}^3\xi \int_W \mathrm{d}\omega\, \mathbf{u}^\dagger(\xi, \omega)\left[\mathbf{G}(\xi, \mathbf{x}, \omega)\mathrm{s}(\omega)\delta\mathrm{S}(\xi, \omega)\right], \tag{18.1.11}$$

where W represents a suitable frequency band. This construction requires the adjoint field $\mathbf{u}^\dagger$ for the governing equation (18.1.2). We can rewrite (18.1.11) in the form

$$\nabla_{\mathrm{S}}\mathcal{X}\, \delta\mathrm{S}(\mathbf{x}, \omega) = \int_V \mathrm{d}^3\xi\, \mathrm{K}_{\mathrm{S}}(\xi, \omega)\, \delta\mathrm{S}(\xi, \omega), \tag{18.1.12}$$

where the noise-source kernel

$$\mathrm{K}_{\mathrm{S}}(\xi, \omega) = \int_W \mathrm{d}\omega\, \mathbf{u}^\dagger(\xi, \omega)\mathbf{G}(\xi, \mathbf{x}, \omega)\mathrm{s}(\omega). \tag{18.1.13}$$

The construction of K_{S} requires the extraction of the adjoint operator $\boldsymbol{\mathcal{L}}^\dagger$ as in Section 16.1 and its solution with excitation by adjoint source-time functions that depend on the explicit form of the misfit functionals employed. These adjoint sources have to be applied at the receiver locations. Fortunately, from step (i) of the forward modelling procedure the Green's tensor has already been stored and so can be reused to build K_{S}. In the time domain, the adjoint equation is solved backwards in time so that the terminal condition that the adjoint wavefield is zero at time t_c acts as an initial condition for numerical calculation.

Structural derivatives

Both governing equations needed to generate the correlation wavefield, (18.1.8) and (18.1.9), depend on the Earth model $\mathbf{m}$. Because we have a two-stage modelling procedure, we need to produce the adjoint fields for each system to be able to express structural derivatives of the overall misfit function $\mathcal{X}(\mathbf{m}, \mathrm{S})$ in a convenient form. The directional derivatives with respect to the model parameters can be cast into the form

$$\nabla_{\mathbf{m}}\mathcal{X}\,\delta\mathbf{m}(\mathbf{x},\omega) = \int_V d^3\xi\, \mathrm{K}_{\mathbf{m}}(\xi,\omega)\delta\mathbf{m}(\xi,\omega), \tag{18.1.14}$$

where the structural sensitivity kernel $\mathrm{K}_{\mathbf{m}}$, in the frequency domain, takes the form

$$\mathrm{K}_{\mathbf{m}} = \int_W d\omega\, \left[\mathbf{u}^{\dagger}(\xi,\omega)\nabla_{\mathbf{m}}\boldsymbol{\mathcal{L}}[\langle C_{ij}(\mathbf{x},\mathbf{x}_r,\omega)\rangle] + \mathbf{c}^{\dagger}(\mathbf{x},\omega)\nabla_{\mathbf{m}}\boldsymbol{\mathcal{L}}[\mathbf{G}(\mathbf{x},\mathbf{x}_r,\omega)\right]. \tag{18.1.15}$$

Here, $\mathbf{u}^{\dagger}$ is the adjoint field for the regular wave equation (18.1.8), and $\mathbf{c}^{\dagger}$ the adjoint wavefield for the correlation equation (18.1.9). The construction of $\mathbf{c}^{\dagger}$ follows a similar path to the forward modelling scenario, first $\mathbf{u}^{\dagger}$ has to be found and then combined with the power-spectral density to provide the excitation term for the adjoint equation to (18.1.9). The correlation adjoint $\mathbf{c}^{\dagger}$ can be expressed in the form

$$\mathbf{c}_i^{\dagger}(\mathbf{x},\omega) = s(\omega)\int_V d^3\xi\, G_{in}^{*}(\mathbf{x},\xi,\omega)\left\{u_m^{\dagger *}(\mathbf{x},\omega)S_{nm}(\xi)\right\}, \tag{18.1.16}$$

where $u_m^{\dagger *}(\mathbf{x},\omega)$ is the regular adjoint field produced from a receiver location by the local data misfit properties.

Practical misfit measures

For the correlation wavefield, the influence of structure is strongest in the zone between a pair of stations, whilst source contributions dominate outside this zone. It is therefore convenient to construct two different misfit measures, $\mathcal{X}_S$ directed at source retrieval and $\mathcal{X}_{\mathbf{m}}$ for structural inference. An overall misfit function $\mathcal{X}$ can then be constructed by combining these two measures, but the relative weighting can affect the convergence of the iterative minimisation scheme. In principle we can use the gradients of $\mathcal{X}_S$, $\mathcal{X}_{\mathbf{m}}$ as starting points for a sub-space inversion (Section 15.6), which avoids the weighting issue. An alternative approach to exploit the significant degree of separation between the source and structural dependence is to use sequential inversion with a switch between misfit criteria once the minimisation gradient diminishes.

The effect of an uneven source distribution on stacked cross-correlograms between the records at a pair of stations is to introduce differences in the amplitudes of the branches with positive and negative time lag. The amplitudes of correlation waveforms depend primarily on the distribution of sources, but can be affected by attenuation and focussing/defocussing due to 3-D structure. To minimise such structural effects, Ermert et al. (2016) have exploited the logarithmic energy ratio between the causal and anti-causal parts of a correlation waveform. Weighting functions centred on the expected time window for, e.g.,

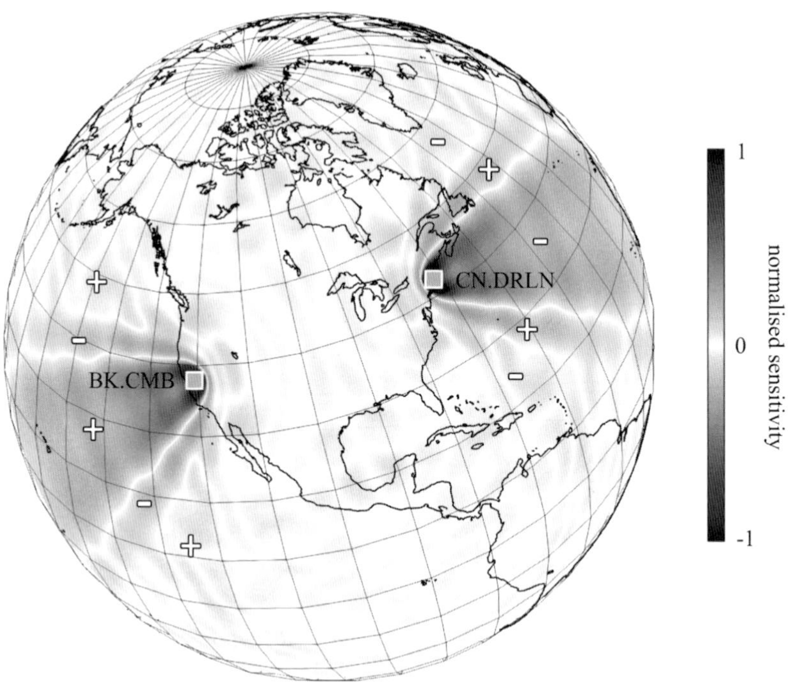

Figure 18.1 Example of the source kernel for the asymmetry measure for the station pair BK.CMB–CN.DRLN using the period band between 100 s and 300 s associated with the Earth's hum.

fundamental mode Rayleigh waves are created symmetrically for both positive and negative time lags. The asymmetry in the correlation waveform between stations i and j is assessed by

$$A_{ij} = \ln E_+ - \ln E_+, \tag{18.1.17}$$

where the energy terms E_+, E_- are defined by

$$E_+ = \int_0^{\infty} dt[w_+(t)C_{ij}(t)]^2, \qquad E_- = \int_{-\infty}^{0} dt[w_-(t)C_{ij}(t)]^2. \tag{18.1.18}$$

Here, $w_+(t)$ and $w_-(t)$ are time windows centred round the surface-wave arrivals on a synthetic correlation function.

We can then construct a data misfit measure based on the discrepancy between synthetic and observed asymmetries in correlation waveforms

$$\mathcal{X}_S = \frac{1}{2}\sum_{n=1}^{N}[A_n - A_n^{obs}]^2, \tag{18.1.19}$$

where the sum is taken over all pairs of receivers for which measurements have been made. Ermert et al. (2016) demonstrate that the noise-source kernel K_S

associated with the pairs of stations at $\mathbf{x}_1$ and $\mathbf{x}_2$ can be expressed as:

$$K_{nm}(\mathbf{y},\omega) = [A_{12} - A_{12}^{obs}]G_{in}(\mathbf{x}_1,\mathbf{y},\omega)G_{jm}^{*}(\mathbf{x}_2,\mathbf{y},\omega) \cdot \frac{1}{\pi}\left[\frac{w_{+}^{2}(\omega) * C_{ij}(\omega)}{E_{+}} - \frac{(w_{-}^{2}(\omega) * C_{ij}(\omega)}{E_{-}}\right], \tag{18.1.20}$$

where $*$ denotes a frequency-domain convolution. In (18.1.20), $w_{+}^{2}(\omega)$ represents the Fourier transform of $w_{+}^{2}(t)$. The propagation term $G_{in}(\mathbf{x}_1,\mathbf{y},\omega)G_{jm}^{*}(\mathbf{x}_2,\mathbf{y},\omega)$ in the kernel $K_{nm}(\mathbf{y},\omega)$ determines the regions where sensitivity to source location is non-zero. The term in square brackets in (18.1.20) controls the sign and amplitude of sensitivity within these regions as a function of frequency.

In Figure 18.1 we illustrate the source kernel (18.1.20) associated with the asymmetry measure for a period range appropriate to study the Earth's hum, using two stations on either side of the North American continent. The logarithmic terms associated with the causal and anti-causal branches have a different sign. Ermert et al. (2016) have shown that a reasonable approximation can be made by employing ray-theoretical kernels along the major arc of the great circle between the stations.

For the measure of data fit associated with structure it is possible, in principle, to use any of the criteria employed in inversion of event data. A simple measure that has only a modest dependence on the source distribution is the use of correlation time shifts as in Section 5.1.1, seeking the time at which the cross-correlation between synthetic and observed surface-wave arrivals is maximised. The time discrepancies $\mathcal{T}_{ij}$ for the surface-wave trains can then be measured on both the causal $\{+\}$ and anti-causal $\{-\}$ branches of the correlation waveforms for each station pair ij. The structural misfit measure is then

$$\mathcal{X}_{\mathbf{m}} = \sum_{n=1}^{N_{+}} \left(\mathcal{T}_{n}^{+}(\mathbf{m})\right)^{2} + \sum_{n=1}^{N_{-}} \left(\mathcal{T}_{n}^{-}(\mathbf{m})\right)^{2}, \tag{18.1.21}$$

where, as in (18.1.19), the sums are taken over all pairs of receivers for which measurements have been made.

In Figure 18.2 we illustrate the structure kernel for the cross-correlation time shift using the same pair of stations and frequency band as in Figure 18.1. This structure kernel depends on the regular adjoint field and that for the correlation field. Its behaviour resembles that for a finite-frequency travel-time kernel between the two stations with little obvious influence from the source distribution.

More complex choices than (18.1.21) for $\mathcal{X}_{\mathbf{m}}$ have a stronger dependence on the amplitude distribution as a function of time for the correlation functions. The result is a tendency to stronger coupling with the source distribution.

18.2 Joint Inversion of Seismic Hum

As an example of inversion of correlation data for both sources and structure we consider the study of Sager et al. (2020) that exploited the Earth's seismic hum in the frequency band 3.3–10 mHz (100–300 s period). This study concentrated

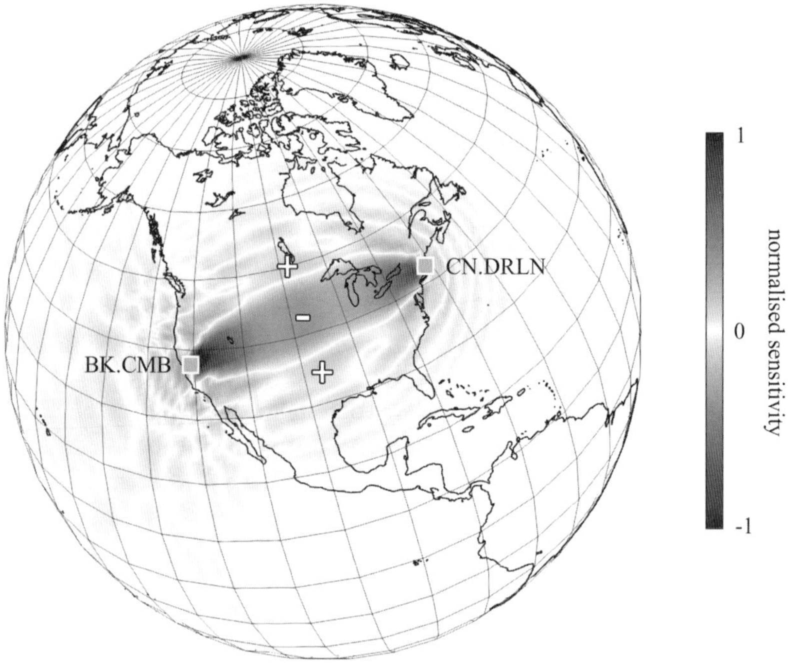

Figure 18.2 Example of the structure kernel using the correlation time shift for the station pair BK.CMB–CN.DRLN using the period band between 100 s and 300 s associated with the Earth's hum.

on just the vertical component of motion, with the simplifying assumption that vertical forcing at the ocean bottom is sufficient to explain the vertical motion, and that horizontal forcing can be neglected. With the relatively narrow frequency band it is also reasonable to assume that there is a partitioning of the spatial and frequency dependence of the source power-spectral density as in (18.1.10): $\mathcal{S}_{zz}(\mathbf{x}, \omega) = s(\omega) S_{zz}(\mathbf{x})$.

The starting point for the joint inversion is the spherically symmetric reference model PREM (Dziewoński & Anderson, 1981) including anisotropy and attenuation, with a spatially homogeneous distribution of sources. Structural inversion is made for the horizontal and vertical P and S wavespeeds, and density ρ. Rather than making a restriction to the oceans the power-spectral density is allowed to be non-zero anywhere at the Earth's surface.

18.2.1 Data and Measurements

The inversion exploited the dataset prepared by Ermert et al. (2016) using data from STS-1 broadband seismometers. The Earth's hum is relatively long period and has low amplitude and so it is advantageous to use very sensitive seismometers. Continuous data were retrieved for the 10-year period from 2004 to 2013 from the IRIS Data Management system in Seattle for 146 stations (see Figure 18.3). For

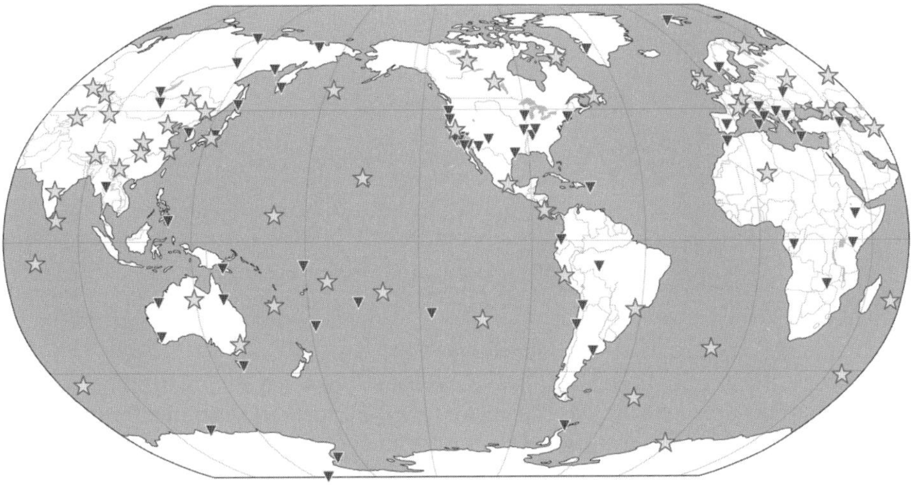

Figure 18.3 Distribution of the 146 STS-1 stations used for correlations across the globe (triangles). The 50 reference stations indicated by stars were used as virtual sources.

each pair of stations, cross-correlations were computed for ~9 hour windows and then stacked. A pre-filter was applied with a pass band between 20 s and 500 s (2–50 mHz). Each correlation function was normalised by the energy of both traces in each window to suppress contributions from large-amplitude (transient) signals such as earthquakes. No further processing was applied, so the forward modelling procedure discussed in Section 18.1.1 can be used directly.

The joint inversion exploits the period band between 100 s and 300 s (3.3–10 mHz), with a zero-phase filter applied to the stacked correlograms. For periods smaller than 100 s (frequency > 0.01 Hz), contamination by earthquakes is more likely. For periods longer than 300 s, effects from atmospheric variation are more important and can only be suppressed by more elaborate processing. With the restricted frequency range it is possible to employ a single spatial distribution for the sources. To concentrate on the northern hemisphere winter, stacks were made for the months of December, January, and February for the 10 year period to enhance the signal-to-noise ratio.

Figure 18.4 shows an example of measurements from the correlation functions using the logarithmic asymmetry and cross-correlation time shifts from the stations BK.CMB, CN.DRLN used for the illustrations of source and structure kernels in Figures 18.1, 18.2. The windows used for the surface wave arrivals are 1200 s wide, centred on the arrival of a Rayleigh wave with a group velocity of 3.7 km/s. The synthetic correlation function is symmetric for a homogeneous distribution of noise sources. However, the observed correlation functions are asymmetric indicating stronger noise sources behind station CN.DRLN in the Atlantic Ocean. The cross-correlation time shifts reveal that the synthetics are too slow, by 8.0 s on the anti-causal branch and 6.5 s on the causal branch.

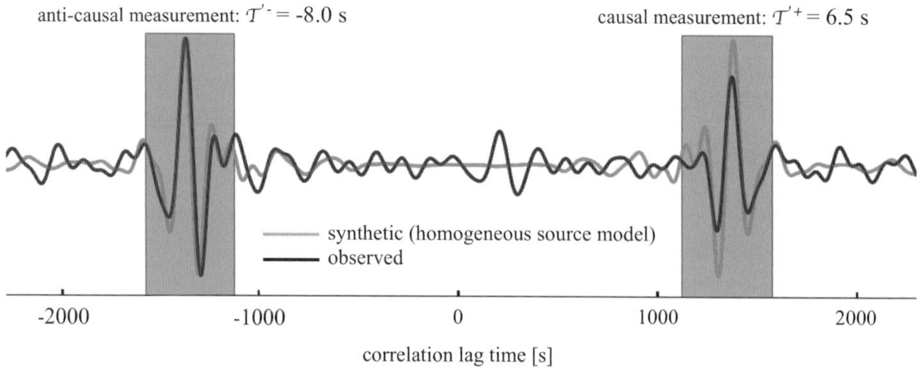

Figure 18.4 Observed (black) and synthetic (grey) correlation functions for the station pair BK.CMB–CN.DRLN in North America for the period band between 100 and 300 s. The windows employed to capture the surface wave arrivals are indicated by light grey boxes. A homogeneous noise source distribution was used to compute the synthetic correlation. Consequently, the synthetic asymmetry measure is 0. For the observed correlation, the asymmetry is −0.8, meaning that amplitudes on the anti-causal side are larger than on the causal side. Travel-time shifts differ in the anti-causal and causal measurement windows, but synthetics are too slow in both cases.

An estimate of the variability in asymmetry measurements for the Earth's hum is available from the work of Ermert et al. (2017), who exploited correlations between broadband stations at the Canadian Yellowknife array (YKA) and all other stations. The broadband station spacing at YKA is around 10 km, much less than the seismic wavelength (~1000 km), so correlation between individual stations and a common reference station should yield similar values. Differences can arise from site-specific effects and instrumental noise. In the 100 – 300 s period band the variance in the asymmetry measurement is rather high, so that a variance reduction of 30% is all that can be expected. Measurements of phase, e.g., by correlation time shift, are more stable and greater variance reduction can be expected.

In view of the significant incoherent noise in the measurements, strong quality controls were applied. Only those correlation functions with a minimum of 200 stacked windows were included. A signal-to-noise ratio of 2 on either the causal or the anti-causal branch was required to accept an asymmetry measurement. For the time shift measurements, shifts of more than half the dominant period are excluded, and a cross-correlation coefficient of at least 0.6 between the windowed synthetics and observations was required.

With the imposition of these quality constraints, there were 2732 asymmetry measurements and 1658 time shifts. In 415 cases, travel-time information can be extracted on both the causal and the anti-causal branch of the same correlation function. Such differences are expected because the distribution of ambient noise sources is not homogeneous.

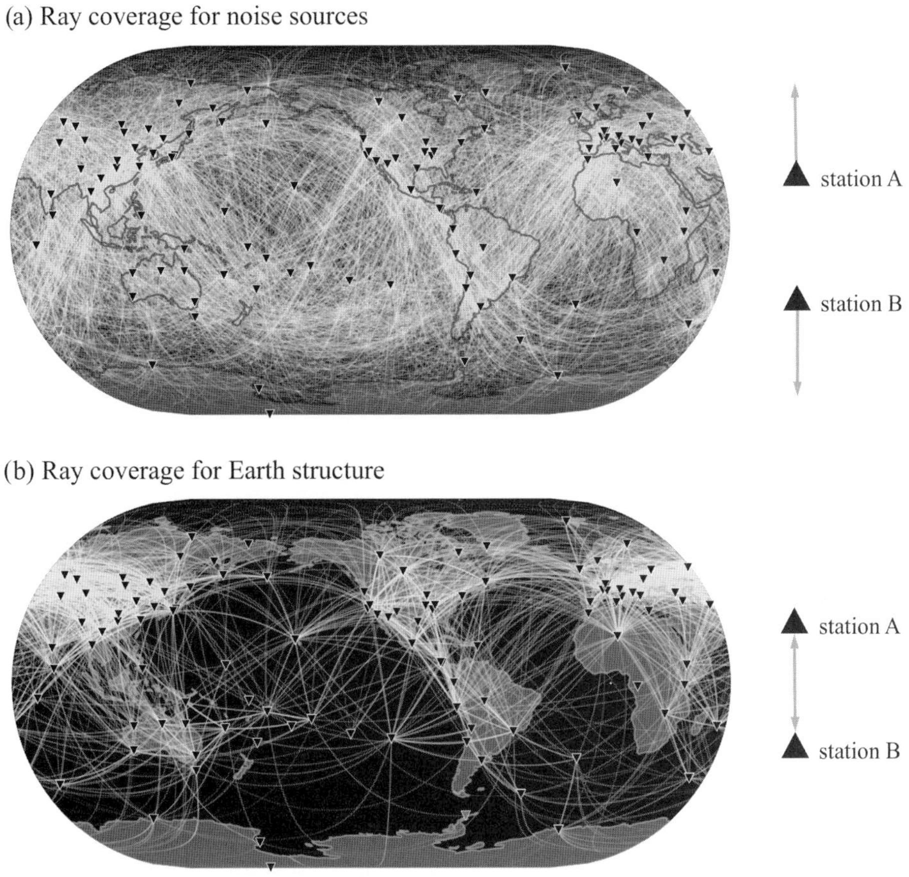

Figure 18.5 Ray coverage for (a) the source inversion and (b) the structure inversion. For source inversion the rays follow the major arc of the great circle, whereas for structural inversion the rays connect the stations along the minor arc. Ray lengths are limited to 10 000 km to mimic attenuation. Source coverage is provided by 5464 rays and 1243 paths are available for structural inversion, in 415 cases travel times could be extracted on both the causal and anti-causal branches.

18.2.2 Computational Aspects

The computational demands for modelling the correlation functions are high and so Sager et al. (2020) have made a number of simplifying assumptions whilst retaining suitable accuracy for matching to the observations. A generating wavefield of about 6 hours duration is needed, and a measurement window corresponding to minor-arc surface waves was used out to 7000 s.

A restricted subset of 50 receivers were used as reference stations (virtual sources), chosen to provide a relatively uniform coverage (Figure 18.4). The patterns of propagation paths, approximated by ray theory, are displayed in Figure (Figure 18.5). These results provide an indication of potential resolution. The concentration of stations in the northern hemisphere is reflected in the structural coverage, and produces banding in the more uniform source coverage.

Since the calculations are made for long-period waves and are dominated by fundamental mode surface waves, the calculation domain is restricted to the outer 1000 km of the globe. The near-surface structure is also simplified so that a single layer is used to 24.4 km depth, the transition to the mantle in PREM.

The source distribution is confined to the Earth's surface and is approximated with a Gaussian spectral basis function $s(\omega)$, and a single spatial distribution $S_{zz}(\mathbf{x})$. Since the receivers are also at the surface the Green's functions needed for step (i) of the forward modelling procedure in Section 18.1.1 are already available for the appropriate frequency range, which significantly simplifies the convolution in step (ii) and the assembly of the source kernel.

Sager et al. (2020) have used a sequential update of source and structural inversions, and so avoid having to make the difficult choice for the appropriate weighting between the different misfit functionals. The asymmetry measure used to infer noise sources is nearly insensitive to unmodelled Earth structure, and the cross-correlation time shift has very weak dependence on the source distribution. As a result, cross-talk between the two aspects of the inversion is limited.

The gradients of the overall misfit functions are used with a limited-memory BFGS (L-BFGS) method (Section 15.5). Sager et al. (2020) apply a diffusion-based smoothing operator to the gradients, with a smoothing length determined by the dominant wavelength. They also impose the physical constraint that power-spectral densities are equal to or larger than zero.

The first stage in the sequential inversion process is a number of source iterations. Once the reduction of the source misfits starts to flatten between iterations a switch is made to the structure inversion, and again, once the return from structural inversion diminishes, a return is made to source inversion. In all, six iterations are made for the distribution of the ambient noise sources and eight iterations for the 3-D structural model (Figure 18.6). For each group of source iterates it is possible to reuse the simulations of the Green's functions.

The error floor for the asymmetry measurements is reached after six iterations. Whereas the misfit associated with the travel times is reduced by 55%, and could probably be taken further. From Figure 18.6 we can see that each sequence of updates has only modest influence (~1%) on the other component of the joint model.

18.2.3 Source and Structural Models

Sager et al. (2020) demonstrate that the joint inversion produces a significant reduction in the variance of the time shifts with a histogram centred around zero. The histogram for the asymmetry measurements shows only a modest improvement from the initial state, but becomes more strongly peaked at zero. The synthetic correlation functions for the final model show a better fit in most cases, even through the details of the waveforms were not used in this joint inversion.

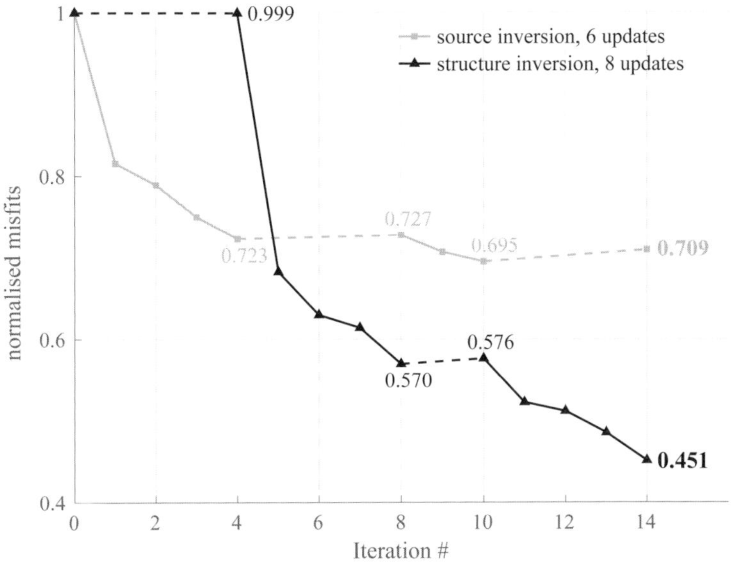

Figure 18.6 Misfit evolution for source inversion (gray) and for structural inversion (black). The dashed lines connect sequences of the same kind, e.g., the first four source updates and the next three after the structural inversion. Misfit values are given at the beginning and end of each sequence.

Source distribution

The model for source power-spectral density for the seismic hum generated by the joint inversion is illustrated in Figure 18.7. Because of the two-sided nature of the source kernels (e.g., Figure 18.1) interpretation is not straightforward, and so it is appropriate to focus on the features with large values of the power-spectral density that represent the major sources of hum for the northern hemisphere winter.

As would be expected, the strongest sources are imaged in the oceans in the northern hemisphere, along the coasts of the northern Pacific and in the northeast Atlantic. There are also strong contributions in the Coral and Tasman Sea near Australia and a broad band of sources in the southern parts of the Pacific, Atlantic, and Indian oceans, which can be linked to the patterns of storms in the austral summer. In general, sources on continents are weak, except for an anomaly in north Asia. The pattern of source distribution is comparable to that of earlier studies (e.g., Nishida & Fukao, 2007; Ermert et al., 2017), but shows stronger localised concentrations. Much, but by no means all, of the excitation of Earth's hum can be linked to excitation in areas of shallower water near continental edges.

Structural model

The combination of the 100–300 s period range and the modest global coverage from the Rayleigh wave segments of the correlation functions means that the

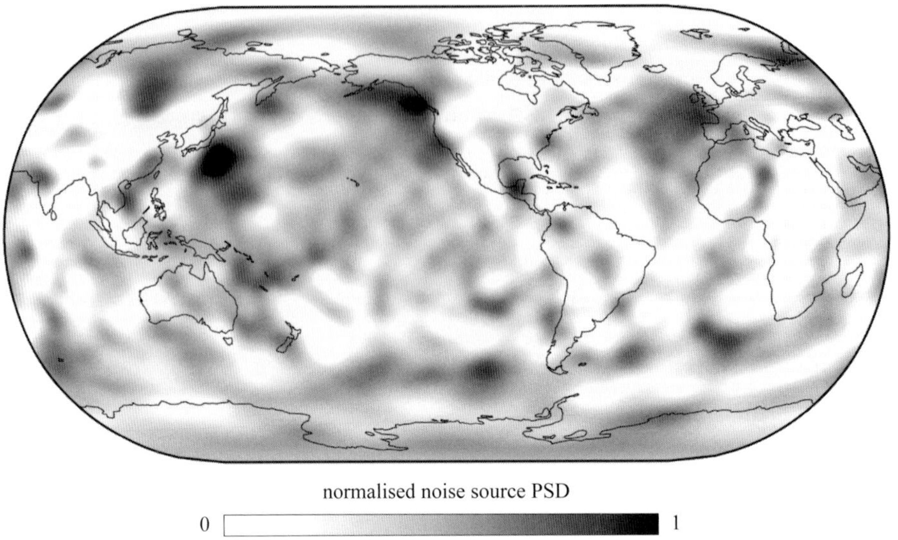

Figure 18.7 Power-spectral density (PSD) distribution of the Earth's hum in the northern hemisphere winter. Light coloured regions represent large values of the power-spectral density and show where the Earth's hum originates.

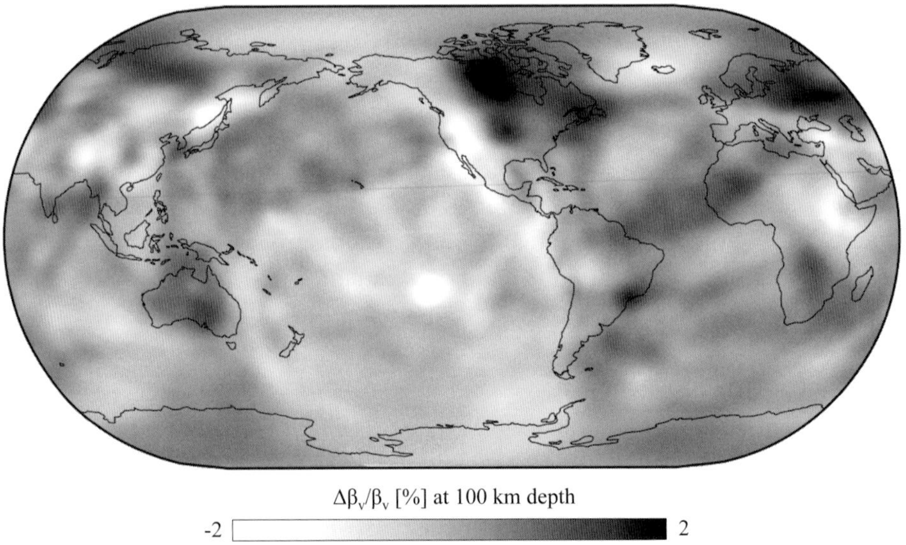

Figure 18.8 Final inversion result for *S* wavespeed β_v at 100 km depth, presented as percentage deviations from the mean.

most robust results are for the vertically polarised *S* wavespeed β_{SV}. Only 1600 time-shift measurements have been employed in the structural inversion, which is orders of magnitude less than usually employed for global-scale tomographic inversions (e.g., Ritsema et al., 1999; Lebedev & Schaeffer, 2013). Nevertheless, there are sufficient data to extract the major features of the wavespeed distribution, with a good correspondence in patterns with, e.g., the model S20RTS of Ritsema et al. (1999) derived from long-period observations.

In Figure 18.8 we plot the wavespeed model at 100 km depth as deviation from the mean. At this depth, there will be a strong imprint from the crust and upper mantle, since the inversion started from a laterally uniform 1-D model (PREM). The cratons typically show elevated wavespeeds, e.g., eastern North America, central and western Australia, the eastern part of Eurasia, eastern South America, as well as western and southern Africa. A slow Pacific belt, the African rift system, and the northern part of the Mid-Atlantic ridge are also clearly imaged. Strong undulations of the crust caused by mountain ranges are imaged as slow anomalies, as in the Himalayas, the Andes, and potentially also the Urals.

For the 100–300 s period band, Nishida et al. (2009) have estimated crustal corrections that correlate strongly with the imaged features. The joint inversion has therefore been able to successfully extract structural information from the correlation functions.

19

New Directions

Seismology can be considered a data-rich science. During the past few decades, large volumes of digital data have become easier to acquire, store, and distribute. As a consequence, major scientific advances in seismology have hardly been impeded by a severe lack of data, though the distribution of data is rather uneven across the globe with the southern hemisphere generally much less well sampled than the northern hemisphere. New data of course promotes new discoveries.

As of November 2019, the volume of freely available seismic waveform data stored by the *Incorporated Research Institutions for Seismology (IRIS)* was around 660 TB. While the current growth of this data volume is around 20% per year, the growth rate is likely to increase sharply, as emerging acquisition technologies such as large-N arrays (e.g., Schmandt & Clayton, 2013; Bowden et al., 2015) and distributed acoustic sensing (DAS) (Lindsey et al., 2017; Walter et al., 2020) become more widespread.

In the light of this wealth of data, one might expect that our knowledge of 3-D Earth structure would have become very detailed. Yet, a comparison of isotropic S-wavespeed models of the Earth's mantle reveals significant discrepancies at length scales of several hundreds to few thousands of kilometres (Schaeffer & Lebedev, 2013). Differences between upper-mantle S-wavespeed models of the western US, covered by USArray stations with an average spacing of around 70 km, have length scales locally exceeding 200 km (Becker, 2012). Global models of shear-wave attenuation up to spherical harmonic degree 8, constructed between the mid 1990s and 2008, were found to have a mutual correlation coefficient of around 0.25, thus revealing major dissimilarities at >5000 km length scale (Dalton et al., 2008).

The reasons for our apparent difficulties to constrain 3-D Earth structure are diverse. (i) While a large volume of data is available in principle, only small subsets of the data are currently being used in tomographic studies. (ii) Instead of systematically using earlier results, new tomographic inversions often start from spherically symmetric Earth models, thereby largely ignoring prior knowledge. (iii) Inversion results tend to be strongly affected by subjective regularisation, which is needed to stabilise the inversion procedure. Regularisation includes explicit damping and smoothing, but also more subtle factors such as the number of iterations in non-linear optimisation. (iv) Artefacts in the inferred Earth models may result from forward modelling errors and simplifications of wave propagation physics.

In many ways, the issues we have just mentioned are only symptoms of a more fundamental limitation of both human and computational power. The sheer volume of available seismic data clearly exceeds the data analysis capacity of a single researcher or a typical academic research group. Thus, further progress in seismic tomography will depend on the development of new modes of collaborative Earth model construction that takes advantage of the distributed human power of the seismic imaging community. An inversion scheme designed to allow the inclusion of new information whilst consistently incorporating prior knowledge is the subject of Section 19.1.

Computational power is often said to grow rapidly, suggesting that we will soon be able to solve entirely new classes of problems. Though it is tempting to repeat such over-simplified statements, it is worthwhile to take a closer look. Firstly, it is important to understand that a rapid increase in global computational resources does not imply that the resources available to individual researchers grow equally fast. In fact, they grow much more slowly because the number of problems that require serious computing also grows quickly. Secondly, we note that the computational requirements for seismic waveform inversion problems roughly scale with frequency to the power of 5. A power of 4 results from the necessary refinements in spatial and temporal discretisation of the forward problem as frequency increases. An additional factor of frequency is due to the fact that the volume sampled by sensitivity kernels is inversely proportional to frequency. Thus, at higher frequencies, more data are needed to fill the Earth with sensitivity. All these considerations imply that merely waiting for computers to become more powerful is likely to be insufficient. Algorithms that are better adapted to circumstances and more efficient are needed for the solution of both the forward and inverse problems. In Section 19.2 we consider an adaptive approach to forward and adjoint modelling, and in Section 19.3 ways of reducing the computational demands of inversion.

We close this chapter in Section 19.4 with an introduction to an emerging Monte Carlo method that is likely to help us to reduce the imprint of subjective choices on inversion results, while also providing more complete and unbiased uncertainty information.

19.1 Multi-Scale Nested Inversion

The Earth is heterogeneous at all scales. Imaging these heterogeneities requires the consistent incorporation of local, regional, and global seismic data across a broad frequency range. This section presents the development of a waveform inversion scheme that evolves on multiple scales through the introduction of regional refinements. The inversion scheme takes advantage of prior information on 3-D Earth structure and enables an evolutionary, collaborative mode of Earth model construction in order to harness distributed human and computational power (Fichtner et al., 2018a).

19.1.1 Evolutionary Multi-Scale Updating

We consider the evolution of a 3-D Earth model by the progressive addition of new information on different scales. This multi-scale update process starts from an initial maximum-likelihood model $\tilde{\mathbf{m}}_0$, associated with the initial data set . Each component of $\tilde{\mathbf{m}}_0$ may represent the continuous distribution of a physical parameter, for instance, a seismic velocity, density, or attenuation. As a function of position $\mathbf{x}$, the model $\tilde{\mathbf{m}}_0$ can be expressed in terms of N_0 basis functions $\mathsf{b}_0^i(\mathbf{x})$,

$$\tilde{\mathbf{m}}_0(\mathbf{x}) = \sum_{i=1}^{N_0} \tilde{\mathbf{m}}_0^i \, \mathsf{b}_0^i(\mathbf{x}) \,. \tag{19.1.1}$$

The basis functions may include (but are not limited to) polynomials, spherical harmonics, constant-property blocks, or some combination of these. The initial maximum-likelihood model vector $\tilde{\mathbf{m}}_0$ contains all the discrete coefficients of $\tilde{\mathbf{m}}_0$, and is defined as the model where the prior probability density function, $\mathsf{P}_0(\mathbf{m}_0)$ attains its global maximum. The construction of a refinement $\tilde{\mathbf{m}}_1$ of the initial model $\tilde{\mathbf{m}}_0$ starts with the addition of N_1 new basis functions $\mathsf{b}_1^i(\mathbf{x})$ to the initial model representation, so that

$$\mathbf{m}_1(\mathbf{x}) = \sum_{i=1}^{N_0} \mathbf{m}_0^i \, \mathsf{b}_0^i(\mathbf{x}) + \sum_{i=1}^{N_1} \mathbf{m}_1^i \, \mathsf{b}_1^i(\mathbf{x}) \,. \tag{19.1.2}$$

The newly added basis functions may, e.g., be intended to describe smaller-scale variations in a region where data have become available with denser coverage than the initial data set. The coefficients $\mathbf{m}_1^i$ are initially constrained by the prior probability density $\mathsf{P}_0(\mathbf{m}_1|\mathbf{m}_0)$. The conditioning on the initial model coefficients $\mathbf{m}_0$ arises from the dependence of the newly added variations $\mathbf{m}_1$ on the structure that is already represented. Through the incorporation of additional data, $\mathbf{d}_1^{obs}$, we obtain a conditional likelihood function $\mathsf{P}_0(\mathbf{d}_1^{obs}, \mathbf{m}_1|\mathbf{m}_0)$. Combining prior and likelihood via Bayes' theorem, yields the conditional posterior for the variations $\mathbf{m}_1$,

$$\mathsf{P}(\mathbf{m}_1|\mathbf{m}_0, \mathbf{d}_1^{obs}) = \mathsf{k}\, \mathsf{P}_0(\mathbf{d}_1^{obs}, \mathbf{m}_1|\mathbf{m}_0) \mathsf{P}_0(\mathbf{m}_1|\mathbf{m}_0) \,, \tag{19.1.3}$$

with normalisation constant k. The joint posterior for the initial model coefficients $\mathbf{m}_0$ and the new variations $\mathbf{m}_1$ can be obtained from the conjunction of the conditional posterior $\mathsf{P}(\mathbf{m}_1|\mathbf{m}_0, \mathbf{d}_1^{obs})$ with the initial model prior $\mathsf{P}_0(\mathbf{m}_0)$,

$$\mathsf{P}(\mathbf{m}_0, \mathbf{m}_1, \mathbf{d}_1^{obs}) = \mathsf{P}(\mathbf{m}_1|\mathbf{m}_0, \mathbf{d}_1^{obs}) \mathsf{P}_0(\mathbf{m}_0) \,. \tag{19.1.4}$$

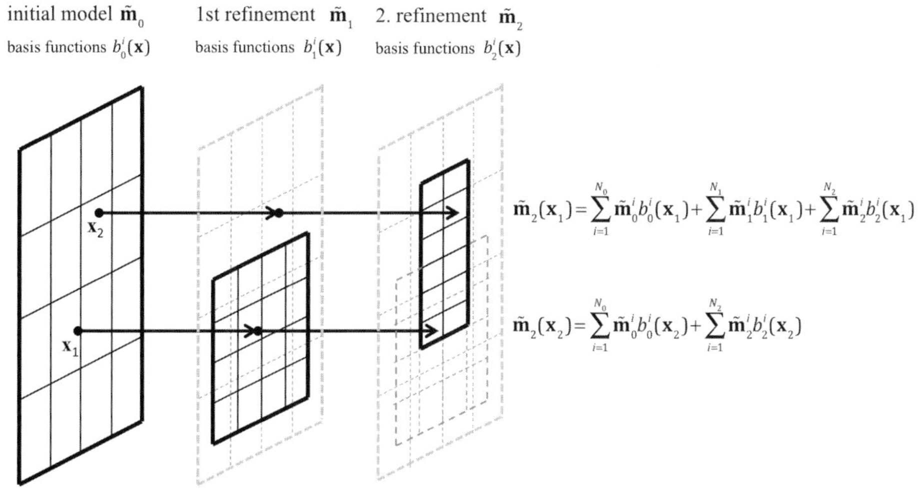
initial model $\tilde{\mathbf{m}}_0$
basis functions $b_0^i(\mathbf{x})$
1st refinement $\tilde{\mathbf{m}}_1$
basis functions $b_1^i(\mathbf{x})$
2. refinement $\tilde{\mathbf{m}}_2$
basis functions $b_2^i(\mathbf{x})$
$\mathbf{x}_2$
$\mathbf{x}_1$
$\tilde{\mathbf{m}}_2(\mathbf{x}_1)=\sum_{i=1}^{N_0}\tilde{\mathbf{m}}_0^i b_0^i(\mathbf{x}_1)+\sum_{i=1}^{N_1}\tilde{\mathbf{m}}_1^i b_1^i(\mathbf{x}_1)+\sum_{i=1}^{N_2}\tilde{\mathbf{m}}_2^i b_2^i(\mathbf{x}_1)$
$\tilde{\mathbf{m}}_2(\mathbf{x}_2)=\sum_{i=1}^{N_0}\tilde{\mathbf{m}}_0^i b_0^i(\mathbf{x}_2)+\sum_{i=1}^{N_2}\tilde{\mathbf{m}}_2^i b_2^i(\mathbf{x}_2)$

Figure 19.1 Illustration of the multi-scale updating scheme. In each refinement step, a new set of basis functions (solid black lines) is added to the previous parameterisation. As a consequence, material parameters at two different points in space, e.g., $\mathbf{x}_1$ and $\mathbf{x}_2$, may be represented by different sets of basis functions.

This process for incorporating new information can be iterated as follows:

(1) The posterior $\mathsf{P}(\mathbf{m}_0, \mathbf{m}_1, \mathbf{d}_1^{\mathrm{obs}})$ serves as the new prior distribution.

(2) New basis functions $\mathsf{b}_2^{\mathsf{i}}$ with coefficients $\mathbf{m}_2$ are added to the continuous model representation (19.1.2).

(3) The new coefficients $\mathbf{m}_2$ are constrained using a new data set $\mathbf{d}_2$, Bayes' theorem, and the conjunction of probability densities to obtain the next posterior, $\mathsf{P}(\mathbf{m}_0, \mathbf{m}_1, \mathbf{m}_2, \mathbf{d}_1^{\mathrm{obs}}, \mathbf{d}_2^{\mathrm{obs}})$.

Depending on the data set, the model parameters of interest, and the geometry of the refinement region, the basis functions for successive stages need not be of the same type. The concept of the development of the successive refinements is illustrated in Figure 19.1.

19.1.2 Pragmatic Simplifications

While the Bayesian updating scheme we have outlined in the previous section has the benefit of being fully consistent, it may require Monte Carlo sampling of a high-dimensional model space with substantial computational effort. Therefore, some pragmatic simplifications can be applied to produce a practical method, at the expense of perfect consistency.

The major simplification is the assumption that all prior probability densities are Gaussian so that posterior probability densities can be calculated explicitly. The maximum-likelihood model $\tilde{\mathbf{m}}_1$ of the conditional posterior $\mathsf{P}(\mathbf{m}_1|\mathbf{m}_0)$ is then approximated by minimising a misfit functional χ, using one of the iterative optimisation methods described in Chapter 15. The previous maximum-likelihood model $\tilde{\mathbf{m}}_0$ can serve as the obvious initial model for the iteration. Repeating the updating process for successive refinements, leads to a new maximum-likelihood model, represented in terms of multi-scale basis functions.

The Hessian matrix $\mathbf{H}$ of χ evaluated at new maximum-likelihood model $\tilde{\mathbf{m}}_1$ equals the inverse covariance, $\mathbf{C}_1^{-1}$ of the conditional posterior distribution $P(\mathbf{m}_1|\mathbf{m}_0)$. While $\mathbf{H}$ can usually not be computed or stored explicitly, useful low-rank approximations may be obtained, for instance, with the help of the L-BFGS method, outlined in Section 15.5. Alternatively, products of the Hessian with arbitrary vectors can be evaluated efficiently with the help of second-order adjoints (Santosa & Symes, 1988; Fichtner & Trampert, 2011a).

An obvious drawback of the simplified Gaussian scheme, is that the coefficients of previous basis functions, $\mathbf{m}_0, \ldots, \mathbf{m}_n$, remain unchanged when a new refinement with coefficients $\mathbf{m}_{n+1}$ is added. Furthermore, the approximation of the maximum-likelihood refinement $\tilde{\mathbf{m}}_{n+1}$ only uses the newly added data $\mathbf{d}_{n+1}$. The resulting approximation error relative to the true joint maximum-likelihood model may have the consequence that previously assimilated data sets, $\mathbf{d}_{i\leq n}$, can no longer be adequately explained.

Avoiding the potential inconsistencies related to our pragmatic simplifications would in principle require re-iterations using data from all previous refinements. Alternatively, we may use the Gaussian approximation in order to replace repeated iterations by the solution of a least-squares problem. For a conceptual illustration, assume that we combine two sets of model parameters, $\mathbf{m}_1'$ and $\mathbf{m}_1''$ into a joint parameter vector $\mathbf{m}_1 = (\mathbf{m}_1', \mathbf{m}_1'')$. These two sets of parameters may, for instance, result from successive updates, or from simultaneous updates that partially overlap. The corresponding conditional prior probability densities are $P_0'(\mathbf{m}_1|\mathbf{m}_0)$ and $P_0''(\mathbf{m}_1|\mathbf{m}_0)$. Combining these with the likelihood functions $\Lambda_0'(\mathbf{m}_1|\mathbf{m}_0)$ and $\Lambda_0''(\mathbf{m}_1|\mathbf{m}_0)$ using Bayes' theorem, yields

$$P(\mathbf{m}_1|\mathbf{m}_0) = k\, P_0'(\mathbf{m}_1|\mathbf{m}_0)\Lambda_0'(\mathbf{m}_1|\mathbf{m}_0)\, P_0''(\mathbf{m}_1|\mathbf{m}_0)\Lambda_0''(\mathbf{m}_1|\mathbf{m}_0)\,. \tag{19.1.5}$$

Within the Gaussian approximation, the maximum-likelihood models, $\tilde{\mathbf{m}}_1'$ and $\tilde{\mathbf{m}}_1''$, and posterior covariances, $\mathbf{C}_1'$ and $\mathbf{C}_1''$, of the individual posterior distributions, $P_0'(\mathbf{m}_1|\mathbf{m}_0)\Lambda_0'(\mathbf{m}_1|\mathbf{m}_0)$ and $P_0''(\mathbf{m}_1|\mathbf{m}_0)\Lambda_0''(\mathbf{m}_1|\mathbf{m}_0)$, can be estimated by independent, iterative misfit minimisations. This implies that the joint maximum-likelihood model, $\tilde{\mathbf{m}}_1$, of $P(\mathbf{m}_1|\mathbf{m}_0)$ is the minimum of

$$\begin{aligned} -\ln P(\mathbf{m}_1|\mathbf{m}_0) &= \tfrac{1}{2}(\mathbf{m}_1 - \tilde{\mathbf{m}}_1')^T \mathbf{C}_1'^{-1}(\mathbf{m}_1 - \tilde{\mathbf{m}}_1') \\ &\quad + \tfrac{1}{2}(\mathbf{m}_1 - \tilde{\mathbf{m}}_1'')^T \mathbf{C}_1''^{-1}(\mathbf{m}_1 - \tilde{\mathbf{m}}_1'') - \ln k\,. \end{aligned} \tag{19.1.6}$$

Since the posterior covariances $\mathbf{C}_1'$ and $\mathbf{C}_1''$ are equal to the Hessians $\mathbf{H}'$ and $\mathbf{H}''$ from the independent misfit minimisations, $\tilde{\mathbf{m}}_1$ can be approximated iteratively from equation (19.1.6) using Hessian–vector products that can be efficiently computed (Santosa & Symes, 1988; Fichtner & Trampert, 2011a). Rearranging (19.1.6) and substituting the minimum $\tilde{\mathbf{m}}_1$, yields the joint (inverse) posterior

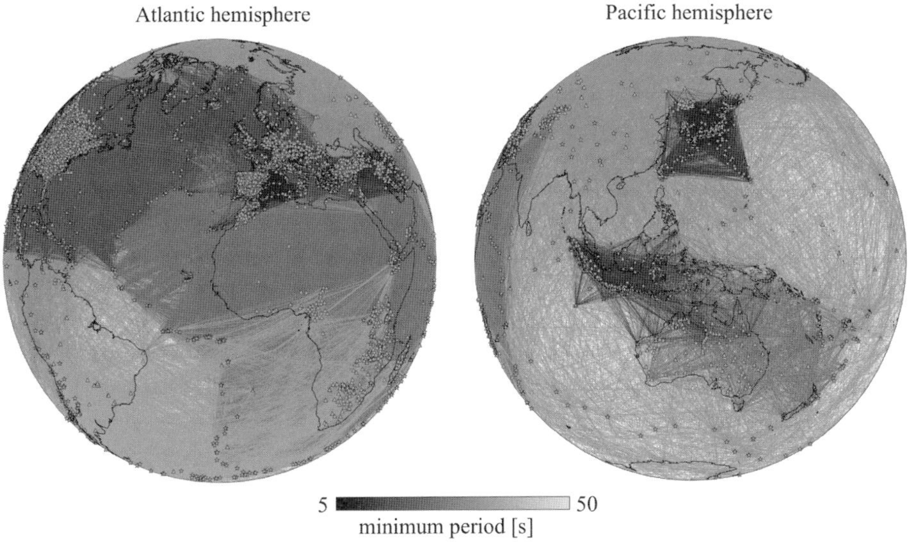

Figure 19.2 Summary of the waveform data included in construction of the 2020 version of the Collaborative Seismic Earth Model (CSEM). Stars indicate the position of (virtual) sources, and triangles mark receiver locations. Great-circle segments connect source–receiver pairs for which measurements of waveform misfit could be made. The depth of tone of the segments represents the minimum period.

covariance,

$$\mathbf{C}_1^{-1} = \mathbf{C}_1'^{-1} + \mathbf{C}_1''^{-1} = \mathbf{H}' + \mathbf{H}'' . \tag{19.1.7}$$

It follows that the action of $\mathbf{C}_1^{-1}$ on a model, needed for resolution analysis and further optimisation, is given by the sum of actions of the individual Hessians on the model. Furthermore, using (19.1.6) and (19.1.7), we can remove potential inconsistency by finding the solution that optimally agrees with the independent updates.

19.1.3 The Collaborative Seismic Earth Model

The multi-scale updating scheme we have outlined, together with its current pragmatic simplifications, enables the construction of a global 3-D seismic Earth model that evolves through successive regional refinements. In contrast to traditional styles of tomography, this Collaborative Seismic Earth Model (CSEM) permits external contributions from multiple researchers or research groups.

The initial model of the CSEM was constructed from the long-wavelength *S*-wavespeed model S20RTS (Ritsema et al., 1999), with *P*-wavespeed variations derived from a depth-dependent *S*-to-*P* scaling relation obtained from body-wave travel-time analyses by Ritsema & van Heijst (2002). An initial crustal model was constructed through vertical smoothing of the global crustal model constructed by Meier et al. (2007).

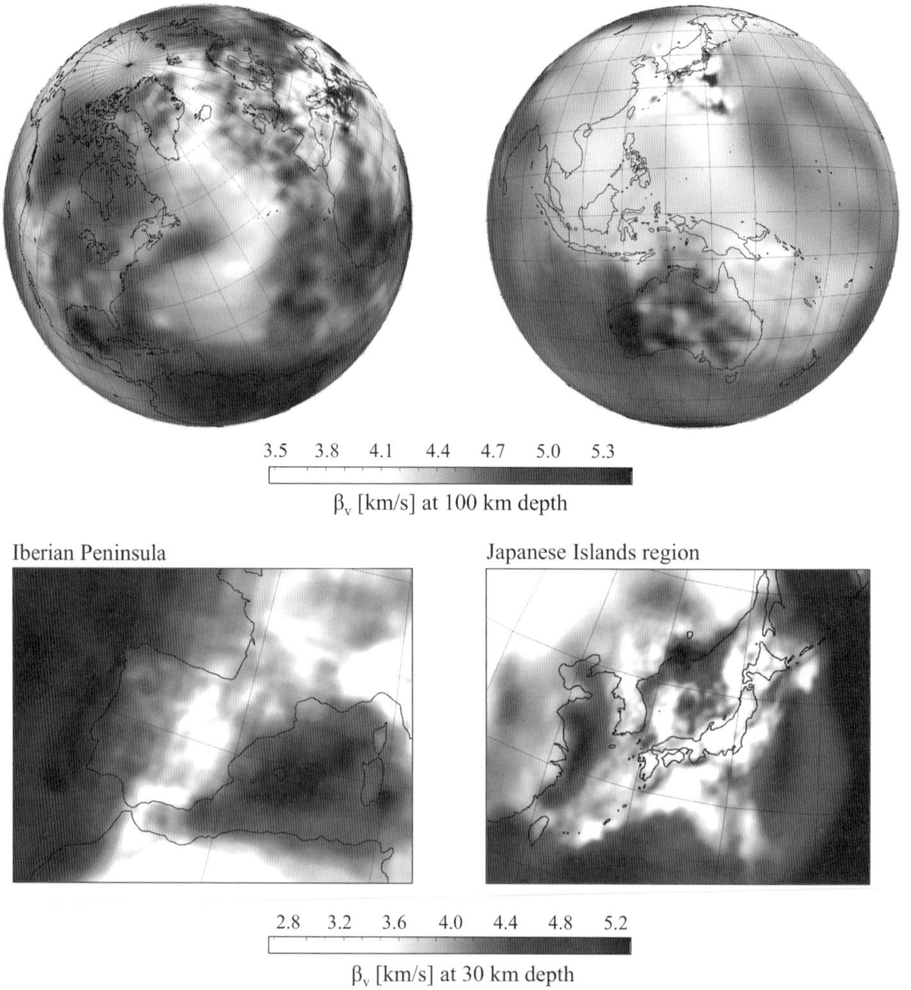

Figure 19.3 The *SV* wavespeed β_v for the 2020 version of the CSEM. Upper panel: structure in the upper mantle at 100 km depth on a global scale. Lower panel: structure at 30 km depth in regions with dense data coverage.

Currently, the CSEM includes a global-scale update at 60 s minimum period, plus 15 updates at variable minimum periods T. These are, in the order in which they have been incorporated, Australasia (T:40 s), Europe (T:50 s), the North Atlantic (T:25 s), Turkey (T:10 s), the South Atlantic (T:55 s), the Western Mediterranean (T:15 s), East Asia (T:20 s), the Iberian Peninsula (T:8 s), Japan (T:15 s), North America (T:30 s), South-East Asia (T:20 s), Western Turkey (T:8 s), the Eastern Mediterranean (T:25 s), the Iranian Plateau (T:20 s), and the African plate (T:40 s).

In total, the 2020 CSEM incorporates seismic waveform data from around 1,200 events recorded by more than 11 000 stations. A summary of the data coverage is presented in Figure 19.2. As a consequence of the construction mode, the resolution of Earth structure in the CSEM is very heterogeneous. While the resolution length in the still poorly covered Pacific Ocean is around 1000 km, details of crustal structure with 10 km length scale are resolved in the Iberian Peninsula, Japan, and the Sea of Marmara region. Some aspects of crustal and upper-mantle structure in the 2020 CSEM are shown in Figure 19.3.

Though all updates of the CSEM have so far been based on waveform inversion of event data, as described in Chapter 17, the generality of the Bayesian updating scheme permits the incorporation of other data types, including derived data, such as travel times of high-frequency body waves, receiver functions, or surface-wave dispersion curves.

While concepts for multi-scale tomography have been known for at least two decades (e.g., Bijwaard et al., 1998), the CSEM constitutes the first collaborative and evolutionary framework for multi-scale seismic Earth model construction. Though simplifications of the pure Bayesian scheme introduced in Section 19.1.2 are still needed to ensure computational feasibility, the CSEM is operational, and further refinements are in progress.

19.2 Wavefield-Adapted Numerical Meshes

Numerical wave propagation constitutes by far the most computationally expensive component of waveform inversion. As a result, numerous strategies have been developed in order to reduce computational cost. These include local time-stepping algorithms (e.g., Rietmann et al., 2017), the use of GPU accelerators (e.g., Gokhberg & Fichtner, 2016), the construction of smooth, long-wavelength equivalent models that permit the use of larger finite elements (e.g., Cupillard & Capdeville, 2018), and simplifications of the wave propagation physics, such as the acoustic approximation widely used in exploration seismology.

For spherically symmetric Earth models, the 3-D seismic wave equation can be simplified to an effective 2-D partial differential equation (Nissen-Meyer et al., 2007). This reduction in dimensionality enables the computation of global wavefields at frequencies as high as 1 Hz, which is currently out of reach for full 3-D simulations. Recently, Leng et al. (2019) generalised this concept to include 3-D variations in elastic parameters and density by adopting a pseudospectral numerical approach in the azimuthal direction. The resulting performance gains, relative to full 3-D simulations, rest on the observation that the complexity of a wavefield, propagating through a smooth Earth model, is high in the vertical and radial directions, but much lower in the azimuthal direction. This difference in complexity is illustrated in Figure 19.4. As discussed later, the precise meaning of 'smooth' in the context of wavefield complexity and numerical accuracy remains to be specified.

In the following paragraphs, we describe a waveform modelling and inversion approach introduced by van Driel et al. (2020) and Thrastarson et al. (2020). This approach builds on the relative smoothness of azimuthal variations, but replaces

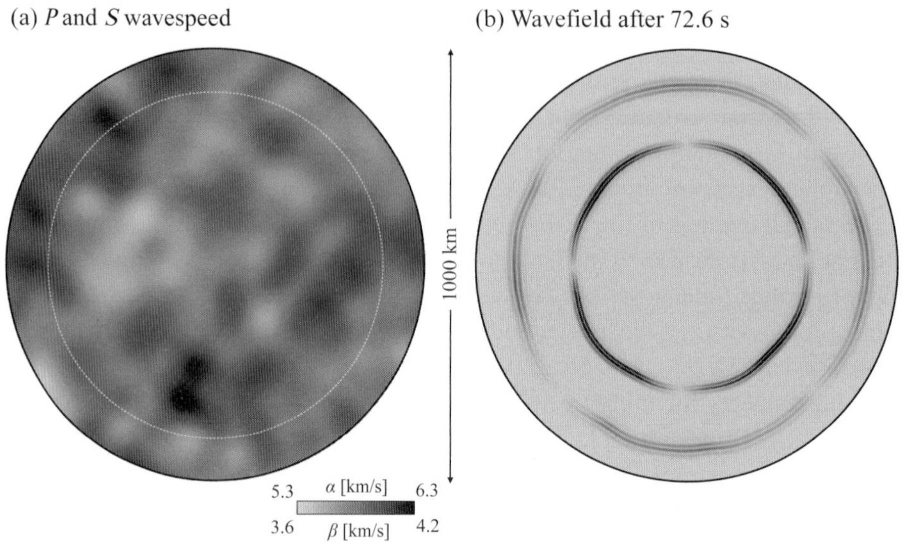

Figure 19.4 Wavefields in smoothly varying media. (a) *P*- and *S*-wavespeed variations in a 2-D circular domain. (b) Wavefield snapshot after 72.6 s. While being distorted relative to a perfectly circular wavefront by the wavespeed variations, the waveform complexity is still much lower in the azimuthal than in the radial direction.

the pseudospectral solution of Leng et al. (2019) by a spectral-element simulation.

19.2.1 Forward Modelling

The 2-D numerical wavefield shown in Figure 19.4 was computed with a spectral-element method (Afanasiev et al., 2019) using the finite-element mesh displayed in Figure 19.5(a). With this conventional meshing approach, elements are distributed in the circular domain such that individual elements are as large as possible, while still sampling the wavefield with around 1.5–2 elements per minimum wavelength. This leads to a total number of 253 700 elements. The relative azimuthal smoothness of the wavefield emanating from the centre of the circular domain, illustrated in Figure 19.4, suggests that elements may be elongated azimuthally. The resulting mesh is shown in Figure 19.5(b). Azimuthal stretching reduces the total number of elements used for this case to 18 500.

As we might expect intuitively, the seismic wavefields computed for the two different meshes are identical for all practical purposes, as illustrated in Figure 19.6. Although there are slight variations between the two styles of simulation, the differences are significantly smaller than the level of match between observed and synthetic waveforms that we would hope to achieve in a waveform inversion. The reduction from 253 700 to 18 500 elements translates, in this specific case, to a reduction in computational requirements by a factor of around 14. Thus, suitably tailored meshing can offer substantial advantages for modelling.

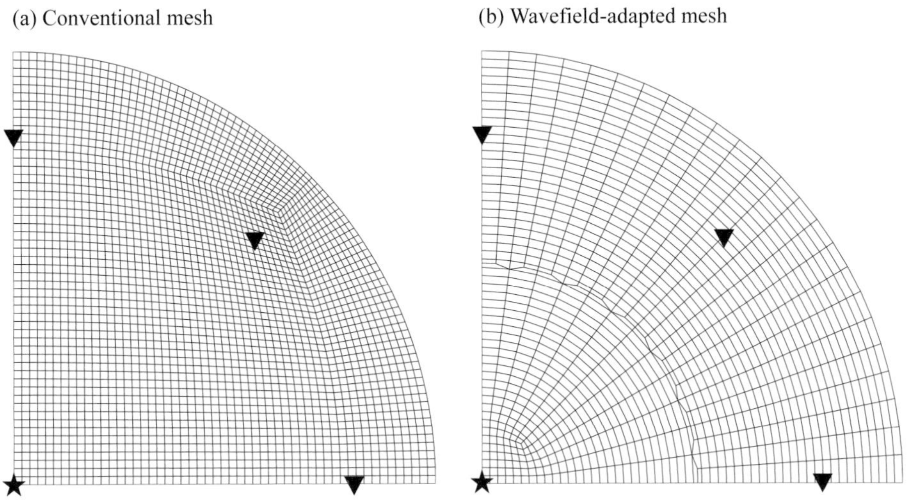

Figure 19.5 One quadrant of a conventional finite-element mesh used for spectral-element simulation (a) and a wavefield-adapted mesh for a source at the location of the black star (b). Both meshes are designed to ensure the sampling of the shortest-wavelength S wave with at least one element at a maximum frequency of 1 Hz. For clarity, only every 25th element is shown for the conventional mesh, meaning that each of the elements shown actually contains an additional 25 elements of similar shape. The total number of elements is 253 700. In the case of the wavefield-adapted mesh, all elements are shown in the azimuthal direction, but only every 5th in the radial direction. In total, there are 18 500 elements in the wavefield-adapted mesh. A comparison of synthetic seismograms for the two meshes, computed for the locations marked by black triangles for all the quadrants, is shown in Figure 19.6.

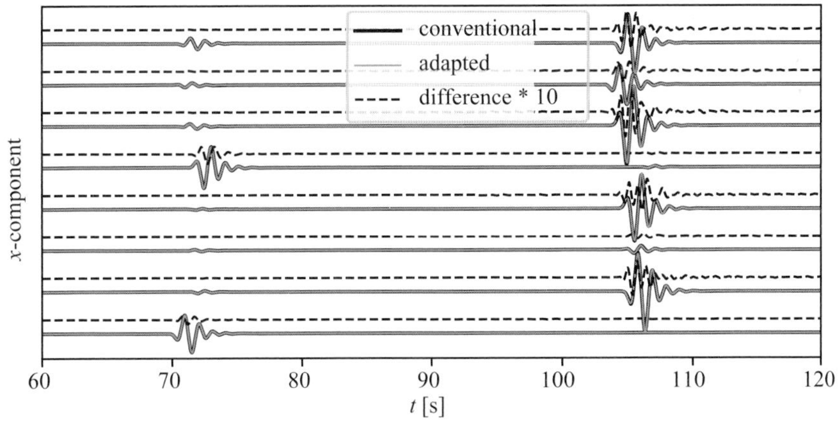

Figure 19.6 Comparison of x-component synthetic seismograms for the two meshes and the receiver locations shown in Figure 19.5, extended to the full mesh. The difference trace between the regular-mesh solution (black) and the adapted-mesh solution (grey) is amplified by a factor of 10 and shown as dashed curve.

19.2.2 Adjoint Modelling

The configuration of a mesh adapted to the nature of the wavefield, as in Section 19.2.1 is source-specific, meaning that a separate mesh must be generated for each source. While this is feasible for regular, physical sources, it appears to present difficulties for the adjoint sources introduced in Chapter 16. After the forward wavefield has been computed and stored on a wavefield-adapted mesh, adjoint sources must be injected at the receiver locations in order to compute the sensitivity kernels needed in iterative, gradient-based optimisation. Therefore, a mesh adapted to the geometry of the forward wavefield would seem to be unsuitable for the propagation of the adjoint wavefield. On the basis of this argument, we might expect that sensitivity kernels computed with the wavefield-adapted meshes would be of poor, and potentially insufficient, quality.

Fortunately, adjoint fields can be suitably modelled and the quality of sensitivity kernels is not degraded. We can demonstrate these results with the aid of the discrete adjoint method (e.g., Pratt et al., 1998), which is a special case of the continuous adjoint method covered in Chapter 16. We consider the wave equation in discretised form that can be expressed, in general, as a vector-matrix equation

$$\mathbf{L}(\mathbf{m})\mathbf{u}(\mathbf{m}) = \mathbf{f}, \tag{19.2.1}$$

where $\mathbf{L}$ is the impedance matrix, $\mathbf{u}$ is a discrete version of the displacement wavefield, and $\mathbf{f}$ is the discretised source. Our interest is in the computation of all partial derivatives of the misfit functional $\chi[\mathbf{u}(\mathbf{m})]$ with respect to the component of the model vector $\mathbf{m}$, that is,

$$\frac{\partial \chi}{\partial m_i} = \nabla_u \chi^T \frac{\partial \mathbf{u}}{\partial m_i}. \tag{19.2.2}$$

As in the continuous case, the difficulty lies in the computation of the partial wavefield derivatives $\partial \mathbf{u}/\partial m_i$, which we will try to eliminate from (19.2.2). We first differentiate the discrete equations of motion (19.2.1) to produce

$$\frac{\partial \mathbf{L}}{\partial m_i}\mathbf{u} + \mathbf{L}\frac{\partial \mathbf{u}}{\partial m_i} = \mathbf{0}. \tag{19.2.3}$$

Now with multiplication of (19.2.3) from the left by an arbitrary vector $\mathbf{v}$, and addition of the result to (19.2.2), we obtain

$$\frac{\partial \chi}{\partial m_i} = \nabla_u \chi^T \frac{\partial \mathbf{u}}{\partial m_i} + \mathbf{v}^T \frac{\partial \mathbf{L}}{\partial m_i}\mathbf{u} + \mathbf{v}^T \mathbf{L}\frac{\partial \mathbf{u}}{\partial m_i}. \tag{19.2.4}$$

Rearranging (19.2.4) allows us to isolate $\partial \mathbf{u}/\partial m_i$,

$$\frac{\partial \chi}{\partial m_i} = \left(\nabla_u \chi^T + \mathbf{v}^T \mathbf{L}\right)\frac{\partial \mathbf{u}}{\partial m_i} + \mathbf{v}^T \frac{\partial \mathbf{L}}{\partial m_i}\mathbf{u}. \tag{19.2.5}$$

Forcing the so-far unspecified vector $\mathbf{v}$ to solve the discrete adjoint equation,

$$\mathbf{L}^T \mathbf{v} = -\nabla_u \chi, \tag{19.2.6}$$

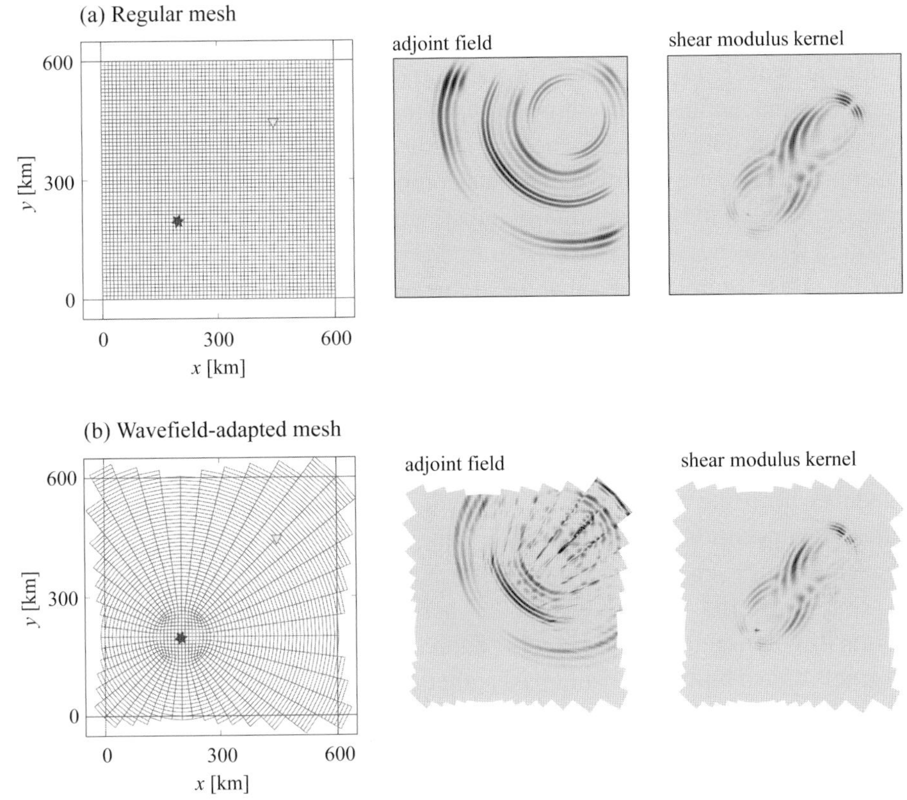

Figure 19.7 Adjoint fields and sensitivity kernels for a regular Cartesian mesh (a) and a wavefield-adapted mesh (b). Source and receiver are marked by a black star and a grey triangle, respectively. The measurement is a cross-correlation time shift, and sensitivity is with respect to the shear modulus. Though the adjoint field on the wavefield-adapted mesh visually appears to be numerically inaccurate, it is in fact the correct adjoint field for this specific discretisation. As a consequence, sensitivity kernels for the two meshes are virtually identical.

indeed eliminates the partial wavefield derivatives, and simplifies the misfit gradient to

$$\frac{\partial \chi}{\partial m_i} = \mathbf{v}^{\mathsf{T}} \frac{\partial \mathbf{L}}{\partial m_i} \mathbf{u} \, . \tag{19.2.7}$$

Equation (19.2.6) constitutes the discrete adjoint equation, with the discrete adjoint source $-\nabla_u \chi$ as right-hand side, and $\mathbf{L}^{\mathsf{T}}$ as discrete adjoint operator. It is the discrete counter-part of the continuous adjoint equation (16.1.11). Finally, inserting the discrete adjoint field into (19.2.7) yields the desired misfit gradient.

The discrete adjoint scheme allows us to compute exact misfit gradients using a discrete forward modelling operator $\mathbf{L}$, which is completely determined by the discretisation of the forward problem. The correct adjoint operator is simply $\mathbf{L}^{\mathsf{T}}$, meaning that a separate mesh to resolve the adjoint wavefield is not necessary. Figure 19.7 illustrates the consequences of this important result. Although the adjoint wavefield computed on the wavefield-adapted mesh appears to be numerically inaccurate when compared with the adjoint field on the regular grid, it is in fact correct with respect to the specific discretisation of the forward problem. As a consequence, the sensitivity kernels computed for a regular mesh and for the wavefield-adapted mesh are virtually indistinguishable.

Though adjoint wavefields are often interpreted as if they are actually physical quantities, this example shows that they represent purely mathematical objects that need not be interpretable within the framework of standard intuitive physics.

19.2.3 Waveform Inversion

In Section 19.2.1 we have introduced forward modelling adapted to the nature of the wavefield and the associated adjoint modelling in Section 19.2.2. These procedures form the foundation of an iterative waveform inversion scheme. In contrast to conventional waveform inversion, separate finite-element meshes must be generated for each source. The corresponding forward and adjoint simulations yield source-specific sensitivity kernels. These are continuous functions expressed in terms of polynomials defined on the source-specific finite elements. Projecting the kernels onto a finite number of basis functions $\mathrm{b}_i(\mathbf{x})$, which define the inversion grid or master model, produces a consistent set of misfit gradients, i.e., partial derivatives of χ with respect to the coefficients m_i of the basis functions $\mathrm{b}_i(\mathbf{x})$. The misfit gradients with respect to the coefficients of the master model allow us to drive an iterative misfit minimisation, using, for instance, one of the methods introduced in Chapter 15.

A proof of concept of the wavefield-adapted mesh approach in 2-D has been presented by Thrastarson et al. (2020), and is summarised in Figure 19.8. The synthetic target model is quasi-random, with quite large *P*- and *S*-wavespeed variations, reaching $\pm 8\%$. To investigate the range of conditions in which meshes adapted to the wavefield can be applied, different levels of azimuthal element stretching are considered. The waveform inversion starts from a homogeneous initial model and uses 30 L-BFGS iterations, after which the L_2 misfit between synthetic and artificial observed waveforms has decreased by 90% or more in all cases.

As expected, the quality of the reconstruction of the target in the different inversions depends critically on the extent to which the finite elements have been stretched. Using 24 or more elements in azimuthal direction produces reconstructions that differ from the target model by less than about 1%. However, with less than 24 elements, it is no longer possible to represent the complexity of the wavefield. Nevertheless, the L-BFGS algorithm successfully reduces the waveform misfit, thereby mapping numerical inaccuracies of the forward solution into reconstruction artefacts. With only 16 azimuthal elements the artefacts are of comparable size to the variations of the original model. It is therefore important to

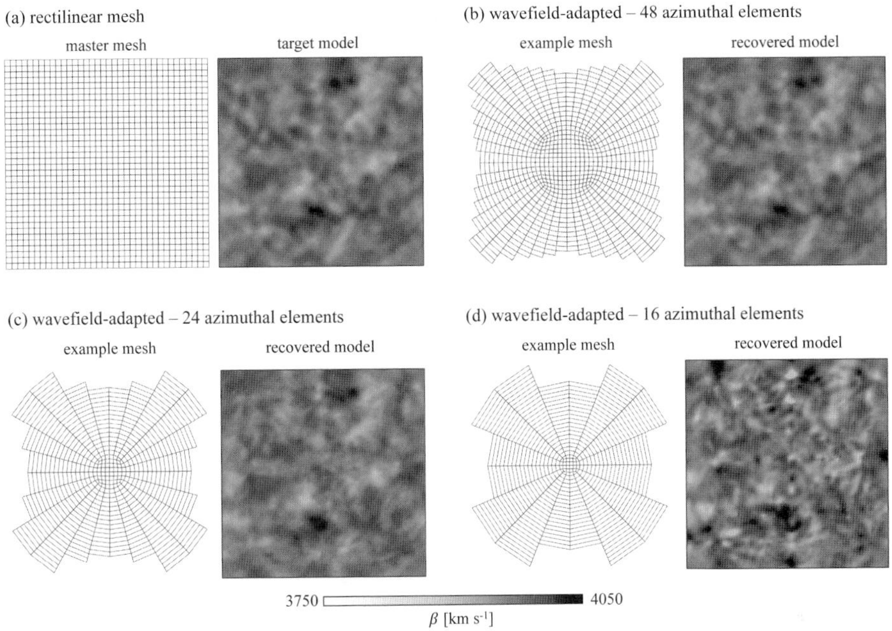

Figure 19.8 Summary of a synthetic 2-D proof-of-concept inversion using wavefield-adapted meshes. (a) S-wavespeed distribution defined on a rectilinear master mesh. (b) Wavefield-adapted mesh with 48 azimuthal elements together with the waveform inversion result for this mesh. Panels (c) and (d) have the same format as (b) but for 24 and 16 azimuthal elements. Using less than 24 azimuthal elements degrades the quality of the forward simulations to the extent that visible artefacts appear in the reconstruction.

ensure that any mesh simplifications are conservative, since excessive reduction is likely to degrade the quality of the inversion result seriously.

Of practical interest is the computational benefit that can be achieved when using meshes geared to the nature of the wavefield meshes instead of a regular mesh. In the specific case considered here the effective reduction in computation time compared with the regular case amounts to a factor of around 10, when using 24 azimuthal elements. This 'speed-up' factor depends primarily on the complexity of the medium, as we discuss below.

19.2.4 Towards Real-Data Applications

The translation of the proof of concept for waveform inversion of 2-D fabricated models from Section 19.2.3 to real-data applications primarily hinges on the extent to which wavefield-adapted meshes can be produced in 3-D. At a global scale, such meshes may be generated by rotating 2-D circular meshes around an axis defined by the location of the source and its antipole. The result of this approach is illustrated in Figure 19.9 for a case designed for low-frequency wavefields.

Numerical accuracy, for any dimensionality, is related to the extent of azimuthal stretching of the elements employed, which, in turn, is limited by the complexity

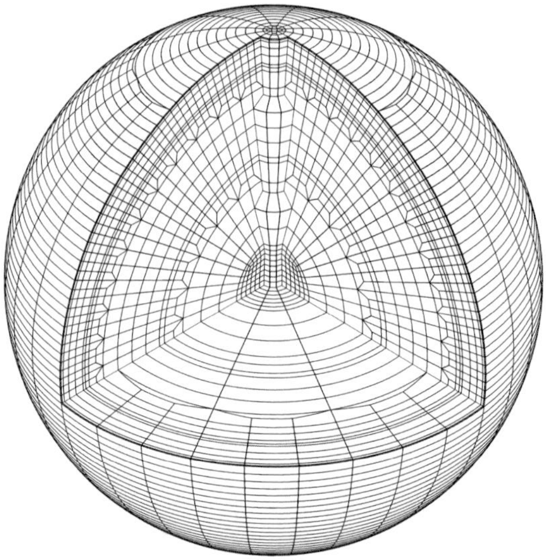

Figure 19.9 Example of a global-scale wavefield-adapted mesh designed for a minimum period of 150 s. The source is assumed to be located along the symmetry axis connecting the north and south poles. Numerical wavefield examples for global meshes can be found in van Driel et al. (2020).

of the medium. Thus, prior to performing a large number of simulations in the context of a waveform inversion, a small number of tests must be conducted in order to determine a suitable degree of azimuthal stretching that still produces acceptably good solutions. The precise meaning of 'acceptable' depends on the specifics of an application and the data quality.

In this sense, waveform inversion based on wavefield-adapted meshes represents a realisation of the No-Free-Lunch theorem, introduced in Section 15.7. This approach derives its efficiency from being highly tuned towards the solution of a very specific class of problems, namely the estimation of medium properties that are sufficiently smooth (or otherwise sufficiently well-known) so that the approximate shape of the wavefield can be anticipated and harnessed in the finite-element mesh design. For media that are too complex, either from the outset or at later stages of an iterative waveform inversion, the admissible azimuthal stretching will be too small to produce significant computational savings.

19.3 Data Redundancy and Parsimonious Inversion

Methods such as homogenisation, local time-stepping, and wavefield-adapted meshes are intended to reduce the computational requirements of forward modelling. Such improvements should, ideally, be complemented by strategies to increase the efficiency of iterative misfit minimisation in waveform inversion. Recent developments in this direction include source stacking (e.g., Capdeville et al., 2005; Tromp & Bachmann, 2019), the use of (quasi-)Newton methods (e.g., Pratt et al., 1998; Métivier et al., 2013), and gradient pre-conditioning based largely on physical intuition (Fichtner et al., 2009; Modrak & Tromp, 2016).

A different family of methods, known as stochastic gradient descent, seems to have emerged originally in the machine learning community where the limiting factor is not forward simulation but the sheer number of data that enters the evaluation of the misfit χ (e.g., Bottou, 2010; Byrd et al., 2011). The fact that several terabytes of data may need to be analysed, in order to compute just one scalar, motivates an approximation of χ using a subset of the data that are actually available. Symbolically, collecting this subset into a batch B gives the misfit approximation

$$\chi \approx \chi(\mathrm{B}) = \frac{1}{|\mathrm{B}|} \sum_{\mathrm{i}} \chi_{\mathrm{i}} \,, \tag{19.3.1}$$

where χ_{i} is the individual misfit of one datum, and $|\mathrm{B}| < \mathrm{N_D}$ is the number of data in the batch. The batch approximation (19.3.1) is used to approximate the gradient of the complete dataset and to update the model. In the next iteration, a new batch is chosen quasi-randomly, leading to the next gradient approximation and model update. In addition to reducing the computational cost for each iteration, the stochastic gradient-descent approach may help to reduce the risk of the inversion being locked into local minima by exploring slightly different misfit surfaces in each iteration. An often quoted disadvantage is slow convergence.

Following successful applications in machine learning, stochastic gradient approaches have received considerable attention in seismic exploration (van Leeuwen & Herrmann, 2013; Matharu & Sacchi, 2019) and medical ultrasound (Boehm et al., 2018). In the following, we describe an adaptation of stochastic gradient descent to seismological problems, characterised by very heterogeneous data coverage (van Herwaarden et al., 2020). We also address the problems of controlling convergence and exploiting curvature information during the inversion.

19.3.1 Adaptive Mini-Batch Optimisation

In the algorithm of van Herwaarden et al. (2020), mini-batches are defined as batches of earthquakes, and so the sum in (19.3.1) is over individual misfits for each earthquake. The composition of the mini-batches varies from one iteration to the next, which we express by a subscript k that indicates iteration number: $\mathrm{B_k}$. In the first iteration, the mini-batch $\mathrm{B_1}$ is constructed from a random choice of an earthquake. The next earthquake is the one furthest from the first; the third has the maximum distance from the first and second, etc. This selection procedure is repeated until the initial maximum mini-batch size has been reached. The approach ensures good coverage of a region with a small number of sources. To avoid the possibility that certain earthquakes are accidentally chosen much more often than others, subsequent iterations can incorporate a selection probability that is inversely proportional to the number of times that earthquake has been selected before.

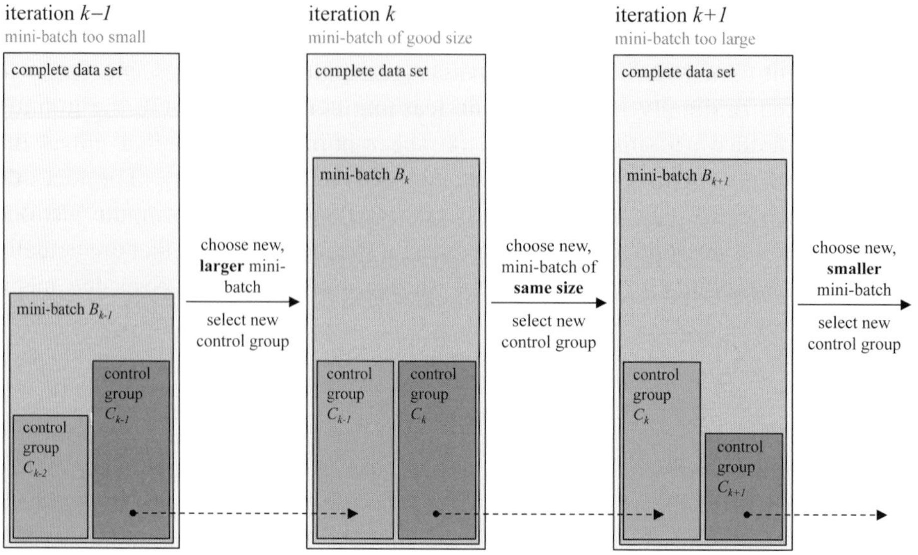

Figure 19.10 Conceptual illustration of the mini-batch scheme for iterative optimisation. The mini-batches $\mathrm{B_k}$ are chosen quasi-randomly so as to ensure good coverage and avoid excessive re-selection of certain events. The control groups $\mathrm{C_k}$ are carried over to the next iteration to ensure misfit reduction. Additionally, the size of the control group is used to adaptively steer the size of the subsequent mini-batch.

A central issue in pure stochastic gradient descent is the assessment of convergence, because $\mathrm{B_{k+1}}$ contains different sources than $\mathrm{B_k}$. To a large extent, this problem can be solved using the concept of control groups, illustrated in Figure 19.10. The control group $\mathrm{C_k}$ is a subset of the mini-batch $\mathrm{B_k}$ that is carried over to the next iteration. Provided that the control group is sufficiently large, its misfit $\chi(\mathrm{C_k})$ will approximate the total misfit χ. Therefore, a model update $\mathbf{m}_{\mathrm{k+1}}$ of $\mathbf{m}_{\mathrm{k}}$ is only accepted when $\chi(\mathrm{C_{k+1}}) < \chi(\mathrm{C_k})$.

The control group $\mathrm{C_k}$ within a mini-batch $\mathrm{B_k}$ is selected as follows. Starting from the complete mini-batch $\mathrm{B_k}$, we remove one earthquake to form a reduced mini-batch $\mathrm{B_k^{(-1)}}$. This earthquake is chosen such that its removal has the least influence on the gradient approximation, in the sense that the angle γ between $\nabla\chi(\mathrm{B_k})$ and $\nabla\chi(\mathrm{B_k^{(-1)}})$ is as small as possible. Subsequently, the subset $\mathrm{B_k^{(-2)}}$ is formed by removing the least influential earthquake from $\mathrm{B_k^{(-1)}}$, etc. This earthquake removal process continues until a pre-defined threshold angle γ_t is reached. At this point the final reduced mini-batch is taken as the control group. The threshold angle may vary as a function of iteration number. During the first few iterations, rough gradient approximations often produce substantial misfit reductions. In later iterations, higher accuracy is often needed in order to ensure continued misfit reductions. This is particularly true for conjugate-gradient and quasi-Newton methods that rely more strongly on accurate gradients than the steepest-descent method.

In addition to monitoring convergence, the control group also serves to adaptively steer the size of the mini-batch. In fact, when the control group is (nearly) as large as the mini-batch itself, this would indicate that the mini-batch is too small. Conversely, a control group that is very small compared to the mini-batch suggests that a smaller number of earthquakes may be sufficient to approximate misfits and gradients with reasonable accuracy. Thus, choosing the next mini-batch B_{k+1} to be a factor of s larger than the current control group C_k, ensures that the mini-batch size adapts to the iteration-dependent accuracy requirements of the misfit minimisation. Typically, a suitable value of the scale factor s is around 2.

In a similar way to the standard L-BFGS method, successive model updates and gradients are used to approximate the inverse Hessian matrix. The Hessian is approximate for two reasons: the unavoidably limited number of models and gradients that enter the L-BFGS updating scheme, and the analysis of different misfit functions in each iteration.

19.3.2 Synthetic and Real Data Illustrations

To illustrate the adaptive mini-batch approach, we consider a data set with 125 earthquakes and 2648 stations located on and around the African Plate, as shown in Figure 19.11(a). We begin with a synthetic waveform inversion, thus deliberately committing an inverse crime in order to assess the functioning of the method under idealised conditions. Using a minimum period of 65 s, we aim to constrain variations in the *S* wavespeeds β_h, β_v, and in the isotropic *P* wavespeed α. For simplicity, we ignore *P*-wave anisotropy, deviations of transverse isotropy from $\eta = 1$, and lateral variations in density. The threshold angle γ_t that steers the size of the control group relative to the mini-batch size is set to 22.5°.

Figure 19.11(b) provides a convergence summary for standard L-BFGS and adaptive mini-batches in terms of the L_2 distance between the target model and the reconstructed model, as a function of simulation count. With L-BFGS, 21 iterations are needed to achieve a model misfit reduction of 50%. Since each iteration uses (artificial) data from all 125 sources, the total number of iterations is 5,250, including both forward and adjoint stages. Using adaptive mini-batches, the same model misfit reduction requires only 1300 simulations, with a batch size gradually increasing from around 10 to around 30 during the iteration. For both methods, the recovered models shown in Figure 19.11(c) are visually indistinguishable.

Since the source–receiver geometry for the synthetic test corresponds to actually available waveform coverage, a real-data inversion with a similar configuration can be performed easily. To illustrate the benefits of the adaptive mini-batch scheme, we *a priori* fix the total number of simulations for forward and adjoint steps combined to 750. Using the standard L-BFGS approach with all 125 sources

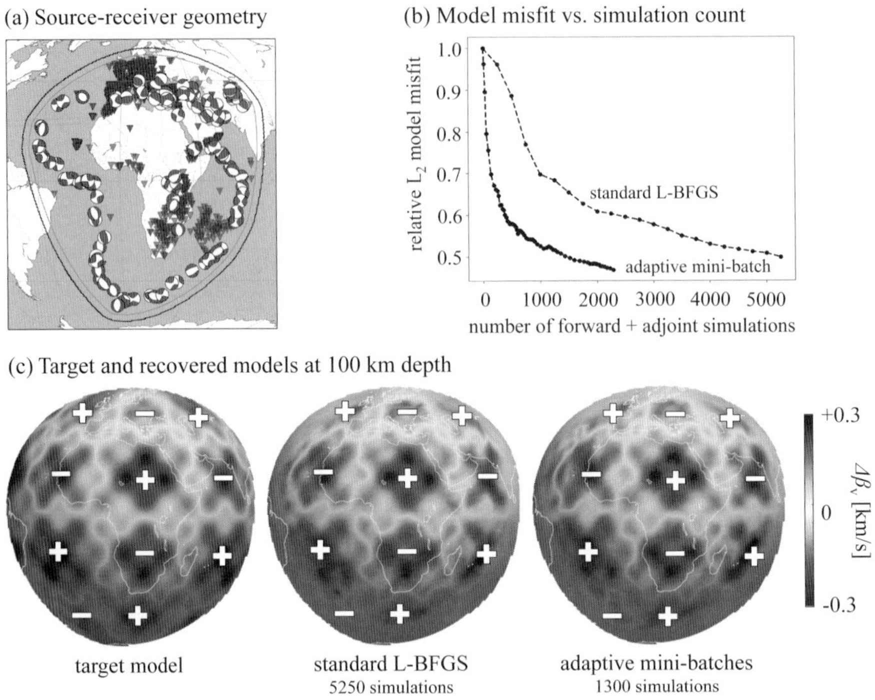

Figure 19.11 Summary of synthetic waveform inversions using the realistic source–receiver geometry shown in panel (a) for the African region. The 125 earthquake epicentres are marked by beachballs, indicating the source mechanism. Receivers are located at the positions of the 2648 black triangles. The black curve outlines the domain used for the wavefield simulation. (b) Difference between the target model $\mathbf{m}_{\text{target}}$ and the reconstructed model $\mathbf{m}_i$ as a function of total wavefield simulation count i, quantified in terms of the L_2 model difference $|\mathbf{m}_{\text{target}} - \mathbf{m}_i|_{L_2}$. In order to reach a 50% model misfit reduction, 5250 simulations with standard L-BFGS, and 1300 simulations with the adaptive mini-batches are needed. (c) Distribution of β_v at 100 km depth in the target model (left), the L-BFGS based recovery after 5250 simulations (centre), and the adaptive mini-batch inversion after 1300 simulations (right).

per iteration, this corresponds to three iterations. In contrast, adaptive mini-batches with around 10 sources per batch enable 40 iterations. The results for both cases are compared in Figure 19.12. While the mini-batch result contains considerable structural detail and has achieved a misfit reduction of 51%, the standard L-BFGS inversion is still close to the initial model with a misfit reduction of 21%. To reach a misfit reduction of nearly 50% with L-BFGS required an additional 10 iterations.

Optimisation schemes are inherently difficult to compare quantitatively because their performance always depends on a specific example, or a family of examples. In this context, it is important to note that the relative efficiency of the adaptive mini-batch approach critically rests on the presence of redundancies in the data set. This makes the method particularly well suited for earthquake tomography because the sources tend to reoccur in similar locations, potentially providing only

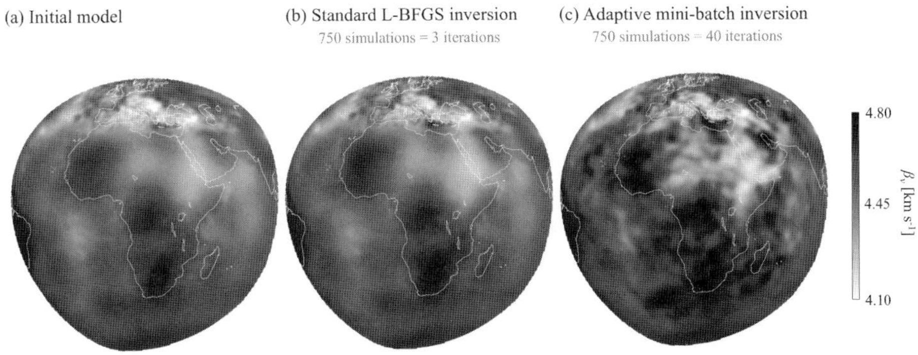

Figure 19.12 Comparison of real-data waveform inversion using a total of 750 forward and adjoint simulations with standard L-BFGS (b) and adaptive mini-batches (c). The initial model in (a) is an earlier version of the CSEM (Fichtner et al., 2018a). The *SV* wavespeed β_v at 100 km depth is shown.

a small amount of new independent information. One may therefore expect that adaptive mini-batches are less efficient in exploration seismology where source and receiver locations are typically chosen such that redundancies are minimised.

The most important tuning parameter in the adaptive mini-batch scheme is the threshold angle γ_t. As is often the case with tuning parameters, a suitable choice requires some experimentation and intuition. Choosing γ_t too small may lead to poor performance because control groups and mini-batches will be very large. On the other hand, choosing γ_t too small may lead to slow convergence because the gradient approximation is too poor.

19.4 Hamiltonian Nullspace Shuttles and Monte Carlo Inversion

Although numerical modelling, nonlinear optimisation, and adjoint-based waveform inversion have progressed significantly during the past decade, our ability to assess the quality of Earth models has hardly been able to keep up with these advances. Methods based on estimates of curvature near the hopefully global optimum derived using the methods discussed in Chapter 15 provide some resolution proxies, but ignore the nonlinear nature of the inverse problem (Bui-Tanh et al., 2003; Fichtner & van Leeuwen, 2015). In the following paragraphs, we introduce a novel class of methods that enable uncertainty quantification via the construction of an artificial Hamiltonian system where a model is interpreted in terms of an n-dimensional particle that orbits through model space.

19.4.1 Nullspace Shuttles

According to Backus & Gilbert (1968), the solution of an inverse problem consists in the description of the 'very' infinite-dimensional space of 'reasonable' models, whose level of misfit is close to the optimum. This space of acceptable models is more commonly referred to as the effective nullspace. In the special case of least-squares problems with a linear forward modelling operator, such as described in Section 14.2, the effective nullspace can be explored using basic linear algebra. In fact, the linear nullspace shuttle (Deal & Nolet, 1996) navigates with the help of singular vectors that correspond to singular values below a certain misfit tolerance.

The nonlinear nullspace shuttle (Fichtner & Zunino, 2019) is suitable for the analysis of nonlinear seismic waveform inversion. This nonlinear approach drops the main limitations of the linear nullspace shuttle, namely that the forward problem is linear and that the misfit χ has the form of a least-squares misfit. To initiate the exploration of the nullspace, we assume that we have estimated some plausible model $\mathbf{m}^{\mathrm{est}}$ that approximately minimises the misfit χ. This minimisation may, for example, have been achieved using one of the nonlinear optimisation methods covered in Chapter 15. As a result of the presence of observational uncertainties, alternative models $\mathbf{m}^{\mathrm{est}}+\Delta\mathbf{m}$ may still be plausible if their associated misfit remains below a given tolerance ε^2,

$$\chi(\mathbf{m}^{\mathrm{est}} + \Delta\mathbf{m}) \leq \chi(\mathbf{m}^{\mathrm{est}}) + \varepsilon^2 \,. \tag{19.4.1}$$

The space spanned by those $\Delta\mathbf{m}$ that satisfy (19.4.1) defines the effective nullspace.

This nonlinear nullspace shuttle approach rests on the interpretation of the model $\mathbf{m}$ in terms of an n-dimensional position vector that marks the position of an imaginary particle in n-dimensional space. The position is assumed to be a function of an artificially introduced time variable t, and so each time corresponds to a new model $\mathbf{m}(t)$. To guide the movement of the particle through model space, we construct artificial equations of motion, borrowing the basic concept from Hamilton's equations in classical mechanics.

We first define the potential energy U of the particle to equal the misfit,

$$U(\mathbf{m}) = \chi(\mathbf{m}) \,. \tag{19.4.2}$$

This potential energy induces an artificial force proportional to $-\nabla U[\mathbf{m}(t)]$, which effectively pulls the model $\mathbf{m}(t)$ in the direction of steepest descent, towards a model $\mathbf{m}(t+\delta t)$ with lower misfit. The movement in the direction of $-\nabla U[\mathbf{m}(t)]$ is complemented by an additional contribution associated with the presence of an artificial, n-dimensional momentum $\mathbf{p}(t)$. Together with an equally artificial, symmetric, and positive definite $n \times n$ mass matrix, $\mathbf{M}$, the momentum defines a kinetic energy

$$K(\mathbf{p}) = \frac{1}{2}\mathbf{p}^T\mathbf{M}^{-1}\mathbf{p} \,. \tag{19.4.3}$$

The sum of potential and kinetic energies defines the artificial Hamiltonian of the system,

$$H(\mathbf{m},\mathbf{p}) = U(\mathbf{m}) + K(\mathbf{p}). \tag{19.4.4}$$

Equipped with the artificial energies U, K and $H = U + K$, the movement of the particle is fully described by Hamilton's equations,

$$\frac{dm_i}{dt} = \frac{\partial H}{\partial p_i}, \qquad \frac{dp_i}{dt} = -\frac{\partial H}{\partial m_i}, \qquad i = 1, ..., n. \tag{19.4.5}$$

Along a trajectory in model-momentum space, the Hamiltonian is preserved. Hence, starting at some approximate minimum $\mathbf{m}^{est}$ of χ, the solution of the Hamiltonian equations (19.4.5) produces a continuous sequence of models $\mathbf{m}(t)$ and associated momenta $\mathbf{p}(t)$ that satisfy

$$\begin{aligned} H[\mathbf{m}(t), \mathbf{p}(t)] &= \chi[\mathbf{m}(t)] + \frac{1}{2}\mathbf{p}(t)^T\mathbf{M}^{-1}\mathbf{p}(t) = H(\mathbf{m}^{est}, \mathbf{p}^{init}) \\ &= \chi(\mathbf{m}^{est}) + \frac{1}{2}\left(\mathbf{p}^{init}\right)^T \mathbf{M}^{-1}\mathbf{p}^{init}. \end{aligned} \tag{19.4.6}$$

When the initial momentum $\mathbf{p}^{init}$ is chosen such that

$$K\left(\mathbf{p}^{init}\right) = \frac{1}{2}\left(\mathbf{p}^{init}\right)^T \mathbf{M}^{-1}\mathbf{p}^{init} = \varepsilon^2, \tag{19.4.7}$$

equation (19.4.6) implies

$$\chi[\mathbf{m}(t)] \le \chi(\mathbf{m}^{est}) + \varepsilon^2, \tag{19.4.8}$$

because the positive definiteness of the mass matrix $\mathbf{M}$ ensures $\mathbf{p}(t)^T\mathbf{M}^{-1}\mathbf{p}(t)$ for all momenta $\mathbf{p}(t)$. Hence, all models $\mathbf{m}(t)$ along the Hamiltonian trajectory are within the effective nullspace.

In the vast majority of interesting cases, Hamilton's equations (19.4.5) must be solved numerically, using, for example, the leapfrog scheme with time step Δt:

$$p_i(t + \Delta t/2) = p_i(t) - \frac{1}{2}\Delta t \left.\frac{\partial U}{\partial m_i}\right|_t, \tag{19.4.9}$$

$$m_i(t + \Delta t) = m_i(t) + \Delta t \left.\frac{\partial K}{\partial p_i}\right|_{t+\Delta t/2}, \tag{19.4.10}$$

$$p_i(t + \Delta t) = p_i(t + \Delta t/2) - \frac{1}{2}\Delta t \left.\frac{\partial U}{\partial m_i}\right|_{t+\Delta t}. \tag{19.4.11}$$

Although other numerical integrators may equally well be used, the leap-frog scheme allows us to illustrate how the specific choices for initial momentum and mass matrix control the orbital properties of the trajectories.

In the extreme case where an increase in the misfit beyond the value $\chi(\mathbf{m}^{est})$ cannot be tolerated, we set $\varepsilon = 0$. Equation (19.4.1) then implies vanishing initial momentum, $\mathbf{p}^{init} = \mathbf{0}$, and the first iteration of the leapfrog scheme from $t = 0$ to $t = \Delta t$ becomes

$$m_i(\Delta t) = m_i^{est} - \frac{1}{2}\Delta t^2 \sum_{k=1}^{n} M_{ik}^{-1} \frac{\partial \chi(\mathbf{m}^{est})}{\partial m_k}. \tag{19.4.12}$$

Equation (19.4.12) reveals an interesting relation between effective nullspace exploration and the gradient-descent methods from Chapter 15. The special case when the mass matrix is chosen to be equal to the identity matrix, $\mathbf{M} = \mathbf{I}$, transforms (19.4.12) into a steepest-descent step with step length $\Delta t^2/2$. Similarly, choosing $\mathbf{M}$ to equal the Hessian $\mathbf{H}$ of χ, we retrieve a variant of Newton's method with step length $\Delta t^2/2$. Since the mass matrix is positive definite, the first step of the zero-tolerance scenario will be in a descent direction, as discussed in Section 15.1.

If, in addition, the estimated model $\mathbf{m}^{\text{est}}$ is sufficiently close to the minimum of χ that the first derivative $\partial\chi(\mathbf{m}^{\text{est}})/\partial m_i = \partial U(\mathbf{m}^{\text{est}})/\partial m_i$ is negligible, the first step of the leapfrog scheme specialises to

$$m_i(\Delta t) = m_i^{\text{est}} + \sum_{k=1}^{n} \Delta t M_{ik}^{-1} p_k^{\text{init}} . \tag{19.4.13}$$

Equation (19.4.13) provides an opportunity to add a specific model perturbation $\Delta\mathbf{m}$ to the estimated model $\mathbf{m}^{\text{est}}$. For the choice of initial momentum

$$\mathbf{p}^{\text{init}} = \gamma \mathbf{M}\, \Delta\mathbf{m} , \tag{19.4.14}$$

with the factor γ set to satisfy $K\left(\mathbf{p}^{\text{init}}\right) = \varepsilon^2$, the first leapfrog iteration adds a scaled version of $\Delta\mathbf{m}$ to $\mathbf{m}^{\text{est}}$. Subsequent iterations will add more of $\Delta\mathbf{m}$ until an increasing misfit derivative starts to dominate the evolution equations, thereby deflecting the trajectory towards a different direction.

A slight rearrangement of equation (19.4.10) shows that the mass matrix has the effect of modifying the direction of the trajectory,

$$\mathbf{m}(t + \Delta t) = \mathbf{m}(t) + \Delta t\, \mathbf{M}^{-1}\, \mathbf{p}(t + \Delta t/2) . \tag{19.4.15}$$

For example, when $\mathbf{M}^{-1}$ is taken as a smoothing operator, the evolution of the Hamiltonian equations will lead to a preferential exploration of smoother models in the effective nullspace. A roughening operator, in contrast, will produce a sequence of rougher alternative models.

The concept of the nonlinear Hamiltonian nullspace shuttle can be illustrated most easily with a 1-D model space and an analytically defined misfit function $\chi(m)$. An example is shown in Figure 19.13 for the misfit function

$$\chi(m) = U(m) = 1 - \cos(0.2m^2) + 0.025m^2 . \tag{19.4.16}$$

By design, the Hamiltonian trajectories are curves of constant total energy H. For a comparatively low tolerance ε, as in Figure 19.13(a), the trajectory remains entirely within the basin of attraction where the estimated model is located. Increasing the tolerance, as shown in Figure 19.13(b), still produces constant-energy trajectories in the combined model-momentum space (phase space), but the projection of the trajectory in model space is able to transition between different basins of attraction.

A more realistic example in the form of a synthetic 2-D nonlinear travel-time tomography is shown in Figure 19.14 (Fichtner et al., 2018b). The misfit $\chi = U$

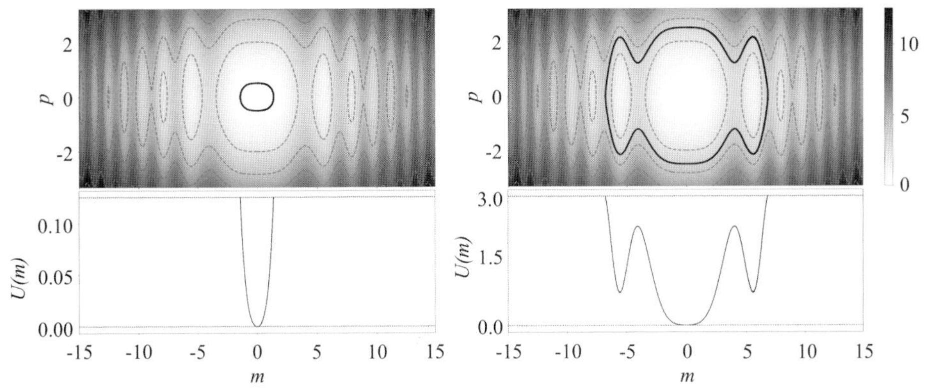

Figure 19.13 Hamiltonian trajectories in model-momentum space (top) and their projection in model space (below). The total energy H is shown in grey scale, with dashed curves marking levels of constant energy. The trajectory is shown as a solid black curve. For a low tolerance (a), the trajectory remains close to the global optimum, whereas it may sample neighbouring basins of attraction in model space when the tolerance is higher (b).

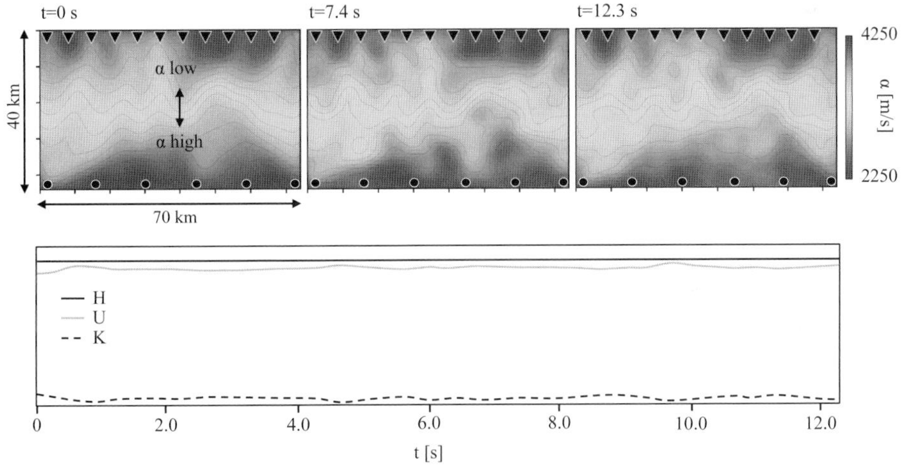

Figure 19.14 Snapshots along the nullspace trajectory of a 2-D nonlinear travel-time tomography problem (top). Sources near the bottom of the domain are marked by black circles, receivers near the surface are shown as black triangles. All models along the trajectories explain the travel times of the waves in passage from the sources to the receivers within a given tolerance. The evolution of the total energy H, the potential energy (misfit) U, and the kinetic energy K is shown in the lower panel. Note that the potential energy U is, by design, always smaller than H.

is defined as the squared sum of travel-time residuals for all source–receiver pairs, and the forward problem is the solution of the eikonal equation for the rays. For a given tolerance, the Hamiltonian trajectory consists of a sequence of P wavespeed models that are, by construction, all consistent with the observed travel

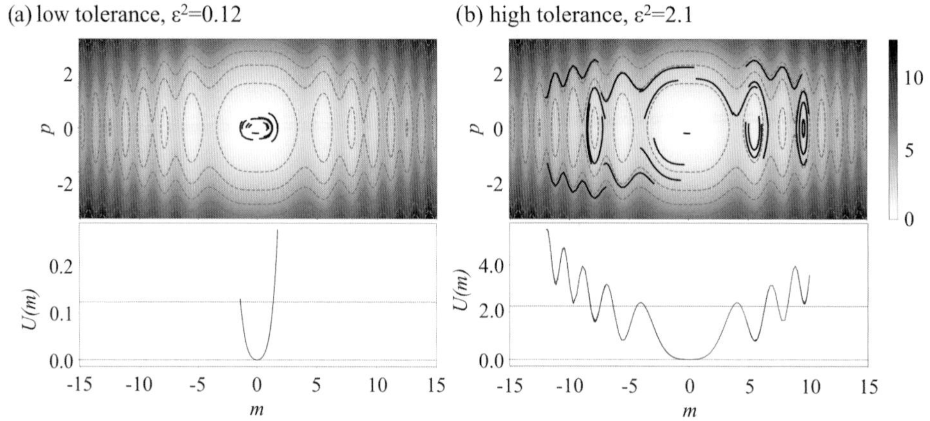

Figure 19.15 Randomised version of the nonlinear Hamiltonian nullspace shuttle. Trajectories are shown in the top panels, superimposed on the total energy H in phase space. The projections in model space are shown below. For sufficiently high tolerances, the shuttle is able to explore the nullspace more exhaustively.

times because their misfit (potential energy) is bounded above by the constant Hamiltonian H.

19.4.2 Hamiltonian Monte Carlo Methods

Instead of being a hard bound, the tolerance itself ε is often uncertain, suggesting that it should be treated as a random variable. Equation (19.4.7) implies that normally distributed tolerances with zero mean and standard deviation ε correspond to random momentum vectors $\mathbf{p}$ drawn from an n-dimensional Gaussian distribution with covariance matrix $2\varepsilon\mathbf{M}$.

Allowing the tolerance ε to exceed a hard bound with some probability gives rise to a randomised version of the nonlinear Hamiltonian nullspace shuttle. After some simulation time, the trajectory is stopped, and a new initial momentum is drawn randomly from the Gaussian distribution with covariance matrix $2\varepsilon\mathbf{M}$. The subsequent trajectory will be on a somewhat lower or higher energy level, thus giving rise to a broader exploration of the nullspace. To see the effect, we return to the analytic example, with the misfit given by equation (19.4.16). The tolerance is now interpreted as a random, normally distributed variable with zero mean and standard deviation ε. The corresponding trajectories are shown in Figure 19.15. For a low tolerance (standard deviation), the nullspace mostly explores the vicinity of the maximum-likelihood model. As the tolerance increases, the shuttle becomes more exploratory and ventures into neighbouring basins of attraction.

The randomised nullspace shuttle is nearly identical to the Hamiltonian Monte Carlo algorithm (e.g., Neal, 2011; Betancourt, 2017) that can be used to sample the posterior probability density

$$P(\mathbf{m}\,|\,\mathbf{d}^{\mathrm{obs}}) = \mathrm{const}\, e^{-\chi(\mathbf{m})}\,, \tag{19.4.17}$$

which results from the application of Bayes' theorem (14.1.7). In the Hamiltonian Monte Carlo approach, the end point of each phase-space trajectory, projected onto model space, is a new sample of $\mathrm{P}(\mathbf{m}\,|\,\mathbf{d}^{\mathrm{obs}})$. The ensemble of many samples provides a fully probabilistic solution of the inverse problem, including any kind of uncertainty information.

The only difference between the randomised nullspace shuttle and Hamiltonian Monte Carlo results from unavoidable inaccuracies in the numerical solution of Hamilton's equations (19.4.5). All explicit time-stepping schemes, including the leapfrog algorithm (19.4.9–19.4.11), do not preserve the total energy H exactly. As a consequence, the numerical approximations of the trajectories tend to deflect towards higher or lower energy levels $\tilde{\mathrm{H}} \neq \mathrm{H}$. To ensure proper importance sampling of the posterior, a correction must be introduced, whereby a trajectory end point is accepted as a new sample only with probability

$$\mathrm{p}^{\mathrm{accept}} = \min\left(1, e^{\mathrm{H}-\tilde{\mathrm{H}}}\right). \tag{19.4.18}$$

Thus, end points with lower energy are generally accepted as a new sample, whereas end points with higher energy may be rejected with probability $1 - e^{\mathrm{H}-\tilde{\mathrm{H}}}$.

The main advantage of the Hamiltonian Monte Carlo approach lies in the fast generation of independent samples, that is, models that are significantly different from models already generated before. While, for example, the numerical cost of generating an independent sample for the Metropolis–Hastings algorithm grows with model space dimension n as $\mathcal{O}(n^2)$ (Creutz, 1988), it only grows as $\mathcal{O}(n^{5/4})$ for Hamiltonian Monte Carlo (Neal, 2011). Intuitively, this attractive property is due to the action of the artificial force $-\nabla \mathrm{U}$, which ensures that models remain within the statistically defined effective nullspace, also referred to as the typical set in the Monte Carlo literature. The efficient generation of independent models makes the Hamiltonian Monte Carlo approach particularly interesting for the solution of tomographic inverse problems, where the necessary misfit derivatives can easily be computed with the help of adjoint techniques.

While the application of Hamiltonian Monte Carlo methods to large-scale 3-D tomographic problems is still the subject of ongoing research, first results for 2-D waveform inversion are promising (Biswas & Sen, 2017). An illustration is shown in Figure19.16, which shows partial results of a probabilistic waveform inversion for *S* wavespeed, *P* wavespeed and density (Gebraad et al., 2020). Each of these physical model parameters is discretised with 180×60 constant-property blocks, resulting in a total of 32 400 model parameters. The prior distributions are constant in the intervals 2000 ± 100 m/s for α, 800 ± 50 m/s for β, and 1500 ± 100 kg/m^3 for ρ. A simple L_2 waveform misfit quantifies the differences between artificial observed and synthetic waveforms.

Numerical experiments indicate that on the order of 10 000 samples are required to achieve reasonable convergence of 1-D and 2-D posterior marginal distributions, though many more samples would certainly be needed to approximate the full

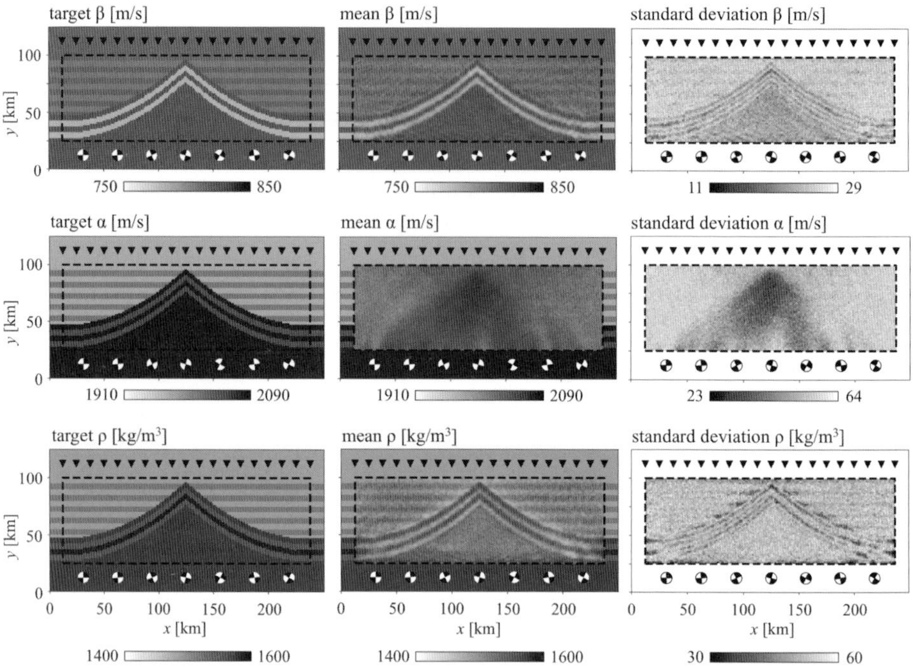

Figure 19.16 Hamiltonian Monte Carlo waveform inversion in 2-D. Target models for *S* wavespeed β, *P* wavespeed α, and density ρ are shown in the left column. Beachballs mark moment-tensor sources, and black triangles are placed at the positions of receivers near the surface. The middle column displays the posterior mean models and the right column shows the posterior standard deviations. While the posterior marginals for β and ρ are approximately Gaussian, the posterior marginals for α are not. This makes the interpretation of the *P* wavespeed mean and standard deviation non-trivial.

32 400-dimensional posterior. The 1-D posterior marginals for β and ρ are nearly Gaussian with small standard deviations relative to the width of the prior distributions. This contributes to the close resemblance of their posterior means and the target model, used to compute the artificial waveform data. In contrast, the posterior 1-D marginals for α are strongly non-Gaussian, mostly because the L_2 waveform misfit emphasises the larger-amplitude *S* waves, while assigning little importance to the lower-amplitude *P* waves. As a consequence, the posterior mean and standard deviation are difficult to interpret intuitively.

In addition to providing a whole ensemble of plausible models and complete uncertainty information, the Hamiltonian Monte Carlo waveform inversion shown in Figure 19.16 also delivers a density model that is free from subjective regularisation, such as artificial correlations between wave speeds and density that are often imposed to stabilise the inversion procedure.

The challenges in future, large-scale applications of Hamiltonian Monte Carlo methods include, but are not limited to, computational resources. A general difficulty for all Monte Carlo methods is the quantification of convergence and the detection of potential locking into local minima. While numerous generic proxies for convergence have been suggested, none of them is universally applicable and failsafe (Cowles & Carlin, 1996), and so it currently seems that the assessment of convergence will continue to require a good dose of physical intuition.

As already indicated in Section 19.4.1, Hamiltonian trajectories are strongly affected by the mass matrix **M**. Though we are in principle free to choose any symmetric and positive definite **M**, this choice has profound effects on the performance of the Hamiltonian Monte Carlo algorithm. The analysis of this algorithm applied to linear least-squares problems reveals that the ideal mass matrix is the Hessian, suggesting that the current manual tuning should be replaced by Hessian approximations, using, for instance, second-order adjoints or quasi-Newton methods.

Appendix: Table of Notation

General

z	–	depth
r	–	radius
t	–	time
ω	–	angular frequency
f	–	frequency
θ, ϕ	–	coordinate angles
$\mathbf{e}_r, \mathbf{e}_\theta, \mathbf{e}_\phi$	–	unit vectors in spherical coordinate system
r_e	–	radius of the Earth
V_e	–	total volume of the Earth
$\mathbf{x}$	–	position vector
x_i	–	position coordinates
$\mathbf{x}_s$	–	source location
$\mathbf{u}$	–	displacement vector
$\mathbf{f}$	–	force vector
τ_{ij}	–	stress tensor
e_{ij}	–	strain tensor
c_{ijkl}	–	elastic modulus tensor
C_{ijkl}	–	viscoelastic relaxation tensor
ψ	–	gravitational potential
ρ	–	density
ρ_0	–	undisturbed density
κ	–	bulk modulus
μ	–	shear modulus
α	–	*P* wavespeed
β	–	*S* wavespeed
ϕ	–	bulk-sound speed
α_h, α_v	–	*P* wavespeeds, transverse isotropy
β_h, β_v	–	*S* wavespeeds, transverse isotropy
η	–	transverse isotropic parameter

General (cont.)

T	– travel time
X	– horizontal distance
p	– horizontal slowness
Δ	– epicentral angle (for propagation in a sphere)
$\wp$	– spherical slowness
p	– phase slowness
g	– group slowness
c	– phase velocity
U	– group velocity
k	– wavenumber
$*$	– complex conjugate
$\star$	– convolution

Stratified media

$\mathfrak{H}$	– stress gradient operator
$\mathbf{u}^e_K$	– Kth modal eigenvector
$\mathbf{e}^e_K$	– strain tensor for Kth mode
$C_K(t)$	– Kth modal time term
Q_K	– quality factor for Kth mode
$\mathbf{x}_s$	– source location
M_{ij}	– point moment tensor
$\mathfrak{M}(t)$	– moment rate tensor
q_α, q_β	– vertical slownesses
$\varepsilon_\alpha, \varepsilon_\beta$	– energy normalisations
U, V, W	– displacement components
P, S, T	– traction components
w_0	– displacement response at surface
$G(p, \omega)$	– medium response
$H(p, \omega)$	– pole contribution
p_c	– slowness hinge point
U_S, D_S	– upgoing, downgoing source radiation
$\mathbf{R}_F$	– free-surface reflection matrix
R_F^{HH}, R_F^{HH}	– free surface reflection coefficients
$\mathbf{W}_F$	– free-surface amplification matrix
$\mathbf{R}_U, \mathbf{R}_D$	– reflection matrices for incident upgoing, downgoing waves
$\mathbf{T}_U, \mathbf{T}_D$	– transmission matrices for incident upgoing, downgoing waves
$\mathbf{X}_R, \mathbf{X}_S, \mathbf{X}_J$	– reverberation terms
$\phi(\omega)$	– incremental phase on path
t^*	– attenuation measure along path
$\mathcal{M}(t)$	– effective source-time function

Stratified media (cont.)

$\tau_M(p)$	–	phase delay for mantle
$\Phi(p, \omega)$	–	phase propagation from source
$\mathbf{Z}(p, \omega)$	–	amplitude propagation from source
$\mathbf{C}_R(p, \omega)$	–	structural effects near receiver
$\mathcal{A}[\cdot]$	–	auto-correlation
$\mathbf{u}_j^e(p, z, \omega)$	–	displacement eigenfunction for surface-wave mode j
$\mathbf{t}_j^e(p, z, \omega)$	–	traction eigenfunction for surface-wave mode j
$\mathcal{S}, \mathcal{R}$	–	source, receiver terms in surface-wave response
L, Q, T	–	rotated components to allow for free-surface *P* incident
V, W, T	–	rotated components to allow for free-surface *S* incident
P, S, H	–	transformed components after removal of free-surface effects

Varying media

$\mathbf{x}_h$	–	horizontal position vector
$\mathbf{G}, G_{ij}$	–	Green's tensor
H_{ijp}	–	stress tensor derived from Green's tensor
$\mathbf{H}, H_{ij}$	–	traction derived from Green's tensor
$\mathbf{n}$	–	normal vector
$\mathbf{e}$	–	equivalent source
$\mathbf{b}$	–	displacement-traction vector
$\mathbf{A}_0$	–	operator matrix for stratified medium
$\mathbf{A}_1$	–	operation matrix for lateral variation
R_F	–	free-surface reflection operator
W_F	–	free-surface amplification operator
$\mathbf{R}_U, \mathbf{R}_D$	–	reflection operator for incident upgoing, downgoing waves
$\mathbf{T}_U, \mathbf{T}_D$	–	transmission operator for incident upgoing, downgoing waves
$\mathbf{X}_R, \mathbf{X}_S, \mathbf{X}_J$	–	reverberation operators
V_U^R, V_D^R	–	receiver operators
Z	–	operator for propagation from source
C_R	–	operator for structural effects near receiver
$A_N(\omega, p)$	-	linear array response
V_U^R, V_U^R	–	receiver operators
$\mathcal{M}$	–	migration operator
$\mathcal{S}$	–	stacking operator
$\mathcal{V}$	–	deconvolution operator
f_U^{AB}, f_D^{AB}	–	focussing functions
G_U^{0J}, G_D^{0J}	–	radiation components
$\mathcal{T}_D^{inv}$	–	direct arrival of inverse transmission
M^{0J}	–	structural coda

Correlation wavefield

$C_{u,v}(t)$	–	cross-correlation of $u(t)$, $v(t)$
$D_{u,v}(t)$	–	convolution of $u(t)$, $v(t)$
$\mathcal{T}_{ij}(t)$	–	correlation time shift between $u_i(t)$ and $u_j(t)$
$V_l(t), V_q(t)$	–	linear and quadratic stacks
$\mathcal{P}$	–	misfit measure
$\mathbf{u}_r$	–	reference seismogram
$\mathbf{u}_c$	–	comparator seismogram
$\boldsymbol{\mathfrak{T}}_{cr}$	–	tensor transfer operator
$\mathfrak{T}_{cr}$	–	transfer operator
$\bar{T}_{cr}(\omega)$	–	transfer function
$\{\mathcal{F}_j\}$	–	filter set
$\mathcal{C}_{ZZ}$	–	normalised correlation
Δ_{12}	–	inter-station distance on sphere
$\mathcal{U}^{12}(\omega)$	–	correlation term
$\mathbf{G}(p, \omega)$	–	medium response
$g_I(p, \omega)$	–	amplitude weight for Ith generalised ray
$\tau_I(p)$	–	phase increment for Ith generalised ray
$\mathcal{I}(\mathbf{x}_i, \mathbf{x}_k, \omega)$	–	interferogram
F	–	generalised force term
$\mathbf{v}$	–	velocity field
$\mathfrak{T}_{ik}$	–	interferogram transfer tensor
$\mathfrak{g}_{ik}$	–	tensor propagation correction
$\mathfrak{f}_{ik}$	–	tensor factorisation residual
$\mathcal{H}$	–	Hilbert transform operator
E	–	envelope operator
$S(p, \omega)$	–	array response

Heterogeneity

$\mathcal{L}$	–	propagation operator
$\mathcal{L}'$	–	heterogeneity operator
$\mathcal{K}$	–	heterogeneity shift operator
$\mathbf{u}^\dagger$	–	adjoint field
$\mathcal{L}^\dagger$	–	adjoint operator
$N(r)$	–	spatial auto-correlation function
$P(k)$	–	power-spectral density
ϵ	–	heterogeneity level
a_x, a_z	–	correlation lengths
ν	–	Hurst number
$\mathbf{D}$	–	finite-difference operator
$\mathbf{R}$	–	discretised density
$\mathbf{M}$	–	discretised modulus

Heterogeneity (cont.)

F_N	–	discrete Fourier transform
L	–	low-wavenumber matrix
H	–	high-wavenumber matrix
χ^{ABC}	–	reverberation operator
$\mathbf{u}(x, z, \omega)$	–	modal displacement field
$\mathbf{t}(x, z, \omega)$	–	modal horizontal traction field
$\mathbf{b}(x, z, \omega)$	–	modal displacement–traction field
$\mathcal{A}_{uu}, \ldots$	–	heterogeneity operators
$\mathbf{B}^{++}, \ldots$	–	modal coupling matrices
c_K	–	modal expansion coefficient
R^{-+}, T^{++}	–	modal reflection and transmission matrices
$\mathbf{r}_U, \mathbf{r}_D$	–	heterogeneity reflection operators
$\mathbf{t}_U, \mathbf{t}_D$	–	heterogeneity transmission operators
$\mathfrak{S}_m$	–	modal power
H_{mn}	–	modal power coupling matrix

Inversion

$\mathfrak{m}$	–	continuous model
$\mathbf{m}$	–	discretised model vector
$\mathbf{d}$	–	data vector
$\Phi(\mathbf{d}_1, \mathbf{d}_2)$	–	measure of data match
$\Psi(\mathbf{m}_1, \mathbf{m}_2)$	–	measure of model match
$\chi(\mathbf{d}, \mathbf{m})$	–	composite misfit function
$P(\mathbf{m})$	–	model probability distribution
$P(\mathbf{m}\|\mathbf{d})$	–	conditional probability
$\Lambda(\mathbf{m}_1\|\mathbf{m}_0)$	–	likelihood function
$\mathbf{C}_D$	–	data covariance
$\mathbf{C}_M$	–	model covariance
$\mathbf{G}[\mathbf{m}]$	–	data functional
$\mathbf{R}_D$	–	data resolution matrix
$\mathbf{R}_M$	–	model resolution matrix
$\mathbf{h}$	–	descent direction
$\mathbf{H}$	–	Hessian matrix
$\mathbf{g}$	–	gradient vector
Υ	–	inversion performance measure
$\mathbf{u}^\dagger$	–	adjoint field
$\mathcal{L}^\dagger$	–	adjoint operator
$\mathbf{f}^\dagger$	–	adjoint source
$\mathbf{G}^\dagger$	–	adjoint Green's function
$\boldsymbol{\sigma}^\dagger$	–	adjoint stress tensor
$\mathbf{c}^\dagger$	–	correlation adjoint

Inversion (cont.)

$\mathcal{T}$	–	correlation time shift
$\mathcal{A}$	–	amplitude measure
$\mathcal{S}_{nm}$	–	source power-spectral density
$s(\omega)$	–	spectral function for power-spectral density
S_{nm}	–	spatial dependence of power-spectral density
$\mathcal{X}$	–	joint misfit function
$\mathcal{X}_S$	–	source power-spectral density misfit function
$\mathcal{X}_M$	–	structural misfit function
H	–	Hamiltonian function
U	–	Hamiltonian potential energy
K	–	Hamiltonian kinetic energy
$:$	–	tensor contraction
$\otimes$	–	tensor product

Mathematical

∇_1	–	gradient on a spherical shell
$\mathbf{e}_r \wedge \nabla_1$	–	surface curl
$J_m(x)$	–	Bessel function
$T_m, T_m^{(1)}$	–	tensor field of vector harmonics
$P_l(x), P_l^m(x)$	–	Legendre functions
$Q_l^{(1),(2)}$	–	travelling wave form of Legendre function
$\mathbf{P}_l^m, \mathbf{B}_l^m, \mathbf{C}_l^m$	–	vector surface harmonics
Y_l^m	–	normalised surface harmonics on a sphere
$Ł$	–	$[l(l+1)]^{\frac{1}{2}}$

Bibliography

Abers G.A., 2000. Hydrated subducted crust at 100–250 km depth, *Earth Planet. Sci. Lett.*, **176**, 323–330.

Abers G.A., Plank T., Hacker B.R., 2003. The wet Nicaraguan slab, *Geophys. Res. Lett.*, **30**, 1098.

Afanasiev M.V., Boehm C., van Driel M., Krischer L., Rietmann M., May D.A., Knepley M.G., Fichtner A., 2019. Modular and flexible spectral-element waveform modelling in two and three dimensions, *Geophys. J. Int.*, **216**, 1675–1692.

Alterman Z., Lowenthal D., 1972. Computer generated seismograms, *Methods in Computational Physics*, **12**, 35–164, ed. Bolt B.A., Academic Press, New York.

Aki K., 1957. Space and time spectra of stationary stochastic waves with special reference to microtremors, *Bull. Earthq. Res. Inst.*, **35**, 415–456.

Aki K., Lee W.H.K., 1976. Determination of three-dimensional velocity anomalies under a seismic array using first *P* arrival times from local earthquakes – 1. A homogeneous initial model, *J. Geophys. Res.*, **81**, 4381–4399.

Aki K., Richards P.G.,2002. *Quantitative Seismology*, 2nd edition, University Science Books, Sausalito.

Aki K., Chouet B., 1975. Origin of coda waves: source, attenuation and scattering effects, *J. Geophys. Res.*, **80**, 3322–3342.

An M., 2012. A simple method for determining the spatial resolution of a general inverse problem, *Geophys. J. Int.*, **191**, 849–864.

Antolik M., Gu Yu-J., Ekström G., Dziewonski A.M., 2003. J362D28: a new joint model of compressional and shear velocity in the Earth's mantle, *Geophys. J. Int.*, **153**, 443–466.

Artemieva I., 2011. *The Lithosphere: An Interdisciplinary Approach*, Cambridge University Press, Cambridge.

Asten M.W., 2006. On bias and noise in passive seismic data from finite circular array data processed using SPAC methods, *Geophysics*, **71**, V153–V162.

Astiz L., Earle P., Shearer P., 1996. Global stacking of broadband seismograms. *Seism. Res. Lett.*, **67**(4), 8–18.

Backus G.E., 1962. Long-wave elastic anisotropy produced by horizontal layering, *J. Geophys. Res.*, **67**, 4427–4440.

Backus G.E., Gilbert F., 1968. The resolving power of gross Earth data, *Geophys. J. Roy. Astr. Soc.*, **16**, 169–205.

Baeten G., Ziolowski A., 1990. *The Vibroseis Source*, Elsevier, Amsterdam.

Bakulin A., Calvert R., 2006, The virtual source method: Theory and case study, *Geophysics*, **71**, SI139–SI150.

Bamberger A., Chavent G., Lailly P., 1977. Une application de la théorie du contrôle à un problème inverse sismique, *Ann. Geophys.*, **33**, 183–200.

Bataille K., Wu R.S., Flatté S.M., 1990. Inhomogeneities near the core–mantle boundary evidenced from scattered waves: A review, *Pure Appl. Geophys.*, **132**, 151–173.

Becker T.W., 2012. On recent seismic tomography for the western United States, *Geochem. Geophys. Geosys.*, **13**, doi:10.1029/2011GC003977.4.

BenMenahem A., Singh S.J., 1981. *Seismic Waves and Sources*, Springer Verlag, New York.

Bensen G.D., Ritzwoller M.H., Barmin M.P., Levshin A.L., Lin F., Moschetti M.P., Shapiro N.M., Yang Y., 2007. Processing seismic ambient noise data to obtain reliable broad-band surface wave dispersion measurements, *Geophys. J. Int.*, **169**, 1239–1260.

Berkhout A.J., 1983. *Seismic Migration: Imaging of Acoustic Energy by Wavefield Extrapolation*, Elsevier, Amsterdam.

Berkhout A.J., 1997. Pushing the limits of seismic imaging, Part I: Prestack migration in terms of double dynamic focusing, *Geophysics*, **62**, 937–954.

Bernauer M., Fichtner A., Igel H., 2014. Optimal observables for multi-parameter seismic tomography, *Geophys. J. Int.*, **198**, 1241–1254.

Betancourt M., 2017. A conceptual introduction to Hamiltonian Monte Carlo, *ArXiv*, arXiv:1701.02434 [stat.ME].

Bijwaard H., Spakman W., Engdahl E.R., 1998. Closing the gap between regional and global travel time tomography, *J. Geophys. Res.*, **103**, 30 055–30 078.

Biswas R., Sen M., 2017. 2D full-waveform inversion and uncertainty estimation using the reversible jump Hamiltonian Monte Carlo, *SEG 2017 Expanded Abstracts*, 1280–1285.

Blom N., Boehm C., Fichtner A., 2017. Synthetic inversions for density using seismic and gravity data, *Geophys. J. Int.*, **209**, 1204–1220.

Blom N., Gokhberg A., Fichtner A., 2020. Seismic waveform inversion of the Central and Eastern Mediterranean upper mantle, *Solid Earth*, **11**, 669–690.

Boehm C., Hanzich M., de la Puente J., Fichtner A., 2016. Wavefield compression for adjoint methods in full-waveform inversion, *Geophysics*, **81**, R385–R397.

Boehm C., Korta N., Vinard N., Balic I., Fichtner A., 2018. Time-domain spectral-element ultrasound waveform tomography using a stochastic quasi-Newton method, *Medical Imaging 2018, Ultrasonic Imaging and Tomography 10580*, doi:10.1117/12.2293299.

Bolton H., Masters G., 2001. Travel times of *P* and *S* from the global digital seismic networks: Implications for the relative variation of *P* and *S* velocity in the mantle, *J. Geophys. Res.*, **106**, 13 527–13 540.

Boore D.M., 1970. Love waves in nonuniform waveguides: Finite difference calculations, *J. Geophys. Res.*, **75**, 1512–1527.

Bottou L., 2010. Large-scale machine learning with stochastic gradient descent, *COMPSTAT 2010*, 177–186.

Boué P., Poli P., Campillo M., Pedersen H., Briand X., Roux P., 2013. Teleseismic correlations of ambient seismic noise for deep global imaging of the Earth, *Geophys. J. Int.*, **194**, 844–848.

Boué P., Poli P., Campillo M., Roux P., 2014. Reverberations, coda waves and ambient noise: Correlations at the global scale and retrieval of the deep phases, *Earth Planet. Sci. Lett.*, **391**, 137–145.

Bowden D.C., Tsai V.C., Lin F.-C., 2015. Site amplification, attenuation, and scattering from noise correlation amplitudes across a dense array in Long Beach, CA, *Geophys. Res. Lett.*, **42**, 1360–1367.

Bozdağ E., Trampert J., Tromp J., 2011. Misfit functions for full waveform inversion based on instantaneous phase and envelope measurements, *Geophys. J. Int.*, **185**, 845-870.

Bozdağ E., Peter D., Lefebvre M., Komatitsch D., Tromp J., Hill J., Podhorszki N., Pugmire D., 2016. Global adjoint tomography: first-generation model, *Geophys. J. Int.*, **207**, 1739–1766.

Brittan J., Jones I., 2019. FWI evolution – from a monolith to a toolkit, *The Leading Edge*, **38**, 178–184.

Brossier R., Operto S., Virieux J., 2009. Robust elastic frequency-domain full waveform inversion using the L_1 norm, *Geophys. Res. Lett.*, **36**, L20319.

Broyden C.G., 1970. The convergence of a class of double-rank minimization algorithms, *J. Inst. Math. Appl.*, **6**, 76–90.

Bui-Tanh T., Ghattas O., Martin J., Stadler G., 2003. A computational framework for infinite-dimensional Bayesian inverse problems part I: The linearized case, with application to global seismic inversion, *SIAM J. Sci. Comp.*, **35**, A2494–A2523.

Bungum H., Husebye E.S., 1971. Errors in time delay measurements, *Pure Appl. Geophys.*, **91**, 56–70.

Bunks C., Saleck F.M., Zaleski S., Chavent G., 1995. Multiscale seismic waveform inversion, *Geophysics*, **69**, 1457–1473.

Burgos G., Montagner J.-P., Beucler E., Capdeville Y., Mocquet A., Drilleau M., 2014. Oceanic lithosphere/asthenosphere boundary from surface wave dispersion data, *J. Geophys. Res. Solid Earth*, **119**, 1079–1093.

Byrd R., Chin G., Neveitt W., Nocedal J., 2011. On the use of stochastic Hessian information in optimization methods for machine learning, *SIAM J. Opt.*, **21**, 977–995.

Cagniard L., 1939. *Réflexion et Réfraction des Ondes Séismiques Progressives*, Gauthier-Villars, Paris.

Campillo M., Paul A., 2003. Long-range correlations in the diffuse seismic coda, *Science*, **299**, 547–549.

Cao S.-H., Kennett B.L.N., 1989. Reflection seismograms in a 3-D elastic model: an isochronal approach, *Geophys. J. Int.*, **99**, 63–80.

Capdeville Y., Gung Y., Romanowicz B., 2005. Towards global Earth tomography using the spectral element method: A technique based on source stacking *Geophys. J. Int.*, **162**, 541–554.

Capdeville Y., Marigo J.-J., 2013. A non-periodic two-scale asymptotic method to take account of rough topographies for 2-D elastic wave propagation, *Geophys. J. Int.*, **192**, 163–189.

Cerjan C., Kosloff D., Kosloff R., Reshef M., 1985. A nonreflecting boundary condition for discrete acoustic and elastic wave equations, *Geophysics*, **50**, 705–708.

Červený V., Popov M., Pšenčík I., 1982. Computation of wave fields in inhomogeneous media – Gaussian beam approach, *Geophys. J. Int.*, **70**, 109–128.

Chapman C.H., Drummond R., 1982. Body wave seismograms in inhomogeneous media using Maslov asymptotic theory, *Bull. Seism. Soc. Am.*, **72**, S277–S317.

Chapman C.H., 2004. *Fundamentals of Seismic Wave Propagation*, Cambridge University Press, Cambridge.

Chen M., Tromp J., Helmberger D.V., Kanamori H., 2007. Waveform modelling of the slab beneath Japan. *J. Geophys. Res.*. **112**, B02305.

Chen P., Zhao L., Jordan T.H., 2007. Full 3D tomography for the crustal structure of the Los Angeles region, *Bull. Seis. Soc. Am.*, **97**, 1094–1120.

Chen Y., Saygin E., 2020. Empirical Green's function retrieval using cross-correlation of noise correlations (C^2) *J. Geophys. Res. Solid Earth*, **125**, e2019JB01826.

Chernov L.A., 1960. *Wave Propagation in a Random Medium*, (Engl. trans. by R.A. Silverman), McGraw-Hill, New York.

Christie P.F., Hughes V., Kennett B.L.N., 1983. Velocity filtering of seismic reflection data, *First Break*, **1**(3), 9–24.

Claerbout J., 1968. Synthesis of a layered medium from its acoustic transmission response, *Geophysics*, **33**, 264–269.

Clayton R.W., Stolt R.H., 1981. A Born-WKBJ inversion method for acoustic reflection data, *Geophysics*, **46**, 1559–1567.

Clayton R.W., 2020. Imaging the subsurface with ambient noise autocorrelations, *Seismol. Res. Lett.*, **91**, 1–6.

Cleary J.R., Haddon R.A.W, 1972. Seismic wave scattering near the core-mantle boundary: A new interpretation of precursors to *PKP*, *Nature*, **240**, 549–551.

Cowles M.K., Carlin B.P., 1996. Markov chain Monte Carlo convergence diagnostics: A comparative review, *J. Am. Stat. Ass.*, **91**, 883–904.

Creutz, M., 1988. Global Monte Carlo algorithms for many-fermion systems, *Phys. Rev. D*, **38**, 1228-1238.

Cupillard P., Capdeville Y., 2018. Non-periodic homogenization of 3-D elastic media for the seismic wave equation, *Geophys. J. Int.*, **213**, 983–1001.

da Costa Filho C.A., Ravasi M., Curtis A., Meles G.A., 2014, Elastodynamic Green's function retrieval through single-sided Marchenko inverse scattering. *Phys. Rev. E*, **90**, 063201.

Dahlen F.A., Tromp J., 1998. *Theoretical Global Seismology*, Princeton University Press, Princeton.

Dahlen F.A., Hung S.-H., Nolet G., 2000. Fréchet kernels for finite-frequency traveltimes – I. Theory, *Geophys. J. Int.*, **141**, 157–174.

Dainty A.M., Toksöz M.N., Anderson K.R., Pines P.J., Nakamura Y., Latham G., 1974. Seismic scattering and shallow structure of the Moon in Oceanus Procellarum, *Moon*, **91**, 11–29.

Dainty A.M., 1990. Studies of coda using array and three-component processing. *Pure Appl. Geophys.*, **132**, 221–244.

Dainty A.M., Toksöz M.N., 1990. Array analysis of seismic scattering, *Bull. Seism. Soc. Am.*, **80**, 2242–2260.

Dalton C.A., Ekström, G., Dziewonski A.M., 2008. The global attenuation structure of the upper mantle, *J. Geophys. Res.*, **113**, doi:10.1029/2007JB005429.

Deal M.M., Nolet G., 1996. Nullspace shuttles, *Geophys. J. Int.*, **124**, 372–380.

Debayle E., Kennett B., Priestley K., 2005. Global azimuthal seismic anisotropy: the unique plate-motion deformation of Australia, *Nature*, **433**, 509–512.

de Hoop A.T., 1958. *Representation theorems for the displacement in an elastic solid and their application to elastodynamic diffraction theory*, D.Sc thesis, Technische Hogeschool Delft.

Doornbos D.J., Vlaar N.J., 1973. Regions of seismic wave scattering in the Earth's mantle and precursors to *PKP*, *Nature Phys. Sci.*, **243**, 58–61.

Doornbos D.J., 1974. Seismic wave scattering near caustics: observation of *PKKP* precursors, *Nature*, **274**, 352–353.

Doornbos D.J., 1976. Characteristics of lower mantle heterogeneities from scattered waves, *Geophys. J. R. Astr. Soc.*, **44**, 447–470.

Draganov D., Wapenaar K., Mulder W., Singer J., Verdel A., 2007. Retrieval of reflections from seismic background-noise measurements, *Geophys. Res. Lett.*, **34**, L04305.

Dziewonski A.M., Hales A.L., Lapwood E.R., 1975. Parametrically simple Earth models consistent with geophysical data, *Phys. Earth Planet. Int.*, **10**, 12–48.

Dziewonski A.M., Hager B.H., O'Connell R.J., 1977. Large-scale heterogeneities in the lower mantle, *J. Geophys. Res.*, **82**, 239–255.

Dziewoński A.M., Anderson D.L. 1981. Preliminary reference Earth model, *Phys. Earth Planet. Inter.*, **25**, 297–356.

Dziewoński A.M., Woodhouse J.H., 1987. Global images of the Earth's interior, *Science* **236**, 37–48.

Ekström G., Abers G.A., Webb S.C., 2009. Determination of surface-wave phase velocities across USArray from noise and Aki's spectral formulation, *Geophys. Res. Lett.*, **36**, L18301.

Ekström G., 2011. A global model of Love and Rayleigh surface wave dispersion and anisotropy, 25–250 s, *Geophys. J. Int.*, **187**, 1668–1686.

Engdahl E.R., van der Hilst R.D., Buland R., 1998. Global teleseismic earthquake relocation with improved travel times and procedures for depth determination, *Bull. Seism. Soc. Am.*, **88**, 722–743.

Ermert L., Villaseñor A., Fichtner A., 2016. Cross-correlation imaging of ambient noise sources. *Geophys. J. Int.*, **204**, 347–364.

Ermert L., Sager K., Afanasiev M., Boehm C., Fichtner A., 2017. Ambient seismic source inversion in a heterogeneous Earth: Theory and application to the Earth's Hum. *J. Geophys. Res. Solid Earth*, **122**, 9184–9207.

Ewing, W. M., Jardetsky, W. S. & Press, F., 1957. *Elastic Waves in Layered Media*, McGraw-Hill, New York.

Faccioli E., Maggio F., Paolucci R., Quarteroni A., 1997. 2D and 3D elastic wave propagation by a pseudospectral domain decomposition method, *J. Seism.*, **1**, 237–251.

Feng J., Yao H., Poli P., Fang L., Wu Y., Zhang P., 2017. Depth variations of 410 km and 660 km discontinuities in eastern North China Craton revealed by ambient noise interferometry, *Geophys. Res. Lett.*, **44**, 8328–8335.

Fichtner A., 2006. The adjoint method in seismology – I. Theory, *Phys. Earth Planet. Int.*, **157**, 86–104.

Fichtner A., Igel H., 2008. Efficient numerical surface wave propagation through the optimization of discrete crustal models - a technique based on non-linear dispersion curve matching (DCM), *Geophys. J. Int.*, **173**, 519–533.

Fichtner A., Kennett B.L.N., Igel H., Bunge H.-P., 2008. Theoretical background for continental and global scale full-waveform inversion in the time-frequency domain, *Geophys. J. Int.*, **175**, 665–685.

Fichtner A., Kennett B.L.N., Igel H., Bunge H.-P., 2009. Full seismic waveform tomography for upper-mantle structure in the Australasian region using adjoint methods, *Geophys. J. Int.*, **179**, 1703–1725.

Fichtner A., Kennett B.L.N., Igel H., Bunge H.-P., 2010. Full seismic waveform tomography for radially anisotropic structure: New insights into the past and present states of the Australasian upper mantle, *Earth. Planet. Sci. Lett.*, **290**, 270–280.

Fichtner A., 2011. *Full Seismogram Waveform Modelling and Inversion*, Springer-Verlag, Heidelberg.

Fichtner A., Trampert J., 2011a. Hessian kernels of seismic data functionals based upon adjoint techniques, *Geophys. J. Int.*, **185**, 775–798.

Fichtner A., Trampert J., 2011b. Resolution analysis in full waveform inversion, *Geophys. J. Int.*, **187**, 1604–1624.

Fichtner A., Trampert J., Cupillard P., Saygin E., Taymaz T., Capdeville Y., Villasenor A., 2013. Multi-scale full-waveform inversion, *Geophys. J. Int.*, **194**, 534–556.

Fichtner A., Kennett B.L.N., Trampert J., 2013. Separating intrinsic and apparent anisotropy, *Phys. Earth Planet. Int.*, **219**, 11–22.

Fichtner A., 2014. Source and processing effects on noise correlations, *Geophys. J. Int.*, **197**, 1527–1531.

Fichtner A., van Leeuwen T., 2015. Resolution analysis by random probing, *J. Geophys. Res.*, **120**, 5549–5573.

Fichtner A., Hanasoge S.M., 2017. Discrete wave equation upscaling, *Geophys. J. Int.*, **209**, 353–357.

Fichtner A., Stehly L., Ermert L., Boehm C., 2017. Generalized interferometry – I: theory for interstation correlations, *Geophys. J. Int.*, **208**, 603–638.

Fichtner A., van Herwaarden D.-P., Afanasiev M., Simute S., Krischer L., Cubuk-Sabuncu Y., Taymaz T., Colli L., Saygin E., Villasenor A., Trampert J., Cupillard P., Bunge H.-P., Igel H., 2018a. The Collaborative Seismic Earth Model: Generation I, *Geophys. Res. Lett.*, **45**, 4007–4016.

Fichtner A., Zunino A., Gebraad L., 2018b. Hamiltonian Monte Carlo solution of tomographic inverse problems, *Geophys. J. Int.*, **216**, 1344–1363.

Fichtner A., Tsai V., 2019. Theoretical foundations of noise interferometry, 109–143, in *Seismic Ambient Noise*, eds. Nakata N., Gualtieri L., Fichtner A., Cambridge University Press, Cambridge.

Fichtner, A., Zunino, A., 2019. Hamiltonian nullspace shuttles, *Geophys. Res. Lett.*, **46**, 644–651.

Fishwick S., Kennett B.L.N., Reading A.M., 2005. Contrasts in lithospheric structure within the Australian Craton, *Earth Planet. Sci. Lett.*, **231**, 163–176.

Flatté S.M., Wu R.S., 1988. Small scale structure in the lithosphere and asthenosphere deduced from arrival time and amplitude fluctuations at NORSAR, *J. Geophys. Res.*, **93**, 6601–6614.

Fletcher R., 1970. A new approach to variable metric algorithms, *Comp. J.*, **13**, 317–322.

Ford H.A., Fischer K.M., Abt D.L., Rychert C.A., Elkins-Tanton L.T., 2013. The lithosphere–asthenosphere boundary and cratonic lithospheric layering beneath Australia from *Sp* wave imaging, *Earth Planet. Sci. Lett.*, **300**, 299–310.

Forsyth D.W., Li A., 2005. Array-analysis of two-dimensional variations in surface wave phase velocity and azimuthal anisotropy in the presence of multi-pathing interference, in *Seismic Earth: Array Analysis of Broadband Seismograms*, eds. Levander A., Nolet G., AGU Geophysical Monograph, **157**, 81–98.

Frasier C.W., 1970. Discrete time solution of plane P–SV waves in a plane layered medium, *Geophysics*, **35**, 197–219.

French S.W., Romanowicz B.A., 2014. Whole-mantle radially anisotropic shear velocity structure from spectral-element waveform tomography, *Geophys. J. Int.*, **199**, 1303–1327.

Frenkel, A., Clayton, R.W., 1986. Finite difference simulation of seismic scattering: implications for the propagation of short-period seismic waves in the crust and models in crustal heterogeneity, *J. Geophys. Res.*, **91**, 6465–6489.

Friederich W., Wielandt E., Stange S., 1993. Multiple forward scattering of teleseismic surface waves: Comparison with an exact solution and Born single-scattering methods, *Geophys. J. Int.*, **112**, 264–275.

Friederich W., Wielandt E., Stange S., 1994. Non-plane geometries of seismic surface wavefields and their implications for regional surface-wave tomography, *Geophys. J. Int.*, **119**, 931–948.

Friederich W., Wielandt E., 1995. Interpretation of seismic surface waves in regional networks: Joint estimation of wavefield geometry and local phase velocity, *Geophys. J. Int.*, **120**, 731–744.

Fuchs, K., 1968. Das Reflexions- und Transmissionsvermögen eines geschichteten Mediums mit belieber Tiefen-Verteilung der elastischen Moduln und der Dichte für schragen Einfall Ebener Wellen, *Z. Geophys.*, **34**, 389–411.

Fuchs K., Müller G., 1971. Computation of synthetic seismograms with the reflectivity method and comparison with observations, *Geophys. J. R. Astr. Soc.*, **23**, 417–433.

Fukao Y., Nishida K., Kobayashi N., 2010. Sea floor topography, ocean infragravity waves, and background Love and Rayleigh waves, *J. Geophys. Res.*, **115**, B04302.

Furumura T., Kennett B.L.N, Furumura M., 1998. Synthetic seismograms for a laterally heterogeneous whole earth models by the pseudospectral method, *Geophys. J. Int.*, **135**, 845–860.

Furumura T., Kennett, B.L.N., 2005. Subduction zone guided waves and the heterogeneity structure of the subducted plate – intensity anomalies in northern Japan, *J. Geophys. Res.*, **110**(B10), B10302.

Furumura T., Kennett B.L.N., 2008. A scattering waveguide in the heterogeneous subducting plate, in *Scattering of Short-Period Seismic Waves in Earth Heterogeneity*, eds. Sato H., Fehler M., *Advances in Geophysics*, **50**, 195–217.

Furumura T., Kennett B.L.N., Padhy S., 2016. Enhanced waveguide effect for deep-focus earthquakes in the subducting Pacific slab produced by a metastable olivine wedge, *J. Geophys. Res. Solid Earth*, **121**, 6779–6796.

Gal M., Reading A.M., Ellingsen S.P., Koper K.D., Burlacu R., 2017. Full wavefield decomposition of high-frequency secondary microseisms reveals distinct arrival azimuths for Rayleigh and Love waves, *J. Geophys. Res. Solid Earth*, **122**, 4660–4675.

Galetti E., Curtis A., 2012. Generalised receiver functions and seismic interferometry, *Tectonophysics*, **532**, 1–26.

Garth T., Rietbrock A., 2014. Downdip velocity changes in subducted oceanic crust beneath Northern Japan : insights from guided waves, *Geophys. J. Int.*, **198**, 1342–1358.

Gauthier O., Virieux J., Tarantola A., 1986. Two-dimensional nonlinear inversion of seismic waveforms: numerical results, *Geophysics*, **51**, 1387–1403.

Gebraad L., Boehm C., Fichtner A., 2020. Bayesian elastic full-waveform inversion using Hamiltonian Monte Carlo, *J. Geophys. Res.*, **125**, e2019JB018428.

Gee L., Jordan T.H., 1992. Generalised seismological data functionals, *Geophys. J. Int.*, **111**, 363–390.

Geli L., Bard P-Y., Jullien B., 1988. The effect of topography on earthquake ground motion: A review and new results, *Bull. Seism. Soc. Am.*, **78**, 42–63.

Giardini D., and 62 others, 2020. The seismicity of Mars, *Nature Geoscience*, **13**, 205–212.

Gilbert F., 1976. The representation of seismic displacements in terms of travelling waves, *Geophys. J. R. Astr. Soc.*, **44**, 275–280.

Gilbert F., Helmberger D.V., 1972. Generalized ray theory for a layered sphere, *Geophys. J. R. Astr. Soc.*, **27**, 57–80.

Gokhberg A., Fichtner A., 2016. Full-waveform inversion on heterogeneous HPC systems, *Comp. Geosci.*, **89**, 260–268.

Goldfarb D., 1970. A family of variable metric updates derived by variational means, *Math. Comp.*, **24**, 23–26.

Gorbatov A., Kennett B.L.N., Saygin E., 2013. Crustal properties from seismic station autocorrelograms, *Geophys. J. Int.*, **192**, 861–870.

Grand S.P., van der Hilst R.D., Widiyantoro S., 1997. Global seismic tomography: a snapshot of convection in the Earth, *Geology Today*, **7**,(4) 1–7.

Grand S.P., 2002. Mantle shear-wave tomography and the fate of subducted slabs, *Phil. Trans. R. Soc. Lond.*, A**360**, 2475–2491.

Gregersen S., 1978. Possible mode conversions between Love and Rayleigh waves at a continental margin, *Geophys. J. R. Astr. Soc.*, **54**, 121–127.

Gregersen S., Alsop L.E., 1974. Amplitudes of horizontally refracted Love waves, *Bull. Seism. Soc. Am.*, **64**, 535–554.

Griewank A., Walther A., 2000. An implementation of checkpointing for the reverse or adjoint mode of computational differentiation, *Trans. Math. Software*, **26**, 19–45.

Gudmundsson O., Kennett B.L.N., Goody A., 1994. Broadband observations of upper mantle seismic phases in northern Australia and the attenuation structure in the upper mantle, *Phys. Earth Planet. Inter.*, **84**, 207–236.

Gutenberg B., 1913. Über die Konstitution des Erdinnern, erschlossen aus Erdbebenbeobachtungen, *Physikalische Zeitschrift*, **14**, 1217–1218.

Halko N., Martinsson P.G., Tropp J.A., 2011. Finding structure with randomness: Probabilistic algorithms for constructing approximate matrix decompositions, *SIAM Review*, **53**, 217–288.

Halliday D., Curtis A., Robertsson J., van Manen D., 2007. Interferometric surface-wave isolation and removal, *Geophysics*, **72**, A69–A73.

Halliday D., Curtis A., 2008. Seismic interferometry, surface waves and source distribution, *Geophys. J. Int.*, **175**, 1067–1087.

Hanasoge S.M., Branicki M., 2013. Interpreting cross-correlations of one-bit filtered noise, *Geophys. J. Int.*, **195**, 1811–1830.

Hansen S.M., Dueker K., Schmandt B., 2015. Thermal classification of lithospheric discontinuities beneath USArray, *Earth Planet. Sci. Lett.*, **431**, 36–47

Harding A.J., 1985. Slowness-time mapping of near offset seismic reflection data, *Geophys. J. R. astr. Soc.*, **80**, 463–492.

Hasselmann K. 1963. A statistical analysis of the generation of microseisms, *Rev. Geophys.*, **1**, 177–210.

Hedlin M.A.H., Shearer P., 2000. An analysis of large scale variations in small-scale mantle heterogeneity using Global Seismographic Network recordings of precursors to *PKP*, *J. Geophys. Res.*, **105**, 13 655–13 673.

Helmberger D. V., 1968. The crust-mantle transition in the Bering Sea, *Bull. Seism. Soc. Am.*, **58**, 179–214.

Helmberger D. V. , Wiggins, R. A., 1971. Upper mantle structure of the midwestern United States, *J. Geophys. Res.*, **76**, 3229–3245.

Hirschmann M.M., 2010. Partial melt in the oceanic low velocity zone, *Phys. Earth Planet. Inter.*, *179*, 60–71.

Hong T.-K., Kennett B.L.N., 2002. On a wavelet-based method for the numerical simulation of wave propagation, *J. Comput. Phys.*, **183**, 577–622.

Hong T.-K., Kennett B.L.N., 2003. Scattering attenuation of 2D elastic waves: theory and numerical modeling using a wavelet-based method, *Bull. Seism. Soc. Am.*, **93**, 922–938.

Hong T.-K., Kennett B.L.N., Wu R.-S., 2004. Effects of the density perturbation in scattering, *Geophys. Res. Lett.*, **31**(13), L13602.

Hong T.-K., Wu R.-S., Kennett B.L.N., 2005. Stochastic features of scattering. *Phys. Earth Planet. Inter.*, **148**, 131–148.

Hopper E., Ford H.A., Fischer K.M., Lekić, V., Fouch, M.J., 2014. The lithosphere-asthenosphere boundary and the tectonic and magmatic history of the northwestern United States, *Earth Planet. Sci. Lett.*, **402**, 69–81.

Hron F., 1972. Numerical methods of ray generation in multilayered media, *Methods in Computational Physics*, **12**, 1–34, ed. Bolt B.A., Academic Press, New York.

Huang H.-H., Lin F.-C., Tsai V.C., Koper K.D., 2015. High-resolution probing of inner core structure with seismic interferometry. *Geophys. Res. Lett.*, **42**, 10622–10630.

Hudson J.A., 1968. The scattering of elastic waves by granular media, *Quart. J. Mech. Appl. Math.*, **21**, 487–502.

Hudson J.A., Douglas A., 1975. Rayleigh wave spectra and group velocity minima, and the resonance of P waves in layered structures, *Geophys. J. R. Astr. Soc.*, **42**, 175–188.

Hudson, J.A., 1982. Use of stochastic models in seismology, *Geophys. J. Int.*, **82**, 649–657.

Hutchinson M.F., 1990. A stochastic estimator of the trace of the influence matrix for Laplacian smoothing splines, *Comm. Stat. Sim.*, **19**, 433–450.

Igel H., Djikpéssé H., Tarantola A., 1996. Waveform inversion of marine reflection seismograms for *P* impedance and Poisson's ratio, *Geophys. J. Int.*, **124**, 363–371.

Igel H., 2016. *Computational Seismology: A Practical Introduction*, Oxford University Press, Oxford.

Ishii M., Tromp J., 2001. Normal mode and free-air gravity constraints on lateral variations in velocity and density of Earth's mantle, *Science*, **285**, 1231–1236.

Ito Y., Shiomi K., Nakajima J., Hino R., 2012. Autocorrelation analysis of ambient noise in northeastern Japan subduction zone, *Tectonophysics*, **572**, 38–46.

Jones I.F., Davison I., 2014. Seismic imaging in and around salt bodies, *Interpretation*, **2**(4), SL1–SL20.

Kárason H., van der Hilst R.D., 2000. Constraints on mantle convection from seismic tomography, in, *The History and Dynamics of Global Plate Motion*, AGU Geophysical Monograph, **121**, 277–288.

Karato S.-I., Karki B.B., 2001. Origin of lateral variation of seismic wave velocities and density in the deep mantle, *J. Geophys. Res.*, **106**, 21 771–21 783.

Kawakatsu H., Watada S., 2007. Seismic evidence for deep-water transportation in the mantle, *Science*, **316**, 1468–1471.

Kawakatsu H., Kumar P., Takei Y., Shinohara M., Kanazawa T., Araki E., Suyehiro K., 2009. Seismic evidence for sharp lithosphere-asthenosphere boundaries of oceanic plates, *Science*, **324**, 499–502.

Kawakatsu H., Utada S., 2017. Seismic and electrical signatures of the lithosphere-asthenosphere system of the normal oceanic mantle, *Ann. Rev. Earth Planet. Sci.*, **45**, 139–167.

Kennett B.L.N., 1972. Seismic waves in laterally inhomogeneous media, *Geophys. J. R. Astr. Soc.*, **27**, 301–325.

Kennett B.L.N., 1975. The effects of attenuation on seismograms, *Bull. Seism. Soc. Am.*, **65**, 1643–1651.

Kennett, B.L.N., Kerry, N. J. & Woodhouse, J. H., 1978. Symmetries in the reflection and transmission of elastic waves, *Geophys. J. R. Astr. Soc.* , **52**, 215–229.

Kennett B.L.N., 1979. The suppression of surface multiples on seismic records, *Geophys. Prospect.*, **27**, 584–600.

Kennett B.L.N., 1983. *Seismic Wave Propagation in Stratified Media*, Cambridge University Press. Second edition 2009, ANU Press.

Kennett B.L.N., Harding A.J., 1984. Guided low-frequency noise from air-gun sources, *Geophys. Prospect.*, **32**, 690–705.

Kennett B.L.N., 1984a. Reflection operator method for elastic waves I – Irregular interfaces and regions, *Wave Motion*, **6**, 407–418.

Kennett B.L.N., 1984b. Reflection operator method for elastic waves II – Composite regions and source problems, *Wave Motion*, **6**, 419–429.

Kennett B.L.N., 1984c. Guided waves in laterally varying media, I: Theoretical development, *Geophys. J. R. Astr. Soc.*, **79**, 235–255.

Kennett B.L.N., 1986. Wavenumber and wavetype coupling in laterally heterogeneous media, *Geophys. J. R. Astr. Soc.*, **87**, 313–331.

Kennett B.L.N., Sambridge M.S., Williamson P.R., 1988. Subspace methods for large scale inverse problems involving multiple parameter classes, *Geophys. J. Int.*, **94**, 237–247.

Kennett B.L.N., 1990. Guided wave attenuation in laterally varying media, *Geophys. J. Int.*, **100**, 415–422.

Kennett, B.L.N., Nolet, G., 1990. The interaction of the *S*-wavefield with upper mantle heterogeneity, *Geophys. J. Int.*, **101**, 751–762.

Kennett B.L.N, Koketsu K., Haines A.J, 1990. Propagation invariants, reflection and transmission in anisotropic, laterally varying media, *Geophys. J. Int.*, **103**, 95–101.

Kennett B.L.N., 1991. The removal of free surface interactions from three-component seismograms, *Geophys. J. Int.*, **104**, 153–163.

Kennett B.L.N., 1993. A two-layer stacking procedure to enhance converted waves, *Geophysics*, **58**, 997–1001.

Kennett B.L.N., Engdahl E.R., Buland R., 1995. Constraints on seismic velocities in the Earth from travel times, *Geophys. J. Int.*, **122**, 108–124.

Kennett B.L.N., 1996. How does the shear-wave structure of the seabed affect the seismic wavefield?, *Geophys. J. Int.*, **124**, 341–348.

Kennett B.L.N., 1998. Guided waves in 3-dimensional structures, *Geophys. J. Int.*, **133**, 159–174.

Kennett B.L.N., Widiyantoro S., van der Hilst R.D., 1998. Joint seismic tomography for bulk-sound and shear wavespeed in the Earth's mantle, *J. Geophys. Res.*, **103**, 12 469–12 493.

Kennett B.L.N., 2001. *The Seismic Wavefield I: Introduction and Theoretical Development*, Cambridge University Press, Cambridge.

Kennett B.L.N., 2002. *The Seismic Wavefield II: Interpretation of Seismograms on Regional and Global Scales*, Cambridge University Press, Cambridge.

Kennett B.L.N., Gorbatov A., 2004. Seismic heterogeneity in the mantle – strong shear wave signature of slabs, *Phys. Earth. Planet. Inter.*, **146**, 88–100.

Kennett B.L.N., Bunge H.-P., 2008. *Geophysical Continua* Cambridge University Press, Cambridge.

Kennett B.L.N., Furumura T., 2008. Stochastic waveguide in the Lithosphere: Indonesian subduction zone to Australian Craton, *Geophys. J. Int.*, **172**, 363–382.

Kennett B.L.N., Fichtner A., 2012. A unified concept for the comparison of seismograms using transfer functions, *Geophys. J. Int.*, **191**, 1403–1416.

Kennett B.L.N. & Furumura, T., 2013. High-frequency Po/So guided waves in the oceanic lithosphere: I – long-distance propagation, *Geophys. J. Int.*, **195**, 1862–1877.

Kennett B.L.N., Fichtner A., Fishwick S., Yoshizawa K., 2013. Australian Seismological Reference Model (AuSREM): mantle component, *Geophys. J. Int.*, **192**, 871–887.

Kennett B.L.N., Furumura T., Zhao Y., 2014. High-frequency *Po/So* guided waves in the oceanic lithosphere: II – Heterogeneity and attenuation, *Geophys. J. Int.*, *199*, 614–630.

Kennett B.L.N., 2015. Lithosphere–asthenosphere *P*-wave reflectivity across Australia, *Earth Planet. Sci. Lett.*, **431**, 225–235.

Kennett B.L.N., Stipčević J., Gorbatov A., 2015a. Spiral arm seismic arrays, *Bull. Seism. Soc. Am.*, **105**, 2109–2116.

Kennett B.L.N., Saygin E., Salmon M., 2015b. Stacking autocorrelograms to map Moho depth with high spatial resolution in southeastern Australia, *Geophys. Res. Lett.*, **42**, 7490–7497.

Kennett B.L.N., Furumura T., 2015. Towards the reconciliation of seismological and petrological perspectives on oceanic lithosphere heterogeneity, *Geochem. Geophys. Geosyst.*, **16**, 3129–3141.

Kennett B.L.N., Furumura T., 2016. Multi-scale seismic heterogeneity in the continental lithosphere, *Geochem. Geophys. Geosys.*, **17**, 791–809.

Kennett B.L.N., Yoshizawa K., Furumura T. 2017. Interactions of multi-scale heterogeneity in the lithosphere: Australia, *Tectonophysics*, **717**, 193–213.

Kennett B.L.N., Pham T.-S., 2018a. The nature of the seismic correlation wavefield: Late coda correlations, *Proc. R. Soc. Lond. A*, **474**, 20180082, doi:10.1098/rspa.2018.0082

Kennett B.L.N., Pham T.-S., 2018b. Evolution of the seismic correlation wavefield for event coda, *Phys. Earth. Planet. Inter.*, **282**, 100–109.

Kennett B.L.N., Sippl C., 2018. Lithospheric discontinuities in central Australia, *Tectonophysics*, **744**, 10–22.

Kennett B.L.N., Furumura T., 2019. Significant *P* wave conversions from upgoing *S* waves generated by very deep earthquakes around Japan, *Prog. Earth Planet. Sci.*, **6**:49.

Kent G.H., Harding A.J., Orcutt J.A., 1993. Distribution of magma beneath the East Pacific Rise between the Clipperton Transform and the $9^\circ 17'$N Deval from forward modeling of common depth point data, *J. Geophys. Res.*, **98**, 13 945–13 969.

Kim N.W., Seriff A.J., 1992. Marine PSSP reflections with a bottom velocity transition zone, *Geophysics*, **57**, 161–170.

Kimman W., Trampert J., 2010. Approximations in seismic interferometry and their effects on surface waves, *Geophys. J. Int.*, **182**, 461–476.

Kind R., Kosarev G.L., Petersen N.V., 1995. Receiver functions at the stations of the German Regional Seismic Network (GRSN), *Geophys. J. Int.*, **121**, 191–202.

King D.W., Haddon R.A.W., Cleary, J. R., 1974. Array analysis of precursors to *PKIKP* in the distance range 128° to 142°, *Geophys. J. R. Astr. Soc.*, **37**, 157–173.

King S.D., Anderson D.L., 1998. Edge-driven convection, *Earth Planet. Sci. Lett.*, **160**, 289–296.

Knopoff L., 1972. Observation and inversion of surface-wave dispersion, *Tectonophysics*, **13**, 497–519.

Koketsu K., Kennett B.L.N., Takenaka H., 1991. 2-D reflectivity method and synthetic seismograms for irregularly layered structures – II. Invariant embedding approach, *Geophys. J. Int.*, **105**, 119–130.

Komatitsch D., Tromp J., 2002a. Spectral-element simulations of global seismic wave propagation – I. Validation, *Geophys. J. Int.*, **149**, 390–412.

Komatitsch D., Tromp J., 2002b. Spectral-element simulations of global seismic wave propagation – II. Three-dimensional models, oceans, rotation and self-gravitation, *Geophys. J. Int.*, **150**, 303–318.

Koper K.D., Franks J.M., Dombrovskaya M., 2004. Evidence for small-scale heterogeneity in Earth's inner core from a global study of PKiKP coda waves. *Earth Planet. Sci. Lett.*, **228**, 227–241.

Korn M., 1988. P-wave coda analysis of short-period array data and the scattering and absorptive properties of the lithosphere. *Geophys. J. Int.*, **93**, 437–449.

Krischer L., Fichtner A., Žukauskaitė S., Igel H., 2015. Large-scale seismic inversion framework, *Seis. Res. Lett.*, **86**, 1198–1207.

Krischer L., Igel H., Fichtner A., 2018. Automated large-scale full seismic waveform inversion for North America and the North Atlantic, *J. Geophys. Res.*, **123**, 5902–5928.

Kunetz A., d'Erceville E., 1962. Sur certaines propriétés d'une onde acoustique plane de compression dans une milieu stratifieé, *Ann. de Geophys.*, **18**, 351–359.

Kuo C., Romanowicz B., 2002. On the resolution of density anomalies in the Earth's mantle using spectral fitting of normal-mode data, *Geophys. J. Int.*, **150**, 162–179.

Lamb H., 1904. On the propagation of tremors over the surface of an elastic solid, *Phil. Trans. R. Soc. Lond.*, **203A**, 1–42.

Lapwood E. R., 1948. The disturbance due to a line source in a semi-infinite elastic medium, *Phil. Trans. R. Soc. Lond.*, **242A**, 63–100.

Lapwood E.R., Usami T., 1981. *Free Oscillations of the Earth*, Cambridge University Press, Cambridge.

Larmat C., Montagner J.-P., Fink M., Capdeville Y., Tourin A., Clévédé E., 2006. Time-reversal imaging of seismic sources and application to the great Sumatra earthquake, *Geophys. Res. Lett.*, **33**, doi:10.1029/2006GL026336.

Laske G., Masters G., 1996. Constraints on global phase velocity maps from long-period polarization data, *J. Geophys. Res.*, **101**, 16 059–16 075.

Laske G., Masters G., Ma Z., Pasyanos M., 2013. Update on CRUST1.0 – A 1-degree global model of Earth's crust, *Geophys. Res. Abstracts*, **15**, EGU2013-2658.

Lebedev S., Schaeffer, A. J., 2013. Global shear speed structure of the upper mantle and transition zone, *Geophys. J. Int.*, **194**, 417–449.

Lehmann I., 1936. P′, *Publ. Bureau Central Seism. Int. Série A*, **14**, 87–115.

Leng K., Nissen-Meyer T., van Driel M., Hosseini K., Al-Attar D., 2019. AxiSEM3D: broadband seismic wavefields in 3-D global Earth models with undulating discontinuities, *Geophys. J. Int.*, **217**, 2125–2146.

Levshin A.L., 1985. Effect of lateral inhomogeneities on surface wave amplitude measurements, *Ann. Geophys.*, **3**, 511–518.

Levshin A.L., Ritzwoller M.H., 2001. Automated detection, extraction, and measurement of regional surface waves, *Pure Appl. Geophys.*, **158**, 1531–1545.

Li X.D., Romanowicz B., 1995. Comparison of global waveform inversions with and without considering cross-branch modal coupling, *Geophys. J. Int.*, **121**, 695–709.

Li L., Boué P., Campillo M., 2020. Observation and explanation of spurious seismic signals emerging in teleseismic noise correlations, *Solid Earth*, **11**, 173–184.

Lin F.-C., Moschetti M.P., Ritzwoller M.H., 2008. Surface wave tomography of the western United States from ambient seismic noise: Rayleigh and Love wave phase velocity maps, *Geophys. J. Int.*, **173**, 281–298.

Lin F.-C., Tsai V.C., Schmandt B., Duputel Z., Zhan Z., 2013. Extracting seismic core phases with array interferometry, *Geophys. Res. Lett.*, **40**, 1049–1053.

Lindsey N.J., Martin E.R., Dreger D.S., Freifeld B., Cole S., James S.R., Biondi B., Ajo-Franklin J., 2017. Fiber-optic network observations of earthquake wavefields, *Geophys. Res. Lett.*, **44**, 11792–11799.

Liu Q., Polet J., Komatitsch D., Tromp, J., 2004. Spectral-element moment tensor inversion for earthquakes in Southern California, *Bull. Seis. Soc. Am.*, **94**, 1748–1761.

Lobkis O.I., Weaver R.L., 2001. On the emergence of the Green's function in the correlations of a diffuse field, *J. Acoust. Soc. Am.*, **110**, 3011–3017.

Lognonné P., and 108 others, 2020. Constraints on the shallow elastic and anelastic structure of Mars from InSight seismic data, *Nature Geoscience*, **13**, 213–220.

Lomas A., Curtis A., 2019. An introduction to Marchenko methods for imaging, *Geophysics*, **84**, F35–F45.

Longuet-Higgins M.S., 1950. A theory of the origin of microseisms, *Phil. Trans. R. Soc. Lond. A*, **243**, 1–35.

Luo Y., Schuster G.T., 1991. Wave-equation traveltime inversion, *Geophysics*, **56**, 645–653.

Love A.E.H., 1911. *Some Problems of Geodynamics*, Cambridge University Press.

Lysmer J., Drake L., 1972. A finite element method for seismology, *Methods in Computational Physics*, **11**, 181–216, ed. Bolt B.A., Academic Press, New York.

Lythgoe K.H., Deuss A., Rudge J.F., Neufeld J.A., 2014. Earth's inner core: Innermost inner core or hemispherical variations?, *Earth Planet. Sci. Lett.*, **385**, 181–189.

Ma S., Beroza G.C., 2012. Ambient-field Green's functions from asynchronous seismic observations, *Geophys. Res. Lett.*, **39**, L06301.

Maggi A., Tape C., Chen M., Chao D., Tromp J., 2009. An automated time-window selection algorithm for seismic tomography, *Geophys. J. Int.*, **178**, 257–281.

Malcolm A.E., Scales J., van Tiggelen, B.A., 2004. Extracting the Green function from diffuse, equipartitioned waves, *Phys. Rev. E*, **70**, doi:10.1103/PhysRevE.70.015601.

Malcolm A.E., Trampert J., 2011. Tomographic errors from wave front healing: more than just a fast bias, *Geophys. J. Int.*, **185**, 385–402.

Mancinelli N.J., Shearer P.M., 2013. Reconciling discrepancies among estimates of small-scale mantle heterogeneity from *PKP* precursors. *Geophys. J. Int.* **195**, 1721–1729.

Margerin L., 2004. Introduction to radiative transfer of seismic waves, in *Seismic Earth: Array Analysis of Broadband Seismograms*, eds. Levander A., Nolet G., AGU Geophysical Monograph, **157**, 229–252.

Marson-Pidgeon K., Kennett B.L.N., 2000. Flexible computation of teleseismic synthetics for source and structural studies, *Geophys. J. Int.*, **125**, 229–248.

Martin S., Rietbrock A., Haberland C., Asch G., 2003. Guided waves propagating in subducted oceanic crust, *J. Geophys. Res. Solid Earth*, **108**, 2536.

Martin S., Rietbrock A., 2006. Guided waves at subduction zones: dependencies on slab geometry, receiver locations and earthquake sources. *Geophys. J. Int.*, **167**, 693–704.

Masters G., Laske G., Bolton H., Dziewonski, A., 2000. The relative behaviour of shear velocity, bulk sound speed, and compressional velocity in the mantle: implications for chemical and thermal structure, in *Earth's Deep Interior: Mineral Physics and Tomography from the Atomic to the Global Scale*. AGU Geophysical Monograph **117**, 63–87.

Matharu G., Sacchi M., 2019. A subsampled truncated Newton method for multiparameter full-waveform inversion, *Geophysics*, **84**, R333–R340.

Maupin V., 1988. Surface waves across 2D structure: a method based on coupled local modes, *Geophys. J. Int.*, **93**, 173–185.

Maupin V., 1992. Modelling of laterally trapped surface waves with application to Rayleigh waves on the Hawaiian swell, *Geophys. J. Int.*, **110**, 553–570.

Maupin V., 2001. A multiple scattering scheme for modelling surface wave propagation in isotropic and anisotropic three-dimensional structures, *Geophys. J. Int.*, **146**, 332–348.

Maupin V., 2007. Introduction to mode coupling methods for surface waves, *Advances in Geophysics*, **48**, 127–155.

Megnin C., Romanowicz B., 2000. The three-dimensional shear velocity structure of the mantle from the inversion of body, surface and higher-mode waveforms, *Geophys. J. Int.*, **143**, 709–728.

Meier U., Curtis A., Trampert J., 2007. Global crustal thickness from neural network inversion of surface wave data, *Geophys. J. Int.*, **169**, 706–722.

Métivier L., Brossier R., Virieux J., Operto S., 2013. Full-waveform inversion and the truncated Newton method, *SIAM J. Sci. Comp.*, **35**, B401–B437.

Métivier L., Brossier R., Mérigot Q., Oudet E., Virieux J., 2016. Measuring the misfit between seismograms using an optimal transport distance: application to full waveform inversion, *Geophys. J. Int.*, **205**, 345–377.

Miller M.S., Kennett B.L.N., Gorbatov A., 2006. Morphology of the distorted subducted Pacific slab beneath the Hokkaido corner, *Phys. Earth Planet. Inter.*, **156**, 1–11.

Miller M.S., Niu F. 2008, Bulldozing the core-mantle boundary: Localized seismic scatterers beneath the Caribbean Sea. *Phys. Earth Planet. Inter.*, **170**, 89–94.

Moczo P., Kristek J., Halada L., 2000. 3D fourth-order staggered-grid finite-difference schemes: stability and grid dispersion, *Bull. Seis. Soc. Am.*, **90**, 587–603.

Moczo P., Kristek J., Galis M., 2014. *The Finite-Difference Modelling of Earthquake Motions: Waves and Ruptures*, Cambridge University Press, Cambridge.

Modrak R., Tromp J., 2016. Seismic waveform inversion best practices: regional, global and exploration test cases, *Geophys. J. Int.*, **206**, 1864–1889.

Modrak R., Borisov D., Lefebvre M., Tromp J., 2018. SeisFlows – flexible waveform inversion software, *Comp. Geosci.*, **115**, 88–95.

Montagner J.-P., Kennett B.L.N., 1996. How to reconcile body-wave and normal-mode reference Earth models?, *Geophys. J. Int.*, **125**, 229–248.

Montelli R., Nolet G., Dahlen F.A., Masters G., Engdahl E.R., Hung S., 2003. Finite-frequency tomography reveals a variety of plumes in the mantle, *Science* **303**, 338–343.

Mosegaard K., Tarantola A., 1995. Monte Carlo sampling of solutions to inverse problems, *J. Geophys. Res.*, **100**, 12431–12447.

Mosegaard K., 2012. Limits to nonlinear inversion, *Appl. Parallel Sci. Comp*, 11–21.

Morozov, I.B., Morozova, E.A. & Smithson, S.B., 1998. On the nature of the teleseismic *Pn* phase observed in the recordings from the ultra-long-range profile "Quartz", *Bull. Seism. Soc. Am.*, **88**, 62–73.

Müller G., 1970. Exact ray theory and its application to the reflection of elastic waves from vertically inhomogeneous media, *Geophys. J. R. Astr. Soc.*, **21**, 261–283.

Nakata N., Chang J.P., Lawrence J.F., Boué P., 2015. Body-wave extraction and tomography at Long Beach, California, with ambient-noise interferometry, *J. Geophys. Res. Solid Earth*, **120**, 1159–1173.

Nakata N., Gualtieri L., Fichtner A. (eds.), 2019. *Seismic Ambient Noise*, Cambridge University Press, Cambridge.

Nakata N., Nishida K., 2019. Body wave exploration, 239–266, in *Seismic Ambient Noise*, eds. Nakata N., Gualtieri L., Fichtner A., Cambridge University Press, Cambridge.

Neal, R. M., 2011. MCMC using Hamiltonian dynamics, *Handbook of Markov chain Monte Carlo*, 113–162.

Nielsen L., Thybo H., Levander A., Solodilov L.N., 2003. Origin of upper-mantle seismic scattering – Evidence from Russian peaceful nuclear explosion data. *Geophys. J. Int.*, **154**, 196–204.

Nishida K., Fukao Y., 2007. Source distribution of Earth's background free oscillations. *J. Geophys. Res.*, **112**, doi: 10.1029/2006JB004720.

Nishida K., Montagner J.-P., Kawakatsu H., 2009. Global surface wave tomography using seismic hum, *Science*, **326**, 112.

Nishida K., 2013. Global propagation of body waves revealed by cross-correlation analysis of seismic hum, *Geophys. Res. Lett.*, **40**, 1691–1696.

Nissen-Meyer T., Fournier A., Dahlen F.A., 2007. A two-dimensional spectral-element method for computing spherical-Earth seismograms – I. Moment-tensor source, *Geophys. J. Int.*, **168**, 1067–1092.

Nissen-Meyer T., van Driel M., Stähler S.C., Hosseini K., Hempel S., Auer L., Colombi A., Fournier A., 2014. AxiSEM: broadband 3-D seismic wavefields in axisymmetric media, *Solid Earth*, **5**, 425–445.

Nocedal J., 1980. Updating quasi-Newton matrices with limited storage, *Math. Comp.*, **35**, 773–782.

Nocedal J., Wright S.J., 1999. *Numerical Optimization*, Springer, New York.

Nolet G., Kennett B.L.N., 1978. Normal mode representations of multiple ray reflections in a spherical Earth, *Geophys. J. R. Astr. Soc.*, **53**, 219–226.

Nolet G., 1990. Partitioned waveform inversion and two-dimensional structure under the network of autonomously recording seismographs, *J. Geophys. Res*, **95**, 8499–8512.

Nolet G., Grand S., Kennett B.L.N., 1994. Seismic heterogeneity in the upper mantle, *J. Geophys. Res.*, **99**, 23 753–23 766.

Nolet G., 2006. *A Breviary of Seismic Tomography*, Cambridge University Press, Cambridge.

Nussenveig H.M., 1965. High frequency scattering by an impenetrable sphere, *Ann. Phys.*, **34**, 23–95.

Oikawa M., Kaneda K., Nishizawa A., 2010. Seismic structures of the 154–160 Ma oceanic crust and uppermost mantle in the Northwest Pacific Basin, *Earth Planets Space*, **62**(4), e13–e16.

Oldham R.D., 1906. The constitution of the Earth as revealed by earthquakes, *Quart. J. Geol. Soc. London*, **62**, 456–476.

Olver F.W.J., 1974. *Asymptotics and Special Functions*, Academic Press, New York.

Paitz P., Sager K., Fichtner A., 2019. Rotation and strain ambient noise interferometry, *Geophys. J. Int.*, **216**, 1938–1952.

Panning M., Romanowicz B., 2006. A three-dimensional radially anisotropic model of shear velocity in the whole mantle, *Geophys. J. Int.*, **167**, 361–379.

Parker T., Shatalin S., Farhadiroushan M., 2014, Distributed acoustic sensing – A new tool for seismic applications, *First Break*, **32**(2), 61–69.

Parkes G., Hatton L., 1986. *The Marine Seismic Source*, D. Riedel, Dordrecht.

Pedersen H.A., Bruneton M., Maupin V., 2006. Lithospheric and sublithospheric anisotropy beneath the Baltic shield from surface-wave array analysis, *Earth Planet. Sci. Lett.*, **244**, 590–605.

Pedersen H.A., Boué P., Poli P., Colombi A., 2015. Arrival angle anomalies of Rayleigh waves observed at a broadband array: a systematic study based on earthquake data, full waveform simulations and noise correlations, *Geophys. J. Int.*, **203**, 1626–1641.

Pekeris C. L., 1948. Theory of propagation of explosive sound in shallow water, *Geol. Soc. Am. Memoirs*, **27**.

Peng Z., Koper K.D., Vidale J.E., Leyton F., .Shearer P., 2008. Inner-core fine-scale structure from scattered waves recorded by LASA. *J. Geophys. Res.*, **113**, B09312.

Peterson J., 1993. Observations and modeling of seismic background noise. *U.S. Geol. Surv. Tech. Rept.*, **93-322**, 1–95.

Pham T.-S., Tkalčić H., 2017. On the feasibility and use of teleseismic *P* wave coda autocorrelation for mapping shallow seismic discontinuities, *J. Geophys. Res. Solid Earth*, **122**, 3776–3791.

Pham T.-S., Tkalčić H., 2018. Antarctic ice properties revealed from teleseismic *P* wave coda autocorrelation, *J. Geophys. Res. Solid Earth*, **123**, 7896–7912.

Pham T.-S., Tkalčić H., Sambridge M., Kennett B.L.N., 2018. Earth's Correlation Wavefield: Late coda correlation, *Geophys. Res. Lett.*, **45**, 3035–3042.

Piromallo C., Morelli A., 1997. Imaging the Mediterranean upper mantle by *P*-wave travel time tomography, *Ann. Geophys.*, **40**, 965–979.

Poli P., Pedersen H.A., Campillo, M., POLENET/LAPNET working group, 2012a. Emergence of body waves from cross-correlation of short period seismic noise, *Geophys. J. Int.*, **188**, 549–558.

Poli P., Campillo M., Pedersen H., LAPNET working group, 2012b. Body-wave imaging of Earth's mantle discontinuities from ambient seismic noise, *Science*, **338**,1063–1065.

Poli P., Thomas C., Campillo M., Pedersen, H.A., 2015. Imaging the D″ reflector with noise correlations, *Geophys. Res. Lett.*, **42**, 60–65.

Poli P., Campillo M., de Hoop M., 2017. Analysis of intermediate period correlations of coda from deep earthquakes. *Earth Planet. Sci. Lett.*, **477**, 147–155.

Poupinet G., Kennett B.L.N., 2004. On the observation of high frequency *PKiKP* and its coda in Australia, *Phys. Earth. Planet. Inter.*, **146**, 497–511.

Pratt R.G., Shin C., Hicks G.J., 1998. Gauss-Newton and full Newton methods in frequency domain seismic waveform inversion, *Geophys. J. Int.*, **133**, 341–362.

Pratt R.G., 1999. Seismic waveform inversion in the frequency domain, Part 1: Theory and verification in a physical scale model, *Geophysics*, **64**, 888–901.

Pugin A., Yilmaz Ö., 2019. Optimum source-receiver orientations to capture PP, PS, SP and SS reflected wave modes, *Leading Edge*, **38**(1), 45–52.

Quarteroni A., Sacco R., Saleri F., 2000. *Numerical Mathematics*, Springer, New York.

Rawlinson N., Kennett B.L.N., 2004. Rapid estimation of relative and absolute delay times across a network by adaptive stacking, *Geophys. J. Int.*, **157**, 332–340.

Rawlinson N., Sambridge M., 2004. Wave front evolution in strongly heterogeneous layered media using the fast marching method *Geophys. J. Int.*, **156**), 631–647

Rawlinson N., Urvoy M., 2006. Simultaneous inversion of active and passive source datasets for 3-D seismic structure with application to Tasmania, *Geophys. Res. Lett.*, **33**, L24313, doi:10.1029/2006GL028105.

Rawlinson N., Fichtner A., Sambridge M., Young M.K., 2014. Seismic tomography and the assessment of uncertainty, *Adv. Geophys.*, **55**, 1–76.

Rawlinson N., Kennett B.L.N., Salmon M., Glen R.A., 2015. Origin of lateral heterogeneities in the upper mantle beneath southeast Australia from seismic tomography, 47–78, in: *The Earth's Heterogeneous Mantle: A Geophysical, Geodynamical, and Geochemical Perspective*, eds. Khan A., Deschamps F., Springer, Heidelberg.

Reading A., Kennett B., Sambridge M., 2003. Improved inversion for seismic structure using transformed *S*-wavevector receiver functions: Removing the effect of the free surface, *Geophys. Res. Lett.*, **30**, 1981.

Resovsky J., Trampert J., 2003. Using probabilistic seismic tomography to test mantle velocity-density relationships, *Earth Planet. Sci. Lett.*, **215**, 121–134.

Rial J.A., Cormier V.F., 1980. Seismic waves at the epicenter's antipode, *J. Geophys. Res.*, **85**, 2661–2668.

Rietmann M., Grote M., Peter D., Schenk O., 2017. Newmark local time stepping on high-performance computing architectures, *J. Comp. Phys.*, **334**, 308–326.

Ritsema J., van Heijst H.J., Woodhouse J.H., 1999. Complex shear wave velocity structure imaged beneath Africa and Iceland, *Science*, **286**, 1925–1928.

Ritsema J., van Heijst H.J., 2002. Constraints on the correlation of *P*- and *S*-wave velocity heterogeneity in the mantle from *P*, *PP*, *PPP* and *PKPab* traveltimes, *Geophys. J. Int.*, **149**, 482–489.

Ritsema J., Deuss A., van Heijst H.J., Woodhouse J.H., 2011. S40RTS: a degree-40 shear-velocity model for the mantle from new Rayleigh wave dispersion, teleseismic traveltime and normal-mode splitting function measurements, *Geophys. J. Int.*, **184**, 1223–1236,

Ritzwoller M.H., Feng L., 2019. Overview of pre- and post-processing of ambient noise correlations, 144–187, in *Seismic Ambient Noise*, eds. Nakata N., Gualtieri L., Fichtner A., Cambridge University Press, Cambridge.

Rosenbrock H.H., 1960. An automatic method for finding the greatest or least value of a function, *Comp. J.*, **3**, 175–184.

Rost S., Earle P.S., 2010. Identifying regions of strong scattering at the core-mantle boundary from analysis of *PKKP* precursor energy, *Earth Planet. Sci. Lett.*, **297**, 616–626.

Rost S., Earle P.S., Shearer P.M., Frost D.A., Selby N.D., 2015. Seismic detections of small-scale heterogeneities in the deep Earth, 367–390, in: *The Earth's Heterogeneous Mantle: A Geophysical, Geodynamical, and Geochemical Perspective*, eds. Khan A. & Deschamps F., Springer, Heidelberg.

Roth M., Korn M., 1993. Single scattering theory versus numerical modelling in 2-D random media, *Geophys. J. Int.*, **112**, 124–140.

Roux P., Sabra K., Gerstoft, P., Kuperman, W., 2005. P-waves from cross correlation of seismic noise, *Geophys. Res. Lett.*, **32**, L19303.

Roux P., 2009. Passive seismic imaging with directive ambient noise: application to surface waves and the San Andreas Fault in Parkfield, CA, *Geophys. J. Int.*, **179**, 367–373.

Rubin K.H., Sinton J.M., Maclennan J., Hellebrand E., 2009. Magmatic filtering of mantle compositions at mid-ocean-ridge volcanoes, *Nature Geosci.*, *2*, 321–328.

Ruigrok E., Campman X., Draganov D., Wapenaar K., 2010. High-resolution lithospheric imaging with seismic interferometry, *Geophys. J. Int.*, **183**, 339–357.

Ruigrok E., Wapenaar K., 2012. Global-phase seismic interferometry unveils P-wave reflectivity below the Himalayas and Tibet, *Geophys. Res. Lett.*, **39**, L11303.

Sager K., Boehm C., Ermert L., Krischer L., Fichtner A., 2018. Sensitivity of seismic noise correlation functions to global noise sources, *J. Geophys. Res.*, **123**, 6911–6921.

Sager K., Boehm C., Ermert L., Krischer L., Fichtner A., 2020. Global-scale full-waveform ambient noise inversion, *J. Geophys. Res.*, **125**, e2019JB018644.

Saito T., 2010. Love-wave excitation due to the interaction between a propagating ocean wave and the sea-bottom topography, *Geophys. J. Int.*, **182**, 1515–1523.

Salmon M., Kennett B.L.N., Stern T., Aitken, A.R.A., 2013. The Moho in Australia and New Zealand, *Tectonophysics*, **609**, 288–298.

Sambridge M.S., Tarantola A., Kennett B.L.N., 1991. An alternative strategy for the non-linear inversion of seismic waveforms, *Geophys. Propsect.*, **39**, 723–726.

Sambridge M.S., 1999. Geophysical inversion with the neighbourhood algorithm: I. Searching a parameter space, *Geophys. J. Int.*, **138**, 479–494.

Sambridge M.S., Gallagher K., Jackson A., Rickwood P., 2006. Trans-dimensional inverse problems, model comparison, and the evidence, *Geophys. J. Int.*, **167**, 528–542.

Santosa F., Symes W.W., 1988. Computation of the Hessian for least-squares solutions of inverse problems of reflection seismology, *Inverse Problems*, **4**, 211–233.

Sato H., Fehler M.C., Maeda T., 2012. *Seismic Wave Propagation and Scattering in the Heterogeneous Earth*, 2nd edition, Springer, Heidelberg.

Saygin E., Kennett B.L.N., 2010. Ambient noise tomography for the Australian Continent, *Tectonophysics*, **481**, 116–125.

Saygin E., Kennett B.L.N., 2012. Crustal structure of Australia from ambient seismic noise tomography, *J. Geophys. Res.*, **117**, B01304.

Saygin E., McQueen H., Hutton L., Kennett B.L.N., Lister G., 2013. Structure of the Mt. Isa region from seismic ambient noise tomography, *Austral. J. Earth Sci.*, **60**, 707–718.

Saygin E., Kennett B.L.N., 2019. Retrieval of interstation local body waves from teleseismic coda correlations, *J. Geophys. Res. Solid Earth*, **124**, 2957–2969.

Schaeffer A.J., Lebedev S., 2013. Global shear speed structure of the upper mantle and transition zone, *Geophys. J. Int.*, **194**, 417–449.

Schimmel M., Paulssen H., 1997. Noise reduction and detection of weak, coherent signals through phase-weighted stacks, *Geophys. J. Int.*, **130**, 497–505.

Schmandt B., Clayton R., 2013. Analysis of P-waves with a 5200-station array in Long Beach, California: evidence for abrupt boundary for Inner Borderland rifting, *J. Geophys. Res.*, **118**, 1–19.

Schmerr N., 2012. The Gutenberg discontinuity: melt at the Lithosphere–Asthenosphere boundary, *Science*, **335**, 1480–1483.

Schultz P.S., 1982. A method for direct estimation of interval velocities, *Geophysics*, **47**, 1657–1671.

Schuster G., 2009. *Seismic Interferometry*, Cambridge University Press, Cambridge.

Sens-Schönfelder C., Snieder R., Stähler S.C., 2015. The lack of equipartitioning in global body wave coda, *Geophys. Res. Lett.*, **42**, 7483–7489.

Sens-Schönfelder C., Brenguier F., 2018. Noise-based monitoring, 267–301, in *Seismic Ambient Noise*, eds. Nakata N., Gualtieri L., Fichtner A., Cambridge University Press, Cambridge.

Sereno T.J., Orcutt J.A., 1985. Synthesis of realistic oceanic *Pn* wavetrains, *J. Geophys. Res.*, **90**, 12 755–12 776.

Shanno D.F., 1970. Conditioning of quasi-Newton methods for function minimization, *Math. Comp.*, **24**, 647–656.

Shapiro N.M., Campillo M., Stehly L., Ritzwoller, M.H., 2005. High resolution surface-wave tomography from ambient seismic noise. *Science*, **307**, 1615–1618.

Shapiro N., 2019. Applications with surface waves extracted from ambient seismic noise, 218–238, in *Seismic Ambient Noise*, eds. Nakata N., Gualtieri L., Fichtner A., Cambridge University Press, Cambridge.

Shearer P.M., Hedlin M.A.H., Earle P.S., 1998. *PKP* and *PKKP* precursor observations: Implications for the small-scale structure of the deep mantle and core, in: *The Core-Mantle Boundary Region*, AGU Geodynamics Monograph **28**, 37–55.

Shearer P.M., Earle P.S., 2008. Observing and modeling elastic scattering in the deep Earth. *Advances in Geophysics*, **50**, 167–193.

Shearer P.M., Rychert C.A., Liu Q., 2011, On the visibility of the inner-core shear wave phase *PKJKP* at long periods, *Geophys. J. Int.*, **185**, 1379–1383.

Shearer P.M., 2017. Deep Earth structure: Seismic scattering in the deep Earth. *Treatise on Geophysics Vol. 1* (second edition), 759–787, Elsevier, Amsterdam,

Shen W., Ritzwoller M.H., 2016. Crustal and uppermost mantle structure beneath the United States, *J. Geophys. Res. Solid Earth*, **121**, 4306–4342.

Shintaku N., Forsyth D.W., Hajewski C.J., Weeraratne D.S., 2014. *Pn* anisotropy in Mesozoic western Pacific lithosphere, *J. Geophys. Res. Solid Earth*, **119**, 3050–3063.

Shirzad T., Shomali Z.-H., 2015. Extracting seismic body and Rayleigh waves from the ambient seismic noise using the rms-stacking method, *Seism. Res. Lett.*, **86**, 1–8.

Shito A., Suetsugu D., Furumura T., Sugioka H., Ito A., 2013. Small scale heterogeneities in the oceanic lithosphere inferred from guided waves, *Geophys. Res. Lett.*, **40**, 1708–1712.

Sieminski A., Trampert J., Tromp J., 2009. Principal component analysis of anisotropic finite-frequency kernels, *Geophys. J. Int.*, **179**, 1186–1198.

Sigloch K., Nolet G., 2006. Measuring finite-frequency body wave amplitudes and travel times, *Geophys. J. Int.*, **167**, 271–287.

Sippl C., 2016. Moho geometry along a north-south passive seismic transect through Central Australia, *Tectonophysics*, **676**, 56–69.

Sippl C., Brisbout L., Spaggiari C., Gessner K., Tkalčić H., Kennett B.L.N., Murdie, R., 2017a. Crustal structure of a Proterozoic craton boundary: East Albany-Fraser Orogen, Western Australia, imaged with passive seismic and gravity anomaly data, *Precambrian Res.*, **296**, 78–92.

Sippl C., Kennett B.L.N., Tkalčić H., Gessner K., Spaggiari, C., 2017b. Crustal surface-wave velocity structure of the east Albany-Fraser Orogen, Western Australia, from ambient noise recordings, *Geophys. J. Int.*, **210**, 1641–1651.

Slob E., Wapenaar K., Broggini F., Snieder R., 2014. Seismic reflector imaging using internal multiples with Marchenko-type equations, *Geophysics*, **79**, S63–S76.

Snieder R., 1986. 3-D linearized scattering of surface waves and a formalism for surface wave tomography, *Geophys. J. Astr. Soc.*, **84**, 581–605.

Snieder R., 2004. Extracting the Green's function from the correlation of coda waves: A derivation based on stationary phase, *Phys. Rev. E.*, **69**, 046610.

Snieder R., Van Wijk K., Haney M., Calvert R., 2008. Cancellation of spurious arrivals in Green's function extraction and the generalized optical theorem, *Phys. Rev. E*, **78**, 1–8.

Snieder R., Sens-Schönfelder C., 2015. Seismic interferometry and stationary phase at caustics, *J. Geophys. Res. Solid Earth*, **120**, 4333–4343.

Snieder R. Duran A., Obermann A., 2019. Locating velocity changes in elastic media with coda wave interferometry, 188-217, in *Seismic Ambient Noise*, eds. Nakata N., Gualtieri L., Fichtner A., Cambridge University Press, Cambridge.

Soldait G., Boschi L., Piersanti A., 2006. Global seismic tomography and modern parallel computers, *Ann. Geophys.*, **49**, 977–986.

Soomro R.A., Weidle C., Cristiano L., Lebedev S., Meier T., PASSEQ Working Group, 2016. Phase velocities of Rayleigh and Love waves in central and northern Europe from automated, broad-band, interstation measurements, *Geophys. J. Int.*, **204**, 517–534.

Spakman W., Wortel M.J.R., Vlaar N.J., 1988. The Hellenic subduction zone: a tomographic image and its geodynamic implications, *Geophys. Res. Lett.*, **15**, 60–63.

Stehly L., Campillo M., Froment B., Weaver R.L., 2008. Reconstructing Green's function by correlation of the coda of the correlation (C3) of ambient seismic noise, *J. Geophys. Res.*, **113**, B11306.

Stehly L., Fry B., Campillo M., Shapiro N., Guilbert J., Boschi L., Giardini D, 2009. Tomography of the Alpine region from observations of seismic ambient noise, *Geophys. J. Int.*, **178**, 338–350.

Sun D., Miller M.S., Piana Agostinetti N., Asimow P., Li D., 2014. High frequency waves and slab structures beneath Italy, *Earth Planet. Sci. Lett.*, **391**, 212–223.

Sun W., Kennett B.L.N., 2016. Receiver structure from teleseisms: Autocorrelation and cross correlation, *Geophys. Res. Lett.*, **43**, 6234–6242,

Sun W., Fu L.-Y., Saygin E., Zhao L., 2018. Insights into layering in the cratonic lithosphere beneath Western Australia, *J. Geophys. Res. Solid Earth*, **123**, 1405–1418.

Takemura S., Maeda T., Furumura T., Obara K., 2016. Constraining the source location of the 30 May 2015 (Mw7.9) Bonin deep-focus earthquake using seismogram envelopes of high-frequency *P* waveforms: Occurrence of deep-focus earthquake at the bottom of a subducting slab, *Geophys. Res. Lett.*, **43**, 4297–4302.

Tanimoto T., Um J., 1999. Cause of continuous oscillations of the Earth, *J. Geophys. res.*, **104**, 28 723–28 739.

Tanimoto T., 2008. Normal-mode solution for the seismic noise cross-correlation method, *Geophys. J. Int.*, **175**, 1169–1175.

Tape C., Liu Q., Maggi A., Tromp J., 2010. Seismic tomography of the southern California crust based upon spectral-element and adjoint methods, *Geophys. J. Int.*, **180**, 433–462.

Tarantola A., Valette B., 1982. Generalized nonlinear inverse problems solved using the least-squares criterion, *Rev. Geophys.*, **20**, 219–232.

Tarantola A., 1984. Inversion of seismic reflection data in the acoustic approximation, *Geophysics*, **49**, 1259–1266.

Tarantola A., 1988. Theoretical background for the inversion of seismic waveforms, including elasticity and attenuation, *Pure Appl. Geophys.*, **128**, 365–399.

Tatham R.S., Stoffa P./L., 1976. Vp/Vs ; a potential hydrocarbon indicator *Geophysics*, **41**, 837–849

Tatham R.H., Goolsbee D.V., 1984. Separation of S-wave and P-wave reflections offshore western Florida, *Geophysics*, **49**, 493–508.

Tauzin B., Bodin T., Debayle E., Perrillat J.-P., Reynard B., 2016. Multi-mode conversion imaging of the subducted Gorda and Juan de Fuca plates below the North American continent, *Earth Planet. Sci. Lett.*, **440**, 135–146.

Tauzin B., Pham T.-S., Tkalčić H., 2019. Receiver functions from seismic interferometry: A practical guide, *Geophys. J. Int.*, **217**, 1–24.

Thomas C., Kendall J.-M., Helffrich G., 2009. Probing two low-velocity regions with *PKP* b-caustic amplitudes and scattering. *Geophys. J. Int.*, **178**, 503–512.

Thompson D.A., Rawlinson N., Tkalčić H., 2019. Testing the limits of virtual deep seismic sounding via new crustal thickness estimates of the Australian continent, *Geophys. J. Int.*, **218**, 787–800.

Thomson C.J., 2012. On the space-time domain form of the reflection operator for a simple flat-lying interface, *Geophys. Prospect.*, **60**, 49–63.

Thrastarson S., van Driel M., Krischer L., Boehm C., Afanasiev M., van Herwaarden D.-P., Fichtner A., 2020. Accelerating numerical wave propagation by wavefield-adapted meshes, Part II: Full-waveform inversion, *Geophys. J. Int.*, **221**, 1591–1604

Thybo H., 2008. The heterogeneous upper mantle low velocity zone, *Tectonophysics*, **416**, 53–79.

Tibuleac I.M., von Seggern, D., 2012. Crust-mantle boundary reflectors in Nevada from ambient seismic noise autocorrelations, *Geophys. J. Int.*, **189**, 493–500.

Tkalčić H., Flanagan M., Cormier V.F., 2006. Observations of near-podal P′P′ precursors: evidence for back scattering from the 150–220 km zone in Earth's upper mantle, *Geophys. Res. Lett.*, **33**, L03305.

Tkalčić H., Cormier V.F., Kennett B.L.N., He, K., 2010. Steep reflections from the earth's core reveal small-scale heterogeneity in the upper mantle, *Phys. Earth Planet. Inter.*, **178**, 80–91,

Tkalčić H., 2015. *The Inner Core*, Cambridge University Press, Cambridge.

Tkalčić H., Pham T.-S., 2018. Shear properties of the Earth's inner core revealed by a detection of *J* waves in global correlation wavefield, *Science*, **362**, 329–332.

Tonegawa T., Nishida K., Watanabe T., Shiomi M., 2009. Seismic interferometry of teleseismic *S*-wave coda for retrieval of body waves: An application to the Philippine Sea slab underneath the Japanese Islands, *Geophys. J. Int.*, **178**, 1574–1586.

Trampert J., Fichtner A., Ritsema J., 2013. Resolution tests revisited: The power of random numbers, *Geophys. J. Int.*, **192**, 676–680.

Tromp J., Tape C., Liu Q., 2005. Seismic tomography, adjoint methods, time reversal and banana-doughnut kernels, *Geophys. J. Int.*, **160**, 195–216.

Tromp J., Luo Y., Hanasoge S., Peter D., 2010. Noise cross-correlation sensitivity kernels, *Geophys. J. Int.*, **183**, 791–819.

Tromp, J., Bachmann, E., 2019. Source encoding for adjoint tomography, *Geophys. J. Int.*, **218**, 2019–2044.

Trorey A.W., 1970. A simple theory for seismic diffractions, *Geophysics*, **35**, 762–784.

Tsai V.C., Moschetti M.P., 2010. An explicit relationship between time-domain noise correlation and spatial autocorrelation (SPAC) results, *Geophys. J. Int.*, **182**, 454–460.

Tseng T.-L., Chen W.-P., Nowack R.L., 2009. Northward thinning of Tibetan crust revealed by virtual seismic profiles, *Geophys. Res. Lett.*, **36**, L24304.

Ursin B., 1983. Review of elastic and electromagnetic wave propagation in horizontally layered media, *Geophysics*, **48**, 1063–1081.

Utsu T., 1966. Regional difference in absorption of seismic waves in the upper mantle as inferred from abnormal distribution of seismic intensities, *J. Fac. Sci. Hokkaido Univ., Ser. VII*, **2**, 359–374.

Vaccari F., Gregersen S., Furlan M., Panza G.F., 1989. Synthetic seismograms in laterally heterogeneous anelastic media by modal summation of P–SV waves, *Geophys. J. Int.*, **99**, 285–295.

Valentine A. P., Woodhouse J. H., 2010. Reducing errors in seismic tomography: Combined inversion for sources and structure, *Geophys. J. Int.*, **180**, 847–857.

Van Avendonk H.J.A., Harding A.J., Orcutt J.A., McClain J.S., 2001. Contrast in crustal structure across the Clipperton transform fault from travel time tomography. *J. Geophys. Res.*, **106**, 10 961–10 981.

VanDecar J.C., Crosson R.S., 1990. Determination of teleseismic relative phase arrival times using multi-channel cross-correlation and least squares, *Bull. Seism. Soc. Am.*, **80**, 150–169.

van der Hilst R., Kennett B., Christie D., Grant J., 1994. Project Skippy explores lithosphere and mantle beneath Australia, *EOS, Transactions AGU*, 177–181.

van der Hilst R.D., Widiyantoro S., Engdahl E.R., 1997. Evidence for deep mantle circulation from global tomography, *Nature*, **386**, 578–584.

van der Neut, J., Vasconcelos I., Wapenaar K., 2015. On Green's function retrieval by iterative substitution of the coupled Marchenko equations, *Geophys. J. Int.*, **203**, 792–813.

Van der Voo R., Spakman W., Bijwaard H., 1999. Tethyan subducted slabs under India, *Earth Planet. Sci. Lett.*, **171**, 7–20.

van Driel M., Boehm C., Krischer L., Afanasiev M., 2020. Accelerating numerical wave propagation using wavefield-adapted meshes, Part I: Forward and adjoint modelling, *Geophys. J. Int.*, **221**, 1580–1590.

van Herwaarden D.-P., Boehm C., Afanasiev M., Thrastarson S., Krischer L., Trampert J., Fichtner A., 2020. Accelerated full-waveform inversion using dynamic mini-batches, *Geophys. J. Int.*, **221**, 1427–1438.

van Leeuwen T., Mulder W.A., 2010. A correlation-based misfit criterion for wave-equation traveltime tomography, *Geophys. J. Int.*, **182**, 1383–1394.

van Leeuwen T., Herrmann F. J., 2013. Fast waveform inversion without source encoding, *Geophys. Pros.*, **61**, 10–19.

van Vleck J.H., Middleton D., 1966. The spectrum of clipped noise, *Proc. IEEE*, **54**, 2–19.

Vered M., BenMenahem A., 1974. Application of synthetic seismograms to the study of low magnitude earthquakes and crustal structure in the Northern Red Sea region, *Bull. Seism. Soc. Am.*, **64**, 1221–1237.

Verschuur D., 2006. *Seismic Multiple Removal Techniques: Past, Present and Future*, EAGE, Houten, Netherlands.

Vidale J.E., Earle P.S., 2000. Fine-scale heterogeneity in the Earth's inner core, *Nature*, **404**, 273–275.

Vinnik L.P., 1977. Detection of waves converted from P to SV in the mantle, *Phys. Earth. Planet. Inter.*, **15**, 39–45.

Virieux J., Operto S., 2009. An overview of full-waveform inversion in exploration geophysics, *Geophysics*, **74**, WCC1–WCC26.

Wathelet M., 2008. An improved neighborhood algorithm: Parameter conditions and dynamic scaling, *Geophys. Res. Lett.*, **35**, L09301.

Walker D.A., Sutton G.H., 1971. Oceanic mantle phases recorded on hydrophones in the North Western Pacific at distances between 9° and 40°, *Bull. Seism. Soc. Am.*, **61**, 65–78.

Walter F., Gräff D., Lindner F., Paitz P., Köpfli M., Chmiel M., Fichtner A., 2020. Distributed acoustic sensing of microseismic sources and wave propagation in glaciated terrain, *Nat. Comm.*, **11**, 2436.

Wang J., Wu G., Chen, X., 2019. Frequency-Bessel transform method for effective imaging of higher-mode Rayleigh dispersion curves from ambient seismic noise data, *J. Geophys. Res. Solid Earth*, **124**, 3708–3723.

Wang T., Song X., Xia H.H., 2015. Equatorial anisotropy in the inner part of Earth's inner core from autocorrelation of earthquake coda. *Nature Geosci.*, **8**, 224–227.

Wang Z., Dahlen F.A., 1995. Validity of surface-wave ray theory on a laterally heterogeneous earth, *Geophys. J. Int.*, **123**, 757–773.

Wapenaar K., 2004. Retrieving the elastodynamic Green's function of an arbitrary inhomogeneous medium by cross correlation, *Phys. Rev. Lett.*, **93**, 254301.

Wapenaar K., Thorbecke J., Draganov D., 2004. Relations between reflection and transmission responses of three-dimensional inhomogeneous media, *Geophys. J. Int.*, **156**, 179–194.

Wapenaar K., Fokkema J., 2006. Green's function representations for seismic interferometry, *Geophysics*, **71**, SI33–SI46.

Wapenaar K.,Thorbecke J., van Der Neut J., Broggini F. , Slob E., Snieder, 2014. Marchenko imaging, *Geophysics*, **79**, WA39–WA57.

Warner M., Guasch Ll., 2015. Adaptive waveform inversion: Theory, *Geophysics*, **61**, RR429–R445.

Waszek L., Deuss, A., 2011. Distinct layering in the hemispherical seismic velocity structure of Earth's upper inner core, *J. Geophys. Res.*, **116**, B12313.

Webb S.C., 2008. The Earth's 'hum': the excitation of Earth normal modes by ocean waves. *Geophys. J. Int.*, **174**, 542–566.

Widiyantoro S., Kennett B.L.N., van der Hilst R.D., 1999. Seismic tomography with *P* and *S* data reveals lateral variations in the rigidity of deep slabs, *Earth. Planet. Sci. Lett.* **173**, 91–100.

Widiyantoro S., Gorbatov A., Kennett B.L.N., Fukao Y., 2000. Improving global shear wave traveltime tomography using three-dimensional ray tracing and iterative inversion, *Geophys. J. Int.* **141**, 747–758.

Willis J.R., 1981. Variational and related methods for the overall properties of composites, *Advances in Applied Mechanics*, **21**, 1–78.

Wittlinger G., Vergne J., Tapponnier P., Farra V., Poupinet G., Jiang M., Su H., Herquel G., Paul A., 2004. Teleseismic imaging of subducting lithosphere and Moho offsets beneath Western Tibet, *Earth Planet. Sci. Lett.*, **221**, 117–130.

Wolpert D.H., Macready W.G., 1997. No Free Lunch Theorems for optimization, *IEEE Trans. Evolutionary Comp.*, **1**, 67–82.

Woodhouse J.H., 1974. Surface waves in a laterally varying layered structure, *Geophys. J. R. Astr. Soc.*, **37**, 461–490.

Woodard M.F., 1997. Implications of localized, acoustic absorption for heliotomographic analysis of sunspots, *Astrophys. J.*, **485**, 890–894.

Wuestefeld A., Wilks M., 2019. How to twist and turn a fibre: Performance modelling for optimum DAS acquisitions, *The Leading Edge*, **38**, 226–231.

Xia H.H., Song X., Wang T., 2016. Extraction of triplicated *PKP* phases from noise correlations, *Geophys. J. Int.*, **205**, 499–508.

Xie J., Nuttli O., 1988. Interpretation of high frequency coda at large distances: stochastic modelling and method of inversion, *Geophys. J. Int.*, **93**, 579–595.

Yanovskaya T.B., 1984. Solution of the inverse problem of seismology for laterally inhomogeneous media, *Geophys. J. R. Astr. Soc.*, **79**, 293–304,

Yokoi T., Margaryan S., 2008. Consistency of the spatial autocorrelation method with seismic interferometry and its consequence, *Geophys. Prospect.*, **56**, 435–451.

Yomogida K., 1992. Fresnel zone inversion for lateral heterogeneities in the Earth, *Pure Appl. Geophys.*, **138**, 391–406.

Yoshizawa K., Kennett B.L.N., 2002a. Nonlinear waveform inversion for surface waves with a neighbourhood algorithm – application to multi-mode dispersion measurements, *Geophys. J. Int.*, **149**, 440–453.

Yoshizawa K., Kennett B.L.N., 2002b, Determination of the influence zone for surface wave paths, *Geophys. J. Int.*, **149**, 118–133.

Yoshizawa K., Kennett B.L.N., 2005. Sensitivity kernels for finite-frequency surface waves, *Geophys. J. Int.*, **162**, 910–926.

Yoshizawa K., 2014. Radially anisotropic 3-D shear wave structure of the Australian lithosphere and asthenosphere from multi-mode surface waves, *Phys. Earth Planet. Inter.*, **235**, 33–48.

Yoshizawa K., Kennett B.L.N., 2015. The lithosphere–asthenosphere transition and radial anisotropy beneath the Australian continent, *Geophys. Res. Lett.*, **42**, 3839–3846.

Young M., Rawlinson N., Arroucau P., Reading A.M., Tkalčić H., 2011. High-frequency ambient noise tomography of southeast Australia: New constraints on Tasmania's tectonic past, *Geophys. Res. Lett.*, **38**, L13313.

Yu C.-Q., Chen W.-P., van der Hilst, R.D., 2013. Removing source-side scattering for virtual deep seismic sounding (VDSS), *Geophys. J. Int.*, **195**, 1932–1941.

Yuan X., Kind R., Li, X., Wang R., 2006. The S receiver functions: synthetics and data example, *Geophys. J. Int.*, **165**, 555–564.

Zhan Z., Ni S., Helmberger D., Clayton, R.W., 2010. Retrieval of Moho reflected shear wave arrivals from ambient seismic noise, *Geophys. J. Int.*, **182**, 408–420.

Zhang J., Yang X., 2013. Extracting surface wave attenuation from seismic noise using correlation of the coda of correlation, *J. Geophys. Res.*, **118**, 2191–2205.

Zhou Y., Dahlen F.A., Nolet G., 2004. Three-dimensional sensitivity kernels for surface wave observables, *Geophys. J. Int.*, **158**, 142–168.

Zhou W., Paulssen H., 2017. P and S velocity structure in the Groningen gas reservoir from noise interferometry, *Geophys. Res. Lett.*, **44**, 11 785–11 791.

Zhu H., Bozdağ E., Duffy T.S., Tromp J., 2013. Seismic attenuation beneath Europe and the North Atlantic: Implications for water in the mantle, *Earth Planet. Sci. Lett.*, **381**, 1–11.

Index